Gravitational N-Body Simulations

This book discusses in detail all the relevant numerical methods for the classical N-body problem. It demonstrates how to develop clear and elegant algorithms for models of gravitational systems and explains the fundamental methematical tools needed to describe the dynamics of a large number of mutually attractive particles. Particular attention is given to the techniques needed to model astrophysical phenomena such as close encounters and the dynamics of black-hole binaries. The author reviews relevant work in the field and covers applications to the problems of planetary formation and star-cluster dynamics, both of Pleiades-type and globular clusters.

Self-contained and pedagogical, this book is suitable for graduate students and researchers in theoretical physics, astronomy and cosmology.

SVERRE AARSETH received his B.Sc. from the University of Oslo in 1959 and his Ph.D. from the University of Cambridge in 1963. After a few years as research assistant to Professor F. Hoyle, he joined the newly created Institute of Theoretical Astronomy in 1967 (which then became the Institute of Astronomy in 1972). His entire career has been spent at this Institute as a post-doctoral research fellow, giving him complete freedom to devote himself exclusively to all aspects of the modern N-body problem. The stimulating Cambridge environment has been ideal for establishing collaborations with visiting astronomers. Dr Aarseth has developed a unique set of codes that include the latest techniques, and are now publicly available. These codes are suitable for laptops and workstations as well as for the most powerful special-purpose computers.

CAMBRIDGE MONOGRAPHS ON MATHEMATICAL PHYSICS

General editors: P. V. Landshoff, D. R. Nelson, S. Weinberg

S. J. Aarseth *Gravitational N-Body Simulations*
J. Ambjørn, B. Durhuus and T. Jonsson *Quantum Geometry: A Statistical Field Theory Approach*
A. M. Anile *Relativistic Fluids and Magneto-Fluids*
J. A. de Azcárrage and J. M. Izquierdo *Lie Groups, Lie Algebras, Cohomology and Some Applications in Physics*†
O. Babelon, D. Bernard and M. Talon *Introduction to Classical integral Systems*
V. Belinkski and E. Verdaguer *Gravitational Solitons*
J. Bernstein *Kinetic Theory in the Early Universe*
G. F. Bertsch and R. A. Broglia *Oscillations in Finite Quantum Systems*
N. D. Birrell and P. C. W. Davies *Quantum Fields in Curved Space*†
M. Burgess *Classical Covariant Fields*
S. Carlip *Quantum Gravity in 2+1 Dimensions*
J. C. Collins *Renormalization*†
M. Creutz *Quarks, Gluons and Lattices*†
P. D. D'Eath *Supersymmetric Quantum Cosmology*
F. de Felice and C. J. S Clarke *Relativity on Curved Manifolds*†
P. G. O. Freund *Introduction to Supersymmetry*†
J. Fuchs *Affine Lie Algebras and Quantum Groups*†
J. Fuchs and C. Schweigert *Symmetries, Lie Algebras and Representations: A Graduate Course for Physicists*†
Y. Fujii and K. Maeda *The Scalar–Tensor Theory of Gravitation*
A. S. Galperin, E. A. Ivanov, V. I. Orievetsky and E. S. Sokatchev *Harmonic Superspace*
R. Gambini and J. Pullin *Loops, Knots, Gauge Theories and Quantum Gravity*†
M. Göckeler and T. Schücker *Differential Geometry, Gauge Theories and Gravity*†
C. Gómez, M. Ruiz Altaba and G. Sierra *Quantum Groups in Two-dimensional Physics*
M. B. Green, J. H. Schwarz and E. Witten *Superstring Theory, volume 1: Introduction*†
M. B. Green, J. H. Schwarz and E. Witten *Superstring Theory, volume 2: Loop Amplitudes, Anomalies and Phenomenology*†
V. N. Gribov *The Theory of Complex Angular Momenta*
S. W. Hawking and G. F. R. Ellis *The Large-Scale Structure of Space-Time*†
F. Iachello and A. Aruna *The Interacting Boson Model*
F. Iachello and P. van Isacker *The Interacting Boson–Fermion Model*
C. Itzykson and J.-M. Drouffe *Statistical Field Theory, volume 1: From Brownian Motion to Renormalization and Lattice Gauge Theory*†
C. Itzykson and J.-M. Drouffe *Statistical Field Theory, volume 2: Strong Coupling, Monte Carlo Methods, Conformal Field Theory, and Random Systems*†
C. Johnson *D-Branes*
J. I. Kapusta *Finite-Temperature Field Theory*†
V. E. Korepin, A. G. Izergin and N. M. Boguliubov *The Quantum Inverse Scattering Method and Correlation Functions*†
M. Le Bellac *Thermal Field Theory*†
Y. Makeenko *Methods of Contemporary Gauge Theory*
N. H. March *Liquid Metals: Concepts and Theory*
I. M. Montvay and G. Münster *Quantum Fields on a Lattice*†
A. Ozorio de Almeida *Hamiltonian Systems: Chaos and Quantization*†
R. Penrose and W. Rindler *Spinors and Space-time, volume 1: Two-Spinor Calculus and Relativistic Fields*†
R. Penrose and W. Rindler *Spinors and Space-time, volume 2: Spinor and Twistor Methods in Space-Time Geometry*†
S. Pokorski *Gauge Field Theories,* 2nd edition
J. Polchinski *String Theory, volume 1: An Introduction to the Bosonic String*
J. Polchinski *String Theory, volume 2: Superstring Theory and Beyond*
V. N. Popov *Functional Integrals and Collective Excitations*†
R. G. Roberts *The Structure of the Proton*†
H. Stephani, D. Kramer, M. A. H. MacCallum, C. Hoenselaers and E. Herlt *Exact Solutions of Einstein's Field Equations, 2nd edition*
J. M. Stewart *Advanced General Relativity*†
A. Vilenkin and E. P. S. Shellard *Cosmic Strings and Other Topological Defects*†
R. S. Ward and R. O. Wells Jr *Twistor Geometry and Field Theories*†
J. R. Wilson and G. J. Mathews *Relativistic Numerical Hydrodynamics*

†Issued as a paperback

Gravitational N-Body Simulations

SVERRE J. AARSETH
Institute of Astronomy
University of Cambridge

PUBLISHED BY THE PRESS SYNDICATE OF THE UNIVERSITY OF CAMBRIDGE
The Pitt Building, Trumpington Street, Cambridge, United Kingdom

CAMBRIDGE UNIVERSITY PRESS
The Edinburgh Building, Cambridge CB2 2RU, UK
40 West 20th Street, New York, NY 10011–4211, USA
477 Williamstown Road, Port Melbourne, VIC 3207, Australia
Ruiz de Alarcón 13, 28014 Madrid, Spain
Dock House, The Waterfront, Cape Town 8001, South Africa

http://www.cambridge.org

First published 2003

Printed in the United Kingdom at the University Press, Cambridge

A catalogue record for this book is available from the British Library

Library of Congress Cataloguing in Publication data
Aarseth, Sverre, J. 1934–
Gravitational N-body simulation: tools and algorithms / Sverre J. Aarseth.
p. cm. – (Cambridge monographs on mathematical physics)
Includes bibliographical references and index.
ISBN 0 521 43272 3
1. Many-body problem. I. Title. II. Series.
QB362.M3A27 2003 521 – dc21 2003046028

ISBN 0 521 43272 3 hardback

To the world's wild and
magical places

Contents

	Preface	*page* xiii
1	**The N-body problem**	**1**
1.1	Introduction	1
1.2	Historical developments	1
1.3	Basic concepts	6
1.4	The first steps	14
2	**Predictor–corrector methods**	**18**
2.1	Introduction	18
2.2	Force polynomials	18
2.3	Individual time-steps	21
2.4	Alternative formulations	24
2.5	Hermite scheme	27
2.6	Block time-steps	28
2.7	Time-symmetric method	30
3	**Neighbour treatments**	**32**
3.1	Introduction	32
3.2	Ahmad–Cohen method	33
3.3	Cosmological adaptation	37
3.4	Multipole expansion	40
3.5	Grid perturbations	43
3.6	Particle in box	46
3.7	Ring scheme	49
4	**Two-body regularization**	**51**
4.1	Introduction	51
4.2	Principles of regularization	52
4.3	Levi-Civita transformation	54

4.4 Kustaanheimo–Stiefel method 56
4.5 Burdet–Heggie alternative 60
4.6 Hermite formulation 61
4.7 Stumpff functions 63

5 Multiple regularization 66
5.1 Introduction 66
5.2 Aarseth–Zare method 67
5.3 External perturbations 73
5.4 Wheel-spoke generalization 75
5.5 Heggie's global formulation 78
5.6 Mikkola's derivation 80
5.7 Chain treatment 82
5.8 Slow-down procedure 86
5.9 Time-transformed leapfrog scheme 90
5.10 Algorithmic regularization 92

6 Tree codes 94
6.1 Introduction 94
6.2 Basic formulation 94
6.3 Collisional treatment 96
6.4 Flattened systems 103

7 Program organization 105
7.1 Introduction 105
7.2 N-body codes 105
7.3 Flowcharts 107
7.4 Scaling and units 110
7.5 Input parameters and options 112
7.6 Basic variables 114
7.7 Data structure 117

8 Initial setup 120
8.1 Introduction 120
8.2 Initial conditions for clusters 120
8.3 Primordial binaries 124
8.4 Open clusters and clouds 127
8.5 Eccentric planar orbits 131
8.6 Motion in 3D 133
8.7 Standard polynomials 136
8.8 Regularized polynomials 138

9 Decision-making 141
9.1 Introduction 141
9.2 Scheduling 142
9.3 Close two-body encounters 144
9.4 Multiple encounters 147
9.5 Hierarchical configurations 150
9.6 Escapers 153
9.7 Mass loss and tidal interactions 154
9.8 Physical collisions 156
9.9 Automatic error checks 160

10 Neighbour schemes 164
10.1 Introduction 164
10.2 Basic Ahmad–Cohen method 165
10.3 Hermite implementation 169
10.4 Parallel adaptations 173
10.5 Black hole binaries in galactic nuclei 175
10.6 Hybrid formulations 177

11 Two-body algorithms 181
11.1 Introduction 181
11.2 General KS considerations 181
11.3 Stumpff Hermite version 186
11.4 KS termination 188
11.5 Unperturbed two-body motion 190
11.6 Slow-down in KS 192
11.7 Hierarchical mergers 194
11.8 Tidal circularization 200
11.9 Chaotic motions 203
11.10 Roche-lobe mass transfer 204

12 Chain procedures 207
12.1 Introduction 207
12.2 Compact subsystems 207
12.3 Selection and initialization 211
12.4 Time stepping 213
12.5 Slow-down implementation 217
12.6 Change of membership 219
12.7 Hierarchical stability 221
12.8 Termination 223
12.9 Tidal interactions 225
12.10 Black hole binary treatment 229

13 Accuracy and performance **234**
13.1 Introduction 234
13.2 Error analysis 234
13.3 Time-step selection 241
13.4 Test problems 242
13.5 Special-purpose hardware 246
13.6 Timing comparisons 250

14 Practical aspects **252**
14.1 Introduction 252
14.2 Getting started 252
14.3 Main results 254
14.4 Event counters 255
14.5 Graphics 257
14.6 Diagnostics 258
14.7 Error diagnosis 261

15 Star clusters **264**
15.1 Introduction 264
15.2 Core radius and density centre 265
15.3 Idealized models 267
15.4 Realistic models 271
15.5 Stellar evolution 279
15.6 Tidal capture and collisions 282
15.7 Hierarchical systems 285
15.8 Spin–orbit coupling 287
15.9 Globular clusters 292

16 Galaxies **297**
16.1 Introduction 297
16.2 Molecular clouds 298
16.3 Tidal disruption of dwarf galaxies 300
16.4 Interacting galaxies 301
16.5 Groups and clusters 303
16.6 Cosmological models 304

17 Planetary systems **307**
17.1 Introduction 307
17.2 Planetary formation 307
17.3 Planetesimal dynamics 312
17.4 Planetary rings 317
17.5 Extra-solar planets 319

18 Small-N experiments 323
18.1 Introduction 323
18.2 Few-body simulations 324
18.3 Three-body scattering 330
18.4 Binary–binary interactions 334
18.5 Chaos and stability 340

Appendix A: Global regularization algorithms 350
A.1 Transformations and equations of motion 350
A.2 External perturbations 352

Appendix B: Chain algorithms 354
B.1 Transformations and switching 354
B.2 Evaluation of derivatives 356
B.3 Errata 358

Appendix C: Higher-order systems 359
C.1 Introduction 359
C.2 Initialization 359
C.3 Termination 360
C.4 Escape of hierarchies 361

Appendix D: Practical algorithms 363
D.1 Maxwellian distribution 363
D.2 Ghost particles 363
D.3 KS procedures for averaging 364
D.4 Determination of pericentre or apocentre 365
D.5 Partial unperturbed reflection 366

Appendix E: KS procedures with GRAPE 367
E.1 Single particles 367
E.2 Regularized KS pairs 368

Appendix F: Alternative simulation method 369
F.1 N-body treatment 369
F.2 Stellar evolution 370

Appendix G: Table of symbols 371
G.1 Introduction 371

Appendix H: Hermite integration method 374

References 377
Index 408

Preface

This book spans my entire life as a research worker at Cambridge. The circumstances that created this opportunity were based entirely on luck and this aspect played a vital part during subsequent developments. In the following chapters, I have tried to give details of the most relevant methods used in so-called 'direct integration' of the classical N-body problem, a method of attack somewhat analogous to scaling a mountain the hard way. This has been enhanced by an extensive discussion of the main algorithms implemented in the associated computer codes. A comprehensive review of related work in the field over the last 40 years is also presented. Throughout the term *N-body simulations* is used exclusively for methods based on direct summation, in keeping with tradition.

Although a wide range of problems is covered, the emphasis is on the dynamics of star clusters. This involves many aspects of stellar evolution. It is fortuitous that the University of Cambridge has a long tradition in this field that dates back to Eddington and Jeans. Fred Hoyle continued this school, which eventually gave rise to the application of synthetic stellar evolution. This subject was pioneered entirely at the Institute, mainly by the sequential efforts of Peter Eggleton, Christopher Tout and Jarrod Hurley, whose work has been vital for realistic star cluster simulations.

I would like to acknowledge the assistance of colleagues who read and commented critically on various chapters – Raul de la Fuente Marcos, Dougles Heggie, Jarrod Hurley, Pavel Kroupa, Derek Richardson, Rainer Spurzem, Christopher Tout and Mark Wilkinson. Specific suggestions for improvements of the contents were made by Doug Lin, Rosemary Mardling and HongSheng Zhao. My thanks also go to Robert Izzard who did most of the figures.

Among my many collaborators, I am especially indebted to Avishai Dekel, Richard Gott, Douglas Heggie, Jarrod Hurley, Pavel Kroupa, Mike Lecar, Doug Lin, Jun Makino, Rosemary Mardling, Steve McMillan,

Seppo Mikkola, Rainer Spurzem, Christopher Tout and Khalil Zare. Likewise, the pioneers Michel Hénon, Sebastian von Hoerner and Roland Wielen provided impetus and advice in the early days. More recently, Piet Hut has acted as a catalyst for stimulating new developments.

Claude Froeschlé, Douglas Heggie, E.L. Stiefel, Victor Szebenely and Khalil Zare educated and influenced me in the fundamental topic of regularization. Moreover, the contributions of Seppo Mikkola to our collaborations in this subject over the past 15 years have been invaluable.

Lastly, in the scientific field, I have benefited greatly from the technical assistance given to me by Jun Makino, Steve McMillan and Rainer Spurzem. My sincere thanks are due to Jun Makino and Makoto Ito who designed the special-purpose HARP-2 computer that occupied my office from 1994, and likewise to Jun Makino who is the driving force behind GRAPE-6, which has recently become the simulator's dream machine.

I made an auspicious start at the newly created Institute of Theoretical Astronomy, founded in 1967 by Sir Fred Hoyle. He was also my Ph.D. supervisor and directly responsible for independently suggesting N-body simulations as a research topic. For all this I am immensely grateful.

My subsequent career would not have been possible without strong support from the Directors of the Institute of Astronomy since the name change in 1972, Donald Lynden-Bell, Sir Martin Rees, Richard Ellis and Douglas Gough. They allowed me complete freedom to pursue my singular interest in dynamics. I have also depended utterly on continuous post-doctoral funding since 1963 by the Government Research Establishments that have undergone several name changes but lately are known as PPARC.

On the personal side, I would like to express my deepest thanks to Patricia who supported my work and also endured my other obsessions of chess and mountaineering. I am very grateful to my mother and father for their help and encouragement during the difficult formative years in Norway. Most of this book was written at the family mountain retreat near beautiful Lake Reinunga, which provided tranquility and inspiration.

Because of my involvement since the beginning, I have taken the opportunity to review the whole subject of N-body simulations as defined above. In view of the increasing activity this is a daunting task, particularly when it comes to making critical comments. In such circumstances my opinion is often expressed instead of merely quoting published work. I apologize for significant omissions and take full responsibility for any misrepresentations that are bound to occur. This book has been in preparation for a very long time. I would like to thank my editor, Tamsin van Essen, and the staff at Cambridge University Press for their patience and advice. Special thanks are due to my copy editor, Robert Whitelock, for his critical appraisal.

In conclusion, our field has undergone a remarkable development, fuelled by an exponential growth of computing power as well as software advances. Although the beginnings were modest and the developments slow, it has now blossomed into a fully fledged scientific activity. For the longer term, further progress is only possible if we attract the younger generation to seek new challenges and enrich our subject. It is therefore my hope that this book will prove timely and serve a practical purpose. Finally, the dedication reflects my many sources of inspiration, whether it be the awesome beauty of the Atacama Desert or more accessible wildlife environments. May our planet's fragile ecosystem and rich diversity be preserved for future enjoyment.

Sverre Aarseth January 2003

1
The N-body problem

1.1 Introduction

The main purpose of this book is to provide algorithms for direct N-body simulations, based on personal experience over many years. A brief description of the early history is included for general interest. We concentrate on developments relating to collisional direct integration methods but exclude three- and four-body scattering, which will be discussed in a separate chapter. In the subsequent section, we introduce some basic concepts which help to understand the behaviour of self-gravitating systems. The topics covered include two-body relaxation, violent relaxation, equipartition of kinetic energy and escape. Although the emphasis is on collisional dynamics, some of the theory applies in the large-N limit that is now being approached with modern hardware and improved numerical techniques. After these theoretical considerations, we turn to the problem at hand and introduce the general principles of direct integration as a beginner's exercise and also describe the first N-body method.

1.2 Historical developments

Numerical investigations of the classical N-body problem in the modern spirit can be said to have started with the pioneering effort of von Hoerner [1960]. Computational facilities at that time were quite primitive and it needed an act of faith to undertake such an uncertain enterprise.* Looking

---

* The story of how it all began is told in von Hoerner [2001]. Also of historical interest is the early study of gravitational interactions between two model galaxies based on measuring the intensity of 37 light bulbs at frequent intervals [Holmberg, 1941]. Finally, a general three-body integration by Strömgren [1900, 1909] was carried out by hand, albeit for only part of an orbit. More than 100 years ago he anticipated that the method of 'mechanical integration' may be extended to deal with four or more bodies and be of considerable importance for the theory of stellar systems.

back at these early results through eyes of experience, one can see that the characteristic features of binary formation and escape are already present for particle numbers as small as $N = 16$, later increased to 25 [von Hoerner, 1963]. In the beginning, integration methods were to a large extent experimental and therefore based on trial and error. This had the beneficial effect of giving rise to a variety of methods, since every worker felt obliged to try something new. However, by Darwinian evolution it soon became clear that force polynomials and individual time-steps† were important ingredients, at least in the quest for larger N [Aarseth, 1963a,b].

The basic idea of a force fitting function through the past points is to enable a high-order integration scheme, with the corresponding intervals satisfying specified convergence criteria. Consistent solutions are then ensured by coordinate predictions before the force summation on each particle is carried out. At the same time, the lack of a suitable method for dealing with persistent binaries inspired the introduction of a softened interaction potential $\Phi = -Gm/(r^2 + \epsilon^2)^{1/2}$, for the separation r with ϵ the softening parameter, which reduces the effect of close encounters. This potential gives rise to a simple expression for the force between two particles. Hence a large value of the softening scale length ϵ describes the dynamics of a so-called 'collisionless system', whereas smaller values may be used to exclude the formation of significant binaries. Although the application was to galaxy clusters, some general results on mass segregation were obtained for $N = 100$ and a mass spectrum [Aarseth, 1963a,b].

Later the integration method was improved to third order [Aarseth, 1966a] and eventually became a fourth-order predictor–corrector scheme [Aarseth, 1968], which survived for some considerable time and was widely used. The subsequent study of star clusters by Wielen [1967] was actually based on a fifth-order polynomial with special error control [Wielen, 1974]. This work compared the *extrapolated* half-life of simulated star clusters with observations and concluded that median life-times of about 2×10^8 yr could be accounted for. The nature of the numerical errors is of prime concern in such work and will be considered in a later chapter. In this context we mention that exponential error growth was demonstrated by Miller [1964] in an important paper where the short time-scale was emphasized. This fundamental feature was highlighted in a code comparison study for a collapsing 25-body system [Lecar, 1968]. In fact, these results led many people to question the validity of N-body simulations and this took many years to dispel.

At that time, the lack of computational facilities dictated a strategy of performing a few calculations at the largest possible value of N or

† The method of individual time-steps was originally suggested by A. Schlüter [private communication, 1961].

undertaking a systematic study of smaller systems. The latter choice was made by van Albada [1968] and yielded considerable insight into fundamental processes involving binary formation and exchange, as well as the energy of escaping particles. Thus it was demonstrated that a dominant binary containing the heaviest components in systems with up to 24 members sometimes acquires more than 100% of the total initial energy. Some interesting properties of long-lived triples were also presented for the first time, including evidence for the so-called 'Kozai cycle' of induced inner eccentricity (to be discussed later). Small systems are notoriously difficult to integrate but here a special fourth-order predictor–corrector method proved highly accurate, at the expense of *two* force evaluations per step in order to ensure convergence. The same time-step was used for all the particles; however, this becomes expensive above $N \simeq 10$ and the scheme of individual time-steps was never implemented.

By concentrating on just one system and using a dedicated computer, it proved possible to reach $N = 250$ [Aarseth, 1968]. Because of a favourable mass spectrum with two dominant (i.e. factor of 5 in mass) bodies, the final binary acquired some 150% of the *initial* total energy. The softening was still a factor of 10 below the small final semi-major axis, thereby justifying this device which does place a lower limit on binary separation.

The early trend towards greater realism led to the study of two new effects. Since open star clusters move in nearly circular Galactic orbits, the external tidal field can be added to the equations of motion using linearized terms. The first such pure N-body implementation was presented by Hayli [1967, 1969, 1970, 1972]. This work showed the characteristic behaviour of low-energy escaping stars passing near the Lagrange points L_1 and L_2. Again an original method was used called the 'category scheme' [cf. Hayli, 1967, 1969, 1974]. It was never fully developed but has some similarities to the popular Hermite method (to be discussed later).

A second effect relating to open clusters is the perturbation by interstellar clouds. The first attempt for $N = 25$ [Bouvier & Janin, 1970] experienced some technical problems in the boundary treatment, which goes to show that even intuitive selection procedures can be misleading. Moreover, distant particles exaggerated the predicted disruption time based on the change in total energy.[‡] In this case the integration method was again of fourth order with two force evaluations per step.

Among other integration schemes that have served a useful purpose we mention explicit Taylor series based on higher force derivatives. In the context of the N-body problem this idea was implemented by successive differentiations of the Newtonian acceleration [Gonzalez & Lecar, 1968; Lecar, Loeser & Cherniack, 1974]. Although quite accurate, a high-order

[‡] This problem was studied more extensively by Terlevich [1983, 1987].

expansion is too expensive to be practical for $N \geq 10$. On the positive side, the initialization of higher derivatives for standard force polynomials employs the explicit derivative approach to good effect.

In the late 1960s, several efforts were made to take advantage of the two-body regularization formulated by Kustaanheimo & Stiefel [1965; hereafter KS]. It became clear that special treatments of energetic binaries are desirable in order to study the long-term evolution of point-mass systems. One brave attempt to avoid the apparent complications of the KS method for N-body applications was based on the variation of parameters method [Aarseth, 1970]. The dominant central binary that usually emerges was represented by the osculating (or instantaneous) two-body elements. Apart from some problems due to secular perturbations, this method worked quite well.§ It also had the advantage of permitting unperturbed solutions which speed up the calculation. On the debit side, the method must be replaced by direct integration for significant perturbations. Still, much useful experience of algorithmic decision-making was gained by this application of celestial mechanics.

The impetus for introducing KS regularization was inspired by the beautiful three-body solution illustrated graphically by Szebehely & Peters [1967]. However, the Hamiltonian development of Peters [1968a,b] for the three-body problem bypasses the problem of evaluating the changing energy of the dominant two-body motion by an explicit calculation of the $N(N-1)/2$ *regular* terms, which is too expensive in the general case. This was eventually solved by introducing an additional equation of motion for the change in the two-body energy due to perturbations. Thus by the time of IAU Colloquium 10 on the N-body problem in 1970 two general codes were presented which included KS regularization [Aarseth, 1972b; Bettis & Szebehely, 1972]. Sadly, the latter proved too expensive for large systems since it employed a high-order Runge–Kutta integrator and was not developed further. However, it did prove itself in an investigation of high-velocity escapers in small systems [Allen & Poveda, 1972].

On the personal front, the next few years saw some interesting applications. One collaboration adopted hierarchical initial conditions inspired by fragmentation theory [Aarseth & Hills, 1972], which led to some energetic interactions. It is now well established that very young clusters show evidence of subclustering. Another effort examined the depletion of low-mass stars and concluded that the preferential effect was somewhat less than expected on theoretical grounds [Aarseth & Wolf, 1972]. The question of energetic binaries halting core collapse was also discussed [Aarseth, 1972a]. It was shown that a central binary may acquire a

§ The treatment of mixed secular terms was later improved by Mikkola [1984a] who introduced the variation of the epoch.

significant fraction of the total energy even for systems with $N = 500$; a calculation that took some 500 hours to complete [Aarseth, 1974]. It is noteworthy that the dominant binary acquired 50% of the total energy after only 12 crossing times (defined in the next section). Finally, a small contribution contained the first simulation of what we now call primordial binaries [Aarseth, 1975], which has become a major industry.¶

The 1970s brought about two important technical developments which are still being used. First we mention the Ahmad–Cohen [1973] neighbour scheme. The basic idea here is to represent the force acting on a particle by a sum of two polynomials, with the neighbour contribution being updated more frequently. Although there are programming complications due to the change of neighbours, the method is truly collisional and speeds up the calculation significantly even for quite modest values of N. Before the advent of the HARP special-purpose computer (to be described later), this algorithm facilitated the simulation of larger cluster models with $N \simeq 10^4$ where the gain may be a factor of 10.

The second innovation occurred by a happy combination of circumstances which resulted in a three-body regularization method [Aarseth & Zare, 1974]. This was achieved by the introduction of two coupled KS solutions which permit two of the particle pairs to approach each other arbitrarily close, provided this does not take place simultaneously. It turns out that the third interaction modifies the equations of motion in a way that still maintains regularity, as long as the corresponding distance is not the smallest. Following this development, the global formulation by Heggie [1974] was a notable achievement, especially since it was generalized to the N-body problem.

It is perhaps surprising that, for practical purposes, the algorithm based on two separable KS solutions is preferable to the global regularization for $N = 3$. However, the treatment of just four particles in a similar way had to wait for a technical simplification, eventually conceived by Mikkola [1985a].‖ In the event, the Ahmad–Cohen method was combined with standard KS as well as the *unperturbed* three- and four-body regularization methods to form the embryonic NBODY5 code towards the end of the 1970s. Right from the start, the KS treatment was generalized to an arbitrary number of simultaneous particle pairs, necessitating a considerable amount of automatic decision-making.

A comparison of the multiple regularization methods has been carried out for $N = 3$ and $N = 4$ [Alexander, 1986], whereas a general review of integration methods for few-body systems is also available [Aarseth, 1988a]. An early study of core collapse for $N = 1000$ illustrated the

¶ The study of initial hard binaries was urged in the thesis of Heggie [1972b].

‖ The early history of multiple regularization has been recorded by Mikkola [1997b].

usefulness of the new techniques [Aarseth, 1985b]. Finally, we mention a pioneering development of a hybrid code which combined the Fokker–Planck method with direct integration and KS regularization [McMillan & Lightman, 1984a,b].

We end this historical review by noting that ideas for increasing the speed of the calculation were discussed at an early stage [Aarseth & Hoyle, 1964]. At that time an increase in the particle number from 100 to 300 seemed to be the practical limit based on an argument that gave the computing time proportional to N^3 for a given degree of evolution. This analysis also anticipated subsequent developments of introducing a collisionless representation in order to reach much larger values of N. It was estimated that a shell method with up to five spherical harmonics would allow $N \simeq 5000$ to fit the current maximum memory of 64 K.

Modern times have seen some significant advances, both as regards software and hardware. The N-body problem has matured and we are now entering an exciting new area. In this spirit we leave history behind and will attempt to discuss a variety of relevant N-body developments in subsequent chapters.

1.3 Basic concepts

In this book, we are primarily interested in applications of the original Newton's Law of Gravity, as opposed to a modified expression including softening. The equations of motion for a particle of index i in a system containing N particles then take the form

$$\ddot{\mathbf{r}}_i = -G \sum_{j=1;\, j\neq i}^{N} \frac{m_j(\mathbf{r}_i - \mathbf{r}_j)}{|\mathbf{r}_i - \mathbf{r}_j|^3} \,. \tag{1.1}$$

For convenience, we use scaled units in which $G = 1$ and define the left-hand side of (1.1) as the force *per unit mass*, $\mathbf{F}_i$. Given the initial conditions $m_i, \mathbf{r}_i, \mathbf{v}_i$ for the mass, coordinates and velocity of each particle at some instant t_0, the set of $3N$ second-order differential equations (1.1) then defines the solutions $\mathbf{r}_i(t)$ over the time interval $(-\infty, \infty)$. Alternatively, the complete solutions are also specified by $6N$ first-order equations that must be solved in a self-consistent manner, and the latter procedure is in fact usually chosen in practice.

It has been known since Newton's days that the *N-body problem* defined by (1.1) only admits exact solutions for the case of two interacting particles. All that is known with certainty beyond this is that there exist ten integrals of the motion. For completeness, let us introduce these fundamental relations which are often used as a check on accuracy. The total energy and angular momentum (E and $\mathbf{J}$) of the system are defined by

$$E = \frac{1}{2}\sum_{i=1}^{N} m_i \mathbf{v}_i^2 - \sum_{i=1}^{N}\sum_{j>i}^{N} \frac{G m_i m_j}{|\mathbf{r}_i - \mathbf{r}_j|}, \tag{1.2}$$

$$\mathbf{J} = \sum_{i=1}^{N} \mathbf{r}_i \times m_i \mathbf{v}_i . \tag{1.3}$$

The two terms of (1.2) represent the total kinetic and potential energy, respectively. Multiplying (1.1) by m_i and performing a summation, we obtain

$$\sum_{i=1}^{N} m_i \ddot{\mathbf{r}}_i = 0 \tag{1.4}$$

by symmetry. Integrating, we find that in the absence of any external forces the centre of mass of the system moves with constant velocity, thus providing an additional six conserved quantities. The demonstration that the total energy and angular momentum are also constant can be left as an exercise [see e.g. Roy, 1988, pp.113–115 for proofs]. We define T, U, W as the total kinetic, potential and external energy, with $U < 0$. The basic energy relation then takes the more general and compact form

$$E = T + U + W , \tag{1.5}$$

which is convenient for discussions. Another quantity useful for numerical algorithms is the Lagrangian energy,

$$L = T - U , \tag{1.6}$$

although the positive sign convention for U is often chosen here.

From the above, it follows that a good numerical scheme for conservative systems needs to maintain satisfactory values for the ten constants of the motion during all times of interest. Unfortunately, errors are always present in any step-wise scheme (as in the simplest numerical computation), hence we speak about the *deviation* from the initial values instead. Since the total energy is the difference between two large numbers, T and $|U|$, experience has shown that this is the most sensitive quantity for monitoring the accuracy. However, if we are unlucky, the errors might still conspire in such a way as to cancel and thereby render energy conservation meaningless. Yet, the general tendency is for such errors to be systematic and hence more readily identified. In order to make progress beyond the basic scheme outlined above, we shall simply take a positive attitude towards obtaining numerical solutions and delay a fuller discussion of this difficult subject until later.

The crossing time is undoubtedly the most intuitive time-scale relating to self-gravitational systems. For a system in approximate dynamical equilibrium it is defined by

$$t_{\rm cr} = 2\, R_{\rm V}/\sigma\,, \tag{1.7}$$

where $R_{\rm V}$ is the virial radius, obtained from the potential energy by $R_{\rm V} = GN^2\bar{m}^2/2|U|$, and σ is the rms velocity dispersion. In a state of approximate equilibrium, $\sigma^2 \simeq GN\bar{m}/2R_{\rm V}$, which gives

$$t_{\rm cr} \simeq 2\sqrt{2}\,(R_{\rm V}^3/GN\bar{m})^{1/2}\,, \tag{1.8}$$

with $\bar{m}$ the mean mass, or alternatively $t_{\rm cr} = G(N\bar{m})^{5/2}/(2|E|)^{3/2}$ from $E = \frac{1}{2}U$. Unless the total energy is positive, any significant deviation from overall equilibrium causes a stellar system to adjust globally on this time-scale which is also comparable to the free-fall time. The close encounter distance is a useful concept in collisional dynamics. It may be defined by the expression [Aarseth & Lecar, 1975]

$$R_{\rm cl} = 2\, G\bar{m}/\sigma^2\,, \tag{1.9}$$

which takes the simple form $R_{\rm cl} \simeq 4\, R_{\rm V}/N$ at equilibrium.

Since much of this book is devoted to star clusters, it may be instructive to introduce some basic parameters for clusters to set the scene for the subsequent numerical challenge. A rich open star cluster may be characterized by $N \simeq 10^4$, $\bar{m} \simeq 0.5\, M_\odot$ and $R_{\rm V} \simeq 4\,{\rm pc}$, which yields $t_{\rm cr} \simeq 5 \times 10^6\,{\rm yr}$. Many such clusters have ages exceeding several Gyr, hence a typical star may traverse or orbit the central region many times, depending on its angular momentum. Another relevant time-scale in N-body simulations is the orbital period of a binary. Let us consider a typical close binary with separation $a \simeq R_{\rm V}/N$. With a period of $\simeq 700\,{\rm yr}$ this would make some 7000 orbits in just one crossing time. Thus, in general, if $a = fR_{\rm V}/N$ there would be $\simeq N/f^{3/2}$ Kepler orbits per crossing time.

The subject of relaxation time is fundamental and was mainly formulated by Rosseland [1928], Ambartsumian [1938, 1985], Spitzer [1940] and Chandrasekhar [1942]. The classical expression is given by

$$t_{\rm E} = \frac{1}{16}\left(\frac{3\pi}{2}\right)^{1/2}\left(\frac{NR^3}{Gm}\right)^{1/2}\frac{1}{\ln(0.4N)}\,, \tag{1.10}$$

where R is the size of the homogeneous system [Chandrasekhar, 1942]. For the purposes of star cluster dynamics, the half-mass relaxation time is perhaps more useful since it is not sensitive to the density profile.

Following Spitzer [1987], it is defined by**

$$t_{\rm rh} = 0.138 \left(\frac{N r_{\rm h}^3}{Gm}\right)^{1/2} \frac{1}{\ln(\gamma N)}, \tag{1.11}$$

where $r_{\rm h}$ is the half-mass radius and $\Lambda = \gamma N$ is the argument of the Coulomb logarithm. Formally this factor is obtained by integrating over all impact parameters in two-body encounters, with a historical value of $\gamma = 0.4$. Some of the most important subsequent determinations are due to Hénon [1975] and Giersz & Heggie [1994a], who obtained the respective values 0.15 and 0.11 for equal masses, with the latter derived from numerical measurements. Although this factor only enters through the term $\ln(\gamma N)$, it can still make a significant difference in numerical comparisons which are now becoming quite reliable when using ensemble averages. As the second authors point out, the corresponding value for a general mass spectrum is reduced considerably. From the numerical example above we then have $t_{\rm rh} \simeq 3 \times 10^8$ yr for $r_{\rm h} \simeq 4$ pc and an equal-mass system with $N = 1 \times 10^4$ stars of half a solar mass. In comparison, $t_{\rm rh} \simeq 3 \times 10^{10}$ yr for a globular cluster with $N \simeq 10^6$ and $r_{\rm h} \simeq 25$ pc.

An alternative viewpoint on the derivation of the two-body relaxation time is promoted in the review by Spurzem [1999]. Based on the pioneering work of Larson [1970] which was continued by Louis & Spurzem [1991] and Giersz & Spurzem [1994], the collisional term in the Fokker–Planck description can be developed to yield unambiguous expressions for the classical types of relaxation discussed here. Now the relaxation time emerges naturally as the consequence of the interaction of two distribution functions and the choice of their form as well as that of the Coulomb logarithm uniquely determines the nature of the different processes. Thus instead of assuming the usual small angle deflections of the orbit, it is inferred directly that the Coulomb integral starts at an angle of 90°.

The expression (1.11) gives an estimate of the time for the rms velocity change arising from small angle deflections at the half-mass radius to become comparable to the initial velocity dispersion. It serves as a useful reference time for significant dynamical changes affecting the whole cluster even though there is no corresponding numerically well-defined quantity. The assumption of approximate equilibrium with the above definition of the crossing time leads to the relation [Spitzer, 1987]

$$\frac{t_{\rm rh}}{t_{\rm cr}} \simeq \frac{N}{22 \ln(\gamma N)}, \tag{1.12}$$

which shows that close encounters become less important for increasing particle number since the potential is smoother. Hence if the relaxation

** Also see Spitzer & Hart [1971a] for an alternative derivation.

time for an equal-mass system exceeds the time interval of interest by a significant factor, the use of the collisionless approximation which neglects close encounters may be justified. However, the approach to the collisionless regime is slow and in any case the central relaxation time may be much shorter.

An equivalent formulation of the relaxation time in terms of the deflection angles suffered by a test star yields comparable values to (1.10) [Williamson & Chandrasekhar, 1941]. This expression has in fact been tested numerically for different velocities [Lecar & Cruz-González, 1972] and particle numbers $N \leq 2500$ [Aksnes & Standish, 1969], providing agreement with theory on the assumption of independent individual encounters.

The concept of dynamical friction was introduced by Chandrasekhar [1942] who elucidated the tendency for a star to be decelerated in the direction of its motion. This refinement reconciled the predicted escape rate with the possible presence of some old open clusters. However, the analysis was not extended to the case of massive stars which later merited considerable interest with the emphasis on mass segregation in stellar systems. In the case of a slow-moving body of mass $m_2 \gg \bar{m}$ but within 20 % of the total mass, the frictional force can be written in the simplified form [Binney & Tremaine, 1987]

$$\frac{d\mathbf{v}_2}{dt} = -\frac{4\pi \ln \Lambda G^2 \rho m_2}{v_2^3} \left[\mathrm{erf}(X) - \frac{2X}{\sqrt{\pi}} \exp(-X^2) \right] \mathbf{v}_2 \,, \tag{1.13}$$

where ρ is the background density and $X = v_2/(2\sigma)^{1/2}$.

Rich star clusters are usually centrally concentrated, with an extended halo. The majority of central stars are strongly bound and therefore experience changes in their orbital elements on shorter time-scales than given by (1.11). A corresponding mean relaxation time can be derived by integrating the general expression [e.g. Chandrasekhar, 1942] for a given cluster model. This was done a long time ago for polytropic models, increasing the classical value by a factor of 4 in the case of $n = 5$ [King, 1958]. On the other hand, the central relaxation time can be much shorter for realistic models with high central densities. This runaway process called core collapse (and its aftermath) has fascinated theoreticians and will be discussed further in another chapter. Let us just remark that the formation of a bound halo, together with a small fraction of escaping particles, is a direct consequence of this process by virtue of energy conservation. In short, the evolution takes place because there is no equilibrium.

So far we have mainly considered equal-mass systems, which are more amenable to analytical treatment and have therefore attracted more attention. However, the general case of a mass spectrum is more relevant for star cluster simulations. The time-scale associated with some aspects of mass segregation is probably better determined than the relaxation times above. Analysis of a two-component system dominated by light particles gave rise to the equipartition time for kinetic energy [Spitzer, 1969]

$$t_{\rm eq} = \frac{(\bar{v}_1^2 + \bar{v}_2^2)^{3/2}}{8(6\pi)^{1/2} G^2 \rho_{01} m_2 \ln N_1}\,, \tag{1.14}$$

where ρ_{01} is the central density of the N_1 light stars of mass m_1. It is envisaged that the heavy particles of mass m_2 lose kinetic energy through encounters with lighter particles of mass m_1 and spiral inwards.

The expression above holds, provided that the heavy particles do not form a self-gravitating system, in which case standard relaxation takes over. The equipartition condition is expressed in terms of the corresponding total masses as $M_2/M_1 < \beta(m_1/m_2)^{3/2}$, where $\beta \simeq 0.16$ for large mass ratios. After a phase of contraction the heavy particles begin to form a self-gravitating system and the evolution rate slows down. To the extent that the expression (1.14) is applicable, it can be seen that the presence of a mass spectrum speeds up the early evolution. Hence, in general, we have that $t_{\rm eq} \simeq t_{\rm E}\bar{m}/m_2$ for the case of two unsegregated populations with comparable velocity dispersions [Spitzer, 1969]. Comprehensive theoretical discussions of time-scales and evolution processes in rich star clusters can be found in several reviews [Meylan & Heggie, 1997; Gerhard, 2000]. However, we emphasize that as yet there is no consistent theory of the relaxation time for a realistic IMF.

Although most old clusters are in a state of approximate virial equilibrium, this may not be the case for very young clusters. Non-equilibrium initial conditions are often chosen in simulations in order to model systems with significant mass motions. Some early simulations that employed a spherical shell model demonstrated that collisionless systems reach overall equilibrium on a relatively short time-scale [Hénon, 1964, 1968]. The concept of *violent relaxation* [Lynden-Bell, 1967] was introduced to describe galaxy formation but is equally relevant for star clusters. Before making some general comments, let us write the virial theorem in the traditional scalar form [Chandrasekhar, 1942, p.219; Fukushige & Heggie, 1995]

$$d^2I/dt^2 = 4T + 2U + 4A - 4W\,, \tag{1.15}$$

where I is the moment of inertia and A represents the angular momentum contribution, $\Omega_z J_z$, for cluster motion in the Galactic plane with angular velocity Ω_z. Hence in this case the virial ratio is defined by

$$Q_{\rm vir} = (T + A)/|U - 2W|)\,. \tag{1.16}$$

Setting $A = 0$ and $W = 0$ for simplicity and choosing initial velocities, collapse takes place if $Q_{\rm vir} < 0.5$, with enhanced mass motions for small values.

A qualitative description of the collapse phase may be made by considering the energy per unit mass of a particle,

$$E_i = \tfrac{1}{2}\mathbf{v}_i^2 + \Phi_i \,, \tag{1.17}$$

with velocity $\mathbf{v}_i$ and potential Φ_i. In the extreme case of starting from rest, all the particles move inwards on radial orbits. These orbits are perturbed by neighbouring particles, acquiring angular momentum. This leads to a dispersion in the collapse times, even for a homogeneous system. Consequently, the early arrivals are decelerated in their outward motion, whereas the late-comers experience a net acceleration. Following the bounce, the core–halo system may also have a significant fraction of particles with positive energy that subsequently escape. The initial collapse therefore leads to a considerable redistribution of the binding energies and the system undergoes violent relaxation. An early investigation of homogeneous N-body systems starting from rest [Standish, 1968a] showed that about 15% of the particles gained enough energy to escape. A variety of one-dimensional experiments made at the time also confirmed that an equilibrium distribution is only reached for the inner part [Lecar & Cohen, 1972].

A much more careful analysis is needed to provide a detailed description of even the simplest collapsing system and is beyond the present scope [Aarseth, Lin & Papaloizou, 1988]. However, it is worth emphasizing that such systems can be studied by numerical methods, which may be used to test theoretical ideas. In the present context, violent relaxation is assumed to be collisionless and is therefore only applicable in the limit of large N. However, the general process is also effective in systems with $N = 500$ which are in fact subject to mass segregation at the same time [Aarseth & Saslaw, 1972; Aarseth, 1974].

Following on from non-equilibrium systems, the analogy with an eccentric binary illustrates some aspects relating to the virial theorem. Consider a plot of the virial ratio, $Q_{\rm vir}$, for collapsing systems that shows several oscillations of decreasing amplitude about the equilibrium value $\bar{Q}_{\rm vir} = 0.5$ [Standish, 1968a].[††] However, small fluctuations are still present even after many crossing times. This behaviour can be understood by examining an isolated binary. Taking the ratio of kinetic and potential energy leads to the simple expression

$$Q_{\rm vir} = 1 - R(t)/2a \,, \tag{1.18}$$

[††] We are not concerned with the excess of kinetic energy due to escaping particles.

where $R(t)$ is the instantaneous separation. Hence an eccentric binary exhibits a varying virial ratio which depends on the phase and eccentricity. Now let such an energetic binary be part of the system. Even if its energy is constant, the contribution to the virial ratio may dominate the whole system near an eccentric pericentre. Needless to say, this feature is not of dynamical significance and because of the special treatment of binaries in the present formulation, such contributions are not included here.

Star clusters orbiting the Galaxy are subject to an external tidal field which tends to increase the disruption rate. In this connection we introduce the classical concept of tidal radius [von Hoerner, 1957; King, 1962]. The simple picture of the tidal radius is that stars that move outside this distance escape from the cluster on a relatively short time-scale. However, actual orbit calculations show that the situation is more complicated even for clusters in circular orbits [Ross, Mennim & Heggie, 1997; Heggie, 2001]. In the case of globular clusters, the process of tidal shocks also needs to be modelled [Ostriker, Spitzer & Chevalier, 1976; Spitzer, 1987].

According to theory, close encounters act to maintain a Maxwellian velocity distribution in equilibrium systems. Thus after one relaxation time, a fraction $Q_{\rm e} \simeq 0.007$ should exceed the escape velocity in an isolated system [Chandrasekhar, 1942] and then be replenished. When discussing escape from stellar systems, we distinguish between ejection due to one close encounter [Hénon, 1969] and evaporation, caused by the cumulative effect of many weak encounters. From general considerations, the former outcome declines in importance with increasing N for systems dominated by single stars, whereas the presence of binaries complicates the issue. Although the process of escape is fundamental, the complexity of the interactions is such that only general statements can be made, especially when different masses are involved. For example, classical theory states that equipartition of kinetic energy will be achieved on a time-scale $t_{\rm eq}$ which is comparable to $t_{\rm rh}$ for $N \simeq 100$ and modest mass ratios. A moment's reflection is enough to show that this argument is fallacious.

In self-gravitating systems the central escape velocity is some factor, $f_{\rm e} \geq 2$, times the rms velocity, where the actual value depends on the density profile. Consequently, the equipartition condition $mv^2 = \text{const}$ can only be satisfied for modest mass ratios, beyond which escape invariably occurs. What actually happens is that the lighter particles occupy a larger volume and hence their relaxation time increases. A better way to look at energy equipartition is to compare $m\bar{v}^2$ for certain mass groups at similar central distances, rather than globally. In any case, the tendency for massive particles to be preferentially concentrated in the central region is a direct consequence of the equipartition process, whereby the loss of kinetic energy leads to inward spiralling [Aarseth, 1974]. These simple considerations show that although theoretical concepts are very useful for a general understanding of dynamics, numerical solutions can often obtain a more consistent picture, albeit for limited particle numbers.

1.4 The first steps

The well-known saying about learning to walk before you can run is highly appropriate for the aspiring N-body simulator, since much play is made of making runs. Hence we start our Odyssey at the most primitive stage in order to illustrate the main principles involved for performing direct numerical integrations.

In order to obtain numerical solutions, we proceed by advancing all coordinates and velocities using sufficiently small intervals, re-evaluating the accelerations by the summation (1.1) after each increment. At the most primitive level we can relate the solutions at time t to the previous solution at time t_0 by a Taylor series expansion to lowest order as

$$\begin{aligned} \mathbf{v}_i(t) &= \mathbf{F}_i \Delta t + \mathbf{v}_i(t_0)\,, \\ \mathbf{r}_i(t) &= \tfrac{1}{2}\mathbf{F}_i \Delta t^2 + \mathbf{v}_i(t_0)\Delta t + \mathbf{r}_i(t_0)\,, \end{aligned} \tag{1.19}$$

where $\Delta t = t - t_0$ is a suitably chosen small time interval and $\mathbf{F}_i$ is evaluated by (1.1) at $t = t_0$. From dimensional considerations, we require that $|\mathbf{v}_i|\Delta t \ll r_{\rm h}$ for meaningful results. A complete solution then involves advancing (1.19) simultaneously for all the particles until some specified condition has been satisfied. This step-by-step method (standard Euler) is clearly very laborious since each force summation includes $O(N)$ operations and Δt needs to be small in order to maintain a reasonable accuracy. However, it does contain the basic idea of obtaining self-consistent solutions for the set of coupled differential equations (1.1).

Numerical solutions of equations (1.19) are readily obtained for the two-body problem. Choosing a circular binary, we find a relative error of the semi-major axis *per orbit* of $\Delta a/a \simeq 8 \times 10^{-3}$ when averaged over ten initial periods. Here the time-step was chosen according to $2\pi\eta R^{3/2}$ from Kepler's Law, with $\eta = 0.0002$, which gives 5000 steps for each revolution.

The errors reduce dramatically by going to the improved Euler method. First provisional coordinates are predicted in the usual way by

$$\tilde{\mathbf{r}}_i(t) = \tfrac{1}{2}\mathbf{F}_i \Delta t^2 + \mathbf{v}_i(t_0)\Delta t + \mathbf{r}_i(t_0)\,, \tag{1.20}$$

whereupon the new force, $\mathbf{F}_i(t)$, is obtained from (1.1). We define the average force during the interval Δt as

$$\bar{\mathbf{F}}_i = \tfrac{1}{2}\left[\mathbf{F}_i(t) + \mathbf{F}_i(t_0)\right]\,. \tag{1.21}$$

The average force is then used to calculate the final values of $\mathbf{v}_i(t)$ and $\mathbf{r}_i(t)$ according to (1.19). Now we obtain $\Delta a/a \simeq -6 \times 10^{-9}$ per revolution, whereas $\eta = 0.002$ gives $\Delta a/a \simeq -6 \times 10^{-6}$, which is considerably more accurate than the standard Euler method above for ten times as many steps. Eccentric orbits require more integration steps because of the smaller pericentre distance and also produce somewhat larger errors. Thus in the case of the improved Euler method an eccentricity $e = 0.75$ leads to $\Delta a/a \simeq -4 \times 10^{-5}$ per revolution with $\eta = 0.002$.

This simple exercise demonstrates an important aspect about numerical integrations, namely that the accuracy may be improved significantly by making better use of existing information at small extra cost. In view of the expensive summation (1.1) for large N, it is worth emphasizing that the improved scheme also uses only one force evaluation per step. This desirable property is exploited in the more sophisticated developments discussed below and in the next chapter.

After illustrating the general principles of direct N-body integration, it may be appropriate to present the basic integration method of von Hoerner [1960] since it is not available in the English literature. For historical reasons, we retain the original notation which does not use vectors. Denoting the coordinates and velocity of a particle i by $x_{\alpha i}$ and $u_{\alpha i}$, respectively, with $\alpha = 1, 2, 3$, the coupled equations of motion for a system of equal masses take the form

$$
\begin{aligned}
\frac{dx_{\alpha i}}{dt} &= u_{\alpha i} \,, \\
\frac{du_{\alpha i}}{dt} &= -Gm \sum_{j=1;\, j\neq i}^{N} \frac{x_{\alpha i} - x_{\alpha j}}{r_{ij}^3} \,,
\end{aligned}
\tag{1.22}
$$

where r_{ij} is the mutual separation. The original derivation adopted the scaling $Gm = 1$ for equal-mass systems but in any case the following discussion is general.

The new time-step is determined from the closest particle pair by taking the harmonic mean of the travel time, $\tau_1 = D_{\rm m}/V_{\rm m}$, and free-fall time, $\tau_2 = D_{\rm m}(2D_{\rm m})^{1/2}$, according to

$$
h_2 = \frac{D_{\rm m}(2D_{\rm m})^{1/2}}{\mu\left[1 + V_{\rm m}(2D_{\rm m})^{1/2}\right]} \,. \tag{1.23}
$$

Here $D_{\rm m}$ is the minimum separation, $V_{\rm m}$ the corresponding relative velocity and μ is an accuracy parameter. Consider the system at an epoch $t_0 = 0$, with h_1 the previous step. Moreover, let u_1 and x_1 denote the velocity at $-\frac{1}{2}h_1$ and coordinates at t_0, respectively, where the subscripts have been omitted for clarity. Assuming a linear dependence over the interval $[-h_1, h_2]$, we write the force as

$$b = b_1 + a_1 t\,, \tag{1.24}$$

where $a_1 = (b_1 - b_0)/h_1$ is the divided force difference over the previous interval, $[-h_1, 0]$. After some algebra we obtain the predicted velocity and coordinates

$$\begin{aligned} u_2^0 &= u_1 + k_0 b_1 + k_2 a_1\,, \\ x_2^0 &= x_1 + h_2 u_2^0 + k_1 a_1\,, \end{aligned} \tag{1.25}$$

with the coefficients $k_0 = \frac{1}{2}(h_2 + h_1),\ k_1 = \frac{1}{24}h_2^3$ and $k_2 = \frac{1}{8}(h_2^2 - h_1^2)$.

The solution can be improved after calculating the new force, b_2, at $t = h_2$. This is achieved[‡‡] by writing a parabolic force fitting function as

$$b = b_1 + a_1 t + d_2(h_1 t + t^2)/(h_2 + h_1)\,. \tag{1.26}$$

Setting $b = b_2$ at the end of the interval h_2 simplifies to

$$d_2 = (b_2 - b_1)/h_2 - a_1\,. \tag{1.27}$$

The contributions from the last term of (1.26) can now be included to yield the corrected solutions for u_2 at $t = h_2/2$ and x_2 at $t = h_2$,

$$\begin{aligned} u_2 &= u_2^0 + k_3 d_2\,, \\ x_2 &= x_2^0 + k_5 d_2\,, \end{aligned} \tag{1.28}$$

where[§§] $k_3 = \frac{1}{24}(h_2^2 + 2h_2h_1 - 2h_1^2)$ and $k_5 = \frac{1}{12}h_2(h_2^2 + h_2h_1 - h_1^2)$.

The employment of a leapfrog method gives rise to enhanced stability for a given integration order [cf. Hut, Makino & McMillan, 1995]. However, the question of the initial velocity requires special attention. Thus it is advantageous to choose a conservative initial step, h_1, and integrate backwards an interval $\Delta t = -\frac{1}{2}h_1$ before beginning the calculation. The subsequent few time-steps, h_2, may then be restricted to grow by a small factor to ensure convergence of the force polynomials. Special care is also needed for evaluating the total energy, since the velocities are known at $t - \frac{1}{2}h_2$ and it is desirable to attain the highest accuracy consistently. Thus for the purpose of calculating the kinetic energy at the end of the current time-step, h_2, the predicted velocity is obtained by integrating (1.26) over $[h_2/2,\ h_2]$ and adding u_2 which finally gives

$$u = u_2 + b_1 h_2/2 + 3a_1 h_2^2/8 + d_2 h_2^2(3h_1/8 + 7h_2/24)/(h_2 + h_1)\,. \tag{1.29}$$

[‡‡] The so-called 'semi-iteration' was also proposed by A. Schlüter [cf. von Hoerner, 1960].

[§§] Corrected for a typographical error in the last term of k_3.

Table 1.1. Integration errors with von Hoerner's method.

N	Steps	$t/t_{\rm cr}$	μ	$\Delta E/E$	ΔJ_z	$a_{\rm min}$
16	6200	5	6	1×10^{-4}	2×10^{-7}	0.046
16	8800	5	10	9×10^{-6}	3×10^{-8}	0.080
16	18 000	10	6	2×10^{-4}	4×10^{-7}	0.022
25	10 000	5	10	1×10^{-6}	1×10^{-8}	0.029
25	16 000	5	20	5×10^{-8}	1×10^{-9}	0.086
25	127 000	10	20	7×10^{-6}	2×10^{-9}	0.007

Unless there are long-lived binaries with short period, test calculations generally give satisfactory energy errors when using $\mu = 10$.

It is instructive to compare von Hoerner's method for the two-body example discussed above. The eccentric orbit with $e = 0.75$ and $\eta = 0.002$ now gives $\Delta a/a \simeq -1.3\times10^{-5}$ per revolution for the case of semi-iteration. This improves to $\simeq -3\times10^{-7}$ when the corrector (1.28) is included. Hence the first N-body method is superior to the improved Euler method for the same number of steps per orbit.

A more general comparison test has also been performed. The initial conditions are generated in the same way as the original paper, which employed virialized velocities inside a homogeneous sphere of radius 1 and $m = 1$. All the calculations are carried out with standard double precision. Table 1.1 gives some characteristic values of relative energy errors and change in the angular momentum about the z-axis for intervals of $0.2t_{\rm cr}$. All deviations are measured with respect to initial values and the last column shows the smallest semi-major axis. Although the relative energy errors are satisfactory in these examples, the presence of a highly eccentric binary introduces a noticeable systematic orbital shrinkage which is expensive to counteract with the present basic treatment.

The device of including the semi-iteration (or corrector) without recalculating the force improves the solutions by almost one full order. It was also adopted in subsequent formulations¶¶ based on high-order force polynomials [cf. Wielen, 1967, 1972; Aarseth, 1968] with $N \leq 250$, whereas the original calculations were performed with up to 16 equal-mass particles. The choice of accuracy parameter $\mu = 6$ led to maximum relative energy errors $\Delta E/E \simeq 4 \times 10^{-3}$ for $t \simeq 9\,t_{\rm cr}$ in spite of only about ten figure machine accuracy combined with a relatively low order. However, the very first general N-body simulation already produced some interesting information on topics such as relaxation time, binary formation and escape that have stood up to the test of time.

¶¶ Already included in a third-order polynomial scheme for $N = 100$ [Aarseth, 1966a].

2
Predictor–corrector methods

2.1 Introduction

In this chapter, we provide the tools needed for standard N-body integration. We first review the traditional polynomial method which leads to increased efficiency when used in connection with individual time-steps. This self-contained treatment follows closely an earlier description [Aarseth, 1985a, 1994]. Some alternative formulations are discussed briefly for completeness. We then introduce the simpler Hermite scheme [Makino, 1991a,b] that was originally developed for special-purpose computers but is equally suitable for workstations or laptops and is attractive by its simplicity. As discussed in a later section, the success of this scheme is based on the novel concept of using quantized time-steps (factor of 2 commensurate), which reduces overheads. Variants of the Hermite method were attempted in the past, such as the low-order scheme of categories [Hayli, 1967, 1974] and the full use of explicit Taylor series derivatives [Lecar, Loeser & Cherniack, 1974]. The former study actually introduced the idea of hierarchical time-steps with respect to individual force calculations using a low-order scheme, whereas the latter formulation is expensive (but accurate) even for modest particle numbers.

2.2 Force polynomials

The force acting on a particle usually varies in a smooth manner throughout an orbit, provided the particle number is sufficiently large. Hence by fitting a polynomial through some past points, it is possible to extend the time interval for advancing the equations of motion and thereby reduce the number of force evaluations. In other words, we can use the past information to predict the future motion with greater confidence. Such a scheme was already introduced in the pioneering work of von Hoerner

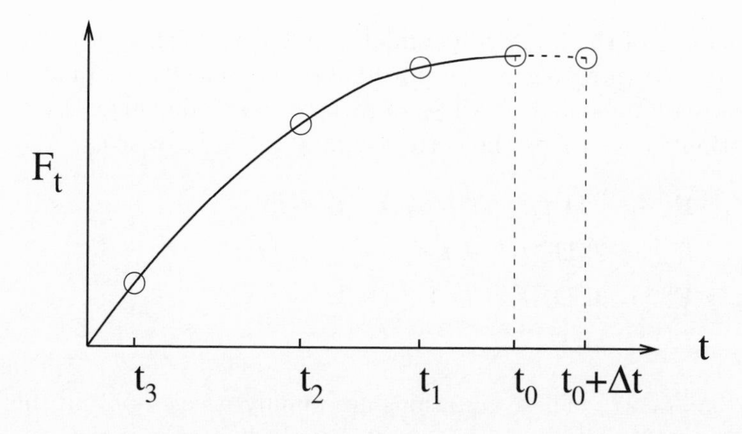

Fig. 2.1. Force polynomial fitting.

[1960], who adopted a quadratic interpolation of the force on each particle. In the following years formulations based on higher orders were employed [Aarseth, 1966a, 1968; Wielen, 1967]. Experience has shown that there is some gain in increasing the order of the integration scheme but the law of diminishing returns applies and it appears that four orders is sufficient for most purposes [Wielen, 1967, 1974]. A subsequent investigation [Makino, 1991a] showed that the fourth-order scheme* is within 30% of the minimum cost for all values of the rms energy error. The present difference formulation is based on the notation of Ahmad & Cohen [1973; hereafter AC] and follows closely an earlier treatment [Aarseth, 1985a]. In the subsequent description we omit the particle subscript in $\mathbf{F}_i$ and related quantities for clarity.

On some time-scale, the force on a particle can be considered to be smoothly varying, as illustrated by Fig. 2.1, and can therefore be approximated by a continuous function. Given the values of $\mathbf{F}$ at four successive past epochs t_3, t_2, t_1, t_0, with t_0 the most recent, we write a fourth-order fitting polynomial at time t valid in the interval $[t_3, t_0 + \Delta t]$ as

$$\mathbf{F}_t = \left\{\left[\left(\mathbf{D}^4(t-t_3) + \mathbf{D}^3\right)(t-t_2) + \mathbf{D}^2\right](t-t_1) + \mathbf{D}^1\right\}(t-t_0) + \mathbf{F}_0\,. \tag{2.1}$$

Using compact notation, the first three divided differences are defined by

$$\mathbf{D}^k[t_0, t_k] = \frac{\mathbf{D}^{k-1}[t_0, t_{k-1}] - \mathbf{D}^{k-1}[t_1, t_k]}{t_0 - t_k}\,, \qquad (k = 1, 2, 3) \tag{2.2}$$

where $\mathbf{D}^0 \equiv \mathbf{F}$ and square brackets refer to the appropriate time intervals, such that $\mathbf{D}^2[t_1, t_3]$, for instance, is evaluated at t_1. The term $\mathbf{D}^4$ is defined similarly by $\mathbf{D}^3[t, t_2]$ and $\mathbf{D}^3[t_0, t_3]$.

* This is two orders more than used by von Hoerner but one order less than Wielen.

Conversion of the force polynomial into a Taylor series provides simple expressions for integrating the coordinates and velocities. Equating terms in the successive time derivatives of (2.1) with an equivalent Taylor series and setting $t = t_0$ yields the corresponding force derivatives

$$
\begin{aligned}
\mathbf{F}^{(1)} &= [(\mathbf{D}^4 t'_3 + \mathbf{D}^3)t'_2 + \mathbf{D}^2]t'_1 + \mathbf{D}^1 \,, \\
\mathbf{F}^{(2)} &= 2![\mathbf{D}^4(t'_1 t'_2 + t'_2 t'_3 + t'_1 t'_3) + \mathbf{D}^3(t'_1 + t'_2) + \mathbf{D}^2] \,, \\
\mathbf{F}^{(3)} &= 3![\mathbf{D}^4(t'_1 + t'_2 + t'_3) + \mathbf{D}^3] \,, \\
\mathbf{F}^{(4)} &= 4!\mathbf{D}^4 \,,
\end{aligned}
\tag{2.3}
$$

where $t'_k = t_0 - t_k$. These equations are mainly used to obtain the Taylor series derivatives at $t = t_0$, when the fourth difference is not yet known. Thus the contribution from $\mathbf{D}^4$ to each order is only added at the end of an integration step, $t_0 + \Delta t$. This semi-iteration, first introduced by von Hoerner [1960], gives increased accuracy at little extra cost (on scalar machines) and no extra memory requirement.

We now describe the initialization procedure, assuming one force polynomial. From the initial conditions, $m_j, \mathbf{r}_j, \mathbf{v}_j$, the respective Taylor series derivatives are formed by successive differentiations of (1.1). Introducing the relative coordinates, $\mathbf{R} = \mathbf{r}_i - \mathbf{r}_j$, and relative velocity, $\mathbf{V} = \mathbf{v}_i - \mathbf{v}_j$, all pair-wise interaction terms in $\mathbf{F}$ and $\mathbf{F}^{(1)}$ are first obtained by

$$
\begin{aligned}
\mathbf{F}_{ij} &= -m_j \mathbf{R}/R^3 \,, \\
\mathbf{F}^{(1)}_{ij} &= -m_j \mathbf{V}/R^3 - 3a\mathbf{F}_{ij} \,,
\end{aligned}
\tag{2.4}
$$

with $a = \mathbf{R} \cdot \mathbf{V}/R^2$. The total contributions are obtained by summation over all N particles. Next, the mutual second- and third-order terms are formed from

$$
\begin{aligned}
\mathbf{F}^{(2)}_{ij} &= -m_j(\mathbf{F}_i - \mathbf{F}_j)/R^3 - 6a\mathbf{F}^{(1)}_{ij} - 3b\mathbf{F}_{ij} \,, \\
\mathbf{F}^{(3)}_{ij} &= -m_j(\mathbf{F}^{(1)}_i - \mathbf{F}^{(1)}_j)/R^3 - 9a\mathbf{F}^{(2)}_{ij} - 9b\mathbf{F}^{(1)}_{ij} - 3c\mathbf{F}_{ij} \,,
\end{aligned}
\tag{2.5}
$$

with

$$
\begin{aligned}
b &= \left(\frac{V}{R}\right)^2 + \frac{\mathbf{R} \cdot (\mathbf{F}_i - \mathbf{F}_j)}{R^2} + a^2 \,, \\
c &= \frac{3\mathbf{V} \cdot (\mathbf{F}_i - \mathbf{F}_j)}{R^2} + \frac{\mathbf{R} \cdot (\mathbf{F}^{(1)}_i - \mathbf{F}^{(1)}_j)}{R^2} + a(3b - 4a^2) \,.
\end{aligned}
\tag{2.6}
$$

A second double summation gives the corresponding values of $\mathbf{F}^{(2)}$ and $\mathbf{F}^{(3)}$ for all particles. This pair-wise *boot-strapping* procedure provides a convenient starting algorithm, since the extra cost is usually small. Here we have employed a compact derivation [Findlay, private communication, 1983] instead of the equivalent but more cumbersome expressions used previously [cf. Aarseth, 1972b].

Appropriate initial time-steps, Δt_i, are now determined, using the general criterion discussed in the next section. Setting $t_0 = 0$, the backwards times are initialized by $t_k = -k\Delta t_i$ $(k = 1, 2, 3)$. Hence this assumes constant time-steps over the past fitting interval. Inversion of (2.3) to third order finally yields starting values for the divided differences,

$$\begin{aligned}
\mathbf{D}^1 &= (\tfrac{1}{6}\mathbf{F}^{(3)}t_1' - \tfrac{1}{2}\mathbf{F}^{(2)})t_1' + \mathbf{F}^{(1)}\,, \\
\mathbf{D}^2 &= -\tfrac{1}{6}\mathbf{F}^{(3)}(t_1' + t_2') + \tfrac{1}{2}\mathbf{F}^{(2)}\,, \\
\mathbf{D}^3 &= \tfrac{1}{6}\mathbf{F}^{(3)}\,.
\end{aligned} \qquad (2.7)$$

It should be remarked that polynomial initialization may also be required at any stage of the calculation, after switching from standard integration to more sophisticated treatments and *vice versa.*

The introduction of a softened potential of the form

$$\Phi = -m/(R^2 + \epsilon^2)^{1/2} \qquad (2.8)$$

is of historical interest [Aarseth, 1963a,b]. This represents a Plummer sphere [Plummer, 1911] with half-mass radius given by $r_{\rm h} \simeq 1.3\epsilon$ [Aarseth & Fall, 1980]. Originally it was used to model galaxies with ϵ representing the characteristic size, and has been employed more generally to reduce the effect of close encounters. Softening may readily be included by modifying all inverse R terms in the denominators of equations (2.4)–(2.6). For some purposes, the corresponding radial force does not fall off sufficiently fast with distance and a steeper r-dependence given by $\Phi = -m/(R^4 + \epsilon^4)^{1/4}$ has been tried [Oh, Lin & Aarseth, 1995]. This representation also has the advantage that outside some distance it can be replaced by the basic point-mass interaction without significant loss of accuracy. Finally, it may be remarked that a softened potential necessitates modifying the virial expression which is often used in simulations. Neglecting external effects, the potential energy is then replaced by a double summation over $m_i\mathbf{r}_{ij} \cdot \mathbf{F}_{ij}$ which gives the virial energy

$$V = -\sum_{i=1}^{N} \sum_{j=1;\, j\neq i}^{N} \frac{m_j|\mathbf{r}_i - \mathbf{r}_j|^2}{(|\mathbf{r}_i - \mathbf{r}_j|^2 + \epsilon^2)^{3/2}}\,. \qquad (2.9)$$

2.3 Individual time-steps

Stellar systems are characterized by a range in density that gives rise to different time-scales for significant changes of the orbital parameters.

In order to exploit this feature, and economize on the expensive force calculation, each particle is assigned its own time-step which is related to the orbital time-scale. Thus the aim is to ensure convergence of the force polynomial (2.1) with the minimum number of force evaluations. Since all interactions must be added consistently in a direct integration method, it is necessary to include a temporary coordinate prediction of the other particles. However, the additional cost of low-order predictions still leads to a significant overall saving since this permits a wide range of time-steps to be used.

Following the polynomial initialization discussed above, the integration cycle itself begins by determining the next particle, i, to be advanced; i.e. the particle, j, with the smallest value of $t_j + \Delta t_j$, where t_j is the time of the last force evaluation. It is convenient to define the present epoch, or 'global' time, t, at this endpoint, rather than adding a small interval to the previous value. The complete integration cycle consists of the sequence given by Algorithm 2.1.

Algorithm 2.1. *Individual time-step cycle.*

1	Determine the next particle: $i = \min_j \{t_j + \Delta t_j\}$
2	Set the new global time by $t = t_i + \Delta t_i$
3	Predict all coordinates $\mathbf{r}_j$ to order $\mathbf{F}^{(1)}$
4	Form $\mathbf{F}^{(2)}$ by the second equation (2.3)
5	Improve $\mathbf{r}_i$ and predict $\mathbf{v}_i$ to order $\mathbf{F}^{(3)}$
6	Obtain the new force $\mathbf{F}_i$
7	Update the times t_k and differences $\mathbf{D}^k$
8	Apply the corrector $\mathbf{D}^4$ to $\mathbf{r}_i$ and $\mathbf{v}_i$
9	Specify the new time-step Δt_i
10	Repeat the calculation at step 1

The individual time-step scheme [Aarseth, 1963a,b] uses two types of coordinates for each particle. We define primary and secondary coordinates, $\mathbf{r}_0$ and $\mathbf{r}_\mathrm{t}$, evaluated at t_0 and t, respectively, where the latter are derived from the former by the predictor. In the present treatment, unless high precision is required, we predict all the coordinates to order $\mathbf{F}^{(1)}$ by

$$\mathbf{r}_j = [(\tfrac{1}{6}\mathbf{F}^{(1)}\delta t'_j + \tfrac{1}{2}\mathbf{F})\delta t'_j + \mathbf{v}_0]\delta t'_j + \mathbf{r}_0\,, \tag{2.10}$$

where $\delta t'_j = t - t_j$ (with $\delta t'_j \leq \Delta t_j$). The coordinates and velocity of particle i are then improved to order $\mathbf{F}^{(3)}$ by standard Taylor series integration (cf. 2.3), whereupon the current force is calculated by direct summation. At this stage the four times t_k are updated (i.e. replacing t_k with t_{k-1}) to be consistent with the definition that t_0 denotes the time of the most recent force evaluation. New differences are now formed (cf. 2.2), including $\mathbf{D}^4$. Together with the new $\mathbf{F}^{(4)}$, these correction terms are combined to improve the current coordinates and velocity to highest order. The coordinate and velocity increments of (2.1) due to the corrector $\mathbf{D}^4$ contain four terms since all the lower derivatives are also modified in (2.3). Consequently, we combine the corresponding time-step factors which yield in compact notation, with $t' = t - t_0$

$$
\begin{aligned}
\Delta \mathbf{r}_i &= \mathbf{F}^{(4)} \left\{ [(\tfrac{2}{3}t' + c)0.6t' + b]\tfrac{1}{12}t' + \tfrac{1}{6}a \right\} t'^3 \,, \\
\Delta \mathbf{v}_i &= \mathbf{F}^{(4)} \left\{ [(0.2t' + 0.25c)t' + \tfrac{1}{3}b]t' + 0.5a \right\} t'^2 \,,
\end{aligned}
\tag{2.11}
$$

where all factorials are absorbed in the force derivatives. The coefficients are defined by $a = t'_1 t'_2 t'_3$, $b = t'_1 t'_2 + t'_1 t'_3 + t'_2 t'_3$, $c = t'_1 + t'_2 + t'_3$, respectively, where the old definition of t'_k still applies. Finally, the primary coordinates are initialized by setting $\mathbf{r}_0 = \mathbf{r}_t$. Hence we have a fourth-order predictor–corrector scheme.

New time-steps are assigned initially for all particles and at the end of each integration cycle for particle i. General considerations of convergence for the corresponding Taylor series (2.1) suggest a time-step of the type

$$
\Delta t_i = \left(\frac{\eta |\mathbf{F}|}{|\mathbf{F}^{(2)}|} \right)^{1/2} , \tag{2.12}
$$

where η is a dimensionless accuracy parameter. Such an expression would have the desirable property of ensuring similar *relative* errors of the force. Moreover, two particles of different mass interacting strongly would tend to have very similar time-steps, which also has certain practical advantages. However, there are situations when this simple form is less satisfactory. After considerable experimentation, we have adopted a more sensitive composite criterion given by

$$
\Delta t_i = \left(\frac{\eta (|\mathbf{F}||\mathbf{F}^{(2)}| + |\mathbf{F}^{(1)}|^2)}{|\mathbf{F}^{(1)}||\mathbf{F}^{(3)}| + |\mathbf{F}^{(2)}|^2} \right)^{1/2} . \tag{2.13}
$$

For this purpose only the last two terms of the first and second force derivatives in (2.3) are included. This expression ensures that all the force derivatives play a role and it is also well defined for special cases (i.e. starting from rest or $|\mathbf{F}| \simeq 0$). Although successive time-steps normally change smoothly, it is prudent to restrict the growth by an inertial factor (e.g. 1.2). Being more sensitive, typical time-steps are about $\sqrt{2}$ times smaller than given by (2.12) for the same value of η.

In summary, the scheme requires the following 30 variables for each particle: m, $\mathbf{r}_0$, $\mathbf{r}_t$, $\mathbf{v}_0$, $\mathbf{F}$, $\mathbf{F}^{(1)}$, $\mathbf{D}^1$, $\mathbf{D}^2$, $\mathbf{D}^3$, Δt, t_0, t_1, t_2, t_3. It is also useful to employ a secondary velocity, denoted $\mathbf{v}_t$, for dual purposes, such as temporary predictions and general evaluations.

2.4 Alternative formulations

Over the years there have been several attempts to determine the optimal order for the individual time-step scheme, as well as other proposals for time-step criteria and polynomial representations. Following the discussion of the divided difference scheme introduced by AC above, it is instructive to consider briefly the original formulation [Aarseth, 1963a,b]. In the improved version with one extra order [Aarseth, 1968], the force polynomial for a given particle is written as an expansion about the reference time $t = 0$ by

$$\mathbf{F} = \mathbf{F}_0 + \sum_{k=1}^{4} \mathbf{B}_k t^k \,. \tag{2.14}$$

The coefficients $\mathbf{B}_k$ for $k = 1, 2, 3$ are obtained by a fitting over three previous times, whereas the fourth coefficient is not evaluated until the end of the current step when its contribution is added as a semi-iteration or corrector. Although this representation is equivalent to the divided difference form (2.1), the coefficients which were derived by Newton's interpolation formula are cumbersome and codes based on (2.14) were phased out in the early 1970s.

A formulation for arbitrary orders was employed by Wielen [1967, 1974] who also introduced divided differences. Following numerical tests, the order $n = 4$ was chosen, compared with $n = 3$ for (2.1) and (2.14).† The time-step is obtained from the difference between the actual and extrapolated values of the particle force. An approximation of the former by a polynomial that includes one more fitting point gives the time-step as a solution of the non-linear equation

$$|\tilde{\mathbf{F}}(t_i + \Delta t_i) - \mathbf{F}(t_i + \Delta t_i)| = \epsilon_{\rm abs} + \epsilon_{\rm rel}|\mathbf{F}| \,. \tag{2.15}$$

Here $\tilde{\mathbf{F}}$ represents the force polynomial of one degree higher than available at the prediction, and the right-hand side consists of absolute and relative error terms. From the known coefficients, the new time-step can be readily obtained by iteration. Provided that $\epsilon_{\rm abs} \ll \epsilon_{\rm rel}|\mathbf{F}|$, this criterion is based on the relative change of the force in a similar manner to (2.13). Although this procedure ensures that the next omitted term is acceptably small, the alternative expression (2.13) involving all the higher force derivatives has proved satisfactory for the particle numbers studied so far. In fact, no specific claim for an improved criterion at this order of integration has been suggested up to now.

Force polynomials can be dispensed with altogether by an explicit calculation of the derivatives. This procedure was introduced by Lecar, Loeser

† With $n = 3$, pair-wise initialization can be done in just two stages (cf. (2.5)).

& Cherniack [1974] who employed up to five force derivatives in the Taylor series solution.[‡] Since evaluation of the higher orders is increasingly expensive without allowing much larger time-steps, this scheme is only suitable for relatively small systems. However, it proved useful in early investigations of Solar System problems since the dominant central body introduces additional errors when integrated by the fourth-order difference method. In the event, the explicit differentiation up to order $\mathbf{F}^{(3)}$ was implemented for initialization of standard force polynomials.

Another early method which was only used by its inventor is the so-called 'category scheme' of Hayli [1967, 1974]. The concept of time quantization was first introduced here and this work may therefore be regarded as a precursor for the Hermite method. Thus the particles are assigned individual time-steps, $t_0/b, t_0/b^2, \ldots, t_0/b^p$, where t_0 is the time unit and b an integer greater than one. Individual forces are evaluated in two parts, with direct summation over all interactions arising from the current and higher levels, whereas the remaining contributions are added as a sum based on linear extrapolation of the first derivative. Finally, appropriate adjustments to the latter are made when a given particle changes its level in the hierarchy. Unfortunately, this scheme was never generalized to higher orders and the relatively large energy errors that result seem a high price to pay for the gain in execution time.

A study by Mann [1987] considered a variety of standard methods. Since only three- and five-body systems were examined, however, the conclusions are not strictly applicable to larger memberships where the force calculation dominates the cost. The absence of close encounters also made the comparison tests less representative of typical situations.

The choice of order and extrapolation method was examined by Press & Spergel [1988] who claimed that considerably higher orders would be beneficial. However, the authors did not take into account that higher order schemes have smaller radii of convergence than lower order ones [Mikkola, private communication]. Thus the theoretical limiting time-step factor of 4 with respect to the standard case for a relative force error of 10^{-4} does not seem a practical proposition. The conclusion of a higher optimal order was also challenged by Sweatman [1994], who proposed that a fixed error per time unit is more appropriate. Addition of a constant part due to the force calculation led to an optimal order $n \leq 5$, in good agreement with Wielen [1967] who adopted $n = 4$. Finally, the investigation by Press & Spergel compared the merits of using rational function extrapolation. It was found that this approach does not offer any significant advantage compared with the standard polynomial representation.

[‡] The original third-order formulation was given by Gonzalez & Lecar [1968].

In a careful analysis of integration schemes, Makino [1991a] suggested a time-step criterion based on the difference between the predicted and corrected velocity change according to

$$\Delta t_{\rm new} = \Delta t_{\rm old} \left(\frac{\epsilon_{\rm v} \Delta t_{\rm old} |\mathbf{F}|}{|\Delta \mathbf{v}_{\rm p} - \Delta \mathbf{v}_{\rm c}|} \right)^{1/(p+1)} , \tag{2.16}$$

where $\epsilon_{\rm v}$ is a parameter controlling the accuracy and p is the step-number ($p = n + 1$). Note that when evaluating $\Delta \mathbf{v}_{\rm p}$, we need the full p-step predictor as opposed to $\mathbf{F}$ and $\dot{\mathbf{F}}$ used in the original scheme. Depending on other factors, such as hardware and particle number, the extra effort of performing a high-order prediction for the particle (or particles) under consideration may well be acceptable in return for a more sensitive criterion if this should prove to be the case.[§] However, the expression (2.16) did not show any significant improvement for the step-number $p = 4$, whereas the unfavourable comparison for $p > 4$ is not justified without including the highest force derivatives by generalizing (2.13).

The question of calculation cost was also addressed by Makino [1991a]. Thus from actual N-body simulations, as opposed to a formal analysis of extrapolation methods, it was concluded that the standard difference formulation $p = 4$ is within 30% of the minimum value for a range of practical accuracies. It should be noted that higher orders would be prone to numerical problems and be more cumbersome both as regards initialization[¶] and use of the corrector.

An investigation of rational extrapolation by Sweatman [2002b] contains some interesting ideas. This work generalizes the scalar extrapolation formalism of Press & Spergel to a vector representation that appears to have considerable merit. One of the expressions considered is of the form

$$\mathbf{f}_{\rm v}(t) = \frac{\mathbf{p}_3 t^3 + \mathbf{p}_2 t^2 + \mathbf{p}_1 t + \mathbf{p}_0}{q_4 t^4 + q_3 t^3 + q_2 t^2 + q_1 t + q_0} , \tag{2.17}$$

where $\mathbf{p}(t)$ is a polynomial with vector coefficients. Although a complete N-body code based on this development has yet to be constructed, the maximum interval of force extrapolations for a given accuracy tends to be larger for the vector rational interpolant. In particular, there is a greater relative increase in extrapolation time for the smallest time-steps, which are also the most time-consuming. Hence after all this time, a new way of looking at our old problem may result in code improvements. This theme is taken up in the next section and has already proved itself in practical applications.

[§] For large N and Δt, (2.16) is now used as a secondary criterion (cf. section 15.9).

[¶] The higher differences may be set to zero at the expense of the time-step criterion.

2.5 Hermite scheme

Although the standard polynomial scheme has proved itself over more than 30 years, the rapid advance in computer technology calls for a critical appraisal and search for alternative formulations. The recent design of special-purpose computers, to be described in chapter 13, poses a particular challenge for software developments. The essential idea is to provide a very fast evaluation of the force and its first derivative by special hardware, and these quantities are then utilized by the integration scheme which is implemented on some front-end machine, such as a standard workstation.

In order to increase the accuracy of integration based on the explicit values of $\mathbf{F}$ and $\mathbf{F}^{(1)}$, it is desirable to include a high-order corrector in the manner of the polynomial formulation. Following Makino [1991a], we write a Taylor series for the force and its first derivative to third order about the reference time t as

$$\mathbf{F} = \mathbf{F}_0 + \mathbf{F}_0^{(1)} t + \tfrac{1}{2}\mathbf{F}_0^{(2)} t^2 + \tfrac{1}{6}\mathbf{F}_0^{(3)} t^3 \,, \tag{2.18}$$

$$\mathbf{F}^{(1)} = \mathbf{F}_0^{(1)} + \mathbf{F}_0^{(2)} t + \tfrac{1}{2}\mathbf{F}_0^{(3)} t^2 \,. \tag{2.19}$$

Substituting $\mathbf{F}_0^{(2)}$ from (2.19) into (2.18) and simplifying, we obtain the third derivative corrector

$$\mathbf{F}_0^{(3)} = 6[2(\mathbf{F}_0 - \mathbf{F}) + (\mathbf{F}_0^{(1)} + \mathbf{F}^{(1)})t]/t^3 \,. \tag{2.20}$$

Similarly, substitution of (2.20) into (2.18) gives the second derivative corrector

$$\mathbf{F}_0^{(2)} = 2[-3(\mathbf{F}_0 - \mathbf{F}) - (2\mathbf{F}_0^{(1)} + \mathbf{F}^{(1)})t]/t^2 \,. \tag{2.21}$$

Using $\mathbf{F}$ and $\mathbf{F}^{(1)}$ evaluated at the beginning of a time-step, the coordinates and velocities are predicted to order $\mathbf{F}^{(1)}$ for all particles by

$$\mathbf{r}_i = [(\tfrac{1}{6}\mathbf{F}_0^{(1)} \delta t_i' + \tfrac{1}{2}\mathbf{F}_0)\delta t_i' + \mathbf{v}_0]\delta t_i' + \mathbf{r}_0 \,,$$

$$\mathbf{v}_i = (\tfrac{1}{2}\mathbf{F}_0^{(1)} \delta t_i' + \mathbf{F}_0)\delta t_i' + \mathbf{v}_0 \,, \tag{2.22}$$

where again $\delta t_i' = t - t_i$. Following the evaluation of $\mathbf{F}$ and $\mathbf{F}^{(1)}$ by summation over all contributions in (2.4), the two higher derivatives are obtained by (2.20) and (2.21). This gives rise to the composite corrector for the coordinates and velocity of a particle with time-step Δt_i,

$$\Delta \mathbf{r}_i = \tfrac{1}{24}\mathbf{F}_0^{(2)} \Delta t_i^4 + \tfrac{1}{120}\mathbf{F}_0^{(3)} \Delta t_i^5 \,,$$

$$\Delta \mathbf{v}_i = \tfrac{1}{6}\mathbf{F}_0^{(2)} \Delta t_i^3 + \tfrac{1}{24}\mathbf{F}_0^{(3)} \Delta t_i^4 \,. \tag{2.23}$$

For the purpose of subsequent predictions, an improved value of the second force derivative at the *end* of the current time-step is obtained by

$$\mathbf{F}^{(2)} = \mathbf{F}_0^{(2)} + \mathbf{F}_0^{(3)} \Delta t_i \,. \tag{2.24}$$

The prediction (2.22) can also be written in an equivalent implicit form (cf. section 2.7) which brings out the time-symmetric nature of the Hermite integrator. We emphasize the simplicity of the scheme which compensates for the extra operations of obtaining the force derivative. The name Hermite is used in numerical analysis to denote a polynomial based on the function and its derivative [see e.g. Stoer & Bulirsch, 1980, p. 52]. The basic Hermite interpolation scheme was first introduced by Makino [1991a], who also compared the convergence characteristics with the divided difference method. In addition to greater simplicity, the coefficient in the leading error term is significantly smaller than for the corresponding order of the polynomial representation.

2.6 Block time-steps

In order to reduce the prediction overheads of the Hermite scheme, it is advantageous to quantize‖ the time-steps, permitting a group of particles to be advanced at the same time [Hayli, 1967, 1974; McMillan, 1986]. In arbitrary units, with the maximum time-step defined by Δt_1, we choose hierarchical levels by the rule

$$\Delta t_n = \Delta t_1/2^{n-1}\,. \tag{2.25}$$

In principle, any level n may be prescribed. However, it is rare for more than about 12 levels to be populated in a realistic simulation with $N \leq 1000$, increasing by a few levels for $N \simeq 10^4$.

At the start of a calculation, the natural time-step given by (2.13), or a suitable low-order expression, is first specified. The nearest truncated value is then selected according to (2.25). At a general time, one of the following three cases apply when comparing the previous time-step, $\Delta t_{\rm p}$, with the new value given by (2.13):

- Reduction by a factor 2 if $\Delta t_i < \Delta t_{\rm p}$
- Increase by 2 if $\Delta t_i > 2\Delta t_{\rm p}$ and t commensurate with $2\Delta t_{\rm p}$
- No change if $\Delta t_{\rm p} < \Delta t_i < 2\Delta t_{\rm p}$

Hence time-steps can be reduced after every application of the corrector, and more than once if necessary, whereas increase by a factor 2 is only permitted every *other* time (or a factor 4 every fourth time). This behaviour reminds us of an orchestra of N members, where all the different players are in harmony. One false note, and the whole performance ends in chaos. A schematic distribution of hierarchical time-steps is illustrated in Fig. 2.2, where the membership at each level is arbitrary. However, the presence of small time-steps does affect the efficiency.

‖ The phrase 'quantization of time' was introduced at an early stage [cf. Hayli, 1974].

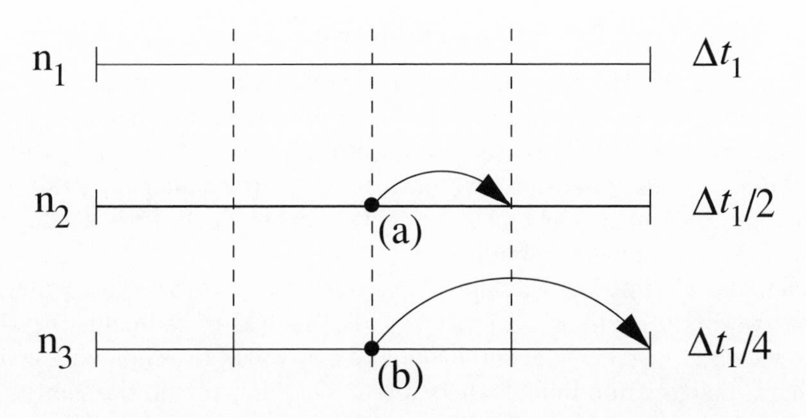

Fig. 2.2. Hierarchical time-step levels. The steps can be reduced at (a) and increased at (b). Each level l contains n_l members with time-steps $\Delta t_1/2^{l-1}$.

Extra care is necessary when initializing Hermite integration at an arbitrary time, which may be required in a sophisticated code. Thus after truncation according to (2.25), the new time-step must now also be commensurate with the current time, t. This frequently involves large reduction factors; i.e. ten or more successive reductions by 2, but is the price one must pay for such a scheme. Note that in this case, *every* time-step can be increased by a factor of 2, and even by 4 every other time, thereby reaching the appropriate value quite quickly. On the other hand, the following advantages may be emphasized:

- The Hermite scheme is self-starting
- New polynomials do not require evaluation of (2.5)
- Stability is increased due to the explicit $\mathbf{F}^{(1)}$
- The corrector is significantly faster
- Hierarchical time-steps reduce predictions
- Special-purpose computers can obtain $\mathbf{F}$ and $\mathbf{F}^{(1)}$

So far, experience with the Hermite integration scheme has been very favourable. Thus even on a standard workstation, the simpler formulation compensates for the extra cost of evaluating $\mathbf{F}^{(1)}$ explicitly, and the corresponding code based on the Ahmad–Cohen [1973] neighbour scheme is slightly faster than the standard code for $N \simeq 1000$ and a similar number of steps. The Hermite scheme may also permit slightly longer time-steps because of the increased stability of the corrector. A detailed discussion is given by Makino [1991b] and Makino & Aarseth [1992] (but see chapter 13). For a listing of the basic method see Appendix H.

2.7 Time-symmetric method

In principle, the Hermite integration scheme can be applied to a wide range of self-gravitating systems. However, in some problems it is possible to take advantage of characteristic features, such as dominant two-body motion or overall expansion. We now discuss a particular adaptation for planetary integration, whereas the next chapter deals with a comoving method for cosmological simulations.

Even the Hermite method produces systematic errors when integrating a two-body orbit over many periods. Instead of reducing the time-step, which is costly, the accuracy can be controlled by employing a time-symmetric algorithm. In fact, there are no secular errors in the semi-major axis for periodic orbits when using a constant time-step here [Quinlan & Tremaine, 1990]. This suggests that an application to a system of minor bodies orbiting the Sun may be beneficial since the planetary perturbations are usually small. Thus the method of Kokubo, Yoshinaga & Makino [1998] achieves a significant improvement in the long-term accuracy by employing almost constant time-steps combined with an iteration for the dominant force. This scheme is quite easy to implement and fits in well with the other predictor–corrector methods presented in this chapter.

The basic integration employs the Hermite scheme which can also be expressed in implicit form [Hut *et al.*, 1995; Makino *et al.*, 1997] as

$$\begin{aligned}\mathbf{r}_1 &= \mathbf{r}_0 + \tfrac{1}{2}(\mathbf{v}_1 + \mathbf{v}_0)\Delta t - \tfrac{1}{10}(\mathbf{a}_1 - \mathbf{a}_0)\Delta t^2 + \tfrac{1}{120}(\dot{\mathbf{a}}_1 + \dot{\mathbf{a}}_0)\Delta t^3\,,\\ \mathbf{v}_1 &= \mathbf{v}_0 + \tfrac{1}{2}(\mathbf{a}_1 + \mathbf{a}_0) - \tfrac{1}{12}(\dot{\mathbf{a}}_1 - \dot{\mathbf{a}}_0)\Delta t^2\,,\end{aligned} \tag{2.26}$$

where the acceleration is denoted by $\mathbf{a}$. The subscripts 0 and 1 at the beginning and end are used symmetrically, thereby making the integrator time-symmetric. The time-steps, Δt, are first obtained by the standard criterion (2.13) before truncation to the nearest block-step value (cf. 2.25). It can be verified by a simple two-body integration that (2.26) yields essentially the same accuracy as the basic Hermite formulation. Although the local truncation error for the coordinates is $O(\Delta t^5)$ compared with $O(\Delta t^6)$ in the latter, the velocity error is $O(\Delta t^5)$ for both schemes. Note that the simplified Hermite scheme (2.26) is expressed in terms of $\mathbf{a}$ and $\dot{\mathbf{a}}$. However, this is not more efficient than the standard form if the two higher derivatives are used for time-step selection.

Comparisons for the two-body problem with modest eccentricity show that the well-known errors in the standard predict–evaluate–correct (PEC) scheme are reduced significantly by going to a P(EC)^2 scheme of the same (i.e. fourth) order. With about 200 steps per orbit for $e = 0.1$ and some 100 orbits, the relative errors *per orbit* change from $\Delta a/a \simeq -7\times 10^{-9}$ to $\simeq 10^{-12}$ and $\Delta e/e$ reduces from $\simeq 10^{-7}$ to $\simeq 10^{-11}$, respectively. The main reason for this improvement is that the PEC corrector step is not time-symmetric [cf. Makino, 1991a]. Moreover, for a longer interval of 10^5 orbits, these errors decrease essentially to zero for a $P(EC)^n$ scheme, with $n = 3$ or 4. Unfortunately, the main improvements only apply to the semi-major axis and eccentricity, which are the most important elements, whereas the argument of pericentre and the pericentre passage time still exhibit a small linear error growth.

When going to a larger system, even the $P(EC)^2$ scheme becomes quite expensive since this involves recalculating all the accelerations. Note that the original time-symmetric formulation [Hut *et al.*, 1995] has only been demonstrated to be viable for hierarchical triples [Funato *et al.*, 1996] and it is difficult to maintain shared time-steps in more complicated situations. However, the application to planetary systems exploits the dominant solar force and hence preserves some of the advantage at modest extra cost. For most of the time, the planetesimals move on nearly circular orbits with small perturbations. Consequently, the time-steps can remain constant over long intervals when using block-steps. The main points of the integration cycle are given by Algorithm 2.2.

Algorithm 2.2. *Time-symmetric integration scheme.*

1	Determine the members of the current block-step at time t
2	Predict coordinates and velocities of all particles to order $\dot{\mathbf{F}}$
3	Improve the coordinates and velocity of particle i to $\mathbf{F}^{(3)}$
4	Obtain the force and first derivative from the planetesimals
5	Add the dominant contributions to $\mathbf{F}$ and $\dot{\mathbf{F}}$
6	Include the Hermite corrector $\mathbf{F}^{(2)}$ and $\mathbf{F}^{(3)}$ for $\mathbf{r}_i$ and $\mathbf{v}_i$
7	Repeat steps 5 and 6 using the corrected values
8	Specify new time-step and update the reference time, t_i
9	Complete the cycle for all other particles with $t_j + \Delta t_j = t$

As demonstrated [Kokubo *et al.*, 1998], the overall conservation of total energy is improved by about two orders of magnitude compared with the standard Hermite integration. In practice, the errors of an $n = 3$ iteration scheme are almost indistinguishable from the case $n = 2$. On the other hand, the improved prediction at step 3 speeds up the convergence significantly and is therefore worthwhile. Finally, it should be emphasized that such simulations can be very expensive on workstations if the number of planetesimals is large, since the evolution time is also rather long. This poses the question of devising approximate methods where the distant interactions are either neglected or included by a tree code approach, as will be discussed in subsequent chapters.

3
Neighbour treatments

3.1 Introduction

Most star clusters are characterized by large memberships that make direct N-body simulations very time-consuming. In order to study such systems, it is therefore necessary to design methods that speed up the calculations while retaining the collisional approach. One good way to achieve this is to employ a neighbour procedure that requires fewer total force summations. The AC neighbour scheme [Ahmad & Cohen, 1973, 1974] has proved very effective for a variety of collisional and collisionless problems. It is particularly suitable for combining with regularization treatments, where dominant particles as well as perturbers can be selected from the corresponding neighbour lists without having to search all the particles.

The AC method can also be used for cosmological simulations. The study of galaxy clustering usually employs initial conditions where the dominant motions are due to the universal expansion. By introducing *comoving* coordinates, we can integrate the deviations from the smooth Hubble flow, thereby enabling much larger time-steps to be used, at least until significant density perturbations have been formed. Naturally, such a direct scheme cannot compete with more approximate methods, such as the P^3M algorithm [Efstathiou & Eastwood, 1981; Couchman, 1991] or the parallel tree code [Dubinski, 1996], which have been developed for large N, but in some problems the ability to perform detailed modelling may still offer considerable advantages.

Several other methods to speed up direct N-body calculations have been tried for a wide range of problems. In subsequent sections, we provide a description of multipole expansion and grid perturbations, as well as the so-called 'particle in box method'. Although the potential representation in spherical harmonics is essentially collisionless for practical

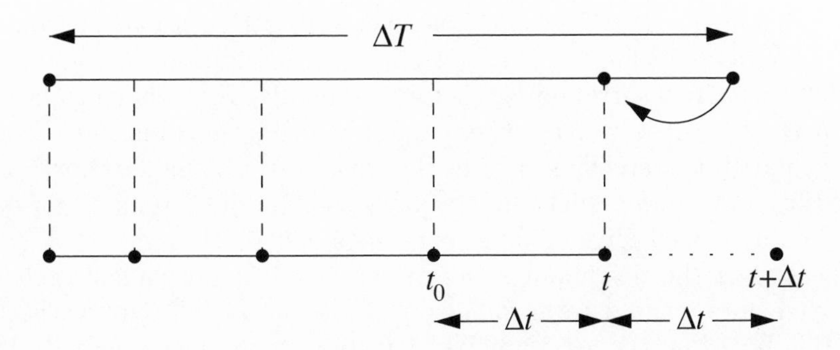

Fig. 3.1. Irregular and regular steps.

purposes, such expansions can be combined with direct summation in regions where encounters are important. In another class of problems, the motions are dominated by a smooth central force field (i.e. the Solar System or interstellar clouds in a disc galaxy), together with an irregular force component. One way to study such problems is to employ a perturbed two-body formulation (p. 45). Finally, we discuss some related methods for studying planetesimal dynamics that are based on the idea of selecting a typical patch or azimuthal ring of particles in differential motion around the Sun while subject to local perturbations and boundary conditions.

3.2 Ahmad–Cohen method

The main idea of the AC scheme is to reduce the effort of evaluating the force contribution from distant particles by combining two polynomials based on separate time-scales. Splitting the total force on a given particle into an *irregular* and a *regular* component by

$$\mathbf{F} = \mathbf{F}_{\mathrm{I}} + \mathbf{F}_{\mathrm{R}}\,, \tag{3.1}$$

we can replace the full N summation in (1.1) by a sum over the n nearest particles, together with a prediction of the distant contribution. Likewise, (3.1) may be differentiated once for the purpose of low-order prediction. This procedure can lead to a significant gain in efficiency, provided the respective time-scales are well separated and $n \ll N$. The two-time-scale scheme for one particle is illustrated in Fig. 3.1.

In order to predict the coordinates of neighbours at the irregular time-steps, we need to extrapolate the regular force and its first derivative. The former can be readily obtained from (2.3), without the term $\mathbf{D}_{\mathrm{R}}^4$ and using appropriate time arguments, $t - T_k$, where T_k refers to the regular force times. Similarly, differentiation of the regular part of (3.1) using T_k yields the corresponding derivative which should be added to $\mathbf{F}_{\mathrm{I}}^{(1)}$.

To implement the AC scheme, we form a list for each particle containing all members inside a sphere of radius R_s. In addition, we include any particles within a surrounding shell of radius $2^{1/3}R_s$ satisfying the condition $\mathbf{R} \cdot \mathbf{V} < 0.1R_s^2/\Delta T_i$, where again $\mathbf{V} = \mathbf{v}_i - \mathbf{v}_j$ and ΔT_i denotes the regular time-step. This ensures that fast approaching particles are selected from the buffer zone. A subsidiary search criterion based on m/r^2 may be introduced for including more distant massive bodies.

The size of the neighbour sphere is modified at the end of each regular time-step when a total force summation is carried out. A selection criterion based on the local number density contrast has proved itself for a variety of problems, but may need modification for interacting subsystems. To sufficient approximation, the local density contrast is given by

$$C = \frac{2n_i}{N}\left(\frac{r_h}{R_s}\right)^3 , \tag{3.2}$$

where n_i is the current membership and r_h is the half-mass radius. In order to limit the range, we adopt a predicted membership

$$n_p = n_{max}(0.04C)^{1/2} , \tag{3.3}$$

subject to n_p being within $[0.2n_{max}, 0.9n_{max}]$, with n_{max} denoting the maximum permitted value.* The new neighbour sphere radius is then adjusted using the corresponding volume ratio, which gives

$$R_s^{new} = R_s^{old}\left(\frac{n_p}{n_i}\right)^{1/3} . \tag{3.4}$$

An alternative and simpler strategy is to stabilize all neighbour memberships on the same constant value $n_p = n_0$ [Ahmad & Cohen, 1973; Makino & Hut, 1988]. Since it is not clear whether this strategy is advantageous, the original density-related expression which has proved itself is still being used. A recent analysis by Spurzem, Baumgardt & Ibold [2003] concluded that the Makino–Hut relation $n_p \propto N^{3/4}$ may not be necessary for large systems with realistic density contrasts. Hence it appears that relatively small neighbour numbers of about 50 are sufficient to take advantage of the AC scheme even for $N \sim 40\,000$. In any case, the actual code performance does not depend sensitively on the neighbour number because of compensating factors: reduction of n_i speeds up the irregular force summation but leads to shorter regular time-steps. A number of refinements are also included as follows:

- To avoid a resonance oscillation in R_s, we use $(n_p/n_i)^{1/6}$ in (3.4) if the predicted membership lies between the old and new values

---

* As a recent refinement, $\tilde{n}_p = n_p(r_h/r)$ is now used outside the half-mass radius.

- R_s is modified by the radial velocity factor $1 + \mathbf{r}_i \cdot \mathbf{v}_i \Delta T_i / r_i^2$ outside the core
- The volume factor is only allowed to change by 25%, subject to a time-step dependent cutoff if $\Delta T_i < 0.01 t_{cr}$
- If $n_i \leq 3$, $r < 3r_h$ and the neighbours are leaving the standard sphere, R_s is increased by 10%

Following the recalculation of the regular force, the gain or loss of particles is recorded when comparing the old and new neighbour list. Regular force differences are first evaluated, assuming there has been no change of neighbours. This gives rise to the provisional new regular force difference

$$\mathbf{D}_R^1 = [\mathbf{F}_R^{new} - (\mathbf{F}_I^{old} - \mathbf{F}_I^{new}) - \mathbf{F}_R^{old}]/(t - T_0)\,, \tag{3.5}$$

where $\mathbf{F}_R^{old}$ denotes the old regular force, evaluated at time T_0, and the net change of irregular force is contained in the middle brackets. In the subsequent discussions of the AC method, regular times and time-steps are distinguished using upper case characters. All current force components are obtained using the *predicted* coordinates (which must be saved), rather than the corrected values based on the irregular term $\mathbf{D}_I^4$ since otherwise (3.5) would contain a spurious force difference. The higher differences are formed in the standard way, whereupon the regular force corrector is applied (if desired).

A complication arises because any change in neighbours requires appropriate corrections of both the force polynomials, using the principle of successive differentiation of (3.1). The three respective Taylor series derivatives (2.4) and (2.5) are accumulated to yield the net change. Each force polynomial is modified by first adding or subtracting the correction terms to the corresponding Taylor series derivatives (2.3) (without the $\mathbf{D}^4$ term), followed by a conversion to standard differences by (2.7).

Implementation of the AC scheme requires the following additional set of regular variables: $\mathbf{F}_R$, $\mathbf{D}_R^1$, $\mathbf{D}_R^2$, $\mathbf{D}_R^3$, ΔT, T_0, T_1, T_2, T_3, as well as the neighbour sphere radius R_s and neighbour list of size $n_{max} + 1$. The corresponding code for softened potentials (NBODY2) has been described in considerable detail elsewhere [Aarseth, 2001b]. In the point-mass case, close encounters are treated by KS regularization, giving rise to an analogous code for the AC scheme [Aarseth, 1985a], as well as a new Hermite AC code with full regularization. As an application of the AC method, we mention a scheme for studying the interaction between a star cluster and a molecular cloud in which the latter is treated by smoothed particle hydrodynamics [Theuns, 1992a].

The combination of the Hermite method with the AC scheme (HACS) gives rise to a very efficient treatment. Following the discussion in chapter

2, the implementation of these two methods is much facilitated. Here we summarize some of the main features for evaluating the derivatives. A full description with actual comparisons is given elsewhere for a softened interaction potential [Makino & Aarseth, 1992].

In HACS, we need to form the current total force for the predictions as well as for the next corrector step by the low-order extrapolation

$$\mathbf{F}(t) = \mathbf{F}_{\rm I}(t) + \mathbf{F}_{\rm R}(T_0) + \mathbf{F}_{\rm R}^{(1)}(T_0)(t - T_0)\,. \tag{3.6}$$

Similarly, the total force derivative is just the sum

$$\mathbf{F}^{(1)}(t) = \mathbf{F}_{\rm I}^{(1)}(t) + \mathbf{F}_{\rm R}^{(1)}(T_0)\,. \tag{3.7}$$

At the end of a regular time-step, we need the equivalent of the difference (3.5), namely the *change* in the regular force assuming there has been no loss or gain of neighbours, which yields

$$\Delta\mathbf{F}_{\rm R} = \mathbf{F}_{\rm R}^{\rm new} - (\mathbf{F}_{\rm I}^{\rm old} - \mathbf{F}_{\rm I}^{\rm new}) - \mathbf{F}_{\rm R}^{\rm old}\,. \tag{3.8}$$

Note that the inner bracket of (2.20) and (2.21) requires the reverse sign of this expression. Likewise, the corresponding new regular force derivative based on the *old* neighbours is given by

$$\mathbf{F}_{\rm R}^{(1)} = \widetilde{\mathbf{F}}_{\rm R}^{(1)} - (\mathbf{F}_{\rm I}^{(1)} - \widetilde{\mathbf{F}}_{\rm I}^{(1)})\,, \tag{3.9}$$

where the quantities $\widetilde{\mathbf{F}}^{(1)}$ denote the respective new derivatives and the terms in brackets contain the net change. Finally, we combine (3.9) with the previous regular force derivative to form the higher derivatives used for the corrector. If required, the corresponding corrected values at the end of the interval may be obtained by including contributions from the change of neighbours.

In the Hermite versions, we employ the code convention that the quantities $\mathbf{D}^k$ denote Taylor series *derivatives*, rather than differences as used in the old polynomial formulation. Consequently, the first derivatives are initialized by

$$\begin{aligned} \mathbf{D}_{\rm I}^1 &= \widetilde{\mathbf{F}}_{\rm I}^{(1)}\,, \\ \mathbf{D}_{\rm R}^1 &= \widetilde{\mathbf{F}}_{\rm R}^{(1)}\,. \end{aligned} \tag{3.10}$$

Construction of the second and third irregular and regular derivatives proceeds according to (2.21) and (2.20), which are still expressed with respect to the old neighbours. If desired, corrections for any change in membership may be carried out by employing the explicit derivatives given by (2.5). However, this procedure may be omitted for single particles (but *not* for c.m. particles), provided the interval for energy checks is commensurate with the largest time-step [cf. Makino & Aarseth, 1992].

The main advantage of HACS is its simplicity and slightly improved efficiency. It has much in common with the basic AC scheme, including a similar number of variables. Thus the backwards times $t_k (k = 1, 2, 3)$ of both types are no longer needed, whereas the separate variables $\mathbf{F}_{\rm I}^{(1)}$ and $\mathbf{F}_{\rm R}^{(1)}$ must be introduced. A discussion of various practical aspects will be delayed until chapter 10.

3.3 Cosmological adaptation

Cosmological N-body simulations are characterized by dominant radial motions, with subsequent formation and growth of clusters due to density inhomogeneities which then develop peculiar velocities. Such systems may be studied more efficiently by a comoving formulation that integrates the deviations from overall expansion rather than the absolute motions. Although there are fast Fourier transform methods and tree codes for large N, it is still of some interest to employ direct summation which can handle more modest particle numbers (say $N \simeq 10^5$). Again the derivation is based on a previous formulation [cf. Aarseth, 1985a].

In standard Newtonian cosmology, a spherical boundary of radius S containing total mass M is governed by the equation of motion

$$\ddot{S} = -GM/S^2 \,. \tag{3.11}$$

Comoving coordinates for each galaxy, $\boldsymbol{\rho}_i = \mathbf{r}_i/S$, are then introduced by scaling the physical coordinates in terms of the expansion factor. We introduce a softened potential of the form $-m/(r^2 + \epsilon_0^2)^{1/2}$, where ϵ_0 may be associated with the half-mass radius of a galaxy. After a combination of the softened version of (1.1) with (3.11), the corresponding comoving equation of motion takes the form

$$\ddot{\boldsymbol{\rho}}_i = -\frac{2\dot{S}}{S}\dot{\boldsymbol{\rho}}_i - \frac{G}{S^3}\left[\sum_{j=1;j\neq i}^{N} \frac{m_j(\boldsymbol{\rho}_i - \boldsymbol{\rho}_j)}{(|\boldsymbol{\rho}_i - \boldsymbol{\rho}_j|^2 + \epsilon^2)^{3/2}} - M\boldsymbol{\rho}_i\right] , \tag{3.12}$$

where $\epsilon = \epsilon_0/S$.

Although this equation may be integrated directly (say by a low-order method), the presence of S^3 in the denominator is inconvenient in the AC scheme because of the explicit derivative corrections during neighbour changes. This problem can be avoided by introducing the time smoothing

$$t' = S^{3/2} \,. \tag{3.13}$$

Hence differentiation with respect to the fictitious time, τ, can be performed using the operator $d/d\tau = S^{3/2} d/dt$. This gives the new velocity

$$\boldsymbol{\rho}_i' = S^{3/2} \dot{\boldsymbol{\rho}}_i \,. \tag{3.14}$$

Similarly, by differentiating $\mathbf{r}_i = \boldsymbol{\rho}_i S$, the physical velocity is recovered from

$$\mathbf{v}_i = S' \boldsymbol{\rho}_i / S^{3/2} + \boldsymbol{\rho}_i' / S^{1/2} \,. \tag{3.15}$$

We note that the comoving peculiar velocities tend to increase in magnitude according to (3.14), whereas the fictitious time-steps are decreasing in a similar manner. A second differentiation of (3.14) with substitution from (3.12) then yields the new equation of motion

$$\boldsymbol{\rho}_i'' = -\frac{S'}{2S}\boldsymbol{\rho}_i' - G \sum_{j=1;j\neq i}^{N} \frac{m_j(\boldsymbol{\rho}_i - \boldsymbol{\rho}_j)}{(|\boldsymbol{\rho}_i - \boldsymbol{\rho}_j|^2 + \epsilon^2)^{3/2}} + GM\boldsymbol{\rho}_i \,. \tag{3.16}$$

The corresponding equation of motion for S is readily derived by applying the rule of differentiation with respect to the fictitious time twice. We introduce scaled units with $G = 1$ and obtain

$$S'' = 3S'^2/2S - MS \,. \tag{3.17}$$

This equation may be integrated using the method of explicit derivatives. For this purpose it is sufficient to include two further Taylor series terms, provided that a conservative time-step is used based on for example some fraction of $(S''/S^{(4)})^{1/2}$. Two successive differentiations of (3.17) then give rise to the corresponding derivatives, which simplify to

$$\begin{aligned} S^{(3)} &= \left(\frac{3S'^2}{S^2} - 4M \right) S' \,, \\ S^{(4)} &= 15 \left(\frac{S'^2}{2S^2} - M \right) \frac{S'^2}{S} + 4M^2 S \,. \end{aligned} \tag{3.18}$$

Particles crossing the boundary may be subject to a mirror reflection in order to conserve the comoving mean density inside S. Any such particle (i.e. $\rho_i > 1$, $\rho_i' > 0$) is assigned an equal negative comoving radial velocity and new polynomials are initialized to avoid discontinuity effects. A corresponding correction to the total energy can also be performed, giving

$$\Delta E_i = 2m_i S'(S'\rho_i + S\rho_i' - S')/S^3 \tag{3.19}$$

when converted to physical units, where ρ_i' denotes the old radial velocity.

The comoving formulation may readily be adapted to the *standard* AC scheme. Thus the velocity-dependent term of (3.16) can be absorbed in the neighbour force. It can be seen that this viscous force acts as a cooling during the expansion to counteract the increase in peculiar velocities. Although the last term of (3.16) would combine naturally with the regular force, experience has shown that it is advantageous to retain it separately; i.e. this procedure provides increased stability at the expense of a few additional operations. This can be readily understood from the experience with the Hermite scheme, where the first force derivative is also obtained by explicit differentiation. We therefore write the total force as a sum of *three* contributions,

$$\mathbf{F} = \mathbf{F}_{\rm I} + \mathbf{F}_{\rm R} + M\boldsymbol{\rho}_i\,, \qquad (3.20)$$

where the last term is treated by explicit differentiation whenever time derivatives are needed.

Initialization of the force polynomials is similar to the case of standard integration, except that the softening parameter must also be differentiated; i.e. $\epsilon' = -\epsilon_0 S'/S^2$. In the differentiation of (3.20) the three required derivatives of $\boldsymbol{\rho}_i$ can be substituted from known quantities. The principle of boot-strapping also applies during the integration when converting to the Taylor series force derivatives (2.3) for prediction purposes.

The integration itself is quite similar to the AC scheme, but now the boundary radius must also be advanced, instead of using the time itself which is less useful here. However, if desired, the physical time may be integrated by using the explicit derivatives of (3.13) which are readily formed. Irregular and regular time-steps are first obtained by (2.13), written in terms of prime derivatives. In comoving coordinates, the regular force tends to remain constant, whereas the irregular component grows in magnitude due to increased clustering. The regular interval, $\delta\tau_{\rm R}$, is therefore incremented iteratively by a small factor (e.g. 1.05), provided that the predicted change of the regular force does not exceed a specified fraction of the *irregular* component by

$$[(\tfrac{1}{6}|\mathbf{F}_{\rm R}^{(3)}|\delta\tau_{\rm R} + \tfrac{1}{2}|\mathbf{F}_{\rm R}^{(2)}|)\delta\tau_{\rm R} + |\mathbf{F}'_{\rm R}|]\delta\tau_{\rm R} < \eta_R|\mathbf{F}_{\rm I}|\,, \qquad (3.21)$$

where η_R is a dimensionless tolerance parameter.

Because of the small peculiar motions of distant particles with respect to a given particle, it is possible to omit the full coordinate prediction when evaluating the total force (3.20). Likewise, the neighbour prediction is performed to order $\mathbf{F}$ only. Also note that the softening length must be updated frequently according to $\epsilon = \epsilon_0/S$; this is done after each integration of the boundary radius. The actual gain in efficiency depends on the expansion rate and degree of clustering or initial peculiar velocities, but may exceed a factor of 2 for a typical hyperbolic model with $N = 4000$ members (Aarseth, Gott & Turner, 1979).

3.4 Multipole expansion

So far, we have been concerned with direct summation methods. This approach inevitably leads to a restriction in the particle number that can be considered. One of the biggest technical challenges is concerned with the simulation of star clusters. Such systems usually develop a core–halo structure with a large range in dynamical time-scale. Thus in the simple theory of relaxation due to stellar encounters, the orbital energy of each star changes at a rate that is related to the inverse square root of the local density. Hence more distant stars move in essentially *collisionless* orbits as far as nearest neighbour interactions are concerned. This behaviour can be modelled by representing the cluster potential in terms of a multipole expansion based on Legendre polynomials. First the basic formulation as applied to the whole system is given [Aarseth, 1967]. This approach is also relevant for other developments in stellar dynamics, such as tree codes [Barnes & Hut, 1986] and several related multipole expansions of lower order [Villumsen, 1982; White, 1983]. In the second part, we describe the (unpublished) implementation of a fully consistent collisional scheme that combines the two techniques.

Given the distribution of K mass-points inside a radius r, the total potential at an external point can be expressed as a sum over the individual contributions,

$$\Phi_{\rm ext} = -\frac{1}{r}\sum_{i=1}^{K} m_i \sum_{n=0}^{l}\left(\frac{r_i}{r}\right)^n P_n(\cos\theta)\,, \tag{3.22}$$

where $P_n(\cos\theta)$ are Legendre polynomials of order n and $\cos\theta = \mathbf{r}\cdot\mathbf{r}_i/rr_i$. This series converges at a rate depending on the ratio r_i/r and the direct summation (1.1) is approached in the limit of large l. To third order, this multipole expansion can be written in compact notation by [Aarseth, 1967; McMillan & Aarseth, 1993]

$$\Phi(r) = -\frac{M}{r} - \frac{D_i x_i}{r^3} - \frac{Q_{ij}x_i x_j}{2r^5} - \frac{S_{ij}x_i^2 x_j + S_{123}x_1 x_2 x_3}{2r^7}\,, \tag{3.23}$$

where the repeated indices i and j run from 1 to 3 and x_j denotes the cyclic components of $\mathbf{r}$. Explicit expressions for the monopole, dipole, quadrupole and octupole coefficients are obtained by a summation over the K internal particles as

$$\begin{aligned}
M &= \sum m_k\,,\\
D_i &= \sum m_k x_{i,k}\,,\\
Q_{ij} &= \sum m_k(3x_{i,k}x_{j,k} - |\mathbf{r}_k|^2\delta_{ij})\,,
\end{aligned}$$

$$S_{ij} = \sum m_k [5(3 - 2\delta_{ij}) x_{i,k}^2 - 3|\mathbf{r}_k|^2] x_{j,k} \,,$$
$$S_{123} = \sum m_k x_{1,k} x_{2,k} x_{3,k} \,. \tag{3.24}$$

The case of an internal point, $r < r_i$, gives rise to an analogous expansion for $\Phi_{\rm int}$ in terms of r/r_i, with the factor $1/r$ appearing as $1/r_i$ inside the first summation of (3.22). Let $\Phi = \Phi_{\rm int} + \Phi_{\rm ext}$ represent the total potential. The equation of motion for a test particle at $\mathbf{r}$ with respect to the reference centre is then

$$\ddot{\mathbf{r}} = -\frac{\partial \Phi}{\partial \mathbf{r}} - \frac{\mathbf{r}}{r}\frac{\partial \Phi}{\partial r} \,, \tag{3.25}$$

where the second term yields only contributions for $r > r_i$. We have also included the dipole term here since it may not be zero in an actual simulation, due to external perturbations or reference to the density centre.

For practical implementations, it is convenient to divide the mass distribution into $n_{\rm shell}$ spherical shells of equal thickness, with membership $N_{\rm shell} \simeq N^{1/2}$. This facilitates using the same moment coefficients for all the particles in one shell. The total force on a particle in shell J (> 1) is then a sum of $J - 1$ internal and $n_{\rm shell} - J$ external (if any) terms of different order. In addition, we need to include contributions from the Jth shell, which may also be divided into an internal and external region.

The gravitational attraction from a given shell, J, to the potential at r arising from the zeroth order is

$$M_J = \frac{1}{r} \sum m_i \,,$$
$$L_J = \sum \frac{m_i}{r_i} \,, \tag{3.26}$$

for the external and internal contributions, respectively. Assuming constant density inside a given shell, we may approximate the corresponding density-weighted integrals within some interval $[r_J, r_{J+1}]$ by

$$M_J = 4\pi \rho_M (r_{J+1}^3 - r_J^3)/3 \,,$$
$$L_J = 2\pi \rho_L (r_{J+1}^2 - r_J^2) \,. \tag{3.27}$$

The potential at $r_J < r < r_{J+1}$ due to these contributions is then

$$\Delta\Phi_J = \frac{4\pi}{r} \int_{r_J}^{r} r^2 \rho_M dr + 4\pi \int_{r}^{r_{J+1}} r \rho_L dr$$
$$= \frac{M_J}{r_{J+1}^3 - r_J^3}(r^2 - r_J^3/r) + \frac{L_J}{r_{J+1}^2 - r_J^2}(r_{J+1}^2 - r^2) \,, \tag{3.28}$$

and the associated force is obtained by differentiation. The six coefficients from the dipole term are treated in a similar way, whereas the simpler linear interpolation is used for higher-order terms.

The total potential, Φ, and corresponding force components can now be obtained consistently at an arbitrary point. However, it is necessary to subtract the self-force from the particle under consideration in order to eliminate a systematic effect. An exact correction for this contribution can be applied using the original expression (3.22), provided that the coordinates $\mathbf{r}_0$ at the previous evaluation are known. Thus, omitting the mass, the term $k = 2$ in the case $r > r_0$ yields a force component

$$\Delta \mathbf{F}_j = \frac{3(\mathbf{r} \cdot \mathbf{r}_0) x_{0,j}}{r^5} + \frac{3}{2} \frac{r_0^2 x_{0,j}}{r^5} - \frac{15}{2} \frac{(\mathbf{r} \cdot \mathbf{r}_0)^2 x_{0,j}}{r^7}, \tag{3.29}$$

which must be subtracted. Special care is needed when r and r_0 lie inside the same shell; i.e. both the internal and external solutions must be used in the interpolation. It can be seen that the approximation $P_2(\cos\theta) \simeq 1$ gives rise to a force $-3x_k/r^3$ when $r \geq r_0$, whereas the corresponding value for $r \leq r_0$ is $2x_k/r^3$. Hence each term of the expansion contributes on average $-x_k/r^3$ to the self-force and it is therefore necessary to include these cumbersome corrections in the basic scheme.

The formulation outlined above was used to study a star cluster containing $N = 1000$ members in the presence of an external tidal field [Aarseth, 1967]. Although this early simulation demonstrated some degree of flattening due to the tidal field, the approximate treatment of the inner regions is not suitable for highly concentrated systems. It was not until about 1988 that the collisionless scheme was combined with the direct approach of *NBODY5* to yield a fully consistent code, which will now be summarized for completeness. In the following we omit any reference to two-body regularization that was also included (to be discussed in subsequent sections) and concentrate on aspects relevant to the standard AC method.

The basic strategy for a full implementation is to divide the cluster into a small number of spherical zones of equal membership. Dictated by the size of characteristic neighbour radii, a choice of only seven zones was made in order to simulate a cluster population of $N = 4000$. Hence larger systems might be divided into slightly more zones. The shell radii are redetermined at least once per crossing time, whereas new moments are evaluated at conservative intervals of $0.004 t_{\rm cr}$ which still exceed the typical regular time-steps and therefore this effort does not add significantly to the overheads.

As might be expected, the neighbour scheme itself is not affected by the new treatment. Before the evaluation of the total force, we determine an inner and outer shell index, I_k ($k = 1, 2$), such that the actual summation is restricted only to the associated particles. These boundaries are chosen by comparing the shell radii $R_{\rm shell}$ with the range $r \pm 2R_{\rm s}$, which in practice represents the maximum for neighbour selection (after including some additional criteria). Here, the distance r is measured with respect to the so-called 'density (or symmetry) centre' of the cluster. Hence, the direct summation is only made over some fraction of the total membership. Finally, the contributions from the moments are added to yield the total force.

The advantages of a full-blown moment scheme can be summarized as follows:

- There is no need to subtract the self-force contribution
- Convergence of (3.22) is increased by avoiding the nearest zones
- The treatment in a dense core requires only evaluation of $\Phi_{\rm int}$
- A considerable efficiency gain is achieved at modest extra cost

3.5 Grid perturbations

Historically, several early N-body simulations were devoted to large scales (i.e. clusters of galaxies) since the time was not ripe for an attack on the technically more challenging star cluster problem. The realization that such an approach might also be fruitful for planetary systems developed more slowly. An early attempt by Hills [1970] investigated the evolution of planetary orbits towards more stable configurations. The work of Cox & Lewis [1980] introduced a greater degree of realism by allowing for physical collisions in an ensemble of 100 planetesimals distributed inside a thin ring. However, in this study the orbits were Keplerian, except when two bodies came within the sphere of influence, whereupon the encounter was treated by a special three-body algorithm obtained from fitting functions. Here we discuss a direct N-body scheme, suitable for a flattened system, where the numerical effort is reduced by including only those perturbations that are significant. The basic method has been described elsewhere [Aarseth, 1985a], together with the results [Lecar & Aarseth, 1986].

In the case of a flattened system of planetesimals orbiting the Sun, it is convenient to perform the calculation in a heliocentric coordinate system. The equation of motion for one body of mass m_i and relative coordinates $\mathbf{r}_i$ is then given by [Brouwer & Clemence, 1961]

$$\ddot{\mathbf{r}}_i = -\frac{M_{\rm s}+m_i}{r_i^3}\mathbf{r}_i - \sum_{j=1;j\neq i}^{N} m_j \left(\frac{\mathbf{r}_i-\mathbf{r}_j}{|\mathbf{r}_i-\mathbf{r}_j|^3} + \frac{\mathbf{r}_j}{r_j^3} \right) , \qquad (3.30)$$

where $M_{\rm s}$ denotes the mass of the Sun. In the planetesimal case, $m_i \ll M_{\rm s}$ and the orbits are subject to a large number of weak encounters with an occasional close approach. Moreover, the system is characterized by differential rotation, such that nearby orbits suffer significant gravitational focusing that induces changes in the eccentricity. The essential features of the evolution are therefore preserved if we exclude the numerous distant interactions from the summation (3.30).

Let us divide the orbital plane into $N_{\rm a}$ heliocentric azimuthal zones. The active perturbers for inclusion in (3.30) are then selected within an angle $2\pi(n_{\rm a}+1)/N_{\rm a}$ centred on the zone associated with m_i, where $n_{\rm a}$ is the number of perturber zones on *either* side, making $2n_{\rm a}+1$ zones (or bins) in all. In order to minimize systematic errors caused by non-symmetrical edge effects, the choice of $N_{\rm a}$ is a compromise between resolution and computational effort. During the early stage when N is relatively large, all indirect terms of (3.30) are also neglected. Thus for a uniform angular distribution, the vectorial contributions in each annular ring tend to cancel and only the small *fluctuating* part of the indirect accelerations is therefore neglected.

For an actual simulation, the membership, N, is determined by the total planetesimal mass as well as the extent of the system. If a narrow ring is chosen, then undesirable boundary effects are introduced. However, we can increase the ring width, and thereby N, by restricting the perturber selection also in the radial direction to speed up the calculation. Let the particles be distributed initially within an annular ring of radius R_1 and R_2. We choose radial zones at $r_k = R_2/k^{1/2}$, with $k = 1, 2, \ldots$, such that the grid cells are smaller in the inner regions. Let $n_{\rm r}$ denote the number of zones to be included on either side of the current particle zone. Each particle j is then assigned a radial zone index $k_j = [R_2^2/r_j^2]$ with $k_j = 0$ for $r_j > R_2$, and an azimuthal index $l_j = 1 + [N_{\rm a}\phi_j/2\pi]$, where ϕ_j is the corresponding phase angle. Active perturbers for particle i must then satisfy the conditions

$$|l_j - l_i| \le n_{\rm a}\,, \tag{3.31}$$

$$|k_j - k_i| \le n_{\rm r}\,. \tag{3.32}$$

The storage requirement for this scheme is quite modest. A matrix L_{aj} contains the list of all particles in each azimuthal zone, whereas k_j and l_j are obtained from lists L_ϕ and $L_{\rm r}$ of size N. These lists are then updated as necessary at the end of each integration cycle. Although the grid cells provide a strictly two-dimensional (2D) description, this representation can also be used for flattened systems. In that case, the third dimension is ignored when assigning perturber zones. This device enables systems with thickness of several typical grid cells to be studied efficiently.

The numerical integrations are performed using the standard differences described in chapter 2, with the coordinates of active perturbers predicted to order $\mathbf{F}^{(1)}$. As a new feature of N-body integrations, we have implemented the energy *stabilization* of Baumgarte [1973] in order to reduce the usual systematic errors associated with elliptic orbits. The concept of energy stabilization was originally introduced by Nacozy [1972] as a least-squares correction procedure involving an expensive recalculation of the total energy at frequent intervals. In the case of dominant two-body motion, we ensure that solutions stay on the correct energy surface by introducing a control term in the equation of motion which now takes the form

$$\ddot{\mathbf{r}}_i = -\frac{M_s + m_i}{r_i^3}\mathbf{r}_i + \mathbf{P}_i - \frac{\alpha(\widetilde{h}_i - h_i)}{v_i^2}\mathbf{v}_i\,, \tag{3.33}$$

where $\mathbf{P}_i$ is the perturbing force, with or without the indirect terms. Here, the last term represents the difference between the explicitly calculated and *integrated* two-body binding energy per unit reduced mass, respectively, and α is a parameter which depends on the time-step. Initially, the binding energy is evaluated by the standard two-body expression

$$h_i = \frac{1}{2}\mathbf{v}_i^2 - (M_s + m_i)/r_i\,, \tag{3.34}$$

and subsequent values are obtained by integrating the equation of motion

$$\dot{h}_i = \mathbf{v}_i \cdot \mathbf{P}_i\,. \tag{3.35}$$

The quantity $\widetilde{h}_i$ is calculated from (3.34) at the start of each integration cycle, when the osculating orbit is known to the highest order. The correction term is then included in the force when predicting the coordinates and velocity of m_i to third order, but is not added explicitly to $\mathbf{F}$ itself (and hence is absent from the force differences). Following some experimentation which depends on the integration scheme, we choose $\alpha = 0.4/\Delta t_i$ [cf. Baumgarte & Stiefel, 1974].

The effectiveness of the stabilization procedure depends on the relative two-body perturbation defined by

$$\gamma_i = |\mathbf{P}_i| r_i^2 / M_s\,. \tag{3.36}$$

Hence it is prudent to omit the stabilizer from (3.33) if the perturbation exceeds some specified value, $\gamma_{\rm cr}$, when the conic reference orbit is not a good approximation. The solution is continued with a new osculating energy once the perturbation becomes sufficiently small again. However, the overheads due to initialization are relatively modest in typical planetesimal systems. We also note that the extra cost of including stabilization is relatively small, since $\mathbf{P}_i$ is already known.

Individual time-steps are again determined by the relative convergence criterion (2.13). In addition, there is a further reduction by a factor

$1 + 4(\gamma_i/\gamma_{\rm cr})^2$ if $\gamma_i \leq \gamma_{\rm cr}$ or by 4 if $\gamma_i > \gamma_{\rm cr}$. The former reduction facilitates the detection of collisions, whereas the latter improves further the treatment of close encounters. A discussion of the collision procedures and other practical details of this scheme will be deferred until the relevant section on applications.

Finally, we include some technical improvements here that resulted from a further investigation of planetary formation [Beaugé & Aarseth, 1990]. In view of the long-range effect of massive bodies that form by accretion, it is desirable to employ a mass-dependent perturber criterion, instead of using a constant number of azimuthal bins. Thus the perturbation from so-called 'embryo planets' ($m_i > 4m_0$, where m_0 is the initial mass) were included out to at least four radial bins on either side. For smaller masses, perturbations were considered when the difference in azimuthal bins by (3.31) were at most an integer determined by $n_{\rm b} = 1 + (m_i/m_0)^{1/2} N_{\rm b}$, with $N_{\rm b}$ a parameter depending on the requirements. This refinement is particularly beneficial when the process of fragmentation is studied, since there may be a large number of bodies with small ($< m_0$) masses, while at the same time the long-range effect of an embryo is treated more carefully. Once this critical mass has been reached, it is also desirable (and not too expensive) to include all the interactions for these emerging planets in order to have more confidence in the final result.

The grid perturbation method was initially devised for studying an interacting ring of molecular clouds in the Galaxy [cf. Aarseth, 1988b], and this application will be discussed in a later chapter.

3.6 Particle in box

In spite of the efficiency gain achieved by the grid perturbation method, direct integration of planetesimals is still quite expensive and therefore only suitable for the later stages. During the earlier stages, when the particle number is large, different local regions tend to evolve in a self-similar manner. Based on this assumption, Wisdom & Tremaine [1988] presented a scheme for studying the collisions of massless particles in a small patch of particles with periodic boundaries in differential rotation about a central planet. Here we are interested in including the mutual gravitational interactions. Because of the long-range nature of such interactions, it is also necessary to add the forces from similar neighbouring regions. Hence this box scheme actually involves force summations over eight surrounding shadow boxes in addition to the usual gravitational interactions, and collisions between ghost particles and real particles across the boundaries are also taken into account.

We adopt a rotating coordinate system with origin at the box centre and the x-axis pointing away from the Sun, which is located at $x = -a$. Let the y-axis be in the direction of circular motion with angular velocity Ω. The linearized equations of motion for a given particle can then be written as [Aarseth, Lin & Palmer, 1993]

$$
\begin{aligned}
\ddot{x} &= 3\Omega^2 x + 2\Omega \dot{y} + F_x + P_x \,, \\
\ddot{y} &= -2\Omega \dot{x} + F_y + P_y \,, \\
\ddot{z} &= -\Omega^2 z + F_z + P_z \,,
\end{aligned} \tag{3.37}
$$

where the total force (per unit mass) is expressed as the sum of the usual internal force $\mathbf{F}$ and the ghost contributions $\mathbf{P}$.

In order to evaluate particle coordinates and velocities in the ghost boxes, we construct a 2D transformation matrix with components $X_{\rm b}, Y_{\rm b}$. Let the central box be contained within an area given by $\pm R_{\rm b}$ in the $x-y$ plane. The eight surrounding boxes are labelled sequentially from 1 to 8, with the respective coordinates of the initial box centres defined by the matrix

$$
(X_{\rm b}, Y_{\rm b}) = (-1,-1), (0,-1), (1,-1), (-1,0), (1,0), (-1,1), (0,1), (1,1) \tag{3.38}
$$

when expressed in units of the box size, $S_{\rm b} = 2R_{\rm b}$, ($\ll a$). The centres of the six boxes for which $X_{\rm b} < 0$ or $X_{\rm b} > 0$ are subject to differential rotation about the central body. This shearing motion is included by introducing the *sliding* distance

$$
D = \tfrac{3}{2}\Omega S_{\rm b} t + R_{\rm b} - (K + 0.5) S_{\rm b} \tag{3.39}
$$

where $K = (\frac{3}{2}\Omega S_{\rm b} t + R_{\rm b})/S_{\rm b}$ denotes the integer part. Figure 3.2 shows the geometry of the sliding box configuration.

In order to obtain the force due to the members of a ghost box of index L^*, we first define appropriate displacements which are added to the corresponding coordinates of a central box particle at x_j, y_j. This yields the ghost particle coordinates

$$
\begin{aligned}
x_{\rm g} &= x_j + X_{\rm b}(L^*) S_{\rm b} \,, \\
y_{\rm g} &= y_j + X_{\rm b}(L^*) D + Y_{\rm b}(L^*) S_{\rm b} \,.
\end{aligned} \tag{3.40}
$$

An additional check is carried out to ensure that $y_{\rm g}$ falls inside the range $[(Y_{\rm b}(L^*) - 0.5)S_{\rm b}]$ to $[(Y_{\rm b}(L^*) + 0.5)S_{\rm b}]$, otherwise it is modified by adding or subtracting the box size, $S_{\rm b}$.

Collisions play an important role in planetesimal systems. For rigid particles we define a collision by the overlapping condition $r < r_i + r_j$, which refers to the sum of the two relevant radii. The outcome of a

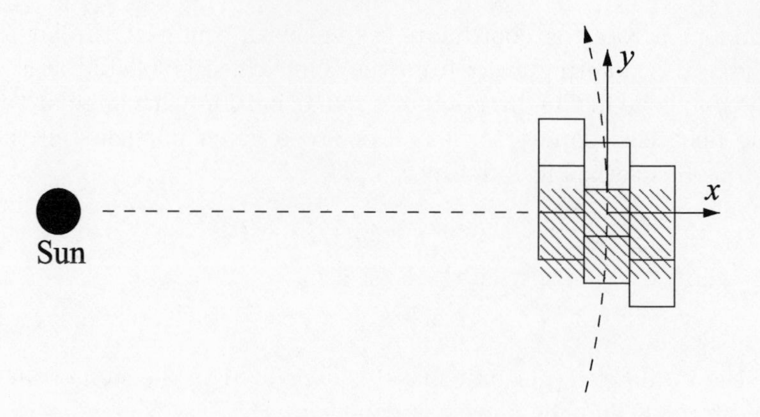

Fig. 3.2. Sliding boxes.

physical collision depends on the impact energy as well as the coefficient of restitution. In the case involving a ghost particle near the boundary we first determine the coordinate and velocity displacements

$$\begin{aligned}\Delta x &= X_{\rm b}(L^*)S_{\rm b}\,,\\ \Delta y &= X_{\rm b}(L^*)D + Y_{\rm b}(L^*)S_{\rm b}\,,\\ \Delta \dot{y} &= -\tfrac{3}{2}\Omega X_{\rm b}(L^*)S_{\rm b}\,,\end{aligned} \tag{3.41}$$

where L^* denotes the relevant box index for the transformation matrix $X_{\rm b}$. Once a collision has been accepted, the second component is predicted to highest order and any ghost displacement terms are added. From the velocity of each particle m_k with respect to the c.m.,

$$\mathbf{w}_k = \frac{(\mathbf{r}_k - \mathbf{r}_{\rm cm})\cdot(\dot{\mathbf{r}}_k - \dot{\mathbf{r}}_{\rm cm})}{|\mathbf{r}_k - \mathbf{r}_{\rm cm}|^2}(\mathbf{r}_k - \mathbf{r}_{\rm cm})\,, \tag{3.42}$$

we determine the new reflected velocity by

$$\dot{\mathbf{r}}'_k = \dot{\mathbf{r}}_k - (1+\epsilon_{\rm r})\mathbf{w}_k\,, \tag{3.43}$$

where $\epsilon_{\rm r}$ is the coefficient of restitution. The case of small impact velocity is treated as an enforced coalescence in order to prevent excessive integration of the sliding phase, and similarly for a semi-major axis smaller than the overlapping distance. After the reflection, force polynomials are initialized for both components in the usual way.

In order to speed up the calculation, a fast square root procedure [cf. Aarseth *et al.*, 1993] was developed† outside a radius $R_{\rm s} = S_{\rm b}/\pi^{1/2}N^{1/4}$. The distance $R_{\rm s}$ also serves to identify neighbours that are predicted to

† The square root function was relatively expensive on older computers.

order $\dot{\mathbf{F}}$ before the force evaluation. Upon obtaining the total force, each planetesimal is advanced a small time interval to ensure convergence of the fourth-order fitting polynomial. Since these time intervals may span many orders of magnitude, it is again beneficial to use individual time-steps. The basic time-step is chosen by a nearest neighbour criterion involving the relative distance, r, and velocity, v, according to

$$\Delta t_i = \min\{\eta r/v,\ \Delta t_{\rm max}\}\,, \tag{3.44}$$

where $\Delta t_{\rm max} = 2\pi\beta/\Omega$ denotes the maximum value. The standard parameters $\eta = 0.02$, $\beta = 0.003$ ensure collision detection. In the case that the closest particle is a ghost, the y-component of the relative velocity is modified by the sliding factor $\Delta\dot{y}$ above.

The last part of the integration cycle treats boundary crossings, defined by $|x| > R_{\rm b}$ or $|y| > R_{\rm b}$. After identifying the relevant box index, L^*, of the corresponding ghost particle, the new coordinates and y-velocity of the injected particle are obtained by

$$\begin{aligned} x' &= x - X_{\rm b}(L^*)S_{\rm b}\,, \\ y' &= y - X_{\rm b}(L^*)D - Y_{\rm b}(L^*)S_{\rm b}\,, \\ \dot{y}' &= \dot{y} + \tfrac{3}{2}\Omega X_{\rm b}(L^*)S_{\rm b}\,. \end{aligned} \tag{3.45}$$

Finally, the increments $\frac{3}{2}\Omega S_{\rm b}X_{\rm b}(L^*)$ and $\Delta J_z = \frac{1}{2}\Omega a S_{\rm b}X_{\rm b}(L^*)$ are added to the y-component of the linear momentum and the z-component of the angular momentum in order to achieve conservation during the boundary crossing.

3.7 Ring scheme

The development of new integration methods sometimes faces subtle problems which may introduce spurious effects. Accordingly, it is also useful to be aware of unsuccessful implementations. In this spirit we discuss a ring formulation that has some similarities with the box method of section 3.6 but is also based on the grid perturbation scheme of section 3.5 [Lin & Aarseth, unpublished, 1997]. Let us consider the dynamical interactions of a thin circular ring of planetesimals moving around the Sun. This description takes into account the curvature of the orbit and therefore would permit the study of more advanced stages of evolution when the collisional mean free path becomes comparable to $2\pi a$. Moreover, the treatment of the physical processes involving collisions can be readily taken over with little modification.

As long as the planetesimal masses remain small, the main effects of the gravitational interactions are due to neighbours within a few per cent of the system size. We therefore employ the 2D perturbation scheme of section 3.5 in which only the nearest members are included as perturbers. In the present formulation, we are mainly concerned with a relatively narrow ring of small planetesimals. For this reason it is sufficient to divide the system into a set of $N_{\rm a}$ azimuthal segments which are used for the perturber selection. The total force on a given planetesimal then consists of contributions from a fixed number of neighbouring perturber bins that are added to the solar term according to (3.37), except that the z-component is included directly to prevent growth after boundary reflections. Here the perturbing force field due to the planetesimals is obtained by summation over the central ring as well as the two adjacent ghost rings, using the sliding box procedure.

Implementation of this neighbour scheme requires a matrix $L_{{\rm a}j}$ (typically of size $N_{\rm a} \times N^{1/2}$) containing the locations of all members in each azimuthal bin together with a reference array $L_{\rm a}$ of size N. Here, the azimuthal index of a planetesimal, j, is given by $l_j = 1 + [N_{\rm a}\phi_j/2\pi]$, where ϕ_j is the phase angle. Active perturbers for particle i must then satisfy the condition (3.31), where $N_{\rm a}$ is a specified parameter.

The centres of the ghost rings are located at different radii from the Sun. Hence the sliding-ring procedure introduces not only different semi-major axes but also eccentricity and inclination distributions for the ghost particles. This prescription can lead to artificial heating and spurious torque due to other rings and similar problems also arise from boundary crossings.

A simplified scheme, based on direct integration of a narrow ring of planetesimals by GRAPE-4, used a procedure in which the eccentricity and inclination were modified to fit the local rms values [Kokubo & Ida, 2000]. Using a selective semi-major axis boundary criterion affects the velocity distribution such that the tangential component becomes larger than the Keplerian circular velocity near the inner boundary, and *vice versa*. This velocity anomaly exerts a torque at the ring boundary that tends to shrink the ring. Hence it was argued that the artificial torque due to the boundary condition is to some extent compensated by the finite ring approximation. Although the most massive protoplanets were confined inside the ring during the later stages, this claim was not substantiated and all that can be said is that the effect of gas drag appears to be more important. In conclusion, the present ring scheme was abandoned after several attempts of introducing consistent ghost-ring prescriptions and boundary crossing conditions which do not lead to artificial heating or cooling.

4

Two-body regularization

4.1 Introduction

Sooner or later during the integration of an N-body system close encounters create configurations that lead to difficulties or at best become very time-consuming if studied by direct methods. On further investigation one usually finds a binary of short period slowing down the calculation and introducing unacceptable systematic errors. Moreover, the eccentricity may attain a large value that necessitates small time-steps in the pericentre region unless special features are introduced. It can be seen that a relative criterion of the type $(\eta R/|\mathbf{F}|)^{1/2}$ for a binary yields an approximate time-step $\Delta t \propto R^{3/2}$, where R is the two-body separation. From this it follows that eccentric orbits require more steps for the same period. Even a relatively isolated binary may therefore become quite expensive to integrate as well as cause a significant drift in the total system energy. It is convenient to characterize the systematic error of a binary integration by the relative change per Kepler orbit, $\alpha = \Delta a/a$. For example, using the basic Hermite method we find $\alpha = -1.3 \times 10^{-6}$ with 270 steps per orbit and an eccentricity $e = 0.9$. At this rate of inward spiralling, the binary energy would be significantly affected after $10^4 - 10^5$ periods. Although better behaved, less eccentric systems are also time-consuming, giving $\alpha = -4 \times 10^{-8}$ for $e = 0.2$ and 135 steps per orbit. These examples highlight the numerical problems associated with the integration of close binaries and point to the need for improved treatments.

There is a long tradition in celestial mechanics for dealing with dominant two-body motion, and we start with some simple developments that are based on time transformations before moving on to more complicated formulations. Naturally, one cannot expect the full treatment by paying half the price, but so-called 'time smoothing' serves as a good introduction to the subject of regularization. It might be considered a bit of an overkill that regularization itself deals with the removal of the collision singularity in the equations of motion. However, experience shows that the transformed equations are much better behaved, even for a circular orbit of the same period. Thus we speak about the benefits of numerical regularization for binaries that may be considered close. The alternative concept of a hard binary [Heggie, 1972b, 1975; Fullerton & Hills, 1982] is slightly more precise, and the corresponding energy may be defined by

$$\epsilon_{\rm hard} = -\bar{m}\sigma^2 \,, \tag{4.1}$$

where σ is the velocity dispersion. Using the virial theorem for a bound system of total mass M we write $\sigma^2 \simeq GM/2r_{\rm h}$, where $r_{\rm h}$ is the half-mass radius. Hence substitution in (4.1) gives $a_{\rm hard} = r_{\rm h}/N$ as a characteristic hard binary semi-major axis for components of mean mass. The dynamical significance of a hard binary is that it has sufficient binding energy to avoid being disrupted by an incoming particle at twice the rms kinetic energy, which is an approximate representation of the central conditions. After these basic considerations we turn to various technical treatments.

4.2 Principles of regularization

The equation of relative motion for a binary with mass components m_k, m_l and separation R is given by

$$\ddot{\mathbf{R}} = -(m_k + m_l)\mathbf{R}/R^3 + \mathbf{F}_{kl} \,, \tag{4.2}$$

where $\mathbf{F_{kl}} = \mathbf{F}_k - \mathbf{F}_l$ is the external perturbation. In the case of dominant two-body motion it is convenient to define the relative perturbation as $\gamma = |\mathbf{F}_{kl}|R^2/(m_k + m_l)$, which measures the strength of the external tidal acceleration. For small γ, the first term of (4.2) is responsible for the numerical problems outlined in the previous section. We now introduce a differential time transformation by

$$dt = R^n d\tau \,, \tag{4.3}$$

where the exponent n can take arbitrary values for experimentation. We denote differentiation with respect to the fictitious time, τ, by primes and construct the operators

$$\frac{d}{dt} = \frac{1}{R^n}\frac{d}{d\tau} \,, \tag{4.4}$$

$$\frac{d^2}{dt^2} = \frac{1}{R^{2n}}\frac{d^2}{d\tau^2} - n\frac{R'}{R^{2n+1}}\frac{d}{d\tau} \,. \tag{4.5}$$

Substitution for the second derivative in (4.2) yields the new equation of motion

$$\mathbf{R}'' = nR'\mathbf{R}'/R - (m_k + m_l)\mathbf{R}/R^{3-2n} + R^{2n}\mathbf{F}_{kl} \,. \tag{4.6}$$

The classical choice $n = 1$ [Sundman, 1912] replaces the R^{-2} type singularity by an undetermined expression, $\mathbf{R}/R$, as $R \to 0$ which is clearly better behaved for small separations. In section 3.3 we introduced the value $n = 3/2$ together with comoving coordinates for cosmological equations of motion. This is also a good compromise for eccentric two-body orbits since using the eccentric anomaly ($n = 1$) is favourable at apocentre and the true anomaly ($n = 2$) is better at pericentre.

Other alternative smoothing functions have been examined [Zare & Szebehely, 1975]. In particular, it was found that the inverse Lagrangian function offers practical advantages. However, for most purposes, integration of the perturbed two-body problem with the time transformation $t' = R$ is preferable when it comes to two-body regularization. For completeness, we also mention a brave early attempt to introduce time smoothing based on the potential or kinetic energy of the most strongly bound cluster members [Heggie, 1972a]. Although the requirement of equal time-steps proved too expensive as far as star cluster simulations are concerned, this method was put to good use in a series of three-body scattering experiments [Hills, 1975].

The essential idea behind regularization is to transform both the time and the coordinates, and the latter proves much harder in the general case of three-dimensional (3D) motion. Since we are discussing the basic concept here, this can be done in 1D, which permits simplifications to be made in the absence of the term $\mathbf{R}/R$ with $n = 1$. If we neglect external perturbations and set $t' = x$, equation (4.6) reduces to

$$x'' = x'^2/x - (m_k + m_l)\,. \tag{4.7}$$

To proceed we make use of the binding energy per unit reduced mass,

$$h = \tfrac{1}{2}\dot{\mathbf{R}}^2 - (m_k + m_l)/R\,, \tag{4.8}$$

which is a constant of the motion evaluated from the initial conditions. Substitution of $\dot{x} = x'/x$ from (4.4) then gives

$$x'' = 2hx + (m_k + m_l)\,. \tag{4.9}$$

As can be readily seen, this equation is regular* for $x \to 0$. However, the displaced harmonic oscillator equation can be simplified by introducing new coordinates from the relation $u^2 = x$. Twice differentiation of u^2, with one more use of the energy equation, finally yields the extremely simple result

$$u'' = \tfrac{1}{2}hu\,. \tag{4.10}$$

We note that the coordinate transformation has reduced the frequency by a factor of 2, as well as eliminated the constant term.

---

* Regularization of two-body collision in 1D is attributed to Euler [cf. Szebehely, 1967].

In summary, the motion of two mass-points on a line can be integrated to high accuracy even though collisions take place. It is useful from a practical point of view to consider 1D as the limiting case for the eccentricity approaching unity. Finally, such an orbit may be calculated with constant time-step,

$$\Delta\tau = \eta_U/|2h|^{1/2}\,, \tag{4.11}$$

where $\eta_U/2\pi$ is a specified fraction of the *physical* period.

4.3 Levi-Civita transformation

An increase of dimension from 1D to 2D first is quite instructive since it deals with the undetermined expression $\mathbf{R}/R$. By analogy with the former development, we write $R = u_1^2 + u_2^2$ which permits the coordinate transformation for the components x and y, denoted R_1 and R_2,

$$\begin{aligned} R_1 &= u_1^2 - u_2^2\,, \\ R_2 &= 2u_1u_2\,. \end{aligned} \tag{4.12}$$

By summation of the squares, it is readily verified that the original relation is recovered. The transformation (4.12) may be written as

$$\mathbf{R} = \mathcal{L}(\mathbf{u})\mathbf{u}\,, \tag{4.13}$$

where the Levi-Civita [1920] matrix is given by

$$\mathcal{L}(\mathbf{u}) = \begin{bmatrix} u_1 & -u_2 \\ u_2 & u_1 \end{bmatrix}. \tag{4.14}$$

The formal derivation of two-body regularization in 2D may be done in terms of manipulations with the Levi-Civita matrix. In the following we follow the presentation of Bettis & Szebehely [1972]. According to Stiefel & Scheifele [1971], the linear matrix (4.14) has the properties

$$\begin{aligned} \mathcal{L}^{\mathrm{T}}(\mathbf{u})\mathcal{L}(\mathbf{u}) &= R\mathbf{I}\,, \\ \mathcal{L}'(\mathbf{u}) &= \mathcal{L}(\mathbf{u}')\,, \\ \mathcal{L}(\mathbf{u})\mathbf{v} &= \mathcal{L}(\mathbf{v})\mathbf{u}\,, \\ \mathbf{u}\cdot\mathbf{u}\mathcal{L}(\mathbf{v})\mathbf{v} - 2\mathbf{u}\cdot\mathbf{v}\mathcal{L}(\mathbf{u})\mathbf{v} + \mathbf{v}\cdot\mathbf{v}\mathcal{L}(\mathbf{u})\mathbf{u} &= 0\,. \end{aligned} \tag{4.15}$$

Here $\mathbf{I}$ is the unit matrix and $\mathbf{u}, \mathbf{v}$ are arbitrary vectors. The second of these equations holds for linear matrices, whereas the third obeys the commutative rule and the fourth defines the so-called 'bilinear relation'.

We now aim to express (4.6) in terms of the new dependent variable $\mathbf{u}$. The first regularized derivative of (4.13) is readily derived by employing the second and third properties (4.15), which yields

$$\mathbf{R}' = 2\mathcal{L}(\mathbf{u})\mathbf{u}'\,. \tag{4.16}$$

From $\mathcal{L}'(\mathbf{u}) = \mathcal{L}(\mathbf{u}')$ we then obtain

$$\mathbf{R}'' = 2\mathcal{L}(\mathbf{u})\mathbf{u}'' + 2\mathcal{L}(\mathbf{u}')\mathbf{u}' \,. \tag{4.17}$$

Using this form for $\mathbf{R}''$, we substitute (4.13) and (4.16) in (4.6) with the choice $n = 1$ to give

$$2\mathbf{u}\cdot\mathbf{u}\mathcal{L}(\mathbf{u})\mathbf{u}'' + 2\mathbf{u}\cdot\mathbf{u}\mathcal{L}(\mathbf{u}')\mathbf{u}' - 4\mathbf{u}\cdot\mathbf{u}'\mathcal{L}(\mathbf{u})\mathbf{u}' + (m_k + m_l)\mathcal{L}(\mathbf{u})\mathbf{u} = (\mathbf{u}\cdot\mathbf{u})^3\mathbf{F}_{kl} \,, \tag{4.18}$$

where $R' = 2\mathbf{u}\cdot\mathbf{u}'$ has been substituted for the scalar radial velocity. By virtue of the last equation (4.15), this expression reduces to

$$2\mathbf{u}\cdot\mathbf{u}\mathcal{L}(\mathbf{u})\mathbf{u}'' - 2\mathbf{u}'\cdot\mathbf{u}'\mathcal{L}(\mathbf{u})\mathbf{u} + (m_k + m_l)\mathcal{L}(\mathbf{u})\mathbf{u} = (\mathbf{u}\cdot\mathbf{u})^3\mathbf{F}_{kl} \,. \tag{4.19}$$

In the next step we multiply by $\mathcal{L}^{-1}(\mathbf{u})$ and introduce the transpose defined by the first relation (4.15), which results in

$$\mathbf{u}'' + \tfrac{1}{2}\{[(m_k + m_l) - 2\mathbf{u}'\cdot\mathbf{u}']/\mathbf{u}\cdot\mathbf{u}\}\mathbf{u} = \tfrac{1}{2}\mathbf{u}\cdot\mathbf{u}\mathcal{L}^{\mathrm{T}}(\mathbf{u})\mathbf{F}_{kl} \,. \tag{4.20}$$

From (4.16) and the definition $\dot{R} = R'/R$ this yields the velocity transformation

$$\dot{\mathbf{R}} = 2\mathcal{L}(\mathbf{u})\mathbf{u}'/R \,. \tag{4.21}$$

Combining (4.21) with $\dot{\mathbf{R}}^{\mathrm{T}} = 2\mathbf{u}'\mathcal{L}^{\mathrm{T}}(\mathbf{u})/R$ and the orthogonality condition, we obtain the simplified square velocity relation

$$\dot{\mathbf{R}}^2 = 4\mathbf{u}'\cdot\mathbf{u}'/R \,. \tag{4.22}$$

With $\mathbf{u}\cdot\mathbf{u} = R$, this enables the final equation of motion to be written as

$$\mathbf{u}'' = \tfrac{1}{2}h\mathbf{u} + \tfrac{1}{2}R\mathcal{L}^{\mathrm{T}}(\mathbf{u})\mathbf{F}_{kl} \,. \tag{4.23}$$

By (4.8) and (4.22), the binding energy per unit reduced mass takes the form

$$h = [(2\mathbf{u}'\cdot\mathbf{u}' - (m_k + m_l)]/R \,. \tag{4.24}$$

The two-body energy h changes with time due to the perturbing force. The scalar product of $\dot{\mathbf{R}}$ and (4.2) leads to the rate of energy change

$$\frac{d}{dt}\left[\tfrac{1}{2}\dot{\mathbf{R}}^2 - \frac{(m_k + m_l)}{R}\right] = \dot{\mathbf{R}}\cdot\mathbf{F}_{kl} \,, \tag{4.25}$$

which reproduces the well-known expression $\dot{h} = \dot{\mathbf{R}}\cdot\mathbf{F}_{kl}$. Conversion to regularized derivative by $h' = \mathbf{R}'\cdot\mathbf{F}_{kl}$ and use of the velocity expression (4.16) then results in the equation of motion

$$h' = 2\mathbf{u}'\cdot\mathcal{L}^{\mathrm{T}}(\mathbf{u})\mathbf{F}_{kl} \,. \tag{4.26}$$

We conclude this derivation by remarking that the beautiful Levi-Civita treatment is seldom used in practical work. It is particularly suitable for educational purposes when introducing students to the subject.

4.4 Kustaanheimo–Stiefel method

The relations (4.12) exploit properties of mapping in the complex plane; hence generalization to 3D is not possible. This stumbling block prevented progress until it was realized by Kustaanheimo & Stiefel [1965] that a 4D formulation could be achieved. Their generalization of the Levi-Civita formalism gave rise to the 4×4 matrix

$$\mathcal{L}(\mathbf{u}) = \begin{bmatrix} u_1 & -u_2 & -u_3 & u_4 \\ u_2 & u_1 & -u_4 & -u_3 \\ u_3 & u_4 & u_1 & u_2 \\ u_4 & -u_3 & u_2 & -u_1 \end{bmatrix} . \tag{4.27}$$

It is convenient to use the same symbol as in (4.14) since it is clear from the context which formulation is intended. Now the basic relation (4.13) still applies but, because of some redundancy, a fourth fictitious coordinate and corresponding velocity is introduced. For completeness, the explicit components of $\mathbf{R}$ are given by

$$\begin{aligned} R_1 &= u_1^2 - u_2^2 - u_3^2 + u_4^2 , \\ R_2 &= 2(u_1 u_2 - u_3 u_4) , \\ R_3 &= 2(u_1 u_3 + u_2 u_4) , \\ R_4 &= 0 . \end{aligned} \tag{4.28}$$

Note that the fourth component of $\mathbf{R}$ is zero by application of (4.13). As can be readily verified, the square root of the sum of the distance squares (4.28) simplifies to

$$R = u_1^2 + u_2^2 + u_3^2 + u_4^2 . \tag{4.29}$$

When we go from a 2D to a 4D treatment, the fourth property (4.15) is not satisfied in general. Thus it can be shown [Stiefel & Scheifele, 1971] that for this to be the case, we require the relation

$$u_4 u_1' - u_3 u_2' + u_2 u_3' - u_1 u_4' = 0 . \tag{4.30}$$

This is called the *bilinear relation* and plays a fundamental role for understanding some theoretical aspects of the development[†] [cf. Stiefel & Scheifele, 1971]. In practical terms, it is a constraint and corresponds to the identity $\dot{R}_4 = 0$. Here we merely remark that (4.30) may also be used as a check on the reliability of the numerical solutions, although this does not seem to have been tried.

Since one of the components of $\mathbf{u}$ is arbitrary, we have some freedom of choice when it comes to specifying the initial components. If $R_1 > 0$, we combine the first equation (4.28) with (4.29), which results in

$$u_1^2 + u_4^2 = \tfrac{1}{2}(R_1 + R) . \tag{4.31}$$

[†] An alternative derivation of the KS variables clarifies its meaning [Yoshida, 1982].

The redundancy choice $u_4 = 0$ then gives rise to the relations

$$\begin{aligned} u_1 &= [\tfrac{1}{2}(R_1 + R)]^{1/2}\,, \\ u_2 &= \tfrac{1}{2}R_2/u_1\,, \\ u_3 &= \tfrac{1}{2}R_3/u_1\,. \end{aligned} \tag{4.32}$$

Likewise, if $R_1 < 0$, we subtract the first equation (4.28) from (4.29) and obtain

$$u_2^2 + u_3^2 = \tfrac{1}{2}(R - R_1)\,. \tag{4.33}$$

Thus by setting $u_3 = 0$ we have

$$\begin{aligned} u_2 &= [\tfrac{1}{2}(R - R_1)]^{1/2}\,, \\ u_1 &= \tfrac{1}{2}R_2/u_2\,, \\ u_4 &= \tfrac{1}{2}R_3/u_2\,. \end{aligned} \tag{4.34}$$

The reason for these alternatives is that the initial values must be numerically well defined, and this is ensured by the above choice.

We also need an expression for the initial regularized velocity, $\mathbf{u}'$. This is achieved by inverting (4.16) and making use of the first relation (4.15), which leads to

$$\mathbf{u}' = \tfrac{1}{2}\mathcal{L}(\mathbf{u})\mathbf{R}'/R\,. \tag{4.35}$$

Using the definition (4.4), we have $\dot{R} = R'/R$. This yields the more convenient expression

$$\mathbf{u}' = \tfrac{1}{2}\mathcal{L}(\mathbf{u})\dot{\mathbf{R}}\,. \tag{4.36}$$

From the above, it can be seen that the 2D formalism carries over to the KS treatment of 3D systems. However, as emphasized by Stiefel & Scheifele [1971], the same line of approach in going from the time-smoothed equations of motion (4.6) to (4.23) cannot be repeated in the general case. The reason is that the KS transformation is ambiguous and there is no unique set of vectors $\mathbf{u}$ for a given $\mathbf{R}$. Accordingly, the only way around this difficulty is to postulate the form (4.23) and verify that the original equations of motion are satisfied. Following this formal exercise, we omit the argument from the Levi-Civita matrix and also represent the perturbing force as $\mathbf{F}$ for simplicity. Hence the equations of motion take the final form

$$\mathbf{u}'' = \tfrac{1}{2}h\,\mathbf{u} + \tfrac{1}{2}R\,\mathcal{L}^{\mathrm{T}}\,\mathbf{F}\,, \tag{4.37}$$

$$h' = 2\,\mathbf{u}' \cdot \mathcal{L}^{\mathrm{T}}\,\mathbf{F}\,, \tag{4.38}$$

$$t' = \mathbf{u} \cdot \mathbf{u}\,. \tag{4.39}$$

It is evident that these equations are well behaved for $R \to 0$. In particular, the relative contribution from the perturbing term is proportional to

R^3 for a circular orbit. Consequently, h tends to a constant rapidly with decreasing perturbation. A total of ten regularized equations are required in order to obtain the solution for the relative motion, compared with just six in the case of direct integration. The method employed here is based on the standard polynomial or Hermite formulations and will be discussed in a subsequent chapter.

In practical applications, the question arises of how to determine the physical time. Notwithstanding the concept of the time element [cf. Stiefel & Scheifele, 1971], which has proved useful in satellite theory, a direct integration of (4.39) to high order has some merit. First, it can also be employed for near-parabolic and hyperbolic motion. A second advantage is that it is suitable during intervals of strong perturbations. Hence the simplicity of such a treatment for general N-body problems outweighs some loss of accuracy for small perturbations on elliptic orbits when phase errors may be tolerated.

For future reference we note that when using the KS variables, the semi-major axis is obtained from $a = -\frac{1}{2}(m_k + m_l)/h$. Another useful expression is the eccentricity, evaluated at an arbitrary phase in the orbit by means of the eccentric anomaly. Hence combining $e\cos\theta = 1 - R/a$ and $e\sin\theta = \mathbf{R}\cdot\dot{\mathbf{R}}/[(m_k + m_l)a]^{1/2}$ we have

$$e^2 = (1 - R/a)^2 + 4(\mathbf{u}\cdot\mathbf{u}')^2/(m_k + m_l)a\,. \tag{4.40}$$

Here we have used the definition $R' = \dot{R}R$ and substituted from $t'' = R'$ by (4.39).

To describe the motion completely, we introduce the centre of mass as a fictitious particle. The corresponding equation of motion is given by the mass-weighted sum of perturbations,

$$\ddot{\mathbf{r}}_{\rm cm} = (m_k\mathbf{F}_k + m_l\mathbf{F}_l)/(m_k + m_l)\,. \tag{4.41}$$

Note that the dominant two-body interaction cancels analytically here. Accordingly, the spatial coordinates of the components are recovered from

$$\begin{aligned} \mathbf{r}_k &= \mathbf{r}_{\rm cm} + \mu\mathbf{R}/m_k\,, \\ \mathbf{r}_l &= \mathbf{r}_{\rm cm} - \mu\mathbf{R}/m_l\,, \end{aligned} \tag{4.42}$$

with $\mu = m_k m_l/(m_k + m_l)$, and similarly for the velocities $\dot{\mathbf{r}}_k, \dot{\mathbf{r}}_l$. Further details of how to obtain the complete solutions are given in a later chapter.

The device of energy stabilization has also been tried with success. Following the discussion of section 3.5, we generalize (4.37) from (3.33) and include a correction term which yields

$$\mathbf{u}'' = \tfrac{1}{2}h\,\mathbf{u} + \tfrac{1}{2}R\mathcal{L}^T\,\mathbf{F} - \alpha(\tilde{h} - h)R\mathbf{u}'/(m_k + m_l)\,. \tag{4.43}$$

Accordingly, $\tilde{h}R$ is calculated explicitly at every step (cf. 4.24) and this quantity is well defined for $R \to 0$. Likewise for the scaling factor α, which we take to be $0.4/\Delta\tau$ on dimensional grounds, with the numerical constant being an optimum choice for the present integration scheme. Although the perturbation may be arbitrarily large in principle because the integration step is shortened, the stabilizing term should only be included for fairly modest values.

The standard KS formulation has also been implemented for a real application to geocentric satellite orbits [Palmer *et al.*, 1998]. Comparisons with direct integration showed that the regularized equations are beneficial for obtaining accurate solutions of highly eccentric orbits.

On a historical note, the first application of KS regularization to the N-body problem was given by Peters [1968a,b]. This development made use of the elegant Hamiltonian formalism which will be exploited in the next chapter. Numerical examples for $N = 3$ [Szebehely & Peters, 1967] and $N = 25$ [Peters, 1968b] were discussed at an early stage. In particular, the former study illustrating a sequence of close encounters did much to publicize the power of regularization in stellar dynamics.

It may also be remarked that the earliest N-body implementations of KS employed the explicit recalculation of (4.24) instead of integrating the perturbation effect according to (4.26). However, very soon thereafter it was realized that the latter procedure is advantageous [Bettis & Szebehely, 1972; Aarseth, 1972b]. One of the reasons for the improved behaviour is that the explicit expression involves the *predicted* velocity which is known to lower order, but in addition the singular nature of (4.24) gives rise to growing oscillations at small separations.

Some advantages in adopting regularization of dominant two-body motion may be summarized as follows:

- The equations of motion are regular and well behaved for $R \to 0$
- A smaller number of integration steps per orbit is required
- The numerical stability of even circular motion is improved
- Further error reduction may be achieved by rectification‡
- Distant contributions can be neglected since $F \propto 1/r^3$
- Two-body elements can be used to study tidal interactions

The main point on the debit side is the need for coordinate transformations. However, the number of operations involved is not very large and no square root is required. In the next section we discuss an alternative regularization that employs the actual physical variables.

‡ Rescaling of $\mathbf{u}$, $\mathbf{u}'$ to the correct value of h will be discussed in chapter 11.

4.5 Burdet–Heggie alternative

Although elegant, the KS formulation involves frequent transformations from the coordinates $\mathbf{u}$ to $\mathbf{r}_k, \mathbf{r}_l$, as well as setting up the initial relations (4.32). Such complications may act as a deterrent to the new practitioner who has to get used to working with fictitious quantities. It is therefore of interest to present an alternative method that is based on physical coordinates, thereby avoiding repetitive transformations. In the following we discuss the independent derivation of Heggie [1973], although an earlier formulation was presented by Burdet [1967, 1968].

We begin by choosing the standard time transformation, $t' = R$, which again gives rise to an equation of motion of the form

$$\mathbf{R}'' = R'\mathbf{R}'/R - M\mathbf{R}/R + R^2\mathbf{F}\,, \tag{4.44}$$

with the notation $M = m_k + m_l$.

Following Heggie [1973], let us define the quantities P and $\mathbf{B}$ by[§]

$$\begin{aligned} P &= -2M/R + \mathbf{R}'^2/R^2\,, \\ \mathbf{B} &= M\mathbf{R}/R - \mathbf{R}'^2\mathbf{R}/R^2 + R'\mathbf{R}'/R\,. \end{aligned} \tag{4.45}$$

Straightforward application of (4.4) results in

$$P = -2M/R + \dot{\mathbf{R}}^2\,, \tag{4.46}$$

$$\mathbf{B} = M\mathbf{R}/R - \dot{\mathbf{R}}^2\mathbf{R} + (\mathbf{R}\cdot\dot{\mathbf{R}})\dot{\mathbf{R}}\,. \tag{4.47}$$

This enables us to write (4.44) as

$$\mathbf{R}'' = P\mathbf{R} + \mathbf{B} + R^2\mathbf{F}\,. \tag{4.48}$$

From (4.2), (4.46) and (4.47), we readily obtain the first-order companion equations for $\dot{P}$ and $\dot{\mathbf{B}}$ which, after conversion to primed derivatives by (4.4), take the final form

$$\begin{aligned} P' &= 2\mathbf{R}'\cdot\mathbf{F}\,, \\ \mathbf{B}' &= -2(\mathbf{R}'\cdot\mathbf{F})\mathbf{R} + (\mathbf{R}\cdot\mathbf{F})\mathbf{R}' + (\mathbf{R}\cdot\mathbf{R}')\mathbf{F}\,. \end{aligned} \tag{4.49}$$

The resulting equations of motion are well behaved for $R \rightarrow 0$, and two-body regularization has therefore been achieved. We also note that, in the absence of perturbations, the quantities P and $\mathbf{B}$ are constants of the motion, and hence in the general case are slowly varying elements.

It remains to determine an equation for the time. By differentiating twice the identity $R^2 = \mathbf{R}\cdot\mathbf{R}$ we obtain, by (4.47) and (4.48), the relation

$$R'' = PR + M + \mathbf{R}\cdot\mathbf{F}\,R\,. \tag{4.50}$$

§ The original notation $\mathbf{Q}$ has been replaced by $\mathbf{B}$ here to avoid confusion with KS.

Higher-order terms may be constructed by further differentiation. However, since this procedure would involve derivatives of the perturbing force, it is advantageous to develop a fitting polynomial for the function (4.50) itself. Hence by (4.3) we have that

$$t^{(3)} = Pt' + M + (\mathbf{R} \cdot \mathbf{F})R\,. \tag{4.51}$$

This function is well behaved near the two-body singularity. Given the first three time derivatives by explicit expressions involving R and $\mathbf{R} \cdot \mathbf{F}$, higher orders may be constructed from divided differences of (4.50) if desired. An alternative solution method is discussed in the next section.

We now turn to the interpretation of the relevant quantities. Since $P = 2h$, the semi-major axis is obtained from

$$a = -M/P\,. \tag{4.52}$$

In the absence of perturbations, the equation of motion (4.48) represents a displaced harmonic oscillator. The frequency is therefore *twice* that given by KS regularization, which is the price to pay for avoiding the coordinate transformation. From $\mathbf{R}'' = 0$ for the centre of the orbit at $\mathbf{R}_0$, we have $\mathbf{B} = -P\mathbf{R}_0$. Since $|\mathbf{R}_0| = ea$ measures the distance from the centre to the focal point, $\mathbf{B}$ is related to the eccentricity vector by

$$\mathbf{B} = -M\mathbf{e}\,. \tag{4.53}$$

Here $-\mathbf{B}/M$ is sometimes known as the Runge–Lenz–Laplace vector, or alternatively it is also named after Hamilton or even Newton.

The Burdet–Heggie scheme has been used successfully for a series of three-body scattering experiments with regularization of the dominant two-body motion [Valtonen, 1974, 1975]. It was also employed for studying linearized departures from a given orbit in theoretical investigations of binary evolution [Heggie, 1975]. Finally, it is pleasing to record that this method proved itself in N-body simulations of core collapse [Giersz & Heggie, 1994a,b].

4.6 Hermite formulation

Hermite integration also lends itself to KS regularization. Although such a scheme has been used successfully in earlier versions of the code[§] NBODY4 [Aarseth, 1996a] and the more recent NBODY6, it has now been replaced by a more sophisticated development, to be presented in the next section. However, this method does represent an attractive alternative for some problems and will therefore be discussed here. Its simplicity also serves as

[§] To be described in subsequent chapters.

a useful introduction to this approach. Before proceeding, we note that the divided difference method of section 2.2 was used to integrate the KS equations of motion since 1969 [cf. Aarseth, 1972b]. The success of the Hermite N-body method [Makino, 1991b; Makino & Aarseth, 1992] soon led to a change-over in order to have a uniform treatment.

Since higher derivatives feature more prominently, we change the notation and define the regularized force (per unit mass) as $\mathbf{F}_{\rm U} = \mathbf{U}''$, with $\mathbf{U}$ replacing $\mathbf{u}$ to conform with the following sections. For simplicity, we also set $\mathbf{Q} = \mathcal{L}^T\mathbf{P}$, where $\mathbf{P}$ from now on is the perturbing force (to avoid confusion with $\mathbf{F}_{\rm U}$). The equations of motion (4.37) and (4.39) then take the Hermite form

$$\begin{aligned}
\mathbf{F}_{\rm U} &= \tfrac{1}{2}h\mathbf{U} + \tfrac{1}{2}R\mathbf{Q}\,,\\
\mathbf{F}'_{\rm U} &= \tfrac{1}{2}(h'\mathbf{U} + h\mathbf{U}' + R'\mathbf{Q} + R\mathbf{Q}')\,,\\
h' &= 2\mathbf{U}'\cdot\mathbf{Q}\,,\\
h'' &= 2\mathbf{F}_{\rm U}\cdot\mathbf{Q} + 2\mathbf{U}'\cdot\mathbf{Q}'\,,\\
t' &= \mathbf{U}\cdot\mathbf{U}\,,
\end{aligned} \tag{4.54}$$

where the standard equation for t' has been included for completeness.

Once again, the energy stabilization term in (4.43) may be included. Note that the stabilization term is only added to $\mathbf{F}_{\rm U}$ for the purpose of predicting $\mathbf{U}$ and $\mathbf{U}'$ and the contribution to $\mathbf{F}'_{\rm U}$ is *not* included. Hence only the basic expression (4.37) is saved for further use by the integration scheme. Following the tradition of *NBODY*5, the stabilized Hermite scheme was tried with low-order prediction (i.e. up to $\mathbf{F}'_{\rm U}$) of the KS variables which gives rise to a fairly simple formulation. However, subsequent experience [Kokubo, Yoshinaga & Makino, 1998; Mikkola & Aarseth, 1998] showed that significant accuracy may be gained by employing higher-order prediction at the expense of some additional operations. Procedures for initialization and integration will be presented in a later section. Finally, we remark that the Burdet–Heggie scheme discussed above could be converted to Hermite form; this might be particularly beneficial for the integration of physical time.

We end this brief section by pointing out an additional equation for R that does not seem to have been employed in actual KS integrations. Twice differentiation of $R = \mathbf{U}\cdot\mathbf{U}$ gives $R'' = 2\,\mathbf{U}''\cdot\mathbf{U} + 2\,\mathbf{U}'\cdot\mathbf{U}'$. Combining the equation of motion with the definition of h from (4.24), we obtain the expression [cf. Bettis & Szebehely, 1972]

$$R'' = 2hR + (m_k + m_l) + \mathbf{U}\cdot\mathbf{Q}\,R\,. \tag{4.55}$$

This differential equation may be solved and used as a check on the accuracy of R or substituted in the higher derivatives of t'.

4.7 Stumpff functions

The emphasis of the present chapter is on the treatment of perturbed binaries. So far, it would appear that the KS method is the ultimate tool for N-body simulations but no stone must be left unturned. Numerical integration of binaries presents formidable problems, mainly because of the time-scale since millions of Kepler periods may be involved. Hence we need to search for methods that combine long-term accuracy with efficiency. In this section, we describe an alternative two-body regularization that builds on the existing KS framework and appears to have significant advantages. The basic idea of the Stumpff KS method [Mikkola & Aarseth, 1998] is to expand the solution in Taylor series, where the higher orders are modified by coefficients representing the truncated terms.

The following discussion is based on the original adaptation. For simplicity, we introduce the notation $\Omega = -h/2$, $\mathbf{V} = \mathbf{U}'$ and $\mathbf{Q} = \mathcal{L}^{\mathrm{T}}(\mathbf{U})\mathbf{P}$ with $\mathbf{P}$ the perturbation. This allows us to write the equations of motion in the shorter form

$$\begin{aligned} \mathbf{U}'' + \Omega\mathbf{U} &= \tfrac{1}{2}R\mathbf{Q}\,, \\ \Omega' &= -\mathbf{V}\cdot\mathbf{Q}\,. \end{aligned} \tag{4.56}$$

The solution for the unperturbed Keplerian case ($\mathbf{Q} = 0$) is given by

$$\begin{aligned} \mathbf{U}_{\mathrm{K}} &= \mathbf{U}_0 G_0(\Omega_0, \tau) + \mathbf{V}_0 G_1(\Omega_0, \tau)\,, \\ \mathbf{V}_{\mathrm{K}} &= \mathbf{V}_0 G_0(\Omega_0, \tau) - \Omega_0 \mathbf{U}_0 G_1(\Omega_0, \tau)\,, \\ t_{\mathrm{K}} &= t_0 + R_0\tau + 2\mathbf{U}_0\cdot\mathbf{V}_0 G_2(4\Omega_0, \tau) + M G_3(4\Omega_0, \tau)\,. \end{aligned} \tag{4.57}$$

Here the functions G_n are defined in Stiefel & Scheifele [1971, pp 141–143]. We have the relations

$$G_n(\Omega, \tau) = \tau^n c_n(\Omega\tau^2)\,, \tag{4.58}$$

where

$$c_n(z) = \sum_{k=0}^{\infty} \frac{(-z)^k}{(n+2k)!} \tag{4.59}$$

are the Stumpff [1962] functions and $z = \Omega\tau^2$. It can be seen from the series expansion that these functions satisfy the recursion relation

$$c_n = \frac{1}{n!} - z\,c_{n+2}\,. \tag{4.60}$$

This expression may be used instead of the power series to evaluate the Stumpff functions for small arguments. The choice of order is a matter for the implementation, to be discussed in a subsequent chapter, but we note here that the expansion converges fast for small z.

In order to discuss the solution of (4.57), we begin by considering the perturbed harmonic oscillator

$$Y'' + \Omega Y = g(\tau)\,, \tag{4.61}$$

where primes denote differentiation with respect to τ and $\Omega = -h/2$ is the square of the frequency if $h < 0$. We obtain a Taylor series solution with derivatives given by

$$Y^{(k+2)} + \Omega Y^{(k)} = g^{(k)}\,. \tag{4.62}$$

Hence if g is a polynomial of at most degree n, the recursion relation above for $k > n$ reduces to the trigonometric one,

$$Y^{(k+2)} + \Omega Y^{(k)} = 0\,. \tag{4.63}$$

With the notation $Y^{(k)}$ for the derivatives at $\tau = 0$, the expansion for Y becomes

$$\begin{aligned} Y = &\sum_{k=0}^{n-2} Y^{(k)} \frac{\tau^k}{k!} \\ &+ Y^{(n-1)} \left(\frac{\tau^{n-1}}{(n-1)!} - \frac{\Omega \tau^{n+1}}{(n+1)!} + \frac{\Omega^2 \tau^{n+3}}{(n+3)!} - + \ldots \right) \\ &+ Y^{(n)} \left(\frac{\tau^n}{n!} - \frac{\Omega \tau^{n+2}}{(n+2)!} + \frac{\Omega^2 \tau^{n+4}}{(n+4)!} - + \ldots \right)\,. \end{aligned} \tag{4.64}$$

The terms associated with the two highest order derivatives are the G-functions defined by (4.58). Hence the Taylor series expansion of a perturbed harmonic oscillator can be made exact for an equation of the type (4.61), provided we modify the coefficients of the last two derivatives of (4.64). Extracting the common terms, we are left with a power series close to unity in each case, with the respective multiplicative factors

$$\begin{aligned} \tilde{c}_{n-1} &= (n-1)!\, c_{n-1}(z)\,, \\ \tilde{c}_n &= n!\, c_n(z)\,. \end{aligned} \tag{4.65}$$

This enables the solution to be obtained by

$$Y(\tau) = \sum_{k=0}^{n-2} Y^k \frac{\tau^k}{k!} + Y^{(n-1)} \frac{\tau^{(n-1)}}{(n-1)!} \tilde{c}_{n-1} + Y^{(n)} \frac{\tau^n}{n!} \tilde{c}_n\,. \tag{4.66}$$

We now proceed to discuss the solution of the equations of motion (4.56). If polynomial approximations are formed for the right-hand sides of the equations for $\mathbf{U}$ and Ω, we can integrate the former with the above method, while for the latter it is sufficient to use the basic polynomial because Ω is slowly varying. Hence we express the three solutions in the form

$$\mathbf{U} = \sum_{k=0}^{n-2} \frac{\mathbf{U}^k}{k!}\tau^k + \frac{\mathbf{U}^{(n-1)}}{(n-1)!}\tau^{n-1}\tilde{c}_{n-1}(z) + \frac{\mathbf{U}^{(n)}}{n!}\tau^n\tilde{c}_n(z)\,,$$

$$\Omega = \Omega_0 + \Omega'\tau + \tfrac{1}{2}\Omega''\tau^2 + \dots\,,$$

$$t = \sum_{k=0}^{n-2} \frac{t^{(k)}}{k!}\tau^k + \frac{t^{(n-1)}}{(n-1)!}\tau^{n-1}\tilde{c}_{n-1}(4z) + \frac{t^{(n)}}{n!}\tau^n\tilde{c}_n(4z), \quad (4.67)$$

where the derivatives are evaluated at $\tau = 0$. The $\tilde{c}$-functions obey the relations

$$\tilde{c}_k(z) = k!\, c_k(z)\,, \quad (4.68)$$

with $z = \Omega_0\tau^2$, and similarly for $\tilde{c}_k(4z)$.

In this formulation it is more meaningful to rewrite the original equation of motion in the form

$$\mathbf{U}'' + \Omega_0\mathbf{U} = (\Omega_0 - \Omega)\mathbf{U} + \tfrac{1}{2}R\mathbf{Q}\,, \quad (4.69)$$

which shows that the expansion has been carried out to sufficient order when the right-hand side has converged.

This concludes the formal development of the KS integration scheme. We remark that the differential term $\Omega_0 - \Omega$ in (4.69) is reminiscent of the classical Encke's method that solves an equation of motion for small displacements from two-body motion [cf. Brouwer & Clemence, 1961]. Algorithms for the implementation of the Stumpff KS method in an N-body code will be described in a later chapter. Some aspects of KS regularization are discussed elsewhere [cf. Aarseth, 1972b, 1985, 1994, 2001c].

For completeness, we mention an original formulation that was also based on Stumpff functions [Jernigan & Porter, 1989]. Although the application employed a recursive binary tree code, the integration method was concerned with weakly perturbed two-body motion of the form (4.56). However, instead of conventional step-wise advancement of the regularized equation of motion, the solutions are expressed as polynomials in terms of Stumpff functions. A key feature is that the perturbing force acting on a binary is given by an expression that takes account of the tidal field and also its time derivative. Provided the perturbing environment of a binary changes slowly, this method allows the solution to be advanced accurately over many Kepler periods in one time-step. Given that most hard binaries in cluster simulations are weakly perturbed, this novel method could be combined with one of the other KS treatments for dealing with more rapidly varying tidal forces.

5
Multiple regularization

5.1 Introduction

In the preceding chapter, we have considered several methods for dealing with the perturbed two-body problem. Formally all these methods work for quite large perturbations, provided the regularized time-step is chosen sufficiently small. However, the selection of the dominant pair in a triple resonance interaction frequently calls for new initializations where the intruder is combined with one of the components. Likewise, one may have situations in which two hard binaries approach each other with small impact parameter. Hence a description in terms of one dominant two-body motion tends to break down during strong interactions, precisely at times when interesting outcomes are likely to occur. Since the switching of dominant components reduces the efficiency and also degrades the quality of the results, it it highly desirable to seek alternative methods for improved treatment.

In this chapter, we discuss several multiple regularization methods that have turned out to be very beneficial in practical simulations. By multiple regularization it is understood that at least two separations in a compact subsystem are subject to special treatment where the two-body singularities are removed. We begin by describing a three-body formulation that may be considered the Rosetta Stone for later developments. The generalization to more members with just one reference body is also included for completeness. A subsequent section outlines the elegant global formulation and is followed by a detailed discussion of the powerful chain regularization. We then present a method for treating binaries of arbitrarily small periods in subsystems by scaling the external perturbation. Finally, we include a new time-smoothing scheme for studying a massive binary in the presence of perturbers. All these methods are invariably somewhat technical but the end result justifies the effort.

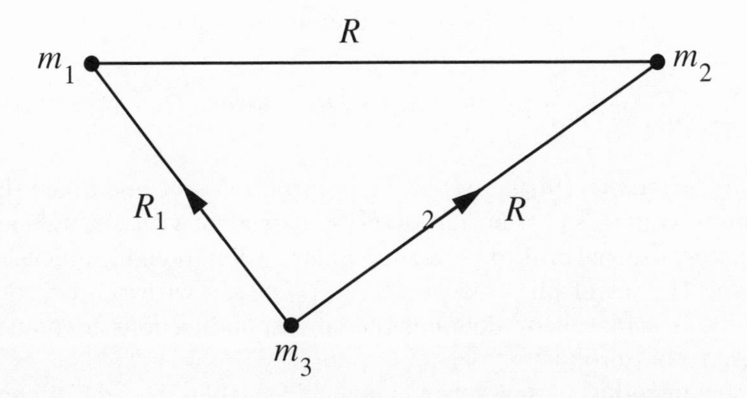

Fig. 5.1. Three-body configuration.

5.2 Aarseth–Zare method

According to Whittaker [1904], the general three-body problem is the most celebrated problem in dynamical astronomy. Hence an improved numerical treatment is likely to be of considerable interest. Whereas there is a vast literature on the restricted three-body problem [Szebehely, 1967], in which one of the masses is zero, relatively little attention has been devoted to the general case. No doubt the reason for the neglect is the intractable nature of this problem, and it is only in modern times that numerical investigations have become the theoretician's tool, requiring more sophisticated treatments.

The basic idea in the following derivation is to introduce two coupled KS regularizations. Let a general three-body configuration be described by the non-zero particle masses m_1, m_2, m_3, with coordinates $\tilde{\mathbf{q}}_k$ and corresponding momenta $\tilde{\mathbf{p}}_k$ $(k = 1, 2, 3)$. This system has the Hamiltonian

$$\widetilde{H} = \sum_{k=1}^{3} \tilde{\mathbf{p}}_k^2/2m_k - m_1 m_3/R_1 - m_2 m_3/R_2 - m_1 m_2/R\,, \tag{5.1}$$

where the distances R_k, R are defined in Fig. 5.1. The corresponding equations of motion form a system of 18 first-order differential equations.

We first reduce the order by six, employing the centre-of-mass (c.m.) integrals; in other words, placing the system at rest in an inertial frame. Using the c.m. condition $\mathbf{p}_3 = -(\mathbf{p}_1 + \mathbf{p}_2)$, we express the kinetic energy in terms of the two independent momenta, $\mathbf{p}_k$ $(k = 1, 2)$, associated with the mass-points m_k. Likewise, from the figure, $\mathbf{q}_k = \mathbf{R}_k$ represents the distance vectors with respect to m_3 which is known as the reference body, whereas $R = |\mathbf{R}_1 - \mathbf{R}_2|$.* Hence the Hamiltonian simplifies to

$$H = \sum_{k=1}^{2} \frac{1}{2\mu_{k3}} \mathbf{p}_k^2 + \frac{1}{m_3} \mathbf{p}_1 \cdot \mathbf{p}_2 - m_1 m_3/R_1 - m_2 m_3/R_2 - m_1 m_2/R \,, \quad (5.2)$$

where $\mu_{k3} = m_k m_3/(m_k + m_3)$ is the reduced mass of m_k. Since the corresponding equations of motion describe a system with six independent coordinates, it is natural to perform regularization in eight dimensions by increasing the set of physical variables in analogy with the KS formulation. This is achieved by defining the corresponding fourth components of $\mathbf{q}_k, \mathbf{p}_k$ to be zero.

We now introduce a new set of canonical variables in eight dimensions, denoted by $\mathbf{Q}$ and $\mathbf{P}$. According to the theory of canonical contact transformations, we choose a generating function $W = W(\mathbf{p}, \mathbf{Q})$. This enables us to obtain the new momenta in terms of the old ones by $\mathbf{P}_k = \partial W/\partial \mathbf{Q}_k$, and likewise the new coordinates by $\mathbf{q}_k = \partial W/\partial \mathbf{p}_k$. At this point we follow the notation of Szebehely [1967] and adopt a separable generating function of the form

$$W = \sum_{k=1}^{2} \mathbf{p}_k \cdot \mathbf{f}_k \,, \quad (5.3)$$

with each $\mathbf{f}_k$ a function of $\mathbf{Q}_k$. This crucial step yields a simple result for the two relative coordinates ($k = 1, 2$),

$$\mathbf{q}_k = \mathbf{f}_k \,. \quad (5.4)$$

The corresponding transformation for the momenta becomes

$$\mathbf{P}_k = \mathbf{A}_k \mathbf{p}_k \,, \quad (5.5)$$

where the 4×4 matrix, $\mathbf{A}_k$, is composed of terms $\partial \mathbf{f}_k/\partial \mathbf{Q}_k$. From a comparison of (5.5) with the regularized velocity transformation (4.36), we see that the matrix $\mathbf{A}_k$ is identical to twice the corresponding *transpose* of the generalized Levi-Civita matrix (4.27). However, in view of the manner in which $\mathbf{A}_k$ is introduced here,† it is more natural to present the derivation based on this quantity.

Since the matrix $\mathbf{A}_k$ is orthogonal, it can be shown that

$$\mathbf{A}_k \mathbf{A}_k^{\mathrm{T}} = 4R_k \mathbf{I} \,, \quad (5.6)$$

where $\mathbf{I}$ is the identity matrix. This allows the physical momenta and relative coordinates to be recovered from standard KS transformations,

* The notation R_3 would cause confusion with the development in the next section.

† Traditionally, the derivation of KS begins with the fundamental relation $\mathbf{R} = \mathcal{L}\mathbf{u}$.

with $\mathbf{p}_k^2 = \mathbf{P}_k^2/4R_k$. Let us introduce the Hamiltonian function in the extended phase space, $\Gamma = H(\mathbf{Q}, \mathbf{P}) - E_0$, where E_0 is the total energy. Regularization is now achieved by employing the time transformation

$$dt = R_1 R_2 d\tau \,. \tag{5.7}$$

With $\Gamma^* = R_1 R_2 \Gamma$, the regularized Hamiltonian takes the form

$$\begin{aligned}\Gamma^* = {} & \frac{1}{8\mu_{13}} R_2 \mathbf{P}_1^2 + \frac{1}{8\mu_{23}} R_1 \mathbf{P}_2^2 + \frac{1}{16m_3} \mathbf{P}_1^{\mathrm{T}} \mathbf{A}_1 \cdot \mathbf{A}_2^{\mathrm{T}} \mathbf{P}_2 \\ & - m_1 m_3 R_2 - m_2 m_3 R_1 - \frac{m_1 m_2 R_1 R_2}{|\mathbf{R}_1 - \mathbf{R}_2|} - E_0 R_1 R_2 \,. \end{aligned} \tag{5.8}$$

From an inspection of (5.2) one can readily identify the terms, with $\mathbf{P}_k^2/\mu_{k3}$ representing the kinetic energy of the relative two-body motions.

The corresponding equations of motion for $k = 1, 2$ are given by

$$\frac{d\mathbf{Q}_k}{d\tau} = \frac{\partial \Gamma^*}{\partial \mathbf{P}_k} \,, \qquad \frac{d\mathbf{P}_k}{d\tau} = - \frac{\partial \Gamma^*}{\partial \mathbf{Q}_k} \,. \tag{5.9}$$

These equations are regular for $R_1 \rightarrow 0$ or $R_2 \rightarrow 0$ since $\mathbf{P}_k \rightarrow$ const as $\mathbf{R}_k \rightarrow 0$, and hence a practical regularization of the three-body problem has been achieved [Aarseth & Zare, 1974].

It may be helpful for implementations to provide the full equations of motion in explicit form. Differentiation according to the first equation (5.9) yields the straightforward result

$$\frac{d\mathbf{Q}_k}{d\tau} = \frac{1}{4\mu_{k3}} R_l \mathbf{P}_k + \frac{1}{16m_3} \mathbf{A}_k \mathbf{A}_l^{\mathrm{T}} \mathbf{P}_l \,, \tag{5.10}$$

where $l = 3 - k$. Likewise, we obtain the more complicated derivative of the momentum in compact notation as

$$\begin{aligned}\frac{d\mathbf{P}_k}{d\tau} = {} & 2 \left(R_l E_0 - \frac{1}{8\mu_{l3}} \mathbf{P}_l^2 + m_l m_3 + \frac{m_1 m_2}{|\mathbf{R}_1 - \mathbf{R}_2|} R_l \right) \mathbf{Q}_k \\ & - \frac{1}{8m_3} \mathbf{P}_k \mathbf{A}_l^{\mathrm{T}} \mathbf{P}_l - \frac{m_1 m_2 R_1 R_2}{|\mathbf{R}_1 - \mathbf{R}_2|^3} \mathbf{A}_k (\mathbf{R}_k - \mathbf{R}_l) \,. \end{aligned} \tag{5.11}$$

With a total of eight components, evaluation of the second equation of motion requires a considerable numerical effort. However, to compensate, the regular structure permits quite large time-steps to be used.

The time transformation (5.7) regularizes all terms in the Hamiltonian associated with the distances R_1 and R_2, leaving one singularity due to the potential energy term $m_1 m_2/R$. After differentiation, equation (5.11) contains two singular terms involving division by R and R^3. In view of the time transformation, these terms have a different origin, with the former having a kinematical interpretation and the latter representing the dynamical interaction. From the geometry of the problem, $R \simeq R_1$ if $R > R_2$ and $R_1 > R_2$; hence in most cases the inequality $R > \min\{R_1, R_2\}$ ensures that the critical terms are *numerically* smaller than the regularized contributions. This desirable property compensates for the divisions and is mainly responsible for the improved treatment of even close triple encounters.

On the debit side, implementation of the switching procedure requires a transformation to physical variables with consequent small loss of accuracy before assigning new particle labels, whereupon the regularized quantities are introduced. However, when the masses are different, the heaviest particle tends to dominate the motion and is a natural reference body, thereby reducing the number of switching transformations.

A comment on the role of the mass ratios may also be in order. Since the present formulation does not include the case of one massless body, a significant loss of accuracy may be anticipated for very large mass ratios; i.e. the contributions from some terms would be downgraded. Moreover, the reason for the inability to deal with one zero mass particle is due to the c.m. condition becoming indeterminate. Equivalently, the Hamiltonian is based on reciprocals of the masses. For a regular treatment of the massless body in the restricted problem see the formulation by Heggie [1974].

In order to make a connection with the standard KS formulation, it is instructive to consider the analogous Hamiltonian equations for just two particles. For this purpose we select the two-body system m_1 and m_3 in the expression (5.8) for Γ^* by omitting all terms connected with m_2, as well as the distance R_2. This simplification leads to a regularized two-body Hamiltonian which was first applied by Peters [1968a,b] to the dominant interaction in the general three-body problem. Introducing the binding energy per unit reduced mass, $h = E_0/\mu_{13}$ and omitting subscripts, we obtain the simple expression

$$\Gamma_2^* = \mathbf{P}^2/8\mu - m_1 m_3 - \mu h \mathbf{Q}^2 . \tag{5.12}$$

The equations of motion then follow by differentiation according to (5.9), which yields

$$\begin{aligned} \frac{d\mathbf{Q}}{d\tau} &= \mathbf{P}/4\mu , \\ \frac{d\mathbf{P}}{d\tau} &= 2\mu h \mathbf{Q} . \end{aligned} \tag{5.13}$$

Finally, the familiar form $\mathbf{Q}'' = \frac{1}{2} h \mathbf{Q}$ is obtained after introducing the regularized velocity $\mathbf{V} = \mathbf{P}/\mu$ in the second equation since the first equation implies that $\mathbf{V} = 4\mathbf{Q}'$.

In the case of 2D problems, (5.10) and (5.11) still apply but we may use the simpler Levi-Civita matrix (4.14). Problems in 1D are ideally suited to the present formulation since no switching is needed by placing the reference body in the middle. Following Mikkola & Hietarinta [1989], we introduce three equal-mass bodies, $m_i = 1$, with coordinates $x_1 < x_0 < x_2$ and take $q_1 = x_0 - x_1$, $q_2 = x_2 - x_0$ as new relative coordinates. From the generating function $S = p_1(x_0 - x_1) + p_2(x_2 - x_0)$, with momenta p_i, the new Hamiltonian becomes

$$H = p_1^2 + p_2^2 - p_1 p_2 - \frac{1}{q_1} - \frac{1}{q_2} - \frac{1}{q_1 + q_2} \,. \tag{5.14}$$

If we adopt the coordinate transformation $q_1 = Q_1^2$, $q_2 = Q_2^2$, the new generating function gives rise to the momenta $P_i = 2\,Q_i p_i$. Combined with the time transformation $t' = q_1 q_2$, this yields the regularized Hamiltonian $\Gamma^* = q_1 q_2 (H - E)$. Thus in explicit form

$$\Gamma^* = \tfrac{1}{4}(P_1^2 Q_2^2 + P_2^2 Q_1^2 - P_1 P_2 Q_1 Q_2) - Q_1^2 - Q_2^2 - \frac{Q_1^2 Q_2^2}{Q_1^2 + Q_2^2} - Q_1^2 Q_2^2 E \,, \tag{5.15}$$

which results in simple equations of motion. Note that the divisor represents the sum of the two distances and is therefore optimal. For data analysis or movie making, the coordinates x_1, x_2 are obtained from substituting the c.m. relation $x_0 = \frac{1}{3}(q_1 - q_2)$. The relative velocities are determined by $\dot{q}_i = \partial H / \partial p_i$, combined with the c.m. condition $\dot{x}_0 = \frac{1}{3}(\dot{q}_1 - \dot{q}_2)$, whereas (5.14) can be used for the energy check. A formulation for unequal masses is given by Mikkola & Hietarinta [1991]. The corresponding Hamiltonian looks similar to the form (5.8), if we note the factor 2 in the definition of $\mathbf{A}_k$ and take account of the reverse sign in the third term due to the different generating function.

To conform with the present style, a discussion of practical details such as decision-making will be postponed to a later section. Next we provide a complete summary of the relevant transformations. Given the initial relative coordinates, $\mathbf{q}_k$, and momenta, $\mathbf{p}_k$, if $q_1 \geq 0$ the regularized coordinates for the first two-body motion are specified by

$$\begin{aligned} Q_1 &= [\tfrac{1}{2}(|\mathbf{q}_1| + q_1)]^{1/2} \,, \\ Q_2 &= \tfrac{1}{2} q_2 / Q_1 \,, \\ Q_3 &= \tfrac{1}{2} q_3 / Q_1 \,, \\ Q_4 &= 0 \,, \end{aligned} \tag{5.16}$$

otherwise the appropriate choice is taken to be

$$Q_2 = [\tfrac{1}{2}(|\mathbf{q}_1| - q_1)]^{1/2} \,,$$

$$
\begin{aligned}
Q_1 &= \tfrac{1}{2}q_2/Q_2\,, \\
Q_3 &= 0\,, \\
Q_4 &= \tfrac{1}{2}q_3/Q_2\,.
\end{aligned}
\tag{5.17}
$$

Likewise, the regularized momenta are given by (5.5). Because of the redundant fourth components of physical coordinates and momenta, numerical work only requires the 4×3 submatrix of $\mathbf{A}_k$, which for $k = 1$ takes the complete form

$$
\mathbf{A}_1 = 2 \begin{bmatrix} Q_1 & Q_2 & Q_3 & Q_4 \\ -Q_2 & Q_1 & Q_4 & -Q_3 \\ -Q_3 & -Q_4 & Q_1 & Q_2 \\ Q_4 & -Q_3 & Q_2 & -Q1 \end{bmatrix}. \tag{5.18}
$$

Inverse transformations yield the physical variables. Thus the relative coordinates, denoted by $\mathbf{R}_k$ $(k = 1, 2)$ in Fig. 5.1, are obtained from

$$
\mathbf{q}_k = \tfrac{1}{2}\mathbf{A}_k^{\mathrm{T}}\mathbf{Q}_k\,, \tag{5.19}
$$

which reproduces the well-known KS transformations, whereas the momenta are inverted by (5.5) and (5.6) according to

$$
\mathbf{p}_k = \tfrac{1}{4}\mathbf{A}_k^{\mathrm{T}}\mathbf{P}_k/R_k\,. \tag{5.20}
$$

Corresponding relations for the second KS pair are derived in a similar way. Finally, the coordinates and momenta expressed in the local c.m. frame are recovered from

$$
\begin{aligned}
\tilde{\mathbf{q}}_3 &= -\sum_{k=1}^{2} m_k \mathbf{q}_k/M\,, \\
\tilde{\mathbf{q}}_k &= \tilde{\mathbf{q}}_3 + \mathbf{q}_k\,, \\
\tilde{\mathbf{p}}_k &= \mathbf{p}_k\,, \\
\tilde{\mathbf{p}}_3 &= -(\mathbf{p}_1 + \mathbf{p}_2)\,, \qquad (k = 1, 2)
\end{aligned}
\tag{5.21}
$$

where $M = m_1 + m_2 + m_3$.

The time transformation (5.7) employed above is not unique and several alternatives have been examined [Aarseth, 1976]. One such case is the choice of the inverse potential energy using $t' = -1/\Phi$, which gives rise to the explicit relation for the time [Baumgarte & Stiefel, 1974] by

$$
t = -(\tau + C)/2E_0 + \sum_{i=1}^{3} \mathbf{r}_i \cdot \mathbf{p}_i/2E_0\,. \tag{5.22}
$$

Here the coordinates and momenta are expressed in the c.m. frame and the constant C is determined initially, when $\tau = 0$. Utilization of the regularized variables developed above, together with the condition (5.6), results in the final regular expression (excepting $E_0 = 0$)

$$t = -(\tau + C)/2E_0 + \sum_{k=1}^{2} \mathbf{Q}_k^{\mathrm{T}} \mathbf{P}_k / 4E_0 \,. \tag{5.23}$$

A change of time transformation requires the equations of motion to be modified. Taking the cue from (5.7), we write

$$dt = g_1 g_2 d\tau \,, \tag{5.24}$$

with $g_1 = R_1 R_2$. Consequently, the potential energy formulation gives

$$g_2 = (m_1 m_3 R_2 + m_2 m_3 R_1 + m_1 m_2 R_1 R_2 / R)^{-1} \,. \tag{5.25}$$

We absorb g_1 in the definition of Γ^* which gives rise to the new equations of motion

$$\begin{aligned} \frac{d\mathbf{Q}_k}{d\tau} &= g_2 \frac{\partial \Gamma^*}{\partial \mathbf{P}_k} \,, \\ \frac{d\mathbf{P}_k}{d\tau} &= -g_2 \frac{\partial \Gamma^*}{\partial \mathbf{Q}_k} - \Gamma^* \frac{\partial g_2}{\partial \mathbf{Q}_k} \,. \end{aligned} \tag{5.26}$$

The existence of an improper integral for the time if both the distances R_1 and R_2 approach zero together has prompted yet another time transformation, given by $g_2 = (R_1 + R_2)^{-1/2}$. Thus the property that $t' \propto R^{3/2}$ near triple collisions represents the limiting case for convergence. Actual experiments show that this choice of time transformation may give better results in practice, together with the condition $\Gamma^* = 0$ [Aarseth, 1976], but $g_2 = (R_1 + R_2)^{-1}$ also merits attention. Note that in the original formulation, with $g_1 = R_1 R_2$, this procedure is not justified since regularity of the equations is lost. Finally, equations of motion for the more recently suggested time transformation $t' = 1/L$ can be readily constructed [cf. Alexander, 1986; Appendix B] and deserves consideration.

5.3 External perturbations

The above method may also be used to study close encounters between binaries and single particles that occur in N-body simulations. In this case we need to include the effect of the external particles in the equations of motion, unless the unperturbed approximation is assumed. Consider a system of N particles and three mass-points m_i with corresponding coordinates $\mathbf{r}_i$ which form a subsystem to be regularized. Consequently, we write the regularized Hamiltonian as

$$\Gamma^* = R_1 R_2 (H_3 + \mathcal{R} - E) \,, \tag{5.27}$$

where H_3 is given by (5.1), $\mathcal{R}$ is the perturbing function expressed in terms of the physical variables $\mathbf{r}$ and $\mathbf{p}$, and E is the total energy. Since $\mathcal{R}$ does not depend on the momenta $\mathbf{P}_k$, the new equations of motion for $k = 1, 2$ take the form

$$\frac{d\mathbf{Q}_k}{d\tau} = \frac{\partial(R_1 R_2 H_3)}{\partial \mathbf{P}_k}, \tag{5.28}$$

$$\frac{d\mathbf{P}_k}{d\tau} = -(H_3 - E_3)\frac{\partial(R_1 R_2)}{\partial \mathbf{Q}_k} - R_1 R_2 \frac{\partial}{\partial \mathbf{Q}_k}(H_3 + \mathcal{R}), \tag{5.29}$$

where $E_3 = E - \mathcal{R}$ is the subsystem energy.

For the present purpose we need only consider the external potential energy part of $\mathcal{R}$, since the kinetic energy of the subsystem is treated independently. Hence we obtain the desired perturbation term from

$$\frac{\partial \mathcal{R}}{\partial \mathbf{Q}_k} = \sum_{i=1}^{3} \frac{\partial \mathcal{R}}{\partial \mathbf{r}_i} \frac{\partial \mathbf{r}_i}{\partial \mathbf{q}_k} \frac{\partial \mathbf{q}_k}{\partial \mathbf{Q}_k}. \tag{5.30}$$

Explicit differentiation yields the perturbing force

$$\partial \mathcal{R}/\partial \mathbf{r}_i = -m_i \mathbf{F}_i, \tag{5.31}$$

where $\mathbf{F}_i$ is defined by

$$\mathbf{F}_i = -\sum_{j=4}^{N} \frac{m_j(\mathbf{r}_i - \mathbf{r}_j)}{|\mathbf{r}_i - \mathbf{r}_j|^3}. \qquad (i = 1, 2, 3) \tag{5.32}$$

The actual coordinates are required for several purposes when studying a perturbed subsystem. Thus from (5.21) and the subsystem c.m. condition $\mathbf{r}_0 = \sum m_i \mathbf{r}_i / M$ we obtain the explicit conversion formulae

$$\begin{aligned}
\mathbf{r}_1 &= \mathbf{r}_0 + (m_2 + m_3)\mathbf{q}_1/M - m_2 \mathbf{q}_2/M, \\
\mathbf{r}_2 &= \mathbf{r}_0 - m_1 \mathbf{q}_1/M + (m_1 + m_3)\mathbf{q}_2/M, \\
\mathbf{r}_3 &= \mathbf{r}_0 - m_1 \mathbf{q}_1/M - m_2 \mathbf{q}_2/M.
\end{aligned} \tag{5.33}$$

Hence the expressions for $\partial \mathbf{r}_i / \partial \mathbf{q}_k$ simplify to mass ratios. From the basic transformation (5.19) it follows that

$$\partial \mathbf{q}_k / \partial \mathbf{Q}_k = \mathbf{A}_k. \tag{5.34}$$

Finally, combining all the terms results in

$$\partial \mathcal{R}/\partial \mathbf{Q}_k = -\mathbf{A}_k[m_1 m_2(\mathbf{F}_2 - \mathbf{F}_1)(-1)^k + m_k m_3(\mathbf{F}_k - \mathbf{F}_3)]/M. \tag{5.35}$$

We remark that differential (or tidal) accelerations appear. Hence only contributions from relatively nearby perturbers need to be taken into account in the summation (5.32), in analogy with standard KS.

It remains to derive an expression for the energy that is affected by the perturbations. From $E_3 = E - \mathcal{R}$, the internal energy change is

$$dE_3/d\tau = -d\mathcal{R}/d\tau\,. \tag{5.36}$$

Consequently, we can also use (5.35) to evaluate E_3' because

$$\frac{d\mathcal{R}}{d\tau} = \sum_{k=1}^{2} \frac{\partial \mathcal{R}}{\partial \mathbf{Q}_k}\frac{d\mathbf{Q}_k}{d\tau}\,. \tag{5.37}$$

Substituting for the equation of motion (5.10) combined with the expression (5.35) and employing the orthogonality condition $\mathbf{A}_k\mathbf{A}_k^{\mathrm{T}} = 4R_k$, we finally arrive at the desired equation

$$\frac{d\mathcal{R}}{d\tau} = -\tfrac{1}{4}\sum_{k=1}^{2} R_l \mathbf{P}_k^{\mathrm{T}} \mathbf{A}_k (\mathbf{F}_k - \mathbf{F}_3)\,. \tag{5.38}$$

Hence when external perturbations are included, the energy, E_3, needs to be updated in a consistent manner for use in (5.29). On the other hand, (5.28) does not contain the external potential. Since the perturbers are advanced separately, the total energy is a sum of two independent terms and can therefore still be used as a check.

This concludes the formal development of the basic three-body regularization. Practical matters relating to code implementation and applications will be discussed in a subsequent chapter.

5.4 Wheel-spoke generalization

The concept of a parallel regularization with respect to one reference body was extended to an arbitrary membership by Zare [1974] at the same time as the method discussed above was developed. This method only appears to have been tried in a binary–binary scattering experiment [Alexander, 1986]. However, since the verdict was favourable and applications to stellar systems with a central black hole seem relevant, the essential points will be presented here. As shown by Fig. 5.2, the generalization to larger systems conjures up the image of a wheel-spoke and we shall employ this name here in the absence of a recognized alternative.

The basic formulation starts with the Hamiltonian for a system of $N+1$ particles containing a subsystem of $n+1$ members. Let the generalized coordinates and momenta be denoted by $\tilde{\mathbf{q}}_i$ and $\tilde{\mathbf{p}}_i$, with $R_{ij} = |\tilde{\mathbf{q}}_i - \tilde{\mathbf{q}}_j|$ representing the inter-particle distances. As before, we reduce the order by employing the six c.m. integrals. Accordingly, the coordinates and momenta are redefined with respect to the reference body, m_0, such that $\mathbf{q}_0$

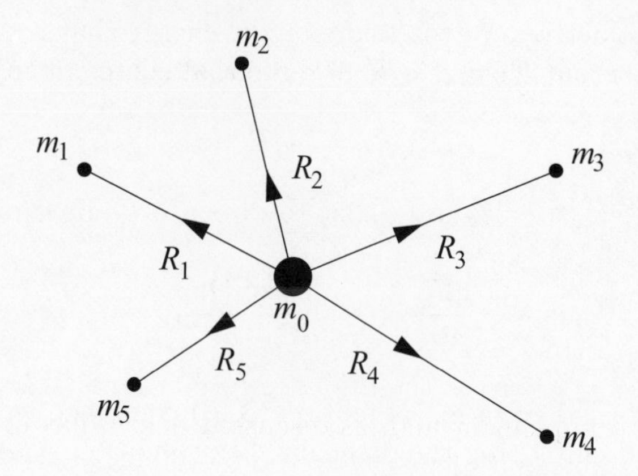

Fig. 5.2. Wheel-spoke geometry.

and $\mathbf{p}_0$ are the corresponding position and momentum. The transformation is achieved by the generating function

$$W(\mathbf{q}_i, \tilde{\mathbf{p}}_i) = \sum_{i=1}^{N} \tilde{\mathbf{p}}_i \cdot \mathbf{q}_i + \left(\sum_{i=0}^{N} \tilde{\mathbf{p}}_i \right) \cdot \mathbf{q}_0 \,. \tag{5.39}$$

Since the reduced Hamiltonian has $\mathbf{q}_0$ as an ignorable coordinate, we can set $\mathbf{p}_0 = 0$ for the c.m. The Hamiltonian then takes the form

$$H = \sum_{i=1}^{N} \frac{\mathbf{p}_i^2}{2\mu_i} + \frac{1}{m_0} \sum_{i<j}^{N} \mathbf{p}_i^{\mathrm{T}} \cdot \mathbf{p}_j - m_0 \sum_{i=1}^{N} \frac{m_i}{R_i} - \sum_{i<j}^{N} \frac{m_i m_j}{R_{ij}} \,, \tag{5.40}$$

where $\mu_i = m_i m_0/(m_i + m_0)$ and $R_i = |\mathbf{q}_i|$.

The system is divided into two subsystems, with masses $m_1, m_2, \ldots, m_n$ in the first one which will be considered for special treatment. The regularization procedure itself is quite similar to the three-body case, and in practice the local Hamiltonian is used instead of (5.40). We introduce n canonical variables $\mathbf{Q}_i$ and $\mathbf{P}_i$ and write the generating function

$$W(\mathbf{p}_i, \mathbf{Q}_i) = \sum_{i=1}^{N} \mathbf{p}_i^{\mathrm{T}} \cdot \mathbf{f}_i(\mathbf{Q}_i) \,, \tag{5.41}$$

where $\mathbf{f}_i(\mathbf{Q}_i)$ represents KS coordinates for $i \leq n$. The corresponding regularized momenta are obtained by

$$\mathbf{P}_i = \mathbf{A}_i \mathbf{p}_i \,, \qquad (i = 1, \ldots, n) \tag{5.42}$$

where the matrix $\mathbf{A}_i$ is defined by (5.18) in the usual way. Moreover, the coordinates and momenta for the remaining system are simply taken as $\mathbf{Q}_i = \mathbf{q}_i$ and $\mathbf{P}_i = \mathbf{p}_i$. Thus the new Hamiltonian consists of two parts, where the equations of motion for $i > n$ are of standard form. By analogy with the preceding section, an additional equation can be derived for the change in the energy of the regularized subsystem due to the perturbations, instead of using the original explicit formulation based on a time-consuming summation if N is large.

A variety of time transformations are available here, as discussed in the original paper. However, from systematic comparisons of binary–binary scattering experiments [Alexander, 1986] it appears that the inverse Lagrangian (i.e. $dt = d\tau/L$) is the best choice in practice. The latter form is also simple to implement and, as discussed below, has been found advantageous in other multiple regularizations. With a suitable time transformation, the equations of motion are regular and permit $R_i \rightarrow 0$ for all the n spokes.

For completeness, the relevant transformations are summarized here. Given the initial conditions $m_i, \tilde{\mathbf{q}}_i, \tilde{\mathbf{p}}_i$, the relative coordinates and absolute momenta are defined by

$$\begin{aligned} \mathbf{q}_i &= \tilde{\mathbf{q}}_i - \tilde{\mathbf{q}}_N \,, \\ \mathbf{p}_i &= \tilde{\mathbf{p}}_i \,, \end{aligned} \tag{5.43}$$

with $i = 1, 2, \ldots, N$. The standard KS coordinate transformations apply for $i \leq n$, whereas for $i > n$ we use the physical coordinates with respect to m_0. Likewise, the regularized momenta are obtained from (5.42). Inverse transformations yield the physical coordinates

$$\begin{aligned} \mathbf{q}_i &= \tfrac{1}{2}\mathbf{A}_i^{\mathrm{T}}\mathbf{Q}_i \,, \qquad & (i = 1, 2, \ldots, n) \\ \mathbf{q}_i &= \mathbf{Q}_i \,, & (i = n+1, \ldots, N) \end{aligned} \tag{5.44}$$

and corresponding momenta

$$\begin{aligned} \mathbf{p}_i &= \tfrac{1}{4}\mathbf{A}_i^{\mathrm{T}}\mathbf{P}_i/R_i \,, \qquad & (i = 1, 2, \ldots, n) \\ \mathbf{p}_i &= \mathbf{P}_i \,. & (i = n+1, \ldots, N) \end{aligned} \tag{5.45}$$

The final coordinates and momenta are then obtained from

$$\begin{aligned} \tilde{\mathbf{q}}_0 &= \sum_{i=1}^{N} m_i \mathbf{q}_i \Big/ \sum_{i=0}^{N} m_i \,, \\ \tilde{\mathbf{q}}_i &= \tilde{\mathbf{q}}_0 + \mathbf{q}_i \,, \\ \tilde{\mathbf{p}}_i &= \mathbf{p}_i \,, \quad (i = 1, 2, \ldots, N) \\ \tilde{\mathbf{p}}_0 &= -\sum_{i=1}^{N} \mathbf{p}_i \,. \end{aligned} \tag{5.46}$$

One can envisage several ways of using the wheel-spoke regularization. However, in order to be efficient, the reference body should contain most of the mass of the regularized subsystem. An attractive feature of this formulation is that no switching of the reference body may be required if this condition is fulfilled. One possible new application would be to a central dominant black hole system, where the closest stars would constitute the regularized part with the external members advanced by (say) the self-consistent field method [Hernquist & Ostriker, 1992]. Such a treatment would permit changes of membership due to collisions and boundary crossings. The method also lends itself to studying collisions in small systems (i.e. $n = N$). In either case, the solutions might be advanced by an individual time-step scheme to speed up the calculation, or the Bulirsch–Stoer [1966] integrator for increased accuracy. Although dominant motions can be integrated more accurately when regularized, any applications to planetary simulations are probably best carried out with the time-symmetric method [Kokubo, Yoshinaga & Makino, 1998] discussed in section 2.7.

5.5 Heggie's global formulation

Although highly efficient, the three-body development discussed above lacks the property of symmetry between the particles. This treatment was extended to a global formulation by Heggie [1974] whose derivation will be followed closely for the case $N = 3$. A somewhat complicated generalization to $N > 3$ is also available but for practical purposes the reformulation of Mikkola [1984a, 1985a] is preferable.

We begin with a Hamiltonian of the form (5.1) but adopt slightly different notation. Let $\tilde{\mathbf{q}}_i$ $(i = 1, 2, 3)$ be the Cartesian coordinates in an inertial frame and $\tilde{\mathbf{p}}_i$ the conjugate momenta. The Hamiltonian then takes the symmetrical form

$$\widetilde{H} = \sum_{i=1}^{3} \tilde{\mathbf{p}}_k^2/2m_i - m_2 m_3/|\tilde{\mathbf{q}}_2 - \tilde{\mathbf{q}}_3| - m_3 m_1/|\tilde{\mathbf{q}}_3 - \tilde{\mathbf{q}}_1| - m_1 m_2/|\tilde{\mathbf{q}}_1 - \tilde{\mathbf{q}}_2| \,. \tag{5.47}$$

We consider the solution of the corresponding Hamiltonian equations

$$\begin{aligned} \dot{\tilde{\mathbf{q}}}_1 &= \tilde{\mathbf{p}}_i/m_i \,, & (i = 1, 2, 3) \\ \dot{\tilde{\mathbf{p}}}_1 &= -\frac{m_1 m_2(\tilde{\mathbf{q}}_1 - \tilde{\mathbf{q}}_2)}{|\tilde{\mathbf{q}}_1 - \tilde{\mathbf{q}}_2|^3} - \frac{m_1 m_3(\tilde{\mathbf{q}}_1 - \tilde{\mathbf{q}}_3)}{|\tilde{\mathbf{q}}_1 - \tilde{\mathbf{q}}_3|^3} \,, & (*) \end{aligned} \tag{5.48}$$

where the asterisk indicates two similar equations by cyclic interchange of indices.

In the following we assume for simplicity that the c.m. is at rest. Let us define the relative distance vectors

$$\mathbf{q}_1 = \tilde{\mathbf{q}}_2 - \tilde{\mathbf{q}}_3 \,. \quad (*) \tag{5.49}$$

From this symmetrical definition we see that the meaning of the associated distances, R_i, is reversed from that of Fig. 5.1. We now introduce new momenta satisfying the relation

$$\tilde{\mathbf{p}}_i = \sum_{j=1}^{3} \mathbf{p}_j^{\mathrm{T}} \partial \mathbf{q}_j / \partial \tilde{\mathbf{q}}_i \,. \tag{5.50}$$

By (5.49) this simplifies to

$$\tilde{\mathbf{p}}_1 = -\mathbf{p}_2 + \mathbf{p}_3 \,. \quad (*) \tag{5.51}$$

Substitution of (5.49) and (5.51) into (5.47) yields the new Hamiltonian

$$H = \frac{\mathbf{p}_1^2}{2\mu_{23}} + \frac{\mathbf{p}_2^2}{2\mu_{31}} + \frac{\mathbf{p}_3^2}{2\mu_{12}} - \left(\frac{\mathbf{p}_2^{\mathrm{T}}\mathbf{p}_3}{m_1} + \frac{\mathbf{p}_3^{\mathrm{T}}\mathbf{p}_1}{m_2} + \frac{\mathbf{p}_1^{\mathrm{T}}\mathbf{p}_2}{m_3} \right)$$
$$- \frac{m_2 m_3}{q_1} - \frac{m_3 m_1}{q_2} - \frac{m_1 m_2}{q_3} \,, \tag{5.52}$$

where $q_i = |\mathbf{q}_i|$ and $\mu_{ij} = m_i m_j/(m_i + m_j)$. This gives rise to the equations of motion

$$\dot{\mathbf{q}}_1 = \mathbf{p}_1/\mu_{23} - \mathbf{p}_3/m_2 - \mathbf{p}_2/m_3 \,,$$
$$\dot{\mathbf{p}}_1 = -m_2 m_3 \mathbf{q}_1/q_1^3 \,. \quad (*) \tag{5.53}$$

Explicit relations for $\mathbf{q}_i$ and $\mathbf{p}_i$ are recovered by similar expressions to those in the previous section and other quantities have the usual meaning.

As above, there is considerable freedom in the choice of time transformations. In the first instance, we adopt the basic relation

$$dt = R_1 R_2 R_3 d\tau \,. \tag{5.54}$$

Denoting the new Hamiltonian by $\widetilde{H}(\mathbf{Q}_i, \mathbf{P}_i)$, we write as before

$$\Gamma^* = R_1 R_2 R_3 (\widetilde{H} - E_0) \,, \tag{5.55}$$

where E_0 is the numerical value of $\widetilde{H}$ along the solution path. From the definition of $\widetilde{H}$, the final regularized Hamiltonian becomes

$$\Gamma^* = \frac{1}{8} \left(\frac{R_2 R_3}{\mu_{23}} \mathbf{P}_1^{\mathrm{T}} \mathbf{P}_1 + \frac{R_3 R_1}{\mu_{31}} \mathbf{P}_2^{\mathrm{T}} \mathbf{P}_2 + \frac{R_1 R_2}{\mu_{12}} \mathbf{P}_3^{\mathrm{T}} \mathbf{P}_3 \right)$$
$$- \frac{1}{16} \left(\frac{R_1}{m_1} \mathbf{P}_2^{\mathrm{T}} \mathbf{A}_2 \mathbf{A}_3^{\mathrm{T}} \mathbf{P}_3 + \frac{R_2}{m_2} \mathbf{P}_3^{\mathrm{T}} \mathbf{A}_3 \mathbf{A}_1^{\mathrm{T}} \mathbf{P}_1 + \frac{R_3}{m_3} \mathbf{P}_1^{\mathrm{T}} \mathbf{A}_1 \mathbf{A}_2^{\mathrm{T}} \mathbf{P}_2 \right)$$
$$- m_2 m_3 R_2 R_3 - m_3 m_1 R_3 R_1 - m_1 m_2 R_1 R_2 - E_0 R_1 R_2 R_3 \,. \tag{5.56}$$

Again the corresponding equations of motion are obtained from the usual expressions (5.9). Differentiation leads to considerable simplification and it can be seen that the resulting equations are regular for collisions between any particle pair.

Alternative time transformations may also be tried here. By analogy with the preceding section, the choice

$$g_2 = (R_1 + R_2 + R_3)^{-3/2} \tag{5.57}$$

leads to regularized equations with the desirable asymptotic behaviour. The modified equations of motion for $i = 1, 2, 3$ then take the form

$$\begin{aligned} \frac{d\mathbf{Q}_i}{d\tau} &= g_2 \frac{\partial \Gamma^*}{\partial \mathbf{P}_i}, \\ \frac{d\mathbf{P}_i}{d\tau} &= -g_2 \left[\frac{\partial \Gamma^*}{\partial \mathbf{Q}_i} - \tfrac{3}{2}\Gamma^* \frac{\partial}{\partial \mathbf{Q}_i} \ln(R_1 + R_2 + R_3) \right]. \end{aligned} \tag{5.58}$$

It has been found [Heggie, 1974] that setting $\Gamma^* = 0$ introduces unstable modes as triple collision is approached. This behaviour is connected with the modified time transformation where growing modes are present without the last term in the momentum equation, which has a stabilizing effect. However, the additional complications introduced by (5.57) may be avoided for less extreme configurations if a lower limit on the system size for non-zero angular momentum can be estimated.

By analogy with the treatment of section 5.3, the global formulation has been extended to the perturbed three-body problem [Heggie, 1974]. This formulation does not appear to have been implemented in any existing N-body code and is given at the end of Appendix A.

5.6 Mikkola's derivation

The previous development was also generalized to arbitrary memberships [Heggie, 1974]. Since this treatment involves a total of $4N(N-1)+1$ equations, the complexity increases rapidly with N, especially as the right-hand side of the second member (5.9) also grows faster than N. The complications of the original formulation are such that $N = 4$ already represents a formidable challenge to the practitioner. Given the importance of binary–binary interactions in star cluster simulations, we therefore present a simpler derivation due to Mikkola [1984a, 1985a] which is easier to implement. This task is accomplished by employing a modified notation while retaining the essential features.

We begin by writing the Hamiltonian in terms of the physical coordinates and momenta,

$$H = \sum_{i=1}^{N} \frac{\mathbf{w}_i^2}{2m_i} - \sum_{i=1}^{N-1} \sum_{j=i+1}^{N} \frac{m_i m_j}{|\mathbf{r}_i - \mathbf{r}_j|}. \tag{5.59}$$

Let $\mathbf{q}_{ij} = \mathbf{r}_i - \mathbf{r}_j$ be the new coordinates and define new momenta by

$$\mathbf{w}_i = \sum_{j=i+1}^{N} \mathbf{p}_{ij} - \sum_{j=1}^{i-1} \mathbf{p}_{ji} \,. \tag{5.60}$$

This transformation introduces $N(N-1)/2$ fictitious particles. A direct substitution shows that a solution of (5.60) is given by

$$\mathbf{p}_{ij} = (\mathbf{w}_i - \mathbf{w}_j)/N \,. \tag{5.61}$$

In order to achieve a simpler notation, we now replace the double indices by a single running index, k. Thus henceforth we work with $\mathbf{q}_k, \mathbf{p}_k$, where $k = (i-1)N - i(i+1)/2 + j$ for $i < j$ is a 1D array that contains $K = N(N-1)/2$ members. This enables (5.60) to be expressed as

$$\mathbf{w}_i = \sum_{k=1}^{K} a_{ik} \mathbf{p}_k \,, \tag{5.62}$$

where by definition $a_{ik} = 1$ and $a_{jk} = -1$ when $k = k(i,j)$ and zero otherwise. We define the mass products $M_k = m_i m_j$. The new Hamiltonian then takes the form

$$H = \sum_{u,v=1}^{K} T_{uv} \mathbf{p}_u \cdot \mathbf{p}_v - \sum_{k=1}^{K} M_k / q_k \,, \tag{5.63}$$

where the matrix elements are given by

$$T_{uv} = \tfrac{1}{2} \sum_{e=1}^{N} a_{eu} a_{ev} / m_e \,. \quad (u = 1, \ldots, K) \quad (v = 1, \ldots, K) \tag{5.64}$$

The standard KS transformations now provide the necessary relations between physical and regularized quantities, where the generalized Levi-Civita matrix $\mathcal{L}$ defined by (4.27) plays the usual role. After introducing the time transformation $dt = g(\mathbf{P}, \mathbf{Q})$, we obtain the desired regularized form Γ^*, with the Hamiltonian function itself in terms of $\mathbf{P}, \mathbf{Q}$ as

$$H = \tfrac{1}{4} \sum_{u,v}^{K} T_{uv} \mathbf{P}_u^{\mathrm{T}} \mathcal{L}_u^{\mathrm{T}} \mathcal{L}_v \mathbf{P}_v / Q_u^2 Q_v^2 - \sum_{e=1}^{N} M_e / Q_e^2 \,. \tag{5.65}$$

Given the final Hamiltonian above, the standard way of carrying out the differentiations after multiplying by the function g of (5.54) leads to complicated expressions for $N > 3$. Instead we revert to the basic form of Γ^* and write the equations of motion as

$$\begin{aligned} \frac{d\mathbf{Q}}{d\tau} &= g \frac{\partial H}{\partial \mathbf{P}} + (H - E_0) \frac{\partial g}{\partial \mathbf{P}} \,, \\ \frac{d\mathbf{P}}{d\tau} &= -g \frac{\partial H}{\partial \mathbf{Q}} - (H - E_0) \frac{\partial g}{\partial \mathbf{Q}} \,. \end{aligned} \tag{5.66}$$

Note that $H - E_0 = 0$ on the solution path but the regularity of the equations is actually lost if this term is omitted.

Several time transformations have been considered. However, use of the Lagrangian by $g = 1/L$ appears to have practical advantages [Zare & Szebehely, 1975; Alexander, 1986]. The paper by Alexander provides a useful systematic comparison of the multiple regularization methods discussed above. Among the few applications of the global regularization method we mention the pioneering work on binary–binary interactions [Mikkola, 1983, 1984b], as well as use of an unperturbed four-body treatment in star cluster simulations with primordial binaries [Aarseth, 1985a; Heggie & Aarseth, 1992]. We also note that the global formulation is simple in one respect, namely there is no decision-making connected with switching of dominant components. Further discussions concerning the properties of the method can be found in the original exposition [Mikkola, 1985a]. Suffice it to state that an examination of the equations shows that two-body collision configurations can be studied without any problems. Appendix A contains the equations of motion and a collection of the relevant formulae, including external perturbations.

5.7 Chain treatment

The basic Aarseth–Zare regularization method has proved itself in scattering experiments and star cluster simulations [Aarseth & Heggie, 1976; Heggie & Aarseth, 1992], as well as in more traditional three-body investigations [Aarseth *et al.*, 1994a,b]. Subsequently, the Heggie–Mikkola global formulation was employed to study binary–binary interactions, likewise for compact subsystems where the external perturbation can be ignored. In view of this activity, it is perhaps remarkable that it took another 17 years for the three-body method to be extended to $N = 4$ [Mikkola & Aarseth, 1990]. This new treatment was reformulated for arbitrary memberships [Mikkola & Aarseth, 1993] with an improved notation, which will be used in the following.

The concept of a chain is very simple and is illustrated in Fig.5.3 for a system of four mass-points. We introduce the dominant two-body forces along a chain of inter-particle vectors, where the pair-wise attractions are treated by the KS formalism. Contributions from the other less dominant interactions are added to the regular terms in a similar way to the original three-body formulation. Hence the number of equations to be integrated is $8(N-1)+1$, compared with $4N(N-1)+1$ in the global implementation.

Consider a system of N particles with inertial coordinates $\mathbf{r}_i$, velocities $\mathbf{v}_i$ and masses m_i, $i = 1, 2, \ldots, N$. Further, take the c.m. to be at rest with the local coordinates $\mathbf{q}_i$ and momenta $\mathbf{p}_i = m_i \mathbf{v}_i$. After selecting the chain vectors connecting the N mass-points, we relabel them $1, 2, \ldots, N$

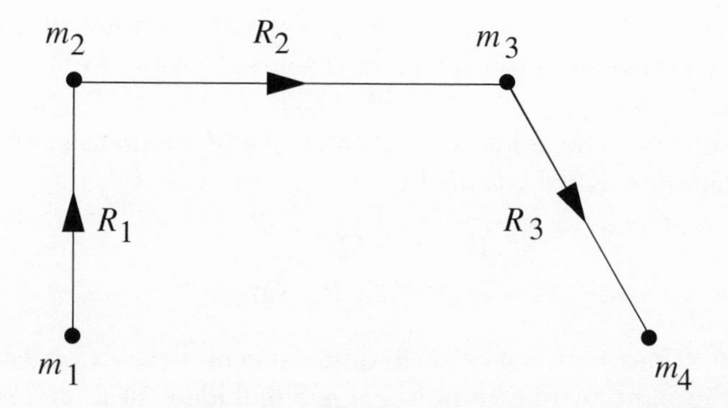

Fig. 5.3. Four-body chain.

as shown in the figure. The generating function

$$S = \sum_{k=1}^{N-1} \mathbf{W}_k \cdot (\mathbf{q}_{k+1} - \mathbf{q}_k) \tag{5.67}$$

then gives the old momenta $\mathbf{p}_k = \partial S/\partial \mathbf{q}_k$ in terms of the new ones. The relative momentum vectors $\mathbf{W}_k$ are obtained recursively by

$$\mathbf{W}_k = \mathbf{W}_{k-1} - \mathbf{p}_k \,, \qquad (k = 2, \ldots, N-2) \tag{5.68}$$

with $\mathbf{W}_1 = -\mathbf{p}_1$ and $\mathbf{W}_{N-1} = \mathbf{p}_N$. These momenta and corresponding relative coordinates, $\mathbf{R}_k = \mathbf{q}_{k+1} - \mathbf{q}_k$, are substituted into the unperturbed Hamiltonian which becomes

$$\begin{aligned} H = \tfrac{1}{2} \sum_{k=1}^{N-1} \left(\frac{1}{m_k} + \frac{1}{m_{k+1}} \right) \mathbf{W}_k^2 - \sum_{k=2}^{N} \frac{1}{m_k} \mathbf{W}_{k-1} \cdot \mathbf{W}_k \\ - \sum_{k=1}^{N-1} \frac{m_k m_{k+1}}{R_k} - \sum_{1 \le i \le j-2}^{N} \frac{m_i m_j}{R_{ij}} \,. \end{aligned} \tag{5.69}$$

We now employ the time transformation $dt = g d\tau$, where $g = 1/L$ is the inverse Lagrangian [cf. Alexander, 1986]. The regularized Hamiltonian $\Gamma^* = g\,(H - E)$ again gives rise to equations of motion of the form (5.9). Inspection of these equations reveals that the two-body solutions are regular for any $R_k \to 0$. Since the time transformation does not introduce terms that cancel analytically, the differentiation of each term is carried out separately. Some care is needed to obtain the partials with respect to $\mathbf{Q}$ of the non-chained part, $U_{\rm nc}$, represented by the last term of (5.69). Let us adopt the traditional notation $\mathcal{L}$ for the generalized Levi-Civita matrix instead of $\widetilde{\mathbf{Q}}$ used in the original paper. Thus if $\mathbf{Q}$ is the KS transform of $\mathbf{R}$ and $U_{\rm nc} = U(\mathbf{R})$, then $\partial U_{\rm nc}/\partial \mathbf{Q} = 2\mathcal{L}^{\rm T}\mathbf{F}$, where $\mathbf{F} = \partial U_{\rm nc}/\partial \mathbf{R}$. Thus a conservative external potential may be treated as an additional part of the non-chained potential. This is discussed below for the perturbed formulation.

When required, the relative coordinates and momenta are recovered from the standard KS relations by

$$\begin{aligned} \mathbf{R}_k &= \mathcal{L}_k\,\mathbf{Q}_k\,, \\ \mathbf{W}_k &= \tfrac{1}{2}\mathcal{L}_k\,\mathbf{P}_k/\mathbf{Q}_k^2\,. \end{aligned} \tag{5.70}$$

Procedures for chain selection require special care in order to ensure that the dominant two-body motions are included. Still, the scheme admits of a certain elasticity at the expense of efficiency, as long as the non-chained distances do not become too small compared with the chained distances. Since this feature forms an important part of the chain algorithms, it may be useful to discuss some relevant aspects.

The chain needs to be constructed *ab initio* before the calculation can begin. First the shortest inter-particle vector is determined. We proceed by searching for the particle which is closest to either end of the known chain. This operation is repeated until all the particles have been included. To facilitate the procedure, it is beneficial to sort all the distances and perform a sequential search for the missing terms.

We readily see that the new chain vectors can be expressed as sums or differences of the old ones. This is achieved by writing physical vectors as sums of chain vectors in a coordinate system where the first particle is located at the origin. Special attention is also required to retain accuracy by carrying out the transformations directly from the old chain vectors, $\mathbf{R}_k$, to the new ones, instead of using the physical coordinates, $\mathbf{r}_i$.

Let the identity of the chain particles defined above be $I_1, I_2, \ldots, I_N$ and denote the identities in the old and new chains by $I_k^{\rm old}$ and $I_k^{\rm new}$. We then express the old partial sums along the chain by

$$\mathbf{q}_{I_k^{\rm old}} = \sum_{\nu=1}^{k-1} \mathbf{R}_\nu^{\rm old}\,, \tag{5.71}$$

with the new chain vectors given by

$$\mathbf{R}_\mu^{\rm new} = \mathbf{q}_{I_{\mu+1}^{\rm new}} - \mathbf{q}_{I_\mu^{\rm new}}\,. \tag{5.72}$$

Hence we need to use the correspondence between the old and new indices to obtain the new chain vectors, $\mathbf{R}_k$. It can be seen that if k_0 and k_1 are such that $I_{k_0}^{\rm old} = I_\mu^{\rm new}$ and $I_{k_1}^{\rm old} = I_{\mu+1}^{\rm new}$, then we have

$$\mathbf{R}_\mu^{\rm new} = \sum_{\nu=1}^{N-1} B_{\mu\nu}\mathbf{R}_\nu^{\rm old}\,, \tag{5.73}$$

where $B_{\mu\nu} = 1$ if ($k_1 > \nu$ and $k_0 \leq \nu$) and $B_{\mu\nu} = -1$ if ($k_1 \leq \nu$ and $k_0 > \nu$), otherwise $B_{\mu\nu} = 0$. In practical applications the chain configuration is rechecked every integration step but the overheads are still modest because the high-order Bulirsch–Stoer [1966] integrator requires many function calls. Moreover, the total number of such switchings by relabelling is relatively small in typical calculations.

The chain method may be used for two purposes. First we have the simple case of a small-N system, in which all the members may be included [Sterzik & Durisen, 1998; Kiseleva *et al.*, 1998]. Distant escapers may be removed from the chain in order to speed up the calculation, but otherwise the procedures outlined above suffice. The alternative usage of studying energetic interactions in a general N-body code requires special considerations. Below we provide the necessary details for including the external perturbations, whereas corresponding algorithms for the interface with an N-body system will be discussed in a later chapter.

Let $\mathbf{F}_j$ be the perturbing acceleration acting on a body of mass m_j in the inertial coordinate system. For consistency with the previous notation, let the chain membership still be denoted by N, with each component subject to an external acceleration, $\mathbf{F}_j$, considered as a known function of time. Hence we assume that all the other bodies are advanced by some direct integration scheme. We now augment the original Hamiltonian (5.69) by adding the perturbing potential

$$\delta U = \sum_{j=1}^{N} m_j \mathbf{q}_j \cdot [\mathbf{F}_j(t) - \mathbf{F}_0(t)] + M\mathbf{r}_0 \cdot \mathbf{F}_0(t)\,, \tag{5.74}$$

where $\mathbf{F}_0 = \sum_{j=1}^{N} m_j \mathbf{F}_j / M$ is the c.m. acceleration in the inertial system defined by $\mathbf{r}_0 = \sum m_j \mathbf{r}_j / M$ and M is the associated mass.

Next we should express the coordinates, $\mathbf{q}_j$, in terms of the chain coordinates, $\mathbf{R}_k$, and substitute these into the expression for δU, which yields the corresponding contributions, $\delta \dot{\mathbf{W}}_k$, to the derivatives of the chain momenta, $\mathbf{W}_k$. However, from the transformation formulae one easily derives the recursive relations

$$\begin{aligned}
\delta \dot{\mathbf{p}}_j &= m_j(\mathbf{F}_j - \mathbf{F}_0)\,, \qquad && (j = 1, \ldots, N) \\
\delta \dot{\mathbf{W}}_1 &= -\delta \dot{\mathbf{p}}_1\,, && \\
\delta \dot{\mathbf{W}}_k &= \delta \dot{\mathbf{W}}_{k-1} - \delta \dot{\mathbf{p}}_k\,, && (k = 2, \ldots, N-2) \\
\delta \dot{\mathbf{W}}_{N-1} &= \delta \dot{\mathbf{p}}_N\,, &&
\end{aligned} \tag{5.75}$$

where δ denotes the perturbative part of the derivative. The corresponding corrections to the derivatives of the KS momenta are then given by

$$\delta \mathbf{P}'_k = 2g\mathcal{L}_k^{\mathrm{T}} \delta \dot{\mathbf{W}}_k\,, \tag{5.76}$$

and the rate of change of the internal energy may be written

$$dE/d\tau = 2\sum_{k=1}^{N-1} \mathbf{Q}_k'^{\mathrm{T}} \mathcal{L}_k^{\mathrm{T}} \delta\dot{\mathbf{W}}_k \,. \tag{5.77}$$

The total energy previously denoted by E_0 may now be integrated as an additional equation. Since only the perturbation with respect to the c.m. enters in the expressions for $\delta\dot{\mathbf{W}}_k$, it is usually sufficient to include tidal contributions from nearby particles. We also note that (5.77) makes use of the term $d\mathbf{Q}_k/d\tau$, which is already known.

A complete collection of formulae for chain regularization is given in Appendix B. The original paper [Mikkola & Aarseth, 1990] makes a comparison with the global method for $N = 3$ and discusses practical aspects.

5.8 Slow-down procedure

Extremely close binaries occur frequently in modern star cluster simulations which usually include a distribution of primordial binaries [McMillan, Hut & Makino, 1990; Heggie & Aarseth, 1992]. Normally such binaries are unperturbed and easily treated but configurations in hierarchical subsystems may give rise to considerable time-scale problems, whether treated by KS or chain regularization. In the case of small perturbations, we may apply the principle of adiabatic invariance and still represent the motions to a good approximation although the orbital phase is lost. Such a scheme has turned out to be very useful [Mikkola & Aarseth, 1996] and will be presented in the following. The question of whether it could also be introduced into the other multiple regularizations described above does not appear to have been addressed.‡

Consider a weakly perturbed binary with the standard equation of relative motion

$$\ddot{\mathbf{r}} = -M_{\mathrm{b}}\mathbf{r}/r^3 + \mathbf{F} \,, \tag{5.78}$$

where M_{b} is the combined mass of the binary components and $\mathbf{F}$ is the external perturbation. One way to solve this equation numerically is to employ the variation of constants method. In that case the equation of motion for a two-body element $q = q(\mathbf{r}, \mathbf{v}, t)$ takes the simple form

$$\dot{q} = \partial q/\partial \mathbf{v} \cdot \mathbf{F} \,, \tag{5.79}$$

where the right-hand side is a small quantity. However, the rapid fluctuations persist and the time-step can only be increased for relatively small perturbations.

‡ A compact triple with superhard inner binary and an eccentric outer orbit could also be partially treated in this way.

The new idea is to slow down the internal dominant two-body motion such that one orbit may represent several Kepler periods. This is achieved by scaling the small perturbation and the physical time-step by a slowly varying factor exceeding unity. In this approximation, we therefore ensure a correct treatment of the secular effects. To demonstrate more carefully what is involved, we write the coupled equations containing the slow-down factor κ as

$$\begin{aligned} \dot{\mathbf{v}} &= -\kappa^{-1}\mathbf{r}/r^3 + \mathbf{F}\,, \\ \dot{\mathbf{r}} &= \kappa^{-1}\mathbf{v}\,. \end{aligned} \tag{5.80}$$

Hence the new period is κ times the original one and the integration is speeded up by the same amount. Let us write the equation for the variation of an element $q = q(t/\kappa, \mathbf{r}, \mathbf{v})$, with t/κ representing the scaling of time. This leads to

$$\begin{aligned} \dot{q} &= \frac{\partial q}{\partial t} + \frac{\partial q}{\partial \mathbf{r}} \cdot \dot{\mathbf{r}} + \frac{\partial q}{\partial \mathbf{v}}\dot{\mathbf{v}} \\ &= \kappa^{-1}\left[\frac{\partial q}{\partial (t/\kappa)} + \frac{\partial q}{\partial \mathbf{r}} \cdot \mathbf{v} - \frac{\mathbf{r}}{r^3} \cdot \frac{\partial q}{\partial \mathbf{v}}\right] + \frac{\partial q}{\partial \mathbf{v}} \cdot \mathbf{F} \\ &= \frac{\partial q}{\partial \mathbf{v}} \cdot \mathbf{F}\,, \end{aligned} \tag{5.81}$$

where the final result simplifies to the earlier form (5.79), because the two-body terms cancel due to q being an element. It should be noted that now the short-period terms are multiplied by the factor κ which therefore tends to counteract the advantage.

In the original Hamiltonian formulation we have

$$H = \tfrac{1}{2}p^2/\mu - M_{\mathrm{b}}/r - U\,, \tag{5.82}$$

where μ is the reduced mass and U is the perturbing force function which yields the external force by $\mathbf{F} = \partial U/\partial \mathbf{r}$. The modified formulation may be described by the new Hamiltonian

$$\tilde{H} = \kappa^{-1}(\tfrac{1}{2}p^2/\mu - M_{\mathrm{b}}/r) - U\,, \tag{5.83}$$

which leads to the above slowed-down equations of motion. This principle can also be applied more generally in an N-body system, and even in complicated situations that may occur in chain regularization [Mikkola & Aarseth, 1993].

The idea may be formulated in the following way. First, separate from the Hamiltonian, H, those terms which give the internal interaction of the pair of particles forming the weakly perturbed binary, denoted H_b. Second, multiply the part H_b by the slow-down factor κ^{-1} to get

$$H_{\mathrm{new}} = \kappa^{-1}H_b + (H - H_b)\,, \tag{5.84}$$

and form the usual Hamiltonian equations of motion

$$\dot{\mathbf{P}} = -\partial H_{\rm new}/\partial \mathbf{Q}\,, \qquad \dot{\mathbf{Q}} = \partial H_{\rm new}/\partial \mathbf{P}\,. \tag{5.85}$$

Alternatively, for a regularized formulation, we first write

$$\Gamma_{\rm new} = H_{\rm new} - E\,, \tag{5.86}$$

and multiply by the time transformation, g, before forming the equations of motion. Since κ changes with time, $H_{\rm new}$ is not constant and for the associated numerical value we may write

$$\dot{E} = \partial \kappa^{-1}/\partial t\, H_b\,. \tag{5.87}$$

In practice, κ is adjusted by a small discrete amount after each step. Hence the corresponding change in E is a delta function, obtained as a fraction $1/\kappa_{\rm new} - 1/\kappa_{\rm old}$ of the instantaneous binding energy, H_b.

The formulation (5.84) may be applied to many dynamical problems featuring weakly perturbed systems and long time-scales. However, in the following we focus on implementation in the chain regularization discussed above.

Given the chain coordinates, $\mathbf{R}_k$, and momenta, $\mathbf{W}_k$, we write the Hamiltonian as

$$H = \sum T_{ij}\mathbf{W}_i \cdot \mathbf{W}_j - \sum M_k/R_k - \sum M_{ij}/R_{ij}\,, \tag{5.88}$$

where the auxiliary quantities affected by the present formulation are given by the original expressions,

$$\begin{aligned} T_{kk} &= \tfrac{1}{2}(1/m_k + 1/m_{k+1})\,, \\ T_{k\,k+1} &= -1/m_k\,, \\ M_k &= m_k m_{k+1}\,. \qquad\qquad (k = 1, \ldots, N-1) \end{aligned} \tag{5.89}$$

To construct an expression for the internal interaction of a binary within the chain, we first note that the relevant term in the potential energy is M_b/R_b if b is the index of the distance between the binary components[§] m_b, m_{b+1}. Hence we must select from the kinetic energy those terms which have $1/m_b$ or $1/m_{b+1}$ as a factor. The resulting combination of terms, however, also contains the c.m. kinetic energy which must be subtracted and added to the rest of the Hamiltonian.

We write the c.m. kinetic energy for the pair m_b, m_{b+1} as

$$T_{\rm cm} = \tfrac{1}{2}(\mathbf{W}_{b+1} - \mathbf{W}_{b-1})^2/(m_b + m_{b+1})\,. \tag{5.90}$$

[§] This notation is employed here for convenience, whereas $m_{\rm b}$ is used generally to define the total binary mass.

This is valid in all cases, provided that the quantities $\mathbf{W}_k$ are defined to be zero whenever the index k is outside the range $[1, N-1]$. Selection of the relevant terms from the Hamiltonian and expansion of the expression for $T_{\rm cm}$ gives the following new expressions for evaluating the matrix T_{ij} and vector M_k in the presence of a slowed-down binary. First we calculate

$$\begin{aligned} T_{k\ k+1} &= -1/m_k/\kappa_k\,, \\ T_{k+1\ k+2} &= -1/m_{k+1}/\kappa_k\,, \qquad (k = 1, 3, 5, \ldots, \leq N) \end{aligned} \tag{5.91}$$

where a slow-down coefficient $\kappa_k = 1$ has been defined for every chain vector, except for $k = b$. Next we evaluate the expressions

$$\begin{aligned} T_{b-1\ b+1} &= -(1 - \kappa_b^{-1})/(m_b + m_{b+1})\,, \\ \delta T_{b-1\ b-1} &= \delta T_{b+1\ b+1} = -\tfrac{1}{2} T_{b-1\ b+1}\,, \end{aligned} \tag{5.92}$$

and finally

$$\begin{aligned} T_{kk} &= -\tfrac{1}{2}(T_{k\ k+1} + T_{k+1\ k+2}) + \delta T_{kk}\,, \\ M_k &= m_k m_{k+1}/\kappa_k\,. \end{aligned} \tag{5.93}$$

Only two modifications remain which affect the evaluation of the equations of motion. First we must change the equation for $\mathbf{A}_k$ that appear in the algorithm for the chain derivatives (cf. B.11) to read

$$\mathbf{A}_k = \tfrac{1}{2} \sum (T_{ki} + T_{ik}) \mathbf{W}_i\,, \qquad (|i - k| \leq 2) \tag{5.94}$$

because there are more non-zero off-diagonal elements in the T-matrix. Hence only the summation limit has changed here. Finally, the total energy that appears explicitly in the usual equations of motion must be modified by the amount

$$\delta E = -\frac{m_b m_{b+1}}{2a} \left(\frac{1}{\kappa_b^{\rm new}} - \frac{1}{\kappa_b^{\rm old}} \right), \tag{5.95}$$

where a is the actual semi-major axis. This contribution should be added to the total energy when the slow-down factor changes from the old value, $\kappa_b^{\rm old}$, to the new one, $\kappa_b^{\rm new}$.

So far we have discussed the slow-down procedure for only one binary. However, we may have several such binaries in a chain subsystem and the operations above may then be repeated for any relevant index b if desired. It may also be remarked that the addition of external perturbations does not affect the slow-down scheme since the internal tidal forces dominate.

Another type of N-body system of current interest consists of one or more planets orbiting a central star which interacts with other cluster members [de la Fuente Marcos & de la Fuente Marcos, 1999] or is perturbed in scattering experiments [Laughlin & Adams, 1998]. In this case the above strategy needs to be modified slightly. Thus the binary energy appearing in (5.95) is replaced by the total subsystem energy which must be calculated explicitly. Moreover, the size of the perturbation is now estimated by considering the outermost planet. The chain regularization is quite effective for studying different types of configurations, provided the condition of approximate isolation is satisfied. Thus the slow-down procedure is applied to a compact subsystem for long time intervals of small perturbations and yet we retain the advantage of treating strong interactions leading to exchange or escape.

5.9 Time-transformed leapfrog scheme

The regularization schemes discussed so far are quite satisfactory for most stellar systems. However, application to the case of very large mass ratios leads to loss of efficiency as well as accuracy. In particular, the problem of black hole binaries in galactic nuclei is currently topical. So far the standard KS treatment discussed above has only been partially successful [Quinlan & Hernquist, 1997; Milosavljević & Merritt, 2001]. On general grounds, extension to chain regularization is also unlikely to be satisfactory because the total subsystem energy that appears explicitly in the equations of motion is dominated by the binary components. In view of the small period and large binding energy of such a binary, some other kind of regularization or smoothing method is therefore desirable.

Below we describe a new time-transformed leapfrog scheme that offers practical advantages for studying systems with large mass ratios [Mikkola & Aarseth, 2002]. Consider first the standard leapfrog equations

$$
\begin{aligned}
\mathbf{r}_{1/2} &= \mathbf{r}_0 + \frac{h}{2}\mathbf{v}_0 \,, \\
\mathbf{v}_1 &= \mathbf{v}_0 + h\,\mathbf{F}(\mathbf{r}_{1/2}) \,, \\
\mathbf{r}_1 &= \mathbf{r}_{1/2} + \frac{h}{2}\mathbf{v}_1 \,,
\end{aligned}
\tag{5.96}
$$

where h is the time-step and $\mathbf{F}$ denotes the acceleration at $t = \frac{1}{2}h$. We adopt a time transformation $ds = \Omega(\mathbf{r})dt$, with Ω an arbitrary function and introduce a new auxiliary quantity $W = \Omega$. The new idea here is to evaluate W by the auxiliary equation

$$
\dot{W} = \mathbf{v} \cdot \frac{\partial \Omega}{\partial \mathbf{r}} \,, \tag{5.97}
$$

rather than explicitly. This allows us to solve the two sets of equations in separate stages; namely (i) $\mathbf{r}' = \mathbf{v}/W$, $t' = 1/W$, $\mathbf{v}' = 0$, $W' = 0$, and (ii)

$\mathbf{v}' = \mathbf{F}/\Omega$, $W' = \dot{W}/\Omega$, $\mathbf{r}' = 0$, $t' = 0$. Consequently, we write

$$\mathbf{r} = \mathbf{r}_0 + s\frac{\mathbf{v}}{W}\,,$$
$$t = t_0 + s\frac{1}{W}\,, \tag{5.98}$$

and

$$\mathbf{v}_1 = \mathbf{v}_0 + s\,\frac{\mathbf{F}(\mathbf{r}_{1/2})}{\Omega(\mathbf{r}_{1/2})}\,,$$
$$W_1 = W_0 + s\,\frac{\mathbf{v} + \mathbf{v}_0}{2\Omega(\mathbf{r}_{1/2})} \cdot \frac{\partial\Omega(\mathbf{r}_{1/2})}{\partial\mathbf{r}_{1/2}}\,. \tag{5.99}$$

Hence the solutions have been combined into a form of leapfrog, taking $s = \frac{1}{2}h$ in (5.98); then $s = h$ in (5.99) and finally again $s = \frac{1}{2}h$ in (5.98), using the midpoint rule. The numerical solutions are obtained by employing the high-order Bulirsch–Stoer method. Thus, with the above leapfrog algorithm, several integrations are performed with gradually decreasing substeps, h, and the results are extrapolated to zero step-length.

Up to now, the choice of the time transformation has not been specified. When considering an application to small subsystems, it is convenient to choose the function

$$\Omega = \sum_{i<j} \frac{\Omega_{ij}}{r_{ij}}\,, \tag{5.100}$$

where Ω_{ij} may be taken as the mass products or simply as unity. Here we make the latter choice which assigns equal weights to all the members. Hence the gradient is simply the force function

$$\frac{\partial\Omega}{\partial\mathbf{r}_k} \equiv \mathbf{G}_k = \sum_{j\neq k} \frac{\mathbf{r}_j - \mathbf{r}_k}{r_{kj}^3}\,. \tag{5.101}$$

The corresponding equations of motion for each particle, k, are given by

$$\mathbf{r}_k' = \frac{\mathbf{v}_k}{W}\,, \qquad t' = \frac{1}{W}\,, \tag{5.102}$$
$$\mathbf{v}_k' = \frac{\mathbf{F}_k}{\Omega}\,, \qquad W' = \frac{1}{\Omega}\sum_k \mathbf{v}_k \cdot \mathbf{G}_k\,, \tag{5.103}$$

whereupon the leapfrog algorithm (5.96) is attained.

This formulation has also been generalized to include separately external perturbations of conservative type as well as relativistic effects. In this case we replace the last pair of equations (5.103) by

$$\mathbf{v}' = (\mathbf{F} + \mathbf{f}(\mathbf{v}))/\Omega\,, \qquad W' = \mathbf{v}\cdot\mathbf{G}/\Omega\,, \tag{5.104}$$

where $\mathbf{f}(\mathbf{v})$ denotes the velocity-dependent part of the acceleration and subscripts have been omitted for simplicity. For small relativistic corrections, the leapfrog velocity integration may be replaced by the implicit midpoint method

$$\mathbf{v}_1 = \mathbf{v}_0 + h(\mathbf{F} + \mathbf{f}(\mathbf{v}_\mathrm{a}))/\Omega\,, \tag{5.105}$$

with $\mathbf{v}_\mathrm{a} = \frac{1}{2}(\mathbf{v}_0 + \mathbf{v}_1)$ the average velocity. Convergent solutions are usually obtained after a few iterations. In either case, the energy of the subsystem is no longer constant but its change can be determined by integration of the additional equation

$$E'(\mathbf{v}_\mathrm{a}) = \sum_k m_k \mathbf{v}_k \cdot \mathbf{f}_k(\mathbf{v}_\mathrm{a})/\Omega\,, \tag{5.106}$$

where $\mathbf{v}_\mathrm{a}$ represents all the average velocities. Finally, this equation can be treated in the same way as (5.104) to yield the energy jump $\Delta E = hE'$.

Tests of small systems ($N = 10$ with two heavy bodies) show that the time-transformed leapfrog (TTL) method is performing well. Thus significantly higher accuracy with about half the number of function evaluations was achieved when including the time transformation. Reliable solutions for coalescence by gravitational radiation have also been obtained involving one or both of the massive binary components. Since the Bulirsch–Stoer integrator is rather expensive when including many interactions, the new method is intended for treating a compact subsystem containing a massive binary and significant perturbers, but intervals of unperturbed motion may also be studied. Finally, it may be noted that integration of the time transforming function W ensures a more well-behaved solution than direct evaluation. A particular advantage is that if any other distances become very large, the time transformation function approaches more closely to the inverse binary separation, which would give exact binary motion by the leapfrog integration.

5.10 Algorithmic regularization

The power of special time transformations was demonstrated in the previous section. Thus when combined with leapfrog integration we essentially achieve a practical regularization scheme where arbitrarily close but non-singular encounters may be studied. This property was first discovered when considering a simplified form of the Hamiltonian [Mikkola & Tanikawa, 1999a,b]. In view of the connection with the TTL method, a brief discussion of the main idea is of interest.¶

¶ Full details of symplectic integration can be found in Mikkola & Saha [2003]. See Mikkola [1997a] for a discussion of time transformations for the few-body problem.

We begin by writing the two-body Hamiltonian as

$$H = T - U\,, \tag{5.107}$$

where T and $U > 0$ define the kinetic and potential energy, respectively. Application of the time transformation

$$ds = U dt \tag{5.108}$$

gives rise to the time-transformed Hamiltonian

$$\Gamma = (T - U + P_t)/U\,. \tag{5.109}$$

Here $P_t = -E_0$ with E_0 the initial total energy. Although this Hamiltonian is not separable, $\Gamma = 0$ on the solution path. This enables us to define the logarithmic function

$$\Lambda = \ln(1 + \Gamma)\,, \tag{5.110}$$

which after simplification leads to the separable form

$$\Lambda = \ln(T + P_t) - \ln U\,. \tag{5.111}$$

A modified leapfrog algorithm can now be introduced that produces correct positions and momenta for an elliptic orbit, albeit with a third-order phase error. Provided the coordinates are not evaluated at the singularity, this treatment also applies to collision orbits. Hence a practical regularization is achieved in the absence of a coordinate transformation. The formulation has been generalized to arbitrary memberships and used successfully to study collisions in a 1D system of six particles.

The numerical solutions are improved by introducing chain coordinates in order to reduce round-off errors. However, comparison with the basic chain method shows that the latter is still more efficient for critical triple encounters. One paper [Mikkola & Tanikawa, 1999b] also contains some useful explanations of the first- and second-order Bulirsch–Stoer method. Finally, for completeness, we mention an analogous derivation of the logarithmic Hamiltonian for symplectic integration [Preto & Tremaine, 1999].

We end this chapter by reviewing two global regularization methods of the general three-body problem in 2D [Lemaître, 1955; Waldvogel, 1972]. The first derivation is based on a complicated Hamiltonian which is not suitable for numerical work. It is also not clear whether a generalization to 3D introduces singular terms [cf. Heggie, 1974]. However, the Hamiltonian of the second formulation does satisfy the requirement of simplicity. A comparison with the three-body methods of sections 5.2 and 5.5 would therefore be of considerable interest.

6
Tree codes

6.1 Introduction

Direct N-body simulations are of necessity expensive because of the need to evaluate all the $N(N-1)/2$ force interaction terms. We have seen that the Ahmad–Cohen [1973] neighbour scheme only alleviates the problem to some extent. However, once the particle number becomes sufficiently large, the dynamical behaviour begins to change because close encounters are less important. This behaviour has inspired methods for collisionless systems to be developed, such as tree codes or multipole expansions. In this chapter, we are concerned with tree codes since some relevant aspects of the latter have already been discussed. First we review the basic features of the pioneering Barnes & Hut [1986] scheme which is widely used in a variety of applications. Since the emphasis in this book is on collisional stellar dynamics, we devote a section to describing a tree code for point-mass interactions [McMillan & Aarseth, 1993] in the hope that it might be revived. The final section deals with an independent development for flattened systems [Richardson, 1993a,b] that has been used to study different stages of planetary formation as well as ring dynamics, where collisions play an important role.

6.2 Basic formulation

In view of the rapid growth in the computational requirements for increasing particle numbers when using direct summation, it is not surprising that several tree-based approaches have been made to speed up the expensive force calculation. The basic idea of employing a tree structure is that the interactions due to a group of distant members can be described by a small number of parameters involving low-order moments of the mass distribution. Depending on the opening angle subtended by each group,

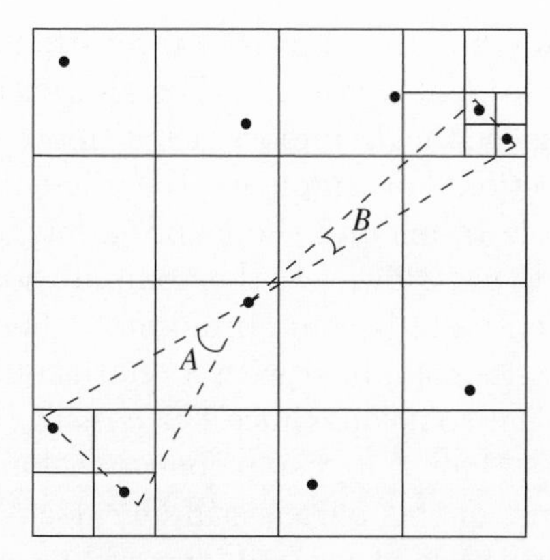

Fig. 6.1. Tree structure.

the total force is then obtained as a sum over these contributions, as well as any individual nearby particles. In the following we outline the basic ideas of the Barnes–Hut [1986] tree-based force algorithm which has been adopted in the methods described in the subsequent sections.

We begin by constructing one empty cubical cell containing the whole system, called the root node. If more than one particle is assigned to any cell, it is split into eight daughter cells, and this procedure is continued recursively until all the members have been allocated single cells. The data structure of each cell provides information about global quantities such as mass and c.m. position accumulated at that level, as well as pointers to the relevant daughter cells that contain further information. Because of the hierarchical subdivision needed to separate two particles, many of the cells will in fact be empty. A typical cell arrangement is shown in Fig. 6.1 for ten particles distributed in 2D.

Given the tree structure, the force on any particle may be obtained by a recursive procedure that starts at the root cell containing the whole system. Let l be the size of the cell under consideration and D the distance from its c.m. to a particle. If $l/D < \theta$, where θ is the opening angle or tolerance, the respective interactions are summed; otherwise the current cell is subdivided into eight cells and each one examined in a similar manner. This procedure continues until all the particles have been included, either in groups or as individual summations. For large N, the total number of such interactions for one particle is of order $\log N$. In some types of simulations involving softened potentials, an opening angle $\theta \simeq 1$ is often used. However, this results in relatively large force errors, albeit of a random nature.

Two strategies may be used to increase the accuracy of the force calculation. By reducing θ below unity more neighbouring particles will be considered individually but the limit $\theta \to 0$ reverts to the direct summation method with additional overheads. The alternative way of providing more information for each cell employs a multipole expansion to be described in the next section. For simplicity, the original formulation did not include the quadrupole terms and the leapfrog integrator used the same time-step for all particles. However, the computational cost of going to higher orders will eventually become prohibitive; hence a compromise of moderate opening angles combined with a relatively low-order expansion offers the best practical solution. Since the cost of rebuilding the whole tree is somewhat less than the cost of one tree force evaluation on all particles, the total cost of this part is still $O(N \log N)$ for large values of N. As will be seen in the next section, the scaling coefficient is surprisingly large if high accuracy is desired; hence special features are needed to achieve an efficient formulation.

Before going on to discuss a code suitable for point-mass calculations, we mention an astrophysical application of the Hernquist [1987] tree code since it pertains to collisions [Arabadjis & Richstone, 1996]. This work included the effects of stellar evolution and collisions, as well as a central black hole and rotation. Star formation was also modelled by reprocessing material ejected through stellar mass loss. The rotationally flattened systems showed a steeper density profile than spherical systems, with preferential radial velocities in the halo.

6.3 Collisional treatment

Several improvements are required in order to achieve satisfactory performance for star cluster simulations. In the first place, it is desirable to increase the accuracy of each force evaluation which can then be used in a high-order polynomial instead of the basic leapfrog integration. The second objective is to introduce an individual time-step scheme both for the particles and the cells. Third, close encounters and persistent binaries may be treated by regularization techniques. The following developments [McMillan & Aarseth, 1993] are based on the Barnes–Hut scheme because of its ease of construction and conceptual simplicity.

Consider a test particle at the position $\mathbf{r} = (r_1, r_2, r_3)$ in the centre-of-mass (c.m.) system of K particles with masses m_k and coordinates $\mathbf{x}_k$. The potential at an external position $\mathbf{r}$ is then given in terms of the multipole expansion [Aarseth, 1967]

$$\Phi(r) = -\frac{M}{r} - \frac{Q_{ij} r_i r_j}{2r^5} - \frac{S_{ij} r_i^2 r_j + S_{123} r_1 r_2 r_3}{2r^7} + O\left(\frac{\Delta x}{r}\right)^5 . \tag{6.1}$$

Here Δx represents the size of the system and repeated indices i and j are summed from 1 to 3. The explicit expressions for the monopole, M, quadrupole, Q, and octupole, S, Legendre coefficients take the form

$$\begin{aligned} M &= \sum m_k \,, \\ Q_{ij} &= \sum m_k (3x_{i,k}x_{j,k} - |\mathbf{x}_k|^2\delta_{ij}) \,, \\ S_{ij} &= \sum m_k [5(3 - 2\delta_{ij})x_{i,k}^2 - 3|\mathbf{x}_k|^2]x_{j,k} \,, \\ S_{123} &= 15 \sum m_k x_{1,k}x_{2,k}x_{3,k} \,, \end{aligned} \tag{6.2}$$

where the summation is over K internal mass-points. Although the dipole term $\mathbf{D} = \sum m_k \mathbf{x}_k$ does not appear in the expansion (6.1) since $\mathbf{r}$ is expressed with respect to the c.m. of the K particles, the value of $\mathbf{D}/M$ for each node must still be known. Analogous expressions are readily constructed for internal points (cf. section 3.4).

The question of expansion order must be decided after appropriate tests. Thus going to one more order than given explicitly here (i.e. hexadecapole) would be quite expensive in view of the rapidly growing number of terms, and it would probably be more efficient to reduce the opening angle slightly. Figure 1 of the original paper illustrates typical relative force errors for different orders and opening angles. For a target median force accuracy of about 10^{-4} considered adequate for the present purpose, we need an opening angle $\theta \leq 0.4$ for quadrupole moments, whereas $\theta \leq 0.5$ is sufficient when octupole terms are included. This general result is in good agreement with other findings [cf. Makino, 1990].

When constructing the cell moments in a tree, we employ a recursive calculation, from the leaves up, such that the moments of each cell are determined by a sum over its eight daughter cells. We begin with the monopole and dipole terms and denote by m_d and x_d the known masses and c.m. of the daughters. The mass M and c.m. coordinates $\mathbf{X}$ of the parents are then simply

$$\begin{aligned} M &= \sum m_d \,, \\ \mathbf{X} &= \frac{1}{M} \sum m_d \mathbf{x}_d \,, \end{aligned} \tag{6.3}$$

where the summation over d is from 1 to 8 and subscripts have been omitted. For the quadrupole moments $\mathbf{Q}$ and $\mathbf{q}_d$, similar expressions to (6.2) again apply, except that the moments of each daughter, $\mathbf{q}_d$, must be included explicitly, which gives

$$Q_{ij} = \sum m_d (3x_{i,d}x_{j,d} - |\mathbf{x}_d|^2\delta_{ij}) + q_{ij,d} \,. \tag{6.4}$$

Lastly, for the octupole terms $\mathbf{S}$ and s_d, we obtain in a similar way

$$S_{ij} = \sum m_d [5(3 - 2\delta_{ij})x_{i,d}^2 - 3|\mathbf{x}_d|^2]x_{j,d} +$$

$$5(1-\delta_{ij})x_{i,d}q_{ij,d} + \tfrac{5}{2}x_{j,d}q_{ii,d} - x_{l,d}q_{jl,d} + s_{ij,d}\,,$$
$$S_{123} = 15\sum m_d[x_{1,d}x_{2,d}x_{3,d} + \tfrac{5}{3}(x_{1,d}q_{23,d} + x_{2,d}q_{31,d} + x_{3,d}q_{12,d})] + s_{123,d}\,, \quad (6.5)$$

where the repeated index, l, in the penultimate term of the first equation is summed. If we define single particles to have zero quadrupole and octupole moments, the above expressions apply whether the daughters of a node are nodes or particles.

Traditionally tree codes have used a low-order integrator and older versions also employed the same time-step for all particles. Since the computational effort of a complete tree construction is $O(N \log N)$, with the same scaling but a smaller coefficient than for determining all the forces, it has been customary to continue this practice. However, star cluster simulations span a much wider range in length- and time-scale which requires the use of high-order integration together with individual time-steps. Hence a different strategy must be devised in order to replace the full reconstruction of the tree at every step.

A solution to the tree construction problem is to consider the deformation of the cells with time and update the current values by prediction. The growing deformation modifies the size of each cell, but this can be controlled by using the effective size in the algorithm, or by reducing θ slightly. Although the particles do not need to fit exactly into the original cubical grid, it is important that the c.m. values for the higher moments of each cell are known to sufficient accuracy. We therefore introduce the idea of individual time-steps for the tree structure, and update the relevant moments by prediction, without explicit reference to the corresponding particle motions. The cells then take on the characteristics of particles, with their own internal derivatives, $\dot{\mathbf{Q}}, \dot{\mathbf{S}}$, as well as time-steps. Now different regions evolve on their own time-scale, and various parts of the tree must be reconstructed at the appropriate time. We delay details of determining the cell time-steps until later and go on to discuss how to extrapolate partial moments.

Suppose that a part of the tree is reconstructed at some time t_0. Moreover, assume that all the relevant particles are synchronized, as in the Hermite scheme, so that any relevant quantity can be evaluated without extrapolation. In addition to $\mathbf{X}_0, \mathbf{Q}_0, \mathbf{S}_0$, we also require the corresponding derivatives $\mathbf{V}_0 = \dot{\mathbf{X}}_0, \mathbf{A}_0 = \ddot{\mathbf{X}}_0$, etc. at time t_0 which are formed by explicit differentiation of (6.2). Cell centroids are predicted to the same order as the particles, i.e.

$$\mathbf{X}(t_0 + \delta t) = \mathbf{X}_0 + \mathbf{V}_0\delta t + \tfrac{1}{2}\mathbf{A}_0\delta t^2 + \tfrac{1}{6}\dot{\mathbf{A}}_0\delta t^3\,, \quad (6.6)$$

where $\delta t < \Delta t$ is the cell time-step. The cell monopole derivatives are obtained from the relevant particle quantities in an analogous manner to the recursive construction of $\mathbf{X}$ from the array $\{\mathbf{x}_d\}$.

Prediction of quadrupole moments also needs to be performed to third order. The corresponding derivatives are evaluated by successive differentiation of (6.4), with $\mathbf{a}_k = \dot{\mathbf{v}}_k$, which gives

$$
\begin{aligned}
\dot{Q}_{ij} &= \sum m_k(3v_{i,k}x_{j,k} + 3x_{i,k}v_{j,k} - 2\mathbf{x}_k \cdot \mathbf{v}_k\delta_{ij}) + \dot{q}_{ij,k}\,, \\
\ddot{Q}_{ij} &= \sum m_k[3a_{i,k}x_{j,k} + 6v_{i,k}v_{j,k} + 3x_{i,k}a_{j,k} \\
&\quad - 2(\mathbf{x}_k \cdot \mathbf{a}_k + |\mathbf{v}_k|^2)\delta_{ij}] + \ddot{q}_{ij,k}\,, \\
Q_{ij}^{(3)} &= \sum m_k[3\dot{a}_{i,k}x_{j,k} + 9a_{i,k}v_{j,k} + 9v_{i,k}a_{j,k} + 3x_{i,k}\dot{a}_{j,k} \\
&\quad - 2(\mathbf{x}_k \cdot \dot{\mathbf{a}}_k + 3\mathbf{v}_k \cdot \mathbf{a}_k)\delta_{ij}] + q_{ij,k}^{(3)}\,.
\end{aligned}
\qquad (6.7)
$$

With multipole prediction, the descent of the tree proceeds almost as previously, except that before the opening criterion is applied to a cell, its c.m. position is updated. If the cell remains unopened, its quadrupole moments are also predicted before use by

$$\mathbf{Q}(t_0 + \delta t) = \mathbf{Q}_0 + \dot{\mathbf{Q}}_0\delta t + \tfrac{1}{2}\ddot{\mathbf{Q}}_0\delta t^2 + \tfrac{1}{6}\mathbf{Q}_0^{(3)}\delta t^3\,. \qquad (6.8)$$

The earlier remark about the cost of the hexadecapole moments also applies to the octupole derivatives which would contain a large number of terms. We neglect the prediction of these derivatives by choosing to limit the time-step instead. However, it seems likely that there would be some benefit by introducing the $\dot{\mathbf{S}}$ terms obtained by differentiating (6.5).

A number of criteria have been used to determine time-steps for the cells. We define three characteristic time-scales for each cell as follows. The first is the crossing time

$$\Delta t_{\rm cross} = \min_k \left[\frac{(\Delta X)^2}{(\mathbf{x}_k - \mathbf{X}) \cdot (\mathbf{v}_k - \mathbf{V})} \right], \qquad (6.9)$$

where the index k refers to all particles in the cell and ΔX is the cell size. Next we introduce a monopole prediction time by

$$\Delta t_{\rm mono} = \left(\frac{\Delta X|\mathbf{A}| + |\mathbf{V}^2|}{|\mathbf{V}||\dot{\mathbf{A}}| + |\mathbf{A}|^2} \right)^{1/2}. \qquad (6.10)$$

Finally, we take the quadrupole prediction time as

$$\Delta t_{\rm quad} = \left(\frac{|\mathbf{Q}||\ddot{\mathbf{Q}}| + |\dot{\mathbf{Q}}^2|}{|\dot{\mathbf{Q}}||\mathbf{Q}^{(3)}| + |\ddot{\mathbf{Q}}|^2} \right)^{1/2}. \qquad (6.11)$$

We now choose the actual cell time-step from

$$\Delta t_{\mathcal{C}} = \eta_{\rm c} \min \{\Delta t_{\rm cross}, \Delta t_{\rm mono}, \Delta t_{\rm quad}\}, \tag{6.12}$$

where the choice $\eta_{\rm c} \simeq 0.1$ for the accuracy parameter gives satisfactory results. In practice, the quadrupole time-scale is the most sensitive. The expressions for the monopole and quadrupole time-steps are based on the original time-step criterion for direct N-body integration.

A cell $\mathcal{C}$ is due to be reconstructed when $t_{0,\mathcal{C}} + \Delta t_{\mathcal{C}} \leq t$, in which case the whole part of the tree below $\mathcal{C}$ is rebuilt. This has the effect of updating all the cells descended from $\mathcal{C}$. Hence a daughter cell whose updating time exceeds that of its parent can simply be excluded from consideration by the time-step scheduling algorithm.

There are several reasons why the standard individual time-step scheme is not well suited to the present tree-based code. First, the cost of frequent particle prediction can become substantial and comparable to a force calculation. Another consideration is that it is difficult to ensure synchronization when both particles and, cells are involved. Moreover, the corrector as well as time-step determinations are scalar operations when particles are advanced one at a time, whereas it is desirable to implement procedures for parallel or vector supercomputers. For these and other reasons we introduce a block time-step algorithm [McMillan, 1986]. This procedure anticipated the later quantization of time-steps in the Hermite integration scheme [Makino, 1991b]. Since the relevant aspects have already been discussed in chapter 2, it suffices to state that the time-steps take values $2^n \Delta t_0$, where Δt_0 is the smallest at a given time.

The scheduling itself is concerned with determining all the members of block n, which are advanced together. A novel algorithm is adopted which does not appear to have been used by others. Thus at the jth step, all particles in or below the n_jth block are updated, where n_j takes the successive values $1, 2, 1, 3, 1, 2, 1, 3, 1, 4, \ldots$ This scheme is quite fast and is implemented by maintaining a sorted list $\{i_\ell\}$ of all the particles such that $\Delta t_{i\ell} \leq \Delta t_{i\ell+1}$ for $1 \leq \ell < N$. The block structure then consists of a second list of pointers $\{p_k\}$ to the ordered time-step list, so that particles $i_1, i_2, \ldots, i_{p1}$ are in block 1, with time-step Δt_0, and particles $i_{p1+1}, i_{p1+2}, \ldots, i_{p2}$ are in block 2, with time-step $2\Delta t_0$, etc. Thus at step j, the particles to be advanced are $i_1, i_2, \ldots, i_{n_j}$, where n_j is defined above. The advantage of this algorithm is that the choice of particles to be integrated does not require any further search.

For the scheduling to work, it is important that the time-step list $\{i_\ell\}$ be ordered such that the block boundaries are correct. Thus particles moving to a lower block are exchanged with the last element in the current block, and the block pointers are adjusted. If necessary, the process is repeated until the correct level has been reached. Likewise, particles moving to a higher block are treated similarly, except that to maintain synchronization at the next block-step, only one adjustment is allowed each time-step. Unlike the present procedure in standard Hermite integration, the natural steps of all particles are retained and are used for limiting the new step in order to ensure convergence. Finally, when a new block is added below block 1, Δt_0 and the step counter j must be replaced by $\frac{1}{2}\Delta t_0$ and $2j$, respectively. Hence at a given time Δt_0 represents the smallest time-step.

Cells also have associated time-steps which must be scheduled in some way. It is convenient to employ the same scheme as above so that when a cell is rebuilt, its time-step and those of any descendants are determined and added to the block list. The block structure itself is defined by the particles and, because two-body regularization is used, the condition $\Delta t < \Delta t_0$ for cells never arises. Hence after a given block of particles has been advanced, we rebuild any cells that are due for updating.

Unlike the particles, cells can contain other cells, and some care is needed to avoid redundancies and inconsistencies, as for example attempting to reconstruct both a cell and its parent at the same time. One can easily avoid this by including a cell in the block list only if its time-step is smaller than that of any ancestor. All the affected cells can then be rebuilt at once, with the entire operation being completely vectorizable, which is desirable for some hardware.

It is quite inexpensive and highly advantageous to ensure that the tree is rebuilt at least as often as the whole system is synchronized; i.e. at the largest time-step. This is achieved by placing the root node in or below the top particle block. Since the tree is predicted to the same accuracy as the particles, we do not impose that the time-step of a particle be smaller than that of its parent cell. The situation can easily be remedied if it were desirable to construct the tree from synchronized particles, namely by requiring that no particle has a time-step exceeding that of its parent node. As can be seen, the present scheme is very flexible and can be modified to suit particular requirements.

Even if there were no numerical problems connected with hard binaries, their presence would cause complications in the construction and maintenance of the tree and block structure. It is therefore natural to implement two-body regularization in analogy with standard N-body codes, and there is no reason why multiple regularization cannot be included. Since practical details of the KS method will be discussed elsewhere, it suffices to deal with some aspects relating to the tree code formulation.

As far as the tree structure is concerned, the c.m. of a regularized pair is treated as one particle since its internal motion is advanced separately. Regularization therefore places a lower limit of $a_{\rm hard} \simeq r_{\rm h}/N$ on the particle separation, where $r_{\rm h}$ is the half-mass radius, and consequently this also prevents the time-step Δt_0 from reaching rather small values. Although the tree sees a binary as a mass-point which is resolved into individual components when necessary, its quadrupole and octupole moments are included correctly in its parent cell.

A bonus of the tree formulation is that it dispenses with the perturber lists used in direct N-body codes. This feature may not lead to much gain in efficiency since the AC method provides lists of relevant particles for perturber selection. However, there is the question of soft binaries for which the neighbour list allocation may not be sufficiently large. This problem may be partially circumvented by restricting regularization to hard binaries and, since the number of perturbers actually decreases with increasing N, it is convenient to retain the standard perturber lists which are calculated from the tree.

Since regularization employs a non-linear internal time, the corresponding physical time-steps cannot be incorporated into the block scheme. However, regularized solutions are distinct from the other treatment and we simply perform all the necessary operations before beginning each block-step. Hence each KS solution is advanced until its next treatment time exceeds the current block time. This part may be vectorized if a large number of perturbed KS pairs require attention simultaneously.

Two particles are selected for KS treatment if their two-body motion is dominant, with time-steps $\Delta t_i < \Delta t_{\rm cl}$, where the close encounter time-step is derived from the corresponding distance $R_{\rm cl} \simeq 4r_{\rm h}/N$, slightly modified by the density contrast. Use of this time-step criterion normally implies that the two particles are approaching each other with $R < R_{\rm cl}$, unless other particles are involved. Because a relative time-step criterion is used, two such particles are normally in the same block, otherwise synchronization can be enforced at the previous step.

The presence of a KS pair requires minor modification of the cell-opening algorithm in order to maintain continuity of the force. Since the KS pair will normally be resolved into its components, we require that the same be true of the separate components in the tree. This is achieved by opening any cell whose diameter D and distance R from the particle in question satisfies $D + \alpha a > \theta R$, where $\alpha \simeq 1$ and a is the relevant semi-major axis.

The collisional tree code described above has reached a state of development where it may be used in large-scale simulations. Based on comparison tests with a fairly modest particle number (i.e. $N = 1024$), the cross-over point with respect to standard summation is estimated to be in the region of $N \simeq 10^4$. One reason is the rather large scaling coefficient required for a high-order scheme. Even at this particle number, the tree code will barely reach its asymptotic $N \log N$ regime; hence a fairly powerful computer will be needed in order to exploit the performance. It is also very encouraging that the results of a core collapse calculation shows excellent agreement with direct integration. Further details of accuracy and timing can be found in the original paper [cf. McMillan & Aarseth, 1993]. Finally, the present formulation is suitable for both workstations and conventional supercomputers, and the time is now ripe for a more serious application to be attempted on faster hardware.

An earlier tree code development which was not pursued [Jernigan & Porter, 1989] also employed two-body regularization. This novel formulation is based on a recursive binary tree data structure and an accurate KS integration method for every level of the binary tree. Given that the computing time scaled as $T_{\rm comp} \propto N^{1.62}$ for $N \leq 32\,{\rm K}$ compared with $N^{1.63}$ for $\theta = 0.5$ above, this method deserves to be resurrected.

6.4 Flattened systems

Planetesimal systems exhibit two characteristic features that facilitate a tree-code approach. First, the dominance of the central body, be it a star or a planet, reduces the effects of the mutual interactions which can therefore be considered as perturbations during most of the motion. Such systems also tend to be disc-like and hence may require less computational effort for dealing with the vertical dimension than fully 3D systems. In the following we describe a unique tree code method [Richardson, 1993a,b] which has also been applied to the problem of planetary rings [Richardson, 1994]. More recently, the cosmological tree code *PKDGRAV* (discussed in a later chapter) was adapted to studying planetary formation as well as ring simulations [Richardson *et al.*, 2000].

The planetesimal method combines two techniques in order to study large particle numbers efficiently. In the first place, the particle in box scheme [Wisdom & Tremaine, 1988] discussed in section 3.6 has been adopted to represent a small self-similar patch, thereby increasing the dynamical range significantly. Again the interactions are evaluated by the Barnes–Hut [1986] tree code, with periodic boundary conditions which include the effect of eight surrounding ghost boxes [Aarseth *et al.*, 1993]. By referring the individual coordinates to the centre of a comoving Cartesian coordinate frame, the equations of motion take the linearized form (3.37). The sliding box procedure is carried out according to section 3.6, and particles leaving the central box are replaced by corresponding ghost images such that, apart from collisions, the particle number is preserved.

The multipole expansion includes terms up to quadrupole order as a compromise between complexity and efficiency. Once again, three derivatives of the quadrupole tensor are calculated explicitly and used for updating during the quadrupole prediction [cf. McMillan & Aarseth, 1993]. The integration scheme is based on the divided difference formulation of section 2.2. To speed up the treatment, a 2D tree is used to describe the flattened 3D system. This calls for special procedures, such as 'effective size' for each node that is recalculated during updates. The effective size is defined as the maximum of the actual size of the node and the predicted y- or z-extensions of each child from the c.m., whereas excursions in the radial x-direction are usually small. This device prevents excessive subdivision for two particles that are arbitrarily close in 2D projection.

A variety of special features have been implemented in the planetesimal code. Most original is the concept of tree repair which is very useful in collision-dominated systems with small time-steps, since otherwise tree construction after every time-step would be extremely expensive. Thus if a particle crosses the boundary of its cell, the relevant nodes are updated consistently without affecting other parts of the tree. This entails destroying old nodes which become de-populated, or creating new ones.

The treatment of collisions also requires considerable care. Both dissipation and spin (discussed in section 17.4) are included for greater realism. In order to reduce the amount of collision overshooting, a more sensitive time-step criterion is employed inside small separations $r < 10r_0$, with

$$\Delta t = 2^{-10r_0/r} \eta r/\dot{r} \,, \tag{6.13}$$

where r_0 is a typical particle size and η is the tolerance factor ($\simeq 0.02$). This results in significant improvement of angular momentum conservation at little extra effort since collisions are relatively rare. The choice of opening angle, $\theta = 0.6$, shows factors of 2–3 speed-up with respect to the direct method for $N = 250$. However, the asymptotic approach to the theoretical $N \log N$ algorithm may require even larger particle numbers [cf. McMillan & Aarseth, 1993].

The planetesimal tree code called *BOX_TREE* has also been applied to a simulation of Saturn's B ring [Richardson, 1993a, 1994], which demonstrated its versatility. Here collisions are a main feature and the time-step criterion (6.13) proved inadequate at high densities. Instead the general expression (2.13) was found to be satisfactory, although it is relatively expensive in the tree code formulation. Another feature of general interest concerns the determination of post-collision velocities and spin rates when normal and transverse restitution coefficients appropriate to small bodies are prescribed. More details of this treatment are presented in section 17.4. Finally, we note that test results for low particle densities are in good agreement with the original box formulation, as well as an analytical model [Goldreich & Tremaine, 1978] when self-gravity is not included.

7

Program organization

7.1 Introduction

We now make an abrupt transition to a presentation of various algorithms utilized by the direct summation codes. Before proceeding further, it will be useful to include some practical aspects in order to have a proper setting for the subsequent more technical procedures. First we introduce the main codes that have been developed for studying different gravitational N-body problems. Where possible, the same data structure has been employed, except that the most recent versions are formulated in terms of the Hermite integration scheme. Since the largest codes are quite complicated, we attempt to describe the overall organization by tables and a flowchart to provide some enlightenment. Later sections give further details concerning input parameters, variables and data structure; each of these elements play an important role for understanding the general construction. We also discuss a variety of optional features which provide enhanced flexibility for examining different processes.

7.2 N-body codes

Before describing the characteristics of the codes, we introduce some short-hand notation to illustrate the different solution methods employed [cf. Makino & Aarseth, 1992]. Thus by ITS we denote the basic individual time-step scheme, whereas ACS defines the Ahmad–Cohen [1973] neighbour scheme. Likewise, HITS and HACS are used for the corresponding Hermite integration methods. Finally, MREG refers to the implementations of unperturbed three-body [Aarseth & Zare, 1974] and four-body chain regularization [Mikkola & Aarseth, 1990], as well as perturbed chain regularization [Mikkola & Aarseth, 1993]. Also note that the latter contains (optionally) the slow-down procedures discussed earlier. Because of the programming complexities involved, only one multiple regularization of each type is currently permitted, but it is extremely rare for more than one such critical interaction to occur at the same time in standard cluster simulations. However, an extension to several compact subsystems or the introduction of wheel-spoke regularization (cf. section 5.4) would be beneficial for some problems. On the other hand, an arbitrary number of simultaneous KS solutions can be used to study large populations of primordial binaries [Aarseth, 1985a, 2001a; Wilkinson *et al.*, 2003].

Table 7.1 summarizes the combination of integration schemes in the main codes, together with typical particle numbers for applications. The choice of names is historical but indicates an increasing degree of complexity in the solution methods, starting with one force polynomial. The codes *NBODY*1 and *NBODY*2 are of less interest here since they are based on a softened potential and are only suitable for collisionless systems. Moreover, a detailed description of these codes can be found elsewhere [Aarseth, 2001b]. Equivalent Hermite versions of *NBODY*1 and *NBODY*2 also exist [cf. Makino & Aarseth, 1992]. Likewise, the cosmological simulation code *COMOVE* discussed in section 3.3 is based on *NBODY*2. Still, it is very useful to gain experience with simple codes before attempting more complicated tasks.

Star cluster simulations are facilitated by the introduction of two-body regularization [Aarseth, 1972b; Bettis & Szebehely, 1972]. The KS method adopted is of the same type as the corresponding direct N-body integrator for compatibility; i.e. the older codes *NBODY*3 and *NBODY*5 are still entirely based on divided differences. Of the four point-mass codes, one (*NBODY*3) is intended for small-N systems, whereas another one (*NBODY*4) has been adapted for the special-purpose HARP and GRAPE computers,* and is likewise based on brute-force solutions of the gravitational many-body problem. This leaves us with the three last codes for general use on laptops, workstations or conventional supercomputers.

It does not follow that a code for small particle numbers is much more compact since many of the key building blocks are essentially similar and the data structure is preserved. Thus the main technical difference between *NBODY*3 and *NBODY*5 is the lack of neighbour lists in the former, although the latter also contains additional astrophysics. Since N is usually relatively modest when using *NBODY*3 (i.e. $N \leq 100$ for efficiency reasons), the extra cost of obtaining perturber lists for two-body regularization by full summation is not a concern, otherwise one could devise an analogous strategy connected with the centre-of-mass (c.m.) motion.

* The names are derived from 'Hermite AcceleratoR Pipeline' and 'GRAvity piPE' [Makino *et al.*, 1997]. We denote by HARP-2 and HARP-3 the older hardware located at Cambridge, whereas GRAPE is used for the generic family of all such machines.

Table 7.1. Direct N-body codes.

Code	Characteristic description	Membership
*NBODY*1	ITS with softening	$3 - 100$
*NBODY*2	ACS with softening	$50 - 10^4$
*NBODY*3	ITS with KS and MREG	$3 - 100$
*NBODY*4	HITS with KS and MREG	$10 - 10^5$
*NBODY*5	ACS with KS and MREG	$50 - 10^4$
*NBODY*6	HACS with KS and MREG	$50 - 10^4$
*NBODY*7	As *NBODY*6 with BH binary	$50 - 10^4$

In other words, the corresponding neighbour list could be renewed on a somewhat longer time-scale to mimic the regular time-step used by *NBODY*2.

The codes *NBODY*5 and *NBODY*6 are mainly intended for star cluster simulations. The former has been a work-horse for more than 20 years now and the main features have been described elsewhere [cf. Aarseth, 1985a, 1994]. With the subsequent arrival of Hermite integration [Makino, 1991a; Makino & Aarseth, 1992], the code *NBODY*6 appears to be more robust and at least as accurate for the same CPU time. Some general aspects have been discussed elsewhere [Aarseth, 1994, 1996b, 1999a,b] but a complete description is still lacking. After a period of testing and developments, it is now beginning to produce useful results [Kroupa, Aarseth & Hurley, 2001]. Although the actual integration methods differ somewhat, many other aspects of interest are in fact very similar and can be discussed together. The very recent code *NBODY*7 [Aarseth, 2003a] contains the additional implementation of BH binary dynamics discussed in section 5.9 and there is also an equivalent GRAPE-6 version which has proved itself [cf. Aarseth, 2003b]. Finally, the HARP or GRAPE code *NBODY*4 deals with several additional astrophysical processes that are of general interest and can be readily included in *NBODY*6.

7.3 Flowcharts

A general description of complicated code structure is facilitated by recourse to flowcharts, supported by tables defining variables, as well as algorithms. The following schematic illustration shown in Fig. 7.1 provides an overview. The simplicity is deceptive and even the first segment contains a number of special features in the star cluster codes.

For convenience, we concentrate on describing the generally available code *NBODY*6 but remark that *NBODY*4 is broadly similar apart from

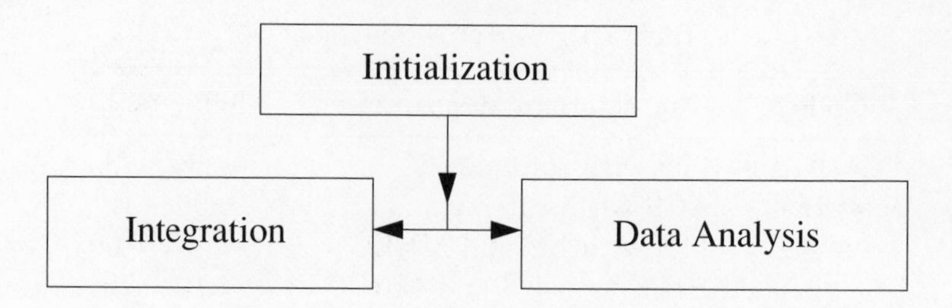

Fig. 7.1. Schematic code structure.

lacking the AC neighbour scheme. Turning first to initialization, the main steps are outlined in Algorithm 7.1 for clarity, whereas the actual scaling from any initial units to N-body units as well as the final astrophysical data conversion is detailed in section 7.4. In view of the variety of initial conditions, there is some provision for explicit generation (i.e. Plummer model); otherwise specially prepared data may be supplied (cf. Table 7.4). Independent optional features such as external tidal fields, primordial binaries (and hierarchical triples), stellar evolution and interstellar clouds are also catered for, as discussed further in the next chapter. Finally, force polynomials and time-steps are assigned and any hard primordial binaries initialized for KS treatment.

Algorithm 7.1. *Initialization procedures.*

1	Initialize useful counters and variables
2	Read input parameters and options
3	Obtain initial conditions (*in situ* or prepared file)
4	Scale all m_i, $\mathbf{r}_i$, $\dot{\mathbf{r}}_i$ to N-body units
5	Define scaling factors for data conversion
6	Introduce external tidal field (optional)
7	Generate a primordial binary distribution (optional)
8	Assign stellar evolution parameters (optional)
9	Add a population of interstellar clouds (optional)
10	Evaluate force polynomials and specify time-steps
11	Regularize any close primordial binaries (optional)

Next we consider the more complicated part which deals with the different ways of obtaining numerical solutions. Since a full discussion of the data structure is deferred until later in this chapter, only some general aspects are included in the flowchart of Fig. 7.2.

Each integration cycle consists of advancing all equations of motion consistently up to the end of the block-step. The new cycle begins by selecting all single and c.m. particles due for treatment by the Hermite

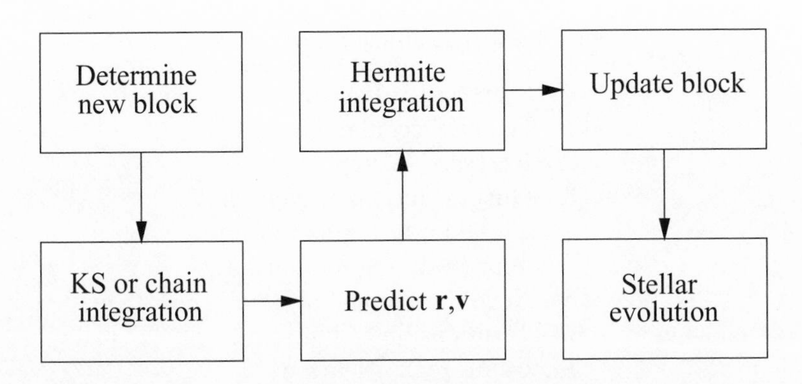

Fig. 7.2. Integration cycle.

scheme (cf. section 9.2). Before dealing with the standard case, any KS or compact subsystem solutions are advanced. If the particle sequence is maintained (e.g. no physical collision), the coordinates and velocities of the relevant neighbours are predicted. Assuming a conventional computer, the irregular and regular integration of each particle is carried out sequentially. However, in order to preserve an invariant outcome, the current coordinates and velocities, $\mathbf{r}_i$, $\mathbf{v}_i$, for particles in the same block are only copied from the corresponding corrected values at the end of the block-step. In the case of KS termination or multiple subsystem initialization, the program flow is interrupted by special procedures discussed later. For completeness, we also mention new KS initialization here, although this task is in fact considered at the beginning of the block-step. This entails setting the time equal to the previous block time, whereupon the cycle starts again with a new sorted sequence. The cycle ends by checking the optional procedures for implementing stellar evolution.

The decision of whether to produce results is also taken just after the new block time has been determined. Thus if the next output time is exceeded, the integration cycle is suspended temporarily and the current time is redefined as the previous block time when all the relevant solutions are updated. Likewise, any additional data saved for later analysis may be initiated at this stage by introducing an appropriate time interval. This requires the prediction of all quantities to highest order, which is also needed for the energy check. At such times, most solutions will usually be known to order $\dot{\mathbf{F}}$ at least. A discussion of any other procedures is less relevant here and is therefore left to a later chapter.

A special indicator is used to facilitate program control. Historically this was introduced to overlay different modules in order to save valuable memory, hence the notation *IPHASE* for this indicator. It may be helpful to consult the list of possible values, given in Table 7.2.

Table 7.2. Indicator for flow control.

0	Standard value
1	New KS regularization
2	KS termination
3	Output and energy check
4	Three-body regularization
5	Four-body regularization
6	New hierarchical system
7	Termination of hierarchy
8	Chain regularization
9	Physical collision
−1	Exceptional cases

All these indicators are used and have the same meaning in the codes *NBODY3* and higher versions, and a few are also found in *NBODY2*. It is convenient to employ a negative value for denoting exceptional cases, such as using only part of a routine, signalling additional procedures or denoting a change of the particle sequence (cf. section 7.7).

7.4 Scaling and units

The use of units in N-body simulations has been rather non-uniform. Ideally, results should be presented in a way that at least facilitates comparison with other models, if not observations. The comparison problems were due to the theoretician's preference for considering a wide variety of scales and reporting the results in terms of natural variables, unlike observers who are constrained by reality. Eventually, it was agreed by the small community of practitioners to employ so-called 'standard N-body units' for the actual calculations [Heggie & Mathieu, 1986]. Briefly stated, the scaled total mass, M, and equilibrium virial radius, r_V, are taken to be unity, with the total energy $E_0 = -0.25$ for bound systems.

Algorithm 7.2. *Scaling to standard units.*

1	Initialize $\mathbf{r}_i$, $\dot{\mathbf{r}}_i$ in the inertial c.m. frame
2	Scale the masses by $\sum_i m_i = 1$, giving $\bar{m} = 1/N$
3	Calculate kinetic, potential and tidal energy T, U, W
4	Define the virial energy expression $V = U + 2W$
4	Multiply the velocities by $Q_v = (Q_{vir}\|V\|/T)^{1/2}$
5	Scale $\mathbf{r}_i$, $\dot{\mathbf{r}}_i$ to $E_0 = -0.25$ using $\beta = (1 - Q_{vir})\, U/E_0$

The steps for achieving the desired scaling are given by Algorithm 7.2. Here $Q_{\rm vir}$ is the specified virial ratio which is a half for overall equilibrium and $V = U + 2W$ is the virial energy. Note that W is not yet determined during the scaling procedure and is therefore only included with zero value to be formally correct. For open star clusters, the z-component of the angular momentum is absorbed in the kinetic energy [cf. Chandrasekhar, 1942, eq. 5.535]. The last operation involves multiplying all coordinates by β and dividing the velocities by $\beta^{1/2}$. As a check, the total energy should now be† $E = -0.25$. In these units we have $\sigma^2 = \frac{1}{2}$ for the mean square equilibrium velocity, which gives a constant crossing time

$$t_{\rm cr} = 2r_{\rm V}/\sigma = 2\sqrt{2}\,. \tag{7.1}$$

Since the central velocity dispersion in a bound cluster exceeds σ by some factor (about 2 for a typical King [1966] model), we define the energy of a hard binary by (4.1). The corresponding scaled binding energy per unit mass is then obtained by the general expression

$$h_{\rm hard} = -4\max\{T, |T + U|\}/M\,, \tag{7.2}$$

which takes the dimensionless value $h_{\rm hard} = -1$ at equilibrium. Note that although $M = 1$ initially, the removal of escapers affects all the global quantities, including the total energy itself. For this reason, a consistent updating of the relevant parameters is desirable as the cluster evolves.

In the case of a cluster with regularized binaries, T represents the sum of contributions from single particles and c.m. motions, whereas the potential energy, U, excludes all internal two-body interactions. Hence the quantity $T+U+W$ is the total energy that binds the cluster and $2Q_{\rm vir}-1$ denotes the fractional departure from overall equilibrium in per cent. With regularized binaries present, the actual total energy is obtained by adding all the predicted two-body binding energies. Note that if the alternative definition of including all the potential energy in U is used, the virial ratio $Q_{\rm vir} = T/|U + 2W|$ may be dominated by a single eccentric binary.

The scaling of astrophysical quantities to internal units is often a source of confusion. The main reason for this is due to taking $G = 1$ for the gravitational constant. Such problems are readily resolved by performing dimensional analysis. Let us now introduce some basic relations in terms of physical units. A cluster is essentially defined by the two global parameters $R_{\rm V}$ and $M_{\rm S}$ together with N, which are usually specified as input (see next section). For most work in stellar dynamics, it is convenient to express the length scale, $R_{\rm V}$, in pc and mean stellar mass, $M_{\rm S}$, in $M_\odot$, from which all quantities of interest can be derived. Distances and masses are then simply obtained from the N-body units by $\tilde{r} = R_{\rm V}\, r$ and $\tilde{m} = M_{\rm S}\, \bar{m}$.

† Subject to a possible small contribution from the optional tidal force.

For the velocity we determine the conversion factor in $\mathrm{km\,s^{-1}}$from

$$\tilde{V}^* = 1 \times 10^{-5}(GM_\odot/L^*)^{1/2}\,, \tag{7.3}$$

where L^* is a length scale. With $G, M_\odot, L^*$ expressed in cgs units and choosing $L^* = 1\,\mathrm{pc}$, this gives $\tilde{V}^* = 6.557 \times 10^{-2}$. Taking account of the total mass and cluster length scale then yields the scaling relation

$$V^* = 6.557 \times 10^{-2}(NM_\mathrm{S}/R_\mathrm{V})^{1/2}\,. \tag{7.4}$$

The scaling of time is also done in two stages as follows. First the fiducial time factor is taken to be $\tilde{T}^* = (L^{*3}/GM_\odot)^{1/2}$, or 14.94 in units of Myr. Converting to the cluster parameters, we obtain

$$T^* = 14.94\,(R_\mathrm{V}^3/NM_\mathrm{S})^{1/2}\,. \tag{7.5}$$

Hence the velocity in units of $\mathrm{km\,s^{-1}}$and the time in Myr are obtained by $\tilde{v} = \tilde{V}^*\,v$ and $\tilde{t} = \tilde{T}^*\,t$, where v and t are in N-body units. Finally, the relation for the crossing time in Myr becomes

$$T_\mathrm{cr} = 2\sqrt{2}\,T^*\,. \tag{7.6}$$

Applying the basic conversion factors $M_\mathrm{S}, R_\mathrm{V}, V^*, T^*$ and dimensional analysis, we can now evaluate all the relevant quantities in convenient physical units. Note that individual masses, m_i, may be expressed in scaled N-body or astrophysical units, depending on the context.

7.5 Input parameters and options

Each code has its own choice of input parameters that can take a range of values according to the desired objective. In practice it is useful to be aware of any limitations and a well-designed code should include consistency checks to prevent inappropriate usage. However, one should also allow a certain flexibility for experimentation; hence any restrictions are often curtailed by compromise. It is useful to distinguish between parameters specifying the model to be investigated and those that control the solutions. Alternatively, an optional facility for including a prepared set of initial conditions can be used to over-ride the choice of model parameters, since only a few explicit models are generated *in situ*.

Table 7.3 gives suggested numerical values for the main variables controlling the integration in the codes NBODY5 and NBODY6 in the case $N = 1000$ and, as can be seen, the simplicity is evident. Given the standard scaling to N-body units discussed above, only a few of these quantities depend on the particle number. First, the central value of the initial neighbour sphere may be taken as $S_0 \simeq 0.3\,(N/1000)^{1/3}$ for the standard scaling $r_V = 1$. At larger distances, the initial neighbour radius is modified according to the expression $S^2 = S_0^2\,(1+r^2)$ if the system is centrally concentrated.

A good choice for the maximum neighbour number is $n_{\max} \simeq 2N^{1/2}$, but this has not been investigated above $N \simeq 1000$.[‡] However, in the case of primordial binaries with a general IMF it is desirable to employ more generous values of $n_{\max}$ since even a hard binary may have a relatively large semi-major axis that requires many perturbers inside the distance $\lambda a(1+e)$, where λ is a dimensionless quantity. From the basic close encounter distance $R_{\rm cl} = 4\,r_{\rm h}/N$ defined by (1.9) and dimensional analysis, the corresponding time-step for typical parabolic motion is determined empirically as $\Delta t_{\rm cl} \simeq 0.04\,(R_{\rm cl}^3/\bar{m})^{1/2}$, where $\bar{m}$ is the mean mass. A slightly smaller value of $R_{\rm cl}$ is used in practice (cf. (9.3)). Note that a binary with mass components $m_i = 10\bar{m}$ would already be hard at a semi-major axis $a \simeq 100/N$.

The codes contain an option for automatic adjustments of all regularization parameters, carried out at intervals $\Delta t_{\rm adj}$. The regularized time-step, to be defined in a later section, is chosen such that there are $2\pi/\eta_U$ steps per unperturbed orbit. The suggested value refers to the Hermite Stumpff integration, whereas $\eta_U = 0.1$ is appropriate for the other KS methods. The value of $h_{\rm hard}$ follows directly from $\epsilon_{\rm hard}/\bar{\mu}$, with $\bar{\mu}$ the mean reduced mass (cf. (4.1)). Most of the remaining quantities are a matter of personal taste but have been included for completeness.

As can be seen, the most important parameters are either dimensionless or have dynamical significance. Hence a change to different particle numbers requires minor modifications of the basic input template. Needless to say, a number of additional features are included in order to permit different effects to be studied, whereas others ensure a smooth running of the code. These are optional and offer facilities for a wide variety of paths to be decided on initially and, if desired, can also be changed during the calculation after a temporary halt which saves the current state. Since many of the relevant procedures will be discussed in later sections, Table 7.4 includes a representative subset (out of 40 options) for inspection. We remark that options offer greater flexibility than logical variables, since the former can be assigned multiple values and defined in order of increasing complexity. Naturally, care must be exercised to ensure that the different options are mutually consistent. This is usually achieved by constructing a template when investigating a specific problem.

[‡] See Spurzem, Baumgardt & Ibold [2003] or Makino & Hut [1988] for different viewpoints regarding large simulations.

Table 7.3. Integration parameters.

η_I	Time-step parameter for irregular force	0.02
η_R	Time-step parameter for regular force	0.03
S_0	Initial radius of the neighbour sphere	0.3
$n_{\rm max}$	Maximum neighbour number	70.0
$\Delta t_{\rm adj}$	Time interval for energy check	2.0
$\Delta t_{\rm out}$	Time interval for main output	10.0
$Q_{\rm E}$	Tolerance for energy check	1.0×10^{-5}
$R_{\rm V}$	Virial cluster radius in pc	2.0
$M_{\rm S}$	Mean stellar mass in solar units	0.8
$Q_{\rm vir}$	Virial theorem ratio $(T/\lvert U+2W\rvert)$	0.5
$\Delta t_{\rm cl}$	Time-step criterion for close encounters	4.0×10^{-5}
$R_{\rm cl}$	Distance criterion for KS regularization	0.001
η_U	Regularized time-step parameter	0.2
$h_{\rm hard}$	Energy per unit mass for hard binary	–1.0
$\gamma_{\rm min}$	Limit for unperturbed KS motion	1.0×10^{-6}
$\gamma_{\rm max}$	Termination criterion for soft binaries	0.01

7.6 Basic variables

All N-body codes require a number of variables for each particle, the minimum being just six if equal masses are chosen. In the following we concentrate on describing *NBODY*6, although *NBODY*5 contains exactly the same direct integration variables. For ease of reference, these are listed in Table 7.5 together with a brief definition. The actual variables employed by the codes are also included for convenience according to the programming style of upper case for all FORTRAN statements.

Given the theoretical exposition of the preceding chapters, the comments to all the entries are self-explanatory. Because of the static memory allocation in older FORTRAN versions, maximum array sizes are specified before compilation. A number of FORTRAN parameters are included in a header file *params.h* and these are defined in Table 7.6. For illustration, Column 3 contains representative values for a cluster simulation that allows up to $N_{\rm s} = 1000$ single particles and $N_{\rm b} = 1000$ hard primordial binaries to be studied. Since every new KS solution also requires a corresponding c.m. particle to be introduced, the storage requirement is for $N_{\rm s} + 3N_{\rm b}$ particle arrays. Moreover, some allocation should also be made for additional regularizations during the early stages since all the original binaries may still be present.

All the important variables are kept together in a global common block *common6.h* to facilitate data communication and restart of a calculation

Table 7.4. Optional features.

1	Manual common save on unit 1 at any time
2	Common save on unit 2 at output time or restart
3	Data bank on unit 3 with specified frequency
5	Different types of initial conditions
7	Output of Lagrangian radii
8	Primordial binaries (extra input required)
10	Regularization diagnostics
11	Primordial triples (extra input required)
13	Interstellar clouds (extra input required)
14	External tidal force; open or globular clusters
15	Multiple regularization or hierarchical systems
16	Updating of regularization parameters $R_{\rm cl}$, $\Delta t_{\rm cl}$
17	Modification of η_I and η_R by tolerance $Q_{\rm E}$
19	Synthetic stellar evolution with mass loss
20	Different types of initial mass functions
22	Initial conditions $m_i, \mathbf{r}_i, \dot{\mathbf{r}}_i$ from prepared data
23	Removal of distant escapers (isolated or tidal)
26	Slow-down of KS and/or chain regularization
27	Tidal circularization (sequential or continuous)
28	Magnetic braking and gravitational radiation
30	Chain regularization (with special diagnostics)
32	Increase of output interval (limited by $t_{\rm cr}$)
34	Roche lobe overflow (only *NBODY4* so far)
35	Integration time offset (cf. section 15.9)

from the saved state. Since the integration variables represent $48N$ double-precision words (cf. section 3.2), this part is of size $96N$ words. In addition, the recommended neighbour list allocation is of size $2(N_{\rm s} + N_{\rm b})^{1/2}$, with some extra locations for algorithmic complications.

The corresponding common variables for the KS scheme are given in Table 7.7. In addition to the 18 basic entries there are another ten arrays pertaining to the Stumpff method. This makes a total of 110 words for each KS solution, compared with 94 words for the difference formulation. As in direct integration, we use the convention 'force' to denote acceleration. Note that standard variables such as $t_0, \Delta t$ and L are also used for KS solutions, with the convention that the first component, m_k, is referred to; i.e. $k = 2I_{\rm p} - 1$ for pair index $I_{\rm p}$. This scheme was introduced before primordial binaries became relevant, but in any case disc space is not usually an issue for direct N-body simulations.

A modern code employs a large number of unique variable names to

Table 7.5. Basic variables.

$\mathbf{x}_0$	X0	Primary coordinates
$\mathbf{v}_0$	X0DOT	Primary velocity
$\mathbf{x}$	X	Coordinates for predictions
$\mathbf{v}$	XDOT	Velocity for predictions
$\mathbf{F}$	F	One half the total force (per unit mass)
$\mathbf{F}^{(1)}$	FDOT	One sixth the total force derivative
m	BODY	Particle mass (also initial mass m_0)
Δt	STEP	Irregular time-step
t_0	T0	Time of last irregular force calculation
$\mathbf{F}_\mathrm{I}$	FI	Irregular force
$\mathbf{D}_\mathrm{I}^1$	FIDOT	First derivative of irregular force
$\mathbf{D}_\mathrm{I}^2$	D2	Second derivative of irregular force
$\mathbf{D}_\mathrm{I}^3$	D3	Third derivative of irregular force
ΔT	STEPR	Regular time-step
T_0	T0R	Time of last regular force calculation
$\mathbf{F}_\mathrm{R}$	FR	Regular force
$\mathbf{D}_\mathrm{R}^1$	FRDOT	First derivative of regular force
$\mathbf{D}_\mathrm{R}^2$	D2R	Second derivative of regular force
$\mathbf{D}_\mathrm{R}^3$	D3R	Third derivative of regular force
R_s	RS	Neighbour sphere radius
L	LIST	Neighbour and perturber list

Table 7.6. FORTRAN parameters.

N_max	Total particle number and c.m. bodies	4010
K_max	KS solutions	1010
L_max	Neighbour lists	100
M_max	Hierarchical binaries	10
M_dis	Recently disrupted KS components	22
M_reg	Recently regularized KS components	22
M_high	High-velocity particles	10
M_cloud	Interstellar clouds	10
N_chain	Chain membership	10

define useful quantities, most of which are scalars. Accordingly, the choice is between local and global allocation. The strategy of the present design is to keep all essential variables for direct and KS integration in one common block, whereas the more complicated chain regularization is described by temporary common variables. Since most chain interactions are of short duration, this usage enables a calculation to be restarted from the

Table 7.7. KS regularization variables.

$\mathbf{U_0}$	U0	Primary regularized coordinates
$\mathbf{U}$	U	Regularized coordinates for predictions
$\mathbf{U}'$	UDOT	Regularized velocity
$\mathbf{F}_{\mathrm{U}}$	FU	One half the regularized force
$\mathbf{F}'_{\mathrm{U}}$	FUDOT	One sixth the regularized force derivative
$\mathbf{F}_{\mathrm{U}}^{(2)}$	FUDOT2	Second derivative of regularized force
$\mathbf{F}_{\mathrm{U}}^{(3)}$	FUDOT3	Third derivative of regularized force
h	H	Binding energy per unit reduced mass
h'	HDOT	First derivative of the specific binding energy
$h^{(2)}$	HDOT2	Second derivative of the binding energy
$h^{(3)}$	HDOT3	Third derivative of the binding energy
$h^{(4)}$	HDOT4	Fourth derivative of the binding energy
$\Delta\tau$	DTAU	Regularized time-step
$t^{(2)}$	TDOT2	Second regularized derivative of physical time
$t^{(3)}$	TDOT3	Third regularized derivative of physical time
R	R	Two-body separation
R_0	R0	Initial value of the two-body separation
γ	GAMMA	Relative perturbation

latest common save by continuing the high-order integration cycle as if no interruption had occurred, so that the end result should be identical to a continuous calculation. This is usually achieved even in the large codes and it is highly desirable that a given outcome be reproducible, at least over short intervals in order to deal with any technical problems. Provisions are also included for changing a variety of input variables and options at restart time. Such a procedure makes it possible to study the effect of varying some condition at a given stage in the evolution, or simply extend the simulation beyond the prescribed termination time.

7.7 Data structure

Elegant code design assists greatly in delineating the complex paths that are an inevitable feature of any such large undertaking. To this end, the data structure itself plays a vital role. Thus we need to allow for a changing situation in which new KS solutions appear or are terminated and escaping particles are removed during the calculation. These requirements make it desirable to abandon a rigid data structure in favour of a flexible scheme. There are essentially two ways to overcome this problem. We can either adopt the style of C-programming and introduce pointers, or

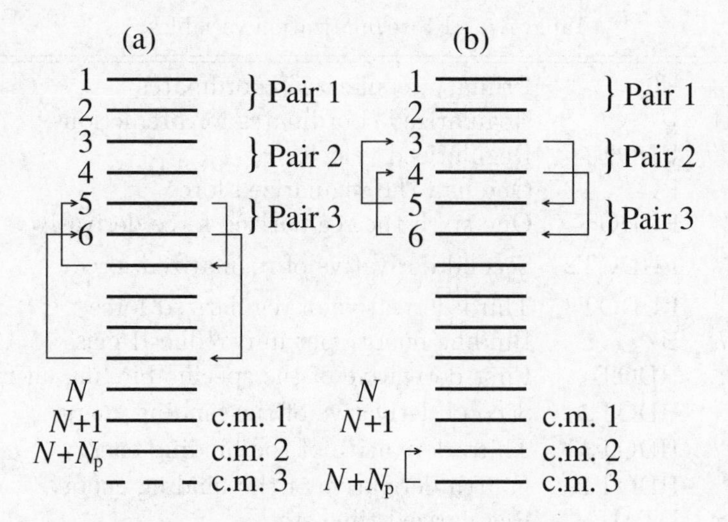

Fig. 7.3. Ordering of KS pairs. The figure shows the data structure for (a) creation of new pair, and (b) KS termination.

modify the relevant particle arrays according to the changing situation.[§] Apart from questions of hardware and efficiency, a FORTRAN program is easier to follow if the arrays can be maintained in an ordered fashion. However, the latter strategy does involve extra cost as well as programming complexities. Fortunately the cost of the relabelling scheme is relatively modest in practice since the number of new KS and other regularized solutions is usually small even with primordial binaries.

The present data organization is based on updating the particle arrays to give a sequential representation [cf. Aarseth, 1985a]. This is achieved by re-ordering all the relevant variables at each new change of configuration. For each KS pair, $I_{\rm p}$, the corresponding c.m. location is situated at $N + I_{\rm p}$. Hence given the particle number, N, and number of pairs, $N_{\rm p}$, the first single particle is located at $I_{\rm s} = 2N_{\rm p} + 1$ and the last c.m. at $N + N_{\rm p}$.

Figure 7.3 illustrates the general case of increasing or reducing the number of regularized solutions with other pairs present. At a given time, all the KS components occupy the $2N_{\rm p}$ first sequential locations, followed by the single particles. In Fig. 7.3(a), the particle pair k, l has been selected for regularization. Consequently, all particle arrays at locations $2N_{\rm p} + 1$ and $2N_{\rm p} + 2$ are exchanged with those of the regularization candidates and the corresponding c.m. is introduced at $N + N_{\rm p} + 1$. At an appropriate stage during the initialization, the current pair index, $N_{\rm p}$, is increased by one. In the reverse case of terminating an existing KS solution, as in Fig 7.3(b), all more recent pairs are moved up to liberate

[§] An alternative tree-type data structure is discussed in Appendix F.

the locations $2N_{\rm p} - 1$ and $2N_{\rm p}$ for the terminated components at $2I_{\rm p} - 1$ and $2I_{\rm p}$, which now become the first single particles, whereupon the corresponding c.m. is removed by compression.

The data structure enables a sequential force summation to be carried out from $I_{\rm s}$ to $N_{\rm tot} = N + N_{\rm p}$, provided due care is exercised for $i > N$. In this case, if $r_{ij}^2 > \lambda^2 R^2$, the force is obtained by including the mass-point contribution, otherwise the two components are first resolved. The c.m. approximation parameter is related to the unperturbed two-body limit for consistency and given by $\lambda = \gamma_{\rm min}^{-1/3}$.

The advantage of having a well-ordered scheme is evident. Given that the cost of a new initialization is $O(N)$, the required switching of arrays does not add significantly since the effort of obtaining the force polynomials themselves dominates on a conventional computer. However, there are additional costs of updating all the internal neighbour and perturber lists due to the changed locations of a few particles. Fortunately this effort can be speeded up in many ways by taking advantage of the sequential ordering, and in any case these operations are fast because only integer arithmetic is involved. Hence the cost of all KS-related initializations in a large simulation is essentially negligible since the total number of such episodes is relatively small even when the number of binaries is large.

In the case of escaper removal, the common arrays of bound cluster members are compressed. Now there is no polynomial initialization but again all relevant lists must be made consistent with the new sequence. Since both the particle number and relevant locations change with time, it is convenient to define a unique label or so-called 'name', $\mathcal{N}_i$, at the beginning. Thus we adopt $\mathcal{N}_i = i$ for all initial binary components and single particles and $\mathcal{N}_j = \mathcal{N}_{2k-1} + N_0$ for the c.m. of each KS pair, k, where N_0 is the initial particle number. This is particularly useful for identification of binary components that may be involved in exchange or collisions. Further details of reconfiguration and escape procedures will be given in later sections, together with a discussion of the data structure for hierarchies.

8
Initial setup

8.1 Introduction

A variety of procedures need to be carried out before the calculation proper can begin. The prescriptions for input parameters and options are discussed in chapter 7. Here we concentrate on different types of initial conditions for star cluster simulations, whereas planetary systems are described elsewhere. The cluster models are first generated for single stars with a specified initial mass function (hereafter IMF) and scaled to internal units. Since a variety of distributions may be considered, we provide several detailed algorithms. Next we present some procedures for including a realistic distribution of primordial binaries. Modelling of star clusters also requires external effects to be added. We distinguish between the motion of open clusters in circular orbits and globular clusters in 3D, with the galactic tidal force truncating the outer parts. Interstellar clouds form another perturbing agent which may be taken into account. Finally, with these procedures completed, the force polynomials for direct solutions as well as for any dominant two-body motions can be initialized.

8.2 Initial conditions for clusters

Although the choice of starting configurations for star cluster simulations is extremely wide, we may be guided by certain principles and observational constraints. Such models are usually represented by a smooth IMF and centrally concentrated density distribution. Depending on the objectives, the velocities may represent approximate equilibrium or initial collapse, whereas cosmological models are characterized by expansion.

The subject of the IMF has a long and chequered history. On the computational side, it has been known for a long time [Wielen, 1967; Aarseth, 1973] that a general mass spectrum increases the evolution rate significantly. The question of the upper and lower mass limits is still a matter for debate. However, the consensus is emerging that the low-mass stars do not form a significant proportion of the total mass [Kroupa, 2001a]. As far as small clusters are concerned, a few massive stars may play an important role due to mass loss on short time-scales. Going to larger clusters ($N \simeq 10^4$), this effect is probably less and we may consider a power-law distribution where the maximum stellar mass is a stochastic variable.

Guided by analysis of observational data, the simplest choice for open clusters is a classical Salpeter-type IMF, given by the power law

$$f(m) \propto m^{-\alpha} \tag{8.1}$$

for a specified mass range $[m_1, m_N]$. The exponent is traditionally taken as $\alpha = 2.3$ [Salpeter, 1955]. The corresponding distribution for each member i is readily obtained by the expression

$$m_i^{-(\alpha-1)} = m_1^{-(\alpha-1)} - (i-1)g_N\,, \tag{8.2}$$

with

$$g_N = (m_1^{-(\alpha-1)} - m_N^{-(\alpha-1)})/(N-1)\,. \tag{8.3}$$

A more realistic alternative is provided by the mass generating function [Kroupa, Tout & Gilmore, 1993]

$$m(X) = 0.08 + \frac{\gamma_1 X^{\gamma_2} + \gamma_3 X^{\gamma_4}}{(1-X)^{0.58}}\,, \tag{8.4}$$

with the random number X uniform in $[0,1]$. The best-fit coefficients for the solar neighbourhood are given by $\gamma_1 = 0.19, \gamma_2 = 1.55, \gamma_3 = 0.05, \gamma_4 = 0.6$. If desired, values outside a specified range may be rejected, with $m_N > 0.08$ as the lower limit. A more flexible form of the fitting function (8.4) has also been developed to allow for any five-part power-law in specified mass intervals [Kroupa, 2001a,b]. As an example, a smaller effective power-law index is indicated below $0.5\,M_\odot$; i.e. $\alpha = 0.3$ for the mass range 0.01–$0.08\,M_\odot$ and $\alpha = 1.3$ for 0.08–$0.5\,M_\odot$. Consequently, the mass fraction in brown dwarfs below $0.08\,M_\odot$ would be less than 6% with a corresponding membership fraction of 37%. Instead of using $\alpha = 2.3$ for all masses above $0.5\,M_\odot$, there is some evidence for $\alpha \simeq 2.5-2.7$ above $1\,M_\odot$. Note that in the case of the standard IMF given by (8.2) it is convenient to specify the mean mass (in solar units) as an input parameter, whereas this quantity is redetermined from the actual distribution when using the relation (8.4).

The initial density distribution is usually taken to be fairly spherical with some degree of central concentration. Traditionally, the Plummer model [Plummer, 1911] has served this purpose well. We therefore begin this exercise by giving the complete algorithm before turning to more versatile models [Aarseth, Hénon & Wielen, 1974].

The space density of the Plummer model is given by

$$\rho(r) = \frac{3M}{4\pi r_0^3} \frac{1}{[1 + (r/r_0)^2]^{5/2}}, \tag{8.5}$$

where r_0 is a scale factor related to the half-mass radius by $r_h \simeq 1.3r_0$ [Aarseth & Fall, 1980]. In the following we adopt the scaling $M = 1$, $r_0 = 1$, which gives the mass inside a sphere of radius r as

$$M(r) = r^3(1 + r^2)^{-3/2}. \tag{8.6}$$

First a radius, r, is chosen by setting $M(r) = X_1$, where X_1 is a random number in [0, 1]. Substituting into (8.6) and simplifying we obtain

$$r = (X_1^{-2/3} - 1)^{-1/2}. \tag{8.7}$$

A rejection may be applied in rare cases of large distances (e.g. $r > 10r_h$). The three spatial coordinates x, y, z are now selected by choosing two normalized random numbers, X_2, X_3, and writing

$$\begin{aligned} z &= (1 - 2X_2)r, \\ x &= (r^2 - z^2)^{1/2} \cos 2\pi X_3, \\ y &= (r^2 - z^2)^{1/2} \sin 2\pi X_3. \end{aligned} \tag{8.8}$$

Let us assume isotropic velocities. From the corresponding potential $\Phi = -(1 + r^2)^{-1/2}$ in scaled units, the escape velocity is given by $v_e = 2^{1/2}(1 + r^2)^{-1/4}$. Since the system is assumed to be in a steady state, we have $f(\mathbf{r}, \mathbf{v}) \propto (-E)^{7/2}$ for the distribution function [Binney & Tremaine, 1987], where E is the specific energy. Hence the probability distribution of the dimensionless velocity ratio $q = v/v_e$ is proportional to

$$g(q) = q^2(1 - q^2)^{7/2}. \tag{8.9}$$

To obtain the velocities we use von Neumann's rejection technique and note that $g(q) < 0.1$ with q in [0, 1]. Let X_4, X_5 be two normalized random numbers. If $0.1X_5 < g(X_4)$ we take $q = X_4$; otherwise a new pair of random numbers is chosen. The isotropic velocity components v_x, v_y, v_z are obtained by employing the principle of (8.8) above, using two further random numbers, X_6, X_7. If a general mass function is chosen, the individual masses can also be assigned sequentially since the system has some discreteness. This introduces further statistical fluctuations in the density distribution but, as can be verified, departures from overall equilibrium during the first crossing time are small if $Q_{\rm vir} = 0.5$.

Next we turn to the more general case of King [1966] models with varying central density contrast. The basic models are characterized by isotropic velocities, subject to a local truncation at the escape velocity. However, the King–Michie [1963] models include a velocity anisotropy and are more appropriate for realistic cluster simulations. The distribution function can be written in the convenient form [Binney & Tremaine, 1987]

$$f(E, \mathbf{J}) = \frac{\rho_1}{(2\pi\sigma^2)^{3/2}} \exp\left(-\frac{J^2}{2r_a^2\sigma^2}\right)\left[\exp\left(-\frac{E}{\sigma^2}\right) - 1\right], \quad (8.10)$$

where $\mathbf{J}$ here is the specific angular momentum, r_a is the anisotropy radius and ρ_1 is the normalization of the density obtained by integrating over velocities with dispersion σ. Hence in the limit $r_\mathrm{a} \to \infty$ this expression reduces to the standard King model. Moreover, from the separable form, it can be seen that at large distances orbits of high angular momentum are deficient, thereby giving rise to velocity anisotropy for isolated systems. More than 30 years ago, preferentially eccentric orbits were also observed outside the half-mass radius of the Pleiades [Jones, 1970].

Until recently, anisotropic velocity distributions have been little used in simulations although detailed models with rotation for the orbit–averaged Fokker–Planck method have been presented [Einsel & Spurzem, 1999]. An early study employed King–Michie models to analyse 111 radial velocities in the globular cluster M3 [Gunn & Griffin, 1979]. Evidence for velocity anisotropy and mass segregation was obtained by a consistent iteration technique. A more general formulation introduced the additional factor $\exp(-\gamma J_z)$ in (8.10) to describe solid-body rotation in the core [Lupton & Gunn, 1987]. In this treatment, E, J^2 and J_z are assumed to be integrals of the motion but, because of the non-spherical potential, J^2 is only an approximate integral. This work also contains various procedures for constructing initial conditions.

More detailed algorithms for generating King–Michie models are now available [Spurzem & Einsel, private communication, 2002]. These procedures are based on generalized two-integral models with rotation that yield density distributions $\rho(r)$ or $\rho(r, z)$ of a continuous system by Runge–Kutta integration. A Monte Carlo rejection technique is then used to create the desired N-body realization. The first simulations of axisymmetric N-body models based on this approach have been reported [Boily, 2000; Boily & Spurzem, 2000].

A sequence of isotropic models can be constructed for different concentration parameters, $W_0 = |\Phi_0|/\sigma_0^2$, where Φ_0 is the central potential and σ_0 the corresponding velocity dispersion. Typical values range from $W_0 = 3$ for extended systems to $W_0 = 9$ for highly concentrated systems with small core radii. The King-type models are more dynamically relevant than the Plummer model which has a relatively large core when used with a tidal cutoff. Detailed algorithms for the generation of initial conditions including tidal truncation are available [Heggie & Ramamani, 1995] and will therefore not be repeated here. These models are approximately self-consistent, with the Galaxy represented by either a point mass or an extended disc potential which are suitable for globular cluster and open cluster simulations, respectively (to be described below). A number of investigations have employed the tidally truncated multi-mass King models which are convenient for systematic studies [cf. Aarseth & Heggie, 1998; Portegies Zwart *et al.*, 1999, 2001; Hurley *et al.*, 2001a].

8.3 Primordial binaries

Already the earliest N-body simulations of point-mass interactions with $N \leq 25$ [von Hoerner, 1960, 1963; van Albada, 1968] demonstrated that binaries are an important feature of self-gravitating systems. Later studies confirmed that this process is also effective for systems with $N = 500$ [Aarseth, 1972a]. At this time, few binaries were known in open clusters and essentially none in globulars. Although this situation was consistent with theoretical expectations of small formation rates, some early N-body and Monte Carlo simulations included the effect of primordial binaries [Aarseth, 1975, 1980; Spitzer & Mathieu, 1980]. In spite of detection difficulties [Trimble, 1980], evidence in favour of a significant distribution accumulated [Mathieu, 1983; van Leeuwen, 1983; Pryor *et al.*, 1989], and close binaries are now observed by the Hubble Space Telescope (HST) at the very centres of globulars* [Grindlay, 1996]. These observational developments gave rise to more systematic explorations of this fascinating subject [McMillan, Hut & Makino, 1990, 1991; Heggie & Aarseth, 1992).

After these historical remarks we turn to the problem at hand, namely the initialization of primordial binaries. This procedure will be illustrated by choosing a flat distribution in $\log(a)$, where a is the semi-major axis [Kroupa, 1995a]. Such a distribution is consistent with the classical result for low-mass stars in the solar neighbourhood [Duquennoy & Mayor, 1991]. Let us assume that the minimum and maximum periods have been specified, either on astrophysical or dynamical grounds. Next we must address the delicate question of the mass ratio. There is currently much uncertainty about this aspect even if we limit our attention to open star clusters. For the present purpose it suffices to combine component masses obtained separately from an unbiased IMF, although the possibility of some correlation for massive binaries may be justified observationally [Eggleton, private communication, 1997].

* For an early review of binaries in globular clusters see Hut *et al.* [1992].

The following algorithmic steps are now carried out for an initial population of N_b primordial binaries. This explicit notation is equivalent to the conventional definition of the binary fraction, $f_\mathrm{b} = N_\mathrm{b}/(N_\mathrm{b} + N_\mathrm{s})$, with N_s being the number of single stars, and is actually used by the codes. Accordingly, we define a provisional particle number, $N = N_\mathrm{s} + N_\mathrm{b}$, which is specified as an input parameter. After generating the $2N_\mathrm{b}$ component masses independently, we form the sum of pair-wise masses in decreasing order and record the individual values separately. If $N_\mathrm{s} > 0$, the single bodies are also generated from the IMF and ordered sequentially, such that the most massive single body is at location $2N_\mathrm{b} + 1$. From the initial coordinates and velocities of these $N_\mathrm{b} + N_\mathrm{s}$ objects we now perform the scaling in the usual way, which produces a system with total energy $E_0 = -0.25$ (modified slightly by any external potential) satisfying the desired virial ratio. Because of the sampling involved in generating the IMF by (8.4), the maximum binary mass usually only exceeds the heaviest single mass by a modest factor for $f_\mathrm{b} \simeq 0.5$, even though the average binary mass is twice that of the singles. Hence a relatively large membership is needed in order for the upper mass limit to be approached.

At this stage, the component masses are introduced by splitting the relevant centre-of-mass (c.m.) body according to the initialization. Let us define the range ψ (say $\psi = 1000$) in semi-major axis, starting from an upper limit a_0. Here a_0 is an input parameter, already scaled by the length unit, R_V. The semi-major axis for pair index i is chosen by

$$a_i = a_0/10^\beta \,, \tag{8.11}$$

with $\beta = X_i \log \psi$ and X_i a normalized random number. Corresponding eccentricities are assigned from a thermal distribution [Jeans, 1929] by $e_i = Y_i^2$, with Y_i another random number, or given a constant value, as required. Using classical expressions [Brouwer & Clemence, 1961, p. 35], we form two-body elements by randomizing the perihelion, node and inclination angles appropriately for each binary. For computational convenience, the initial separation is taken as the apocentre, $R_i = a_i(1+e_i)$. Combining the c.m. coordinates, $\mathbf{r}_\mathrm{cm}$, with the relative separation, $\mathbf{R}_i = \mathbf{r}_{2i-1} - \mathbf{r}_{2i}$, we obtain the global coordinates

$$\begin{aligned} \mathbf{r}_{2i-1} &= \mathbf{r}_\mathrm{cm} + m_{2i}\mathbf{R}_i/(m_{2i-1} + m_{2i}) \,, \\ \mathbf{r}_{2i} &= \mathbf{r}_\mathrm{cm} - m_{2i-1}\mathbf{R}_i/(m_{2i-1} + m_{2i}) \,, \end{aligned} \tag{8.12}$$

and similarly for the velocities. Upon completion of the binary initialization, the total particle number is assigned the final value $N = N_\mathrm{s} + 2N_\mathrm{b}$.

Procedures for so-called 'eigenevolution' [Kroupa, 1995b] have also been included as an alternative to the above. The basic idea is to modify the two-body elements for small pericentre distances, R_p, due to pre-main-sequence evolution when stellar radii are larger. Although this scheme should be considered experimental, the essential steps are given below. First the period is assigned from a distribution with specified minimum period, $P_{\rm min}$ [cf. eq. 11b of Kroupa, 1995b], whereas the eccentricity is again sampled from a thermal distribution. Given the period and mass, the pericentre distance is derived via the semi-major axis. A pre-main-sequence radius of $5\,(m/M_\odot)^{1/2}$ is used to reflect the earlier contraction stage. Using tidal circularization theory [Mardling & Aarseth, 2001], we obtain the modified eccentricity $\tilde{e}_i$ during a characteristic time interval of 10^5 yr. Angular momentum conservation then yields the new semi-major axis by $a_i = R_{\rm p}/(1 - \tilde{e}_i)$. Finally, any case of overlapping enlarged radii is defined as collision and rejected. Depending on the minimum period, only a small fraction of the binaries are affected, with typically a dozen circularized orbits for $P_{\rm min} = 5\,{\rm d}$ and $N_{\rm b} = 2000$.

One useful feature of the data structure is that the primordial binary components have neighbouring labels, $\mathcal{N}_i, \mathcal{N}_j$, referred to as 'names' in the codes, which are helpful for distinguishing original and exchanged binaries. In this connection, we remark that in subsequent data analysis the definition of a primordial binary does not necessarily imply that $|\mathcal{N}_i - \mathcal{N}_j| = 1$, since exchanged descendants still possess a certain binding energy and should therefore also be considered as primordial. On the other hand, new binaries form from single stars even if some of these may originally have been members of disrupted primordials. We now have a cluster of $N_{\rm b}$ binaries with total internal binding energy

$$E_{\rm b} = -\tfrac{1}{2}\sum_{i=1}^{N_{\rm b}} m_{2i-1}m_{2i}/a_i \,. \tag{8.13}$$

Finally, any binary satisfying the close encounter condition $R_i < R_{\rm cl}$ is initialized as a KS solution before the calculation starts. This explicit initialization provides a convenient dataset for subsequent comparison.

Given the preponderance of binaries and even multiple systems in a variety of clusters, it is natural to extend the algorithm outlined above to include hierarchical configurations. We concentrate on the case of primordial triples which can be readily generalized to higher-order systems. The basic idea is to split the mass of the primary into an inner binary, with suitably chosen two-body parameters. Using the outer pericentre and eccentricity, we test the stability (to be defined in the next chapter) before accepting the new initial conditions, otherwise a new inner semi-major axis is generated. Since the first sequential binaries are selected for this procedure, the combined mass will also tend to be slightly larger than for the standard binaries, but other mass distributions may be chosen if desired. As in the case of primordial binaries, the total particle number is increased accordingly. Moreover, the outer components are placed sequentially at $j > 2N_b$ after all the inner binaries have been assigned. This arrangement therefore results in a well-defined data structure that facilitates KS initialization and membership identification. A popular alternative data structure is discussed in Appendix F.

8.4 Open clusters and clouds

A cluster orbiting in the Galaxy is subject to an external tidal force that affects the motions in the outer parts and, indeed, often determines the spatial extent. We consider in turn three different perturbing components, i.e. circular motion near the Galactic plane, an irregular force due to passing interstellar clouds and a constant external potential. A discussion of the latter is left for subsequent sections.

We begin with the simple case of a circular orbit in the solar neighbourhood and adopt a right-handed coordinate system whose origin rotates about the Galaxy with constant angular velocity, Ω_z, and the x-axis pointing away from the centre. Since typical cluster radii are small compared with the size of the Galaxy, it is sufficient to expand the perturbing potential to first order. The well-known equations of motion for the coordinates x_i, y_i, z_i take the form [Aarseth, 1967; Hayli, 1967]

$$\begin{aligned} \ddot{x}_i &= F_x + 4A(A-B)x_i + 2\Omega_z \dot{y}_i \,, \\ \ddot{y}_i &= F_y - 2\Omega_z \dot{x}_i \,, \\ \ddot{z}_i &= F_z + \frac{\partial K_z}{\partial z}\, z_i \,. \end{aligned} \tag{8.14}$$

The first terms represent the sum over all the internal interactions, A and B are Oort's constants of Galactic rotation and $\partial K_z/\partial z$ is the local force gradient normal to the plane of symmetry.

The equations (8.14) induce a tidal torque since K_z is directed towards the plane of symmetry, leading to flattening as well as velocity anisotropy. Taking $\dot{\mathbf{r}}_i \cdot \ddot{\mathbf{r}}_i$ and summing over the masses, we readily obtain the Jacobi energy integral

$$E_{\rm J} = T + U - 2A(A-B)\sum m_i x_i^2 - \frac{1}{2}\frac{\partial K_z}{\partial z}\sum m_i z_i^2 \,, \tag{8.15}$$

since the Coriolis terms cancel. An equivalent constant of motion also applies to individual stars in the absence of encounters and significant mass motions or mass loss.

The concept of tidal radius [von Hoerner, 1957; King, 1962] has played an important role towards the understanding of cluster dynamics. For a star at rest on the x-axis, the central attraction of the cluster is balanced by the tidal force at a distance

$$R_{\rm t} = \left[\frac{GM}{4A(A-B)}\right]^{1/3}, \tag{8.16}$$

where all quantities are now expressed in physical units. According to classical perceptions, the energy per unit mass for a star to escape from an idealized cluster model is given by [Wielen, 1972]

$$C_{\rm crit} = -\tfrac{3}{2}[4G^2M^2A(A-B)]^{1/3}. \tag{8.17}$$

Although this energy limit is less stringent compared with an isolated system, a star satisfying this condition may still take a long time to reach one of the Lagrange points, L_1 or L_2, and some excess is usually required to ensure escape. We shall see later that there exist some orbits that return to the cluster after moving well outside the tidal radius [Ross, Mennim & Heggie, 1997].

The tidal parameters are scaled to N-body units in the following way. First we choose the Oort's constants $A = 14.4$, $B = -12.0\,{\rm km\,s^{-1}\,kpc^{-1}}$ [Binney & Tremaine, 1987]. After converting A and B to ${\rm cm\,s^{-1}\,pc^{-1}}$ and comparing dimensions, star cluster units of solar mass and pc are introduced into the x-component of the tidal coefficient by the scaling

$$\tilde{T}_1 = 4\tilde{A}(\tilde{A}-\tilde{B})\beta\,, \tag{8.18}$$

with $\beta = L^*/GM_\odot$ and L^* in pc. Hence the final scaling to the appropriate length unit and total mass gives

$$T_1 = \tilde{T}_1(R_{\rm V}^3/NM_{\rm S})\,, \tag{8.19}$$

where $\tilde{T}_1 \simeq 0.3537$.

For the z-component, denoted by $\tilde{T}_3$, we have according to the theory of stellar dynamics [Oort, 1965]

$$T_3^* = -4\pi G\rho_0 - 2(A-B)(A+B)\,, \tag{8.20}$$

where $\rho_0 = 0.11\,M_\odot\,{\rm pc}^{-3}$ is the local density [Kuijken & Gilmore, 1989]. Converting to the above units and transforming to $\tilde{T}_3$ as above, we obtain

$$T_3 = \tilde{T}_3(R_{\rm V}^3/NM_{\rm S})\,, \tag{8.21}$$

with $\tilde{T}_3 \simeq -1.4118$. Note that the vertical force gradient is a factor of 4 greater than the x-component for the assumed Galactic model.

The final conversion concerns the angular velocity, given by $\Omega_z = A - B$. Let us define

$$\tilde{T}_4 = 2(\tilde{A}-\tilde{B})\beta^{1/2} \tag{8.22}$$

as *twice* the actual value for computational convenience, with $\tilde{T}_4 \simeq 0.8052$. Accordingly, the final scaled value becomes

$$T_4 = \tilde{T}_4 (R_V^3 / N M_S)^{1/2} \,. \tag{8.23}$$

The corresponding Galactic rotation period is readily obtained using (7.5). For completeness, the scaled tidal radius takes the form

$$r_t = (\tilde{M}/T_1)^{1/3} \,, \tag{8.24}$$

with $\tilde{M}$ denoting the total mass in N-body units.† We remark that one may study tidal effects for circular orbits at different locations in the Galaxy by using appropriate differential rotation constants and vertical force gradients derived from models and/or observation. Moreover, since the coefficient T_3 is approximately constant for $z \leq 0.5\,\mathrm{kpc}$, the present treatment is also valid for open clusters at moderate heights above the Galactic plane, i.e. tidal shocks are not important here.

As far as the AC integration scheme is concerned, the additional force terms in (8.14) must be assigned to the appropriate component of (3.1). Since the tidal contributions arising from T_1 and T_3 change smoothly with position, it is natural to combine them with the regular force. However, the Coriolis terms vary with the velocity and are therefore added to the neighbour force. Note that expressions for the corresponding derivatives are readily obtained for Hermite integration.

The perturbation by interstellar clouds may also be modelled by N-body simulations [Bouvier & Janin, 1970; Terlevich, 1983, 1987]. Consider a spherical region of size R_b surrounding an open cluster and containing a number of individual clouds. These clouds exert an irregular tidal force on the cluster members. By ignoring the contributions from clouds outside the boundary, it is shown below that only the external *fluctuating* force component is neglected. The perturbing effect of each cloud, represented as a polytrope of index 5 (or Plummer model), is included in the equations of motion and provides a regular force

$$\mathbf{F}_c = -GM_c \left[\frac{\mathbf{r}_i - \mathbf{r}_c}{(|\mathbf{r}_i - \mathbf{r}_c|^2 + \epsilon_c^2)^{3/2}} - \frac{\mathbf{r}_i - \mathbf{r}_d}{R_b^3} \right] . \tag{8.25}$$

Here $M_c, \mathbf{r}_c, \epsilon_c$ is the cloud mass, vectorial position and softening length, and $\mathbf{r}_d$ is the density (or symmetry) centre of the cluster. The repulsive force is added to cancel a spurious net inward attraction that would otherwise reduce the escape rate; i.e. the local fast-moving clouds contribute an average density distribution and resulting retardation at large

† In the following, M by itself will be in N-body units unless stated otherwise.

distances. Again this treatment is suitable for implementation in the Hermite codes since taking the time derivative of (8.25) is straightforward.

The clouds are integrated in the rotating frame, with the cluster attraction omitted from their equations of motion (8.14). When the boundary is crossed, a new cloud is introduced at random position angles with isotropic velocities [cf. Terlevich, 1987]. The resulting small force discontinuity is alleviated by employing a mass smoothing function near $R_{\rm b}$. Although the actual force from a given cloud may be large, the differential effect is relatively small due to the high velocity dispersion. Even so, the regular time-steps are reduced by a significant factor because the higher derivatives are affected by the cloud velocities and masses.

The employment of a relatively small boundary radius can be justified by the following considerations. Let us write the equation of motion in an inertial coordinate system as

$$\ddot{\mathbf{r}}_i = \mathbf{F}_i + \mathbf{A}_{\rm out} + \mathbf{A}_{\rm in}\,, \tag{8.26}$$

where $\mathbf{A}_{\rm out}$ and $\mathbf{A}_{\rm in}$ represent all additional accelerations outside and inside the cloud boundary, respectively, and the first term is due to the cluster members. We now assume that $\mathbf{A}_{\rm out} = \bar{\mathbf{A}}_{\rm out}$; i.e. we neglect irregular effects *outside* the boundary. Adding and subtracting $\bar{\mathbf{A}}_{\rm in}$ then results in

$$\ddot{\mathbf{r}}_i = \mathbf{F}_i + \bar{\mathbf{A}} + \mathbf{A}_{\rm in} - \bar{\mathbf{A}}_{\rm in}\,, \tag{8.27}$$

which is readily converted to rotating coordinates. The combined term $\mathbf{A}_{\rm in} - \bar{\mathbf{A}}_{\rm in}$ refers to the clouds and $\bar{\mathbf{A}}$ is the usual total galactic tidal acceleration. Hence this exercise also provides the justification for subtracting the average cloud force in the basic equation (8.25).

The effect of interstellar clouds on the internal dynamics of star clusters was examined in a classical paper by Spitzer [1958], who based his conclusions of reduced life-times on somewhat higher cloud densities and masses than employed in the numerical simulations. From the impulse approximation, the energy change of the cluster due to a cloud with impact parameter $p > r_h$, relative velocity V and mass M_2 is given by

$$\Delta E = \frac{4G^2 M_2^2 M \bar{r^2}}{3b^4 V^2}\,, \tag{8.28}$$

where $\bar{r^2}$ is the mean square radius of the cluster members with mass M and b the impact parameter. This compares with $\Delta E = G^2 M^3/3V^2 a^2$ for the case of a head-on encounter between two equal-mass Plummer models with scale factor a [Binney & Tremaine, 1987].

An independent derivation, based on the so-called 'extended impulse approximation', gave rise to a convergent expression for close encounters between two Plummer spheres [Theuns, 1992b]. Here the magnitude of the velocity change in the orbital plane, $\Delta v \simeq 2M_2/a^2 v$, replaces the singular expression $\Delta v \simeq 2M_2/b^2 v$, in agreement with the head-on encounter result. It was also pointed out that the relative energy gain from the cluster heating does not scale in a simple manner with the parameters defining the encounter.

Numerical simulations [Aguilar & White, 1985] show good agreement with theory for $b \geq 5 \max\{r_{\rm h}, \epsilon_{\rm c}\}$. The heating is a small effect and the actual energy change will be less because the cluster size is usually smaller than typical impact parameters. Finally, we remark that, given the field star density, there may be several hundred such stars inside the tidal radius. Their disrupting effect on open clusters is likely to be small due to the high velocity dispersion but a quantitative investigation is lacking.

8.5 Eccentric planar orbits

Let us now turn to modelling general motions of bound subsystems in a galactic environment. Some examples are globular clusters and dwarf spheroidal galaxies, both of which have received considerable attention [Grillmair, 1998; Oh, Lin & Aarseth, 1995]. We begin by deriving equations of motion suitable for the divided difference formulation which are easier to implement and restrict the treatment to motion in the plane. The next section deals with the case of 3D motion, using a formulation that is also suitable for Hermite integration.

Consider a system of self-gravitating particles orbiting the Galaxy at a distance R_0. We define right-handed coordinates centred on the cluster, with the x-axis pointing towards the Galactic centre. The acceleration in the inertial reference frame of a local test particle at $\mathbf{r} = \mathbf{r}_{\rm R}$ is given by

$$\ddot{\mathbf{r}}_{\rm I} = \ddot{\mathbf{r}}_{\rm R} + 2\boldsymbol{\Omega} \times \dot{\mathbf{r}}_{\rm R} + \boldsymbol{\Omega} \times (\boldsymbol{\Omega} \times \mathbf{r}_{\rm R}) + \dot{\boldsymbol{\Omega}} \times \mathbf{r}_{\rm R}\,, \tag{8.29}$$

where $\boldsymbol{\Omega}$ is the instantaneous angular velocity and $\mathbf{r}_{\rm R}$ refers to rotating coordinates. This equation contains the well-known Coriolis terms as well as the centrifugal and Euler terms.

First we discuss the case of a point-mass potential. In the inertial frame, the equation of motion for a cluster member is

$$\ddot{\mathbf{r}}_{\rm I} = \boldsymbol{\Sigma} - \frac{M_{\rm g}}{R^3}\mathbf{R} + \frac{M_{\rm g}}{R_0^3}\mathbf{R}_0\,, \tag{8.30}$$

with $\boldsymbol{\Sigma}$ the internal attraction and the last two terms representing the tidal acceleration due to the mass $M_{\rm g}$ located at $\mathbf{R} = \mathbf{R_0} + \mathbf{r}$.

In the following we consider motion in the x–y plane with angular velocity Ω_z. The position and angular velocity of the Galactic centre are then given in terms of the unit vectors $\hat{\mathbf{x}}, \hat{\mathbf{y}}, \hat{\mathbf{z}}$ such that

$$\mathbf{R}_0 = -R_0\hat{\mathbf{x}}\,,$$
$$\boldsymbol{\Omega} = (\mathbf{R}_0 \times \dot{\mathbf{R}}_0)/R_0^2 = \Omega_z\hat{\mathbf{z}}\,. \tag{8.31}$$

This allows the centrifugal term to be expressed as

$$\boldsymbol{\Omega} \times (\boldsymbol{\Omega} \times \mathbf{r}) = -\Omega_z^2 \dot{x}\hat{\mathbf{x}} - \Omega_z^2 \dot{y}\hat{\mathbf{y}}\,. \tag{8.32}$$

Similarly, the angular momentum per unit mass is

$$\mathbf{J} = \mathbf{R}_0 \times \dot{\mathbf{R}}_0 = R_0^2\Omega_z\hat{\mathbf{z}} = J_z\hat{\mathbf{z}}\,. \tag{8.33}$$

Differentiation of (8.31) combined with the above relation results in

$$\dot{\boldsymbol{\Omega}} = -\frac{2\dot{R}_0(\mathbf{R}_0 \times \dot{\mathbf{R}}_0)}{R_0^3} = -\frac{2\dot{R}_0\Omega_z}{R_0}\hat{\mathbf{z}}\,. \tag{8.34}$$

Substituting for Ω_z from (8.33) gives rise to the Euler term

$$\dot{\boldsymbol{\Omega}} \times \mathbf{r} = -\frac{2\dot{R}_0 J_z}{R_0^3}(x\hat{\mathbf{y}} - y\hat{\mathbf{x}})\,. \tag{8.35}$$

The equation of motion in rotating coordinates can now be written down by inverting (8.29) and making use of (8.30) and the definition of $\mathbf{R}_0$. This yields the explicit expressions

$$\ddot{x} = \Sigma_x + \frac{M_{\rm g}}{R_0^3}\left[R_0 - (R_0 + x)\left(\frac{R_0}{R}\right)^3\right] - \frac{2\dot{R}_0 J_z}{R_0^3}y + \Omega_z^2 x + 2\Omega_z\dot{y}\,,$$
$$\ddot{y} = \Sigma_y - \frac{M_{\rm g}}{R^3}y + \frac{2\dot{R}_0 J_z}{R_0^3}x + \Omega_z^2 y - 2\Omega_z\dot{x}\,,$$
$$\ddot{z} = \Sigma_z - \frac{M_{\rm g}}{R^3}z\,. \tag{8.36}$$

If $r \ll R$, which is frequently the case, the evaluation of the tidal force terms may be speeded up by expanding $(R_0/R)^{-3}$ to first order. Taking account of the vectorial relation $R^2 = R_0^2 + 2\mathbf{R}_0 \cdot \mathbf{r} + r^2$, we have

$$R^{-3} = R_0^{-3}\left[1 - \frac{3\mathbf{R}_0 \cdot \mathbf{r}}{R_0^2} + O(\frac{r^2}{R_0^2})\right]\,, \tag{8.37}$$

which can be used to simplify the equations of motion above. The full equations were used to study the tidal evolution of globular clusters in eccentric orbits and also the tidal disruption of dwarf spheroidal galaxies

[Oh, Lin & Aarseth, 1992, 1995]. As can be readily verified, these equations reduce to the form (8.14) with reversed sign of $\ddot{x}$ for circular motion if the point-mass nature of the model is taken into account.

The energy of the cluster centre associated with the rotating frame is

$$E = \tfrac{1}{2}\dot{\mathbf{R}}_0^2 + \tfrac{1}{2}J_z^2/R_0^2 - M_{\rm g}/R_0 \,. \tag{8.38}$$

Since this quantity is conserved, we derive the equation of motion by differentiation. Cancelling $\dot{\mathbf{R}}_0$ from the common scalar products yields

$$\ddot{\mathbf{R}}_0 = \left(-\frac{M_{\rm g}}{R_0^3} + \frac{J_z^2}{R_0^4} \right) \mathbf{R}_0 \,. \tag{8.39}$$

The case of a logarithmic potential, $\Phi = V_0^2 \ln(R/a_0)$ represents another useful galaxy model that has many applications. Since the kinematical terms are the same as before, we need consider only the equation of motion in the inertial frame. The alternative form of (8.30) is then

$$\ddot{\mathbf{r}}_{\rm I} = \mathbf{\Sigma} - \frac{V_0^2}{R^2}\mathbf{R} + \frac{V_0^2}{R_0^2}\mathbf{R}_0 \,. \tag{8.40}$$

Hence the factors $M_{\rm g}/R_0^3$, M_g/R^3, and $(R/R_0)^3$ in (8.36) are replaced by V_0^2/R_0^2, V_0^2/R^2 and $(R/R_0)^2$, respectively. The corresponding energy per unit mass for the cluster centre is now given by

$$E = \tfrac{1}{2}\dot{\mathbf{R}}_0^2 + \tfrac{1}{2}J_z^2/R_0^2 + V_0^2 \ln(R_0/a_0) + C \,, \tag{8.41}$$

where V_0 and a_0 denote velocity and radius scales and C is a constant. From $\dot{E} = 0$ and the procedure above we obtain the equation of motion

$$\ddot{\mathbf{R}}_0 = \left(-\frac{V_0^2}{R_0^2} + \frac{J_z^2}{R_0^4} \right) \mathbf{R}_0 \,. \tag{8.42}$$

All these equations are in exact form and therefore suitable for general use. Moreover, the additional equations for the cluster centre itself, (8.39) or (8.42), are simple and can be integrated to high accuracy. Since the third derivative of $\mathbf{R}_0$ involves only the velocity $\dot{\mathbf{R}}_0$ additionally, the Hermite scheme would be appropriate. Hence the values of Ω_z and $\dot{R}_0 = \mathbf{R}_0 \cdot \dot{\mathbf{R}}_0 / R_0$ which appear in the equations of motion can be updated frequently. Finally, note that these formulations are suitable only for the divided difference method (i.e. *NBODY*5) since the force derivatives needed for Hermite integration are rather cumbersome.

8.6 Motion in 3D

The case of 3D globular cluster orbits requires special care with a Hermite method. Thus, in order for such a scheme to work, we need to obtain explicit force derivatives of all the terms. For this purpose we have a choice of two strategies, both of which exploit the computational advantage of a local reference system when the internal dimensions are small compared with the size of the orbit. By generalizing the procedure of section 8.5 to 3D, the number of terms in the equation of motion is larger and their form less amenable to differentiation. In the alternative and preferable formulation, we take a non-rotating, but accelerating, coordinate system also with the origin at the cluster centre, with axes having fixed directions in space.

The construction of realistic galactic potentials requires a significant disc component as well as a halo. Hence we consider a superposition of different characteristic contributions, with each part having a 3D representation. Let us first introduce a spherical potential due to a point-mass $M_{\rm g}$, adapted for Hermite integration. After linearizing the tidal terms in (8.30) using (8.37), the equation of motion in non-rotating coordinates for a mass-point at $\mathbf{r}$ takes the form

$$\ddot{\mathbf{r}} = \mathbf{\Sigma} - \frac{M_{\rm g}}{R_0^3}\left(\mathbf{r} - \frac{3\mathbf{r}\cdot\mathbf{R}_0}{R_0^2}\mathbf{R}_0\right). \tag{8.43}$$

By differentiation the corresponding force derivative becomes

$$\frac{d^3\mathbf{r}}{dt^3} = \dot{\mathbf{\Sigma}} - \frac{M_{\rm g}}{R_0^3}\left(\dot{\mathbf{r}} - \frac{3\dot{\mathbf{r}}\cdot\mathbf{R}_0}{R_0^2}\mathbf{R}_0 - \frac{3\mathbf{r}\cdot\dot{\mathbf{R}}_0}{R_0^2}\mathbf{R}_0 - \frac{3\mathbf{R}_0\cdot\dot{\mathbf{R}}_0}{R_0^2}\mathbf{r}\right) + \frac{3M_{\rm g}}{R_0^5}\left[\dot{\mathbf{R}}_0 - \frac{5(\mathbf{R}_0\cdot\dot{\mathbf{R}}_0)}{R_0^2}\mathbf{R}_0\right](\mathbf{r}\cdot\mathbf{R}_0). \tag{8.44}$$

In the absence of encounters, (8.43) yields a conserved Jacobi integral,

$$E_J = \tfrac{1}{2}\dot{\mathbf{r}}^2 - \mathbf{\Omega}\cdot(\mathbf{r}\times\dot{\mathbf{r}}) + \tfrac{1}{2}\Omega^2 r^2 - \tfrac{3}{2}\Omega^2\left(\frac{\mathbf{r}\cdot\mathbf{R}_0}{R_0}\right)^2 + \Phi_{\rm c}\,, \tag{8.45}$$

with the angular velocity $\Omega^2 = M_{\rm g}/R^3$ and $\Phi_{\rm c}$ the cluster potential at r [Heggie, 2001].

From the definition of the total energy or virial theorem, the external tidal field contribution is given by

$$W = \frac{M_{\rm g}}{2R_0^3}\sum_i m_i\left(r^2 - 3\frac{(\mathbf{r}\cdot\mathbf{R}_0)^2}{R_0^2}\right). \tag{8.46}$$

For the special case of circular cluster motion, the Jacobi energy in the rotating frame given by (8.15) is conserved. This quantity can also be expressed in the non-rotating frame as

$$E = T + U + W - \mathbf{\Omega}\cdot\mathbf{J}\,, \tag{8.47}$$

where **J** is the standard angular momentum. For some purposes, it may be desirable to construct initial conditions without rotation in a rotating frame. This is achieved by adding the quantity

$$\Delta \dot{\mathbf{r}} = -\frac{1}{R_0^2}\left[(\mathbf{r}\cdot\dot{\mathbf{R}}_0)\mathbf{R}_0 - (\mathbf{r}\cdot\mathbf{R}_0)\dot{\mathbf{R}}_0\right] \tag{8.48}$$

to the velocity of each star. Finally, the integration of the cluster centre can be carried out according to the prescription of the previous section. As before, the equation of motion (8.39) can be integrated accurately using explicit derivatives of higher order.

At this stage we introduce a disc potential which breaks the spherical symmetry. Given the requirement of an analytical force derivative, the choice is somewhat limited. In the following we derive the relevant expressions for the Miyamoto–Nagai [1975] potential that in cylindrical coordinates R, z takes the form

$$\Phi = -\frac{M_{\rm d}}{\{R^2 + [a + (b^2 + z^2)^{1/2}]^2\}^{1/2}} \,. \tag{8.49}$$

This expression can represent the full range, from infinitely thin discs to a spherical system by the choice of the coefficients a, b. Taking the gradient gives rise to the radial and vertical force components

$$\begin{aligned} \mathbf{F}_R &= -\frac{M_{\rm d}}{A^3}\mathbf{R}\,, \\ F_z &= -\frac{M_{\rm d} z[a + (b^2 + z^2)^{1/2}]}{A^3(b^2 + z^2)^{1/2}}\,, \end{aligned} \tag{8.50}$$

with $\mathbf{R} = \mathbf{R}_0 + x\hat{\mathbf{x}} + y\hat{\mathbf{y}}$

and $A = \{R^2 + [a + (b^2 + z^2)^{1/2}]^2\}^{1/2}\,. \quad (8.51)$

Differentiation with respect to time and collection of terms yield

$$\begin{aligned} \dot{\mathbf{F}}_R &= -\frac{M_{\rm d}}{A^3}\left\{\dot{\mathbf{R}} - \frac{3}{A^2}\left[R\dot{R} + \frac{a + (b^2 + z^2)^{1/2}}{(b^2 + z^2)} z\dot{z}\right]\mathbf{R}\right\}, \\ \dot{F}_z &= -\frac{3M_{\rm d}[a + (b^2 + z^2)^{1/2}]z}{A^5(b^2 + z^2)^{1/2}}\left[R\dot{R} + \frac{a + (b^2 + z^2)^{1/2}}{(b^2 + z^2)} z\dot{z}\right] \\ &\quad - \frac{M_d[ab^2 + (b^2 + z^2)^{3/2}]}{A^3(b^2 + z^2)^{3/2}\dot{z}}\,. \end{aligned} \tag{8.52}$$

The disc potential contains expressions which do not appear amenable to linearization, although analytical derivation by special software is feasible. Alternatively, we may obtain the smoothly varying differential effect by subtracting the corresponding values for the cluster centre. Note that the latter evaluations are in any case performed at frequent intervals for the purpose of integrating the cluster centre. In view of the many common factors in these expressions, the numerical effort may be considered acceptable even if carried out on the GRAPE-6 host. If necessary, several such components may be superimposed, albeit at extra cost.

As the final part of a realistic galaxy model, we include the logarithmic potential discussed earlier. An expansion to second order gives

$$\Phi = V_0^2 \left[\ln \left(\frac{R_0}{a_0} \right) + \frac{\mathbf{R}_0 \cdot \mathbf{r}}{R_0^2} + \tfrac{1}{2} \left(\frac{r^2}{R_0^2} - \frac{2(\mathbf{r} \cdot \mathbf{R}_0)^2}{R_0^4} \right) \right] . \quad (8.53)$$

In the inertial frame, the corresponding linearized tidal force and first derivative are given by

$$\begin{aligned} \mathbf{F}_{\rm L} &= -\frac{V_0^2}{R_0^2} \left(\mathbf{r} - \frac{2\mathbf{R}_0 \cdot \mathbf{r}}{R_0^2} \mathbf{R}_0 \right) , \\ \dot{\mathbf{F}}_{\rm L} &= -\frac{V_0^2}{R_0^2} \left(\dot{\mathbf{r}} - \frac{2\mathbf{R}_0 \cdot \mathbf{r}}{R_0^2} \dot{\mathbf{R}}_0 \right) + \frac{2V_0^2}{R_0^4} (\dot{\mathbf{R}}_0 \cdot \mathbf{r} + \mathbf{R}_0 \cdot \dot{\mathbf{r}}) \mathbf{R}_0 \\ &\quad + \frac{2V_0^2}{R_0^4} \left(\mathbf{r} - \frac{4\mathbf{R}_0 \cdot \mathbf{r}}{R_0^2} \mathbf{R}_0 \right) (\mathbf{R}_0 \cdot \dot{\mathbf{R}}_0) . \end{aligned} \quad (8.54)$$

Hence the total force is the sum of the contributions (8.43), (8.50) and (8.54), bearing in mind the use of cylindrical coordinates for the disc component. Alternatively, the halo potential may be included by direct evaluation as for the disc above, with acceptable loss of precision.‡

8.7 Standard polynomials

The last task before beginning the calculation is usually to initialize force polynomials and time-steps. These procedures are essentially identical in the divided difference and Hermite formulations, and we therefore concentrate on the latter in connection with the AC neighbour scheme (i.e. the code *NBODY6*). Note that the case of initializing a single force polynomial has already been described in chapter 2, where all the necessary expressions are given.

Only two main input parameters are needed to create the desired polynomials. This stage begins by forming neighbour lists for all the particles. For Plummer or King–Michie models, the basic neighbour radius S_0 is modified by the central distance according to $R_{\rm s}^2 = S_0^2(1 + r_i^2)$ in order to yield an acceptable membership at different radii. This relation compensates for the density profile but should not be used if subsystems are

‡ For moderately flattened logarithmic potentials see Binney & Tremaine [1987].

present. In any case, the radius R_s is increased if there are no neighbours and reduced if the membership exceeds a prescribed value, $n_{\max}$ (which should be a bit smaller than $L_{\max}$ for technical reasons).

Having determined sequential neighbour lists, $L_{j,i}$, with the membership $n_i = L_{1,i}$, we now evaluate the force and its derivative by explicit summation over all particles. If the summation index is identified in the neighbour list, the corresponding contributions are added to the irregular components $\mathbf{F}_{\rm I}$, $\mathbf{D}^1_{\rm I}$; otherwise they are included in the regular terms. After all the particles have been considered, any external contributions are added to the respective components as discussed above. Lastly, the total force and first derivative for each particle are formed by

$$\begin{aligned} \mathbf{F} &= \mathbf{F}_{\rm I} + \mathbf{F}_{\rm R} \,, \\ \dot{\mathbf{F}} &= \mathbf{D}^1_{\rm I} + \mathbf{D}^1_{\rm R} \,, \end{aligned} \tag{8.55}$$

where the index i has been suppressed and $\mathbf{D}$ now denotes a derivative.

Another full N summation is performed in order to obtain the second and third force derivatives. This boot-strapping procedure requires all $\mathbf{F}$ and $\dot{\mathbf{F}}$ to be known, as can be seen from the basic derivation in chapter 2. Again the neighbour list is used to distinguish the irregular and regular components, $\mathbf{D}^2_{\rm I}$, $\mathbf{D}^3_{\rm I}$ and $\mathbf{D}^2_{\rm R}$, $\mathbf{D}^3_{\rm R}$. Since distant particles do not contribute significantly to the regular derivatives, their effect may be neglected outside some distance (say $5R_s$). Higher derivatives of any external forces are added consistently according to type, using boot-strapping.

It now remains for new irregular and regular time-steps, Δt_i, ΔT_i, to be assigned from the definition (2.13), using the block-step rule, (2.25), if relevant. At the same time, the primary coordinates and velocity, $\mathbf{x}_0$, $\mathbf{v}_0$, are initialized and the factorials $\frac{1}{2}$ and $\frac{1}{6}$ are absorbed in $\mathbf{F}$ and $\dot{\mathbf{F}}$ for computational convenience. Finally, the quantity $t_{\rm next} = t_i + \Delta t_i$ is introduced for decision-making in the Hermite scheme.

We note that, strictly speaking, the evaluation of the two higher force derivatives can be omitted in the Hermite AC formulation, as is done in *NBODY*4, the only significant draw-back being the need to construct new time-steps from lower derivatives. In the latter case, we employ the conservative expression

$$\Delta t = \tfrac{1}{2}\, \eta \min\{|\mathbf{F}|/|\dot{\mathbf{F}}|, t_{\rm cr}|\} \,, \tag{8.56}$$

with $t_{\rm cr}$ the *initial* crossing time. However, the extra cost for the standard versions is very small, both initially and at subsequent initializations, and consistency with the difference formulation is retained. The initialization of the latter method contains one essential step more, i.e. conversion of all force derivatives to divided differences according to the formulae of section 2.2. On the other hand, the corresponding time-steps are not truncated to commensurate values.

8.8 Regularized polynomials

The case of primordial binaries discussed above also requires special attention initially. Thus it is elegant and numerically more accurate to begin a KS solution at $t = 0$, instead of delaying by up to half a period when the selection would otherwise be made. Since the primordial binary components are already ordered pair-wise, the regularized quantities can be initialized for each pair in turn satisfying the usual close encounter condition $r_{ij} < R_{\rm cl}$ without the time-step test; otherwise direct integration will be used. We shall now concentrate on a given pair with index i.

Initialization of regularized polynomials is also required at subsequent times as a result of close encounters. The following treatment combines both cases if it is assumed that the two selected particles, m_k, m_l, have already been placed in adjacent locations with pair index i and all arrays are updated, such that the situation is similar to the above.

Standard force polynomials are first formed for each component following the procedure outlined in the previous section. In a general situation, neighbour lists contain single particles as well as the c.m. locations corresponding to the KS components. For the irregular contributions the summation is over both components of other perturbed binaries, whereas the point-mass approximation can normally be used for most of the regular interactions. The two individual quantities $\mathbf{F}_{2i-1}$ and $\mathbf{F}_{2i}$ are combined to give the new c.m. force by

$$\mathbf{F}_{\rm cm} = (m_{2i-1}\mathbf{F}_{2i-1} + m_{2i}\mathbf{F}_{2i})/(m_{2i-1} + m_{2i}) \,, \tag{8.57}$$

and likewise for the derivative, $\dot{F}_{\rm cm}$. However, the more expensive second and third derivative are obtained by one summation, using the binary c.m. which is sufficient for the present purpose. The last part of the c.m. initialization proceeds as described in the previous section.

Next it is the turn of the KS polynomials to be constructed. The basic variables $\mathbf{U}$, $\mathbf{U}'$, R, h are initialized in the standard way of section 4.4, whereupon the perturber list is formed from the existing neighbour list if available (i.e. *NBODY*5 and *NBODY*6). A direct selection is carried out if N is small, whereas for *NBODY*4 the HARP and GRAPE provide the desired list of candidates. Perturbers m_j are then selected according to the tidal approximation by

$$r_{ij} < \left[\frac{2m_j}{(m_k + m_l)\gamma_{\rm min}}\right]^{1/3} R \,. \tag{8.58}$$

A more efficient criterion, $r_{ij}^2 > f(m)\lambda^2 R^2$ with $f(m) = 2m_1/m_{N+i}$ involving the maximum mass, m_1, is first used to exclude more distant particles. Recall that $\lambda = \gamma_{\rm min}^{-1/3}$ is a subsidiary c.m. distance parameter.

It is sometimes the case in the AC method that the neighbour list is not sufficiently large to ensure that enough perturbers can be selected when using the apocentre condition $\lambda a_i(1 + e_i) < R_s$.[§] However, this is not usually a problem for large N or hard binaries.

Considerable care is involved in forming the highest derivatives of the KS force polynomials, especially in Hermite formulations that employ additional differentiation (cf. section 4.6). The essential procedure consists of the following steps summarized in Algorithm 8.1.

Algorithm 8.1. *Initialization of KS polynomials.*

1	Predict $\mathbf{r}_j$, $\dot{\mathbf{r}}_j$ for all the perturbers to order $\dot{\mathbf{F}}_j$
2	Obtain $\mathbf{P}$ and $\dot{\mathbf{P}}$ by summation over all perturbers
3	Add any external effects on the relative motion $\mathbf{R}$, $\dot{\mathbf{R}}$
4	Introduce regularized derivative by $\mathbf{P}' = R\dot{\mathbf{P}}$
5	Define relative perturbation $\gamma = \vert\mathbf{P}\vert R^2/(m_{2i-1} + m_{2i})$
6	Scale $\mathbf{P}$ and $\mathbf{P}'$ by the slow-down factor if $\kappa > 1$
7	Construct $\mathbf{Q} = \mathcal{L}^{\mathrm{T}}\mathbf{P}$ and the expression for $\mathbf{F}_{\mathrm{U}}$
8	Form $\mathbf{F}'_{\mathrm{U}}$, $t^{(2)}$, $t^{(3)}$, h', $h^{(2)}$ by explicit differentiation
9	Include $\mathbf{Q} = \mathcal{L}^{\mathrm{T}}\mathbf{P}$ and $\mathbf{Q}'$ in $\mathbf{F}'_{\mathrm{U}}$, $h^{(2)}$, $h^{(3)}$, $t^{(4)}$
10	Save $f_0^{(2)} = \frac{1}{2}R\mathbf{Q}$, $f_0^{(3)} = \frac{1}{2}(R\mathbf{Q})'$, $h_0 = h$ for corrector
11	Derive two of the terms in $\mathbf{Q}^{(2)} = (\mathcal{L}^{\mathrm{T}}\mathbf{P})''$ using $\mathcal{L}(\mathbf{U}'')$
12	Evaluate $\mathbf{F}_{\mathrm{U}}^{(2)}$, $\mathbf{F}_{\mathrm{U}}^{(3)}$, $h^{(4)}$, $t^{(5)}\, t^{(6)}$ by boot-strapping
13	Improve $\mathbf{F}_{\mathrm{U}}^{(2)}$, $\mathbf{F}_{\mathrm{U}}^{(3)}$, $h^{(3)}$, $h^{(4)}$ to order $\mathbf{Q}^{(2)}$
14	Set regularized time-step $\Delta\tau$, including perturbation
15	Generate Stumpff coefficients $\tilde{c}_3$, $\tilde{c}_4$, $\tilde{c}_5$, $\tilde{c}_4(4z)$, $\tilde{c}_5(4z)$
16	Find Δt from $\Delta\tau$ to order $t^{(6)}$ with $\tilde{c}_4(4z)$, $\tilde{c}_5(4z)$
17	Increase Δt by slow-down factor κ if applicable
18	Absorb factorials $\frac{1}{2}$, $\frac{1}{6}$ in $\mathbf{F}_{\mathrm{U}}$ and $\mathbf{F}'_{\mathrm{U}}$ for fast prediction

This condensed algorithm is almost self-explanatory. Note that the derivatives $\mathbf{F}_{\mathrm{U}}^{(2)}$ and $\mathbf{F}_{\mathrm{U}}^{(3)}$ are required for high-order prediction in the Stumpff formulation, with the short-hand notation $\mathbf{F}_{\mathrm{U}} = \mathbf{U}''$. As for the slow-down factor κ, the procedure differs slightly from that in chain regularization and will be discussed subsequently. In any case, the nominal value $\kappa = 1$ is assigned for most perturbed initializations. We recall that the perturbing force is obtained from the expression $\mathbf{P} = \mathbf{F}_k - \mathbf{F}_l$ and $\mathbf{Q} = \mathcal{L}^{\mathrm{T}}\mathbf{P}$ as defined in section 4.7. Being of a tidal nature, this has the advantage that only relatively nearby members (specified in terms of

[§] Note that this expression does not include any mass dependence, whereas more distant massive perturbers are also copied from the neighbour list if relevant.

$\gamma_{\min}$) need to be selected. For the regularized time-step we still use the original criterion [Aarseth, 1972b]

$$\Delta\tau = \eta_U \left(\frac{1}{2\,|h|}\right)^{1/2} \frac{1}{(1+1000\,\gamma)^{1/3}}\,, \tag{8.59}$$

where η_U is an input parameter. Hence this gives $2\pi/\eta_U$ steps per unperturbed orbit, which is reduced by a factor of 10 for $\gamma \simeq 1$. In the case of nearly parabolic orbits, the factor $1/2|h|$ is replaced by $R/(m_k + m_l)$ which has the property of yielding a small relative change in the variables. If desired, an additional condition to ensure convergence of the Taylor series solution may be obtained by considering the predicted change of the orbital energy,

$$\Delta h = h'\Delta\tilde{\tau} + \tfrac{1}{2}h''\Delta\tilde{\tau}^2\,, \tag{8.60}$$

and specifying some tolerance for the relative change (e.g. $0.001|h|$ but taking care near zero). The smallest of the two values given by (8.59) and (8.60) may then be chosen for the regularized step. At present the second algorithm is only used for initializations, whereas subsequent values of $\Delta\tau$ are restricted to an increase by a factor of 1.2.

The theory of the Stumpff coefficients is given in section 4.7 [cf. Mikkola & Aarseth, 1998]. Suffice it to note that $\tilde{c}_4(4z)$, $\tilde{c}_5(4z)$ with argument $4z = -2h\Delta\tau^2$ are used to modify the last two terms in the Taylor series for the physical time interval, while the first three coefficients appear in the predictor and corrector. In comparing this method with the basic Hermite formulation, it can be seen that the Stumpff coefficients have values close to unity for properly chosen time-steps, $\Delta\tau$.

The algorithm above is also relevant for the standard Hermite KS method. Thus only steps 10 and 15 are not needed in this treatment, although the refinement of steps 13 and 16 (to order $t^{(6)}$) was not implemented for some time. A comparison of the two methods (without energy stabilization) can be found elsewhere [Mikkola & Aarseth 1998], together with the original derivation. In this connection we remark that in order to compensate for a larger recommended value of η_U for the Stumpff method (i.e. 0.2 vs 0.1), a slightly steeper perturbation dependence in (8.59) may be beneficial.

We conclude by remarking that even the earlier KS difference formulation [cf. Aarseth, 1985a] has many similarities with the procedures above. Thus the number of equations to be integrated are actually less but on balance the Hermite schemes are preferable, with the Stumpff method being the most efficient at the cost of some extra programming efforts. However, the original difference method may be useful for problems where it is impractical to calculate the time derivative of the perturbing force.

9
Decision-making

9.1 Introduction

N-body simulations involve a large number of decisions and the situation becomes even more complex when astrophysical processes are added. The guiding principle of efficient code design must be to provide a framework for decision-making that is sufficiently flexible to deal with a variety of special conditions at the appropriate time. Since the direct approach is based on a star-by-star treatment at frequent intervals, this prerequisite is usually satisfied. However, we need to ensure that the relevant tests are not performed unnecessarily. The development of suitable criteria for changing the integration method or identifying procedures to be carried out does in fact require a deep understanding of the interplay between many different modes of interactions. Hence building up the network for decision-making is a boot-strapping operation needing much patience and experience. The aim of a good scheme should be that this part of the calculation represents only a small proportion of the total effort.

This chapter discusses several distinct types of decisions necessary for a smooth performance. First we deal with the important task of selecting the next particle, or block of particles, to be advanced in time. The challenge is to devise an optimized strategy in order to reduce the overheads. Another aspect concerns close encounters, either between single particles or where one or more subsystems already consist of binaries. Such interactions may be studied by multiple regularization or procedures for hierarchical systems in the case of stable configurations. Once selected for special treatment, these systems may need to be terminated either due to external perturbations or internal evolution. The escape process is a characteristic feature of star cluster evolution and it is desirable to remove distant members in order to concentrate on the bound system. Other algorithms discussed below are concerned with mass loss from evolving stars as well as physical collisions. A well-tried scheme for error checking by relying on energy conservation is also described. Finally, it is worth emphasizing that all the above procedures are completely automatic. Ideally a large-scale simulation should proceed from start to finish without any external intervention.

9.2 Scheduling

The procedure of determining the next particle(s) for consideration is known as *scheduling*, which is a queueing problem. Although many solutions to this logistical challenge exist, it is often better to develop special procedures by exploiting certain features. In the following we distinguish between small and large N-values. The basic problem can be formulated as follows. Given a wide distribution of N time-steps, Δt_j, and the corresponding times, t_j, of the last force evaluation, the next particle to be advanced is selected by

$$i = \min_j \{t_j + \Delta t_j\}, \tag{9.1}$$

such that $t = t_i + \Delta t_i$ defines the new time. Moreover, in the case of quantized time-steps, a number of particles may satisfy this condition.

Let us begin by outlining a simple scheme as follows. Construct a list L containing all particles satisfying $t_j + \Delta t_j < t_{\rm L}$. Initially $t_{\rm L} = \Delta t_{\rm L}$, where $\Delta t_{\rm L}$ is a suitably small time interval that is modified to stabilize the list membership on $N^{1/2}$ at each updating of $t_{\rm L}$ and L. A redetermination of the list is made as soon as $t > t_{\rm L}$, followed by another search over all the members. The next particle to be advanced is then determined using (9.1) for the list L.

A more sophisticated algorithm based on the above was developed for the code *NBODY*2 [Aarseth, 2001b]. The list is now ordered by sequential sorting such that the next particle due to be advanced is simply given by the next member. A complication arises for repeated small steps in the interval $\Delta t_{\rm L}$. In order to ensure that the global time increases monotonically, we employ an insert procedure which maintains the sequential ordering. Thus the index of a particle satisfying $t + \Delta t_i < t_{\rm L}$, evaluated at the end of the cycle, is inserted at the appropriate sequential location. The interval $\Delta t_{\rm L}$ is stabilized on a membership chosen as a compromise between the cost of sorting the quantities $t_j + \Delta t_j$ and inserting a small number of particles in the sequential list. Thus a target membership of $2N^{1/2}$ is chosen initially but the value is modified according to the number of inserts. In spite of this complication, comparison with the so-called 'heap-sort algorithm' advocated by Press [1986] favoured the present algorithm above $N \simeq 100$, although the cost of both procedures is $\propto N \ln N$.

With Hermite integration, the block-step structure presents a new feature because a number of particles from several levels are usually advanced together. This situation is handled in the following way. Given the array $\{\tilde{t}_j \equiv t_j + \Delta t_j\}$ and the smallest value, $t_{\min}$, the particles in the next block are those for which $\{\tilde{t}_j\} = t_{\min}$. A list of members is maintained for the corresponding interval $\Delta t_{\rm L}$, with the membership stabilized on the square root of the effective population, $(N - N_{\rm p})^{1/2}$, where $N_{\rm p}$ is the number of KS pairs. Unless there are changes in the particle sequence, this simply requires the next value of $t_{\min}$ to be determined during the integration cycle itself. Hence the subsequent block-step is defined by $\min\{\tilde{t}_j\} - t$, where t is the current time. A full N search is also carried out a *second* time after each non-standard termination or initialization of special configurations, such as multiple regularizations, since this may have resulted in smaller time-steps prescribed at the current block time. Profiling tests show that this scheme is efficient at least up to $N \simeq 10^5$; i.e. the overheads arising from (9.1) only account for at most 1.5 % of total CPU.

Modern star cluster simulations often deal with a large number of primordial binaries [McMillan, Hut & Makino, 1990; Heggie & Aarseth, 1992]. If two-body regularization is used, it is necessary to have a separate scheduling algorithm because the corresponding physical time-steps are not quantized.* In a typical situation the majority of KS solutions are unperturbed, with look-up times of many periods which may still be small. Two different algorithms have been tried for determining the next KS solution to be advanced. An earlier version consisted of the following three-level approach. We begin by constructing a list of members for all KS pairs satisfying $t_i + \Delta t_i < t_{\rm b}$, where the epoch, $t_{\rm b}$, is updated by an interval, $\Delta t_{\rm b}$, stabilized on a membership of $N_{\rm p}/4$. At each new block-step, another list is formed by selecting those pairs for which $t_j + \Delta t_j \le t_{\rm block}$, where $t_{\rm block}$ defines the block boundary time for direct integration. This second list is then examined before each new KS step to determine the smallest look-up time. The first list is also initialized when there is a change in the sequence, but this occurs much less frequently than the renewals after the interval $\Delta t_{\rm b}$.

For relatively large values of $N_{\rm p}$, the sorting algorithm described above is a better alternative. Again the principle of sequential selection is maintained by the insert procedure. The sorted list is stabilized on a membership of $2N_{\rm p}^{1/2}$ as a compromise between renewals and list size. The conditions for reconstructing the full time-step list are the same as above, namely at the end of the interval $\Delta t_{\rm L}$ or after each change in the particle sequence. Most of the latter changes occur at the end of a block-step and

* This may be done in principle by truncating to the nearest quantized value but would not serve any purpose with the two separate treatments.

are much less frequent (by a factor of about 10^4) in large simulations. Since the relevant pointer is advanced by one at the beginning of each block-step, the current value must also be reduced by one on exit. Although the number of inserts may be comparable to the total number of KS steps due to some small values of Δt_i, the insert procedure is speeded up with an appropriate initial guess based on the remaining interval. Profiling of the relative costs shows that the second algorithm for KS selection is significantly faster for large memberships and this procedure has in fact been implemented in the Hermite codes.

Finally, for completeness, we also mention the independent and efficient block-step algorithm of McMillan [1986] that was employed in the tree code described in section 6.3 [cf. McMillan & Aarseth, 1993].

9.3 Close two-body encounters

The question of when to regularize a close encounter is based on considerations of numerical accuracy as well as computational efficiency. However, given that the number of such method changes is relatively small in a typical calculation, the decision-making must be aimed at gaining accuracy. In fact, introduction of the relative motion does not alter the cost per step significantly on conventional computers. Now only one particle is advanced by direct integration, albeit with some force summations over both components, whereas the cost of the regularized solution depends on the perturber number together with initialization and termination. Hence in the limit of few perturbers, the cost of integrating a binary essentially reduces to that of a single particle.† Since hyperbolic two-body encounters are relatively infrequent compared with the various types of bound interactions, the main consideration is to define dominant two-body motion. The benefit of switching to KS regularization is two-fold: accuracy is gained while reducing the number of steps per orbit.

Let us describe an encounter between two stars of mass m_k, m_l in terms of the deflection angle [Chandrasekhar, 1942]

$$\cos\delta = \left[1 + \frac{R^2 V_0^4}{G^2(m_k + m_l)^2}\right]^{-1/2}, \tag{9.2}$$

where R is the impact parameter and V_0 is the pre-encounter relative velocity. We now define a close encounter by a total deflection $2\delta = \pi/2$ and replace V_0 by the approximate rms velocity from the virial theorem, $(GN\bar{m}/2r_{\rm h})^{1/2}$. For equal masses, the close encounter distance is given by

† A hard binary has some N periods in a crossing time whereas each typical c.m. step, estimated by $\Delta t_{\rm cm} \simeq t_{\rm cr}/500$, requires N force summations.

$R_{\rm cl} = 4r_{\rm h}/N$ (cf. (1.9)). However, we are dealing with centrally concentrated clusters so it is more prudent to use the density-related expression

$$R_{\rm cl} = 4r_{\rm h}/[N(\rho_{\rm d}/\rho_{\rm h})^{1/3}]\,, \tag{9.3}$$

where $\rho_{\rm d}/\rho_{\rm h}$ is the central density contrast (to be discussed in chapter 15). With both integration schemes, the corresponding time-scale for a typical parabolic encounter is found empirically to be

$$\Delta t_{\rm cl} \simeq 0.04(\eta_I/0.02)^{1/2}(R_{\rm cl}^3/\bar{m})^{1/2}\,, \tag{9.4}$$

where $\bar{m}$ is the mean mass (i.e. $1/N$ in N-body units).

Having defined dynamically relevant parameters for close two-body encounters, we now turn to the actual implementation. If a particle k satisfies the condition $\Delta t_k < \Delta t_{\rm cl}$, a search is made for other nearby particles. For large N, this procedure uses either the neighbour list (*NBODY*5, *NBODY*6) or an existing list containing short steps (*NBODY*4); otherwise all particles need to be considered. Every particle inside a distance $2R_{\rm cl}$ is recorded[‡] and the closest single particle is denoted by index l. The pair k, l is tentatively accepted for regularization, provided the minimum distance condition $R < R_{\rm cl}$ is satisfied and, in the case of other close particles j, the two-body force is dominant which requires

$$(m_k + m_l)/R^2 > (m_l + m_j)/|\mathbf{r}_l - \mathbf{r}_j|^2\,. \tag{9.5}$$

This condition excludes the possibility of particle l being close to another regularized pair. It is also prudent to accept approaching particles only (i.e. $\dot{R} < 0$). Moreover, nearly circular orbits may be included by extending the radial velocity test to angles slightly less than 90°, using

$$\mathbf{R}\cdot(\dot{\mathbf{r}}_k - \dot{\mathbf{r}}_l) < 0.02[R(m_k + m_l)]^{1/2}\,. \tag{9.6}$$

This modification is especially useful in the difference formulation since the velocity $\dot{\mathbf{r}}_l$ is not predicted, in which case a factor 0.1 is used instead. Note that, with Hermite integration, synchronization of time-steps is required but the steps tend to be identical because a relative criterion is employed. If necessary, we achieve synchronization by halving the value of Δt_k, but only if $t_{0,l} + \Delta t_l < t + \Delta t_k$.

If massive binaries with low eccentricity are present, the time-step condition (9.4) may be too conservative for a close encounter search. Such binaries are often quite energetic and may contribute noticeably to systematic errors. A separate search procedure allowing somewhat wider criteria has therefore been included, i.e. $\Delta t_k < 4\Delta t_{\rm cl}$ and $m_k > 2\bar{m}$. The closest

[‡] The square search distance may be modified by $(m_k + m_l)/2\bar{m}$ for massive particles.

single particle is then identified from either the neighbour list (*NBODY6*) or a list of short time-steps (*GRAPE*). Provided we have $E_b < \epsilon_{hard}$, $e < 0.5$ and the estimated nearest neighbour perturbation is relatively small, the binary components are chosen for KS regularization. This procedure can be very beneficial for systems with significant mass dispersion.

Following acceptance of the regularization conditions, the initialization proceeds as described in section 8.8. The next question relating to decision-making is concerned with the possible termination of a regularization. We distinguish between hard and soft binaries, as well as hyperbolic fly-by's, since these cases require different considerations. The latter are simplest and for them the termination conditions $R > R_0$, $\dot{R} > 0$ may be employed, where R_0 is the initial separation (taken as $2a$ for hard binaries). An additional criterion based on the perturbation is also used to prolong the integration in case $R_0 < R_{cl}$ (see below).

Soft binaries require more care since high eccentricities are often involved. Hence termination during the outward motion with $R > R_{cl}$ might be followed by a new initialization after the apocentre passage even if the maximum perturbation is modest. Hence if such a binary is in a low-density region this switching might occur repeatedly unless some precaution is taken. Instead we introduce a secondary perturbation parameter, γ_{max}, (usually 0.01) and employ the termination conditions

$$R > R_0 \, , \quad \gamma > \gamma_{max} \, , \tag{9.7}$$

otherwise $\lambda R > R_s$ is used for the AC scheme.

Hard binaries often experience strong perturbations without suffering exchange. Thus if $\gamma > 0.2$ (say) a search for the dominant perturber is made based on the vectorial force. Termination with subsequent selection of the new component is only carried out if the negative radial velocity condition is also satisfied. In the rare case of no such test being successful before $\gamma = 0.5$ is reached, a direct comparison of dominant force terms among the perturbers usually results in a new companion being identified, whereupon termination is activated.[§]

Termination in the Hermite scheme has been modified slightly to fit in with the existing block-steps. Thus, except for collisions, it is advantageous to switch over to direct integration at the end of the current block-step instead of the need to create a smaller new step. Even if the termination conditions are satisfied, the regularized solution is extended until the corresponding physical time-step exceeds the remaining interval,

$$\Delta t_{2i-1} > t_{block} - t \, , \tag{9.8}$$

[§] The vectorial expression of the perturbation gives $\gamma = 0.5$ for an equilateral triangle and equal-mass bodies.

where i is the pair index. The final solution is then predicted at the time $t_{\rm block}$ by iteration of the regularized time (discussed in chapter 11). This novel procedure works well because the block-steps are usually small compared with the time-scale for significant changes in two-body configurations; i.e. the very nature of a strong perturbation implies a small value of the associated centre-of-mass (c.m.) step.

In conclusion, the identification of dominant two-body motion for KS regularization and its termination is essentially carried out with only two parameters, the close encounter time-step, $\Delta t_{\rm cl}$, and individual dimensionless perturbation, γ_i.

9.4 Multiple encounters

Compact subsystems where binaries are involved in strong interactions with single particles or other binaries may be treated by several types of multiple regularization described previously. Since many aspects of decision-making are similar, we concentrate on the more versatile chain regularization method. Let us first consider the case of a single particle of mass m_j approaching a hard binary with component masses m_k and m_l, semi-major axis a and eccentricity e. A convenient time to check for chain regularization is at each binary apocentre passage, provided $\Delta t_{\rm cm} < \Delta t_{\rm cl}$. Thus a small value of the c.m. time-step ensures a strong interaction. The following conditions for the intruder need to be satisfied before the configuration can be selected for treatment. In addition to a negative radial velocity, we employ the condition of a compact subsystem,

$$|\mathbf{r}_{\rm cm} - \mathbf{r}_j| < \max\{3R_{\rm grav}, R_{\rm cl}\}\,. \tag{9.9}$$

Here $R_{\rm grav}$ is the characteristic gravitational radius defined by

$$R_{\rm grav} = (m_k m_l + m_{\rm cm} m_j)/|E_{\rm b} + E_{\rm out}|\,, \tag{9.10}$$

with $E_{\rm b}$, $E_{\rm out}$ the binding energies of the inner and outer relative motion. If the subsystem is not well bound (i.e. $E_{\rm out} \gg 0$), we employ the expression $R_{\rm grav} = \frac{1}{2}(R + |\mathbf{r}_{\rm cm} - \mathbf{r}_j|)$ instead for decision-making purposes, with R the binary separation. In order to ensure a strong interaction, the condition for a small pericentre distance is also imposed by

$$a_{\rm out}(1 - e_{\rm out}) < |a|(1 + e)\,, \tag{9.11}$$

where $a_{\rm out}$ is the semi-major axis of the intruder with respect to the binary c.m. and $e_{\rm out}$ is the corresponding eccentricity. Moreover, a factor 2 may be included on the right-hand side for small apocentre distances (e.g. $a_{\rm out}(1 + e_{\rm out}) < 0.01R_{\rm cl}$), and even a factor 2.5 might be tried for large outer eccentricities¶ (e.g. $e_{\rm out} > 0.9$).

¶ Such systems still violate the hierarchical stability condition (9.14).

The condition (9.9) ensures that the transition from a perturbed KS system is not made until the interaction can be said to be strong, and chain regularization becomes a more appropriate solution method. For a parabolic intruder orbit and equal masses, $R_{\rm grav} = 6a$ and the perturbation at this distance would only be a few per cent, whereas the remaining time until maximum interaction barely exceeds one orbital period. However, it is more efficient to initiate the chain regularization with the existing KS pair, otherwise another dominant solution may be initialized before the next check. Because orbital phase is a random element here, the desirable change from a binary KS solution to chain regularization is usually achieved with the present criteria, but a small fraction of new hyperbolic KS motions do occur. Hence if a second approaching KS binary is terminated prematurely, both components are in fact subject to a distance test and may be selected, instead of the fourth particle being treated as a strong perturber.

Chain regularization may also be initiated with four members if two hard binaries approach each other closely. In this case the second binary is treated analogously with the single particle above and its internal energy added to $E_{\rm b}$ for use by (9.10). We generalize (9.11) by enlarging the apocentre cross section such that the semi-major axes are added. Once accepted for treatment, both KS solutions are terminated in the usual way. A discussion of aspects pertaining to the actual integration will be presented in another chapter.

A more general chain selection scheme has been added in order to prevent the intruder approaching too closely before being identified, as might be the case between two apocentre passages. Thus if the perturbation is significant (say $\gamma > 0.2$), the configuration is examined more carefully. Before deciding on termination of the KS solution, we consider the suitability for chain regularization. The actual decision for acceptance is then made using the standard conditions (9.9) and (9.11).

Decision-making for termination is much more involved than in KS regularization. A search for escape candidates is performed after each integration step if $\sum_k R_k > 3R_{\rm grav}$. Since this procedure requires velocity information, the basic variables are obtained after every integration step. The distances R_k are ordered and the index of the largest one with mass $m_{\rm esc}$ is noted. First we express the relevant distance and *radial* velocity in the local c.m. frame. Multiplication by the factor $M_{\rm sub}/(M_{\rm sub} - m_{\rm esc})$, with $M_{\rm sub}$ the subsystem mass, then yields the corresponding distance d and radial velocity $\dot{d}$ with respect to the remaining system, which for a chain membership $N_{\rm ch} = 3$ would be the binary c.m. This gives rise to the approximate two-body equation

$$E_{\rm d} = \tfrac{1}{2}\dot{d}^2 - M_{\rm sub}/d\,. \tag{9.12}$$

The simplest case is a single particle being ejected from a three-body system with hyperbolic motion (i.e. $\dot{d} > 0$, $E_{\rm d} > 0$). Such a particle is a candidate for escape provided $d > 3R_{\rm grav}$ [cf. Standish, 1971]. In addition, we employ a subsidiary distance condition which may continue the solution for small external perturbations up to a somewhat larger distance. If $E_{\rm d} < 0$, termination also occurs for an extended orbit with $d > 2R_{\rm cl}$, provided certain conditions are met. In practice it is beneficial to delay KS initialization for small pericentre separations. For this purpose the semi-major axis of the inner two-body motion is evaluated by inverting the expression $E_{\rm b} = -m_k m_l/2a$ in (9.10). Combining with (9.12) we write $E_{\rm out} = \mu_{\rm out} E_{\rm d}$, where $\mu_{\rm out} = m_{\rm esc} m_{\rm cm}/M_{\rm sub}$ and $m_j = m_{\rm esc}$ is the outer mass. Re-arranging terms, we finally obtain

$$a = \frac{R_{\rm grav}}{2 + 2\frac{m_j}{\mu}(1 + \frac{R_{\rm grav}}{M_{\rm sub}} E_{\rm d})}\,, \tag{9.13}$$

with $\mu = m_k m_l/m_{\rm cm}$. Consequently, we delay starting a new two-body solution for any value of (9.12) in the pericentre region $R < a$. The most distant single particle is treated similarly if $N_{\rm ch} > 3$, except for the semi-major axis condition above. Now the remaining subsystem is retained, whereupon initialization takes place as described previously and the ejected particle is prepared for direct integration in the usual way.

Configurations with $N_{\rm ch} > 3$ may also contain two particles escaping together, rather than just one. Provisional close components are identified by examining the smallest particle separation at either end of the chain. Again the procedure above is used, with (9.12) describing the relative motion of the combined c.m. Such particle pairs may be initialized as one KS solution if the separation is suitably small. The alternative case of $E_{\rm d} < 0$ and an extended orbit is also considered. Thus $N_{\rm ch} = 4$ and the potential binary exerting a small perturbation on the two other particles may result in termination with two separate KS solutions. Furthermore, two KS solutions may be initialized if the two smallest distances are well separated; e.g. $R_1 + R_3 < R_2/5$, where R_2 is the middle distance.

Change of membership by the addition of an intruder, followed by escape of another body may lead to long-lived systems that become time-consuming when treated by chain regularization. At the simplest level, stable triples form during strong binary–binary interactions and may persist over long intervals even in the presence of external perturbations. The possibility of more complex structures also needs to be considered. Thus the case $N_{\rm ch} = 4$ with an inner or outer binary of small size can be reduced to a hierarchical triple and likewise a five-body subsystem may be simplified. In the event of approximate stability being satisfied, the chain treatment is replaced by the appropriate combination of KS and direct solutions. Although semi-stable systems of higher orders are relatively rare during chain regularization, it is desirable to include such terminations in the decision-making. Fortunately, the available two-body separations together with the value of $N_{\rm ch}$ can be used to distinguish the different cases (cf. section 11.7).

Finally, we note that procedures exist for initiation of unperturbed treatments of three-body regularization [Aarseth & Zare, 1974] as well as four-body chain regularization [Mikkola & Aarseth, 1990]. The unperturbed global four-body regularization [Heggie, 1974; Mikkola, 1985a] was also used for some time [cf. Aarseth, 1985a] until replaced by the chain method for $N = 4$. Although the selection criteria are essentially similar to those above, the relevant procedures will be discussed later in connection with initialization and termination.

9.5 Hierarchical configurations

The formation and persistence of hierarchical systems poses difficult technical problems for N-body simulations. Such configurations are known to exist in the Galactic field, and it is therefore not surprising that they should appear in cluster models containing a significant binary population. The technical problem arises because the inner binary in a triple invariably has a short period that needs to be integrated as a perturbed KS solution over long time intervals, and likewise the outer orbit by the direct method. It has been known for a long time [cf. Harrington, 1972] that most observed triples appear to be quite stable as indicated by their age. According to perturbation theory, the inner semi-major axis does not exhibit any secular effects for sufficiently large ratios of the outer pericentre to the inner apocentre distance and this feature can be exploited in numerical integrations.

A number of systematic studies of triple systems have been made in order to determine the boundaries of stable configurations. Most of these investigations [cf. Harrington, 1975; Eggleton & Kiseleva, 1995] have specifically addressed the problem of comparable masses. However, only a restricted set of initial conditions can be explored because of the large parameter space; i.e. usually both the inner and outer orbits are taken to be circular. The importance of stable hierarchies in cluster simulations has been recognized for some time and such criteria were used extensively [cf. Aarseth, 1985a] until replaced by a more general expression.

A new approach, based on the binary–tides problem [Mardling, 1995], inspired a semi-analytical stability criterion that holds for a wide range of outer mass ratios and arbitrary outer eccentricities [Mardling & Aarseth, 1999]. The limiting outer pericentre distance, $R_{\rm p}^{\rm crit}$, is expressed in terms of the inner semi-major axis, a_{in}, by the relation

$$R_p^{crit} = C\left[(1+q_{out})\frac{(1+e_{out})}{(1-e_{out})^{1/2}}\right]^{2/5} a_{in}\,, \tag{9.14}$$

where $q_{out} = m_3/(m_1+m_2)$ is the outer mass ratio, e_{out} is the corresponding eccentricity and $C \simeq 2.8$ is determined empirically. This criterion applies to coplanar prograde orbits, which are the most unstable, and ignores a weak dependence on the inner eccentricity and mass ratio m_1/m_2.[‖] Since inclined systems tend to be more stable, we adopt an heuristic linear correction factor of up to 30%, in qualitative agreement with early work [Harrington, 1972] and recent unpublished experiments.

The criterion (9.14) ensures stability against escape of the outermost body. However, the alternative outcome of exchange with one of the inner components also needs to be considered. The latter case is treated by the semi-analytical criterion [Zare, 1976, 1977]

$$(\mathbf{J}^2E)_{crit} = -\frac{G^2 f^2(\rho)g(\rho)}{2(m_1+m_2+m_3)}\,, \tag{9.15}$$

where $\mathbf{J}$ is the total angular momentum, with $f(\rho)$ and $g(\rho)$ algebraic functions of the masses which can be solved by iteration. If $\mathbf{J}^2E < (\mathbf{J}^2E)_{crit}$ (with $E < 0$), no exchange can occur and the inner binary retains its identity. However, this criterion is only sufficient; hence exchange is not inevitable if the criterion is violated, but if it does occur, escape follows. Application of the expression (9.15) shows that for planar prograde motion, the escape boundary lies above the exchange limit for $q_{out} < 5$ [cf. Mardling & Aarseth, 2001], and this is mostly satisfied in star cluster simulations.

We now define a triple that satisfies the outer pericentre condition $a_{out}(1-e_{out}) > R_p^{crit}$ to be stable, in the sense that a_{in} will be assumed constant in spite of small short-term fluctuations. Such a system is then converted to a KS solution, where the combined mass m_1+m_2 plays the role of the inner component. Notwithstanding some current terminology,[**] we refer to this procedure as a *merger* but emphasize that it is only a temporary approximation and the two components are re-activated at the end. Quadruples and higher-order systems may also be treated in an analogous way by modifying the relation (9.14) appropriately, with a correction factor $f = 1 + 0.1a_2/a_{in}$ for the smallest binary of size a_2.

The actual identification of a stable hierarchy is again carried out at the apocentre phase, provided $\Delta t_{cm} < \Delta t_{cl}$. This may appear to be

‖ A complete criterion has now been derived from first principles [Mardling, 2003b].

** The identity of the two high-density cores is retained during the major phase of a long-lived galaxy merger, whereas two colliding stars soon become indistinguishable.

rather conservative but the additional condition $E_{\rm out} < \frac{1}{4}\epsilon_{\rm hard}$ is employed, mainly because a long life-time is still possible. Hence it follows that the time-step for the c.m. will be quite small for such configurations. We also note that even an unperturbed binary may be found inside a stable hierarchy. If we ignore the cube root of the mass ratio, this condition arises when we have

$$a_{\rm out}(1 - e_{\rm out}) > \lambda a_{\rm in}(1 + e_{\rm in}) \,, \tag{9.16}$$

where $\lambda = \gamma_{\rm min}^{-1/3}$. Hence a stability search is also performed during the check for unperturbed two-body motion, provided the time-step is below $\Delta t_{\rm cl}$. As an alternative to the more careful standard procedure, a fast evaluation of the stability parameter and inclination is carried out first if $\Delta t_{\rm cm} < 4\,\Delta t_{\rm cl}$, with existing information of the closest perturber. For a typical hard outer orbit, (9.16) allows for a large range of inner binary sizes and hence stable configurations of higher order are possible.

Occasionally, long-lived hierarchies occur that do not satisfy the stability criterion (9.14), even after the inclination effect has been included. A fraction of such systems exhibit relatively large outer eccentricity. In the case of high outer eccentricity and in analogy with tides in binaries, the amount of energy exchanged during a pericentre passage of the outer body is proportional to $(a_{\rm in}/R_{\rm p})^6$. Hence the small energy exchange at each passage owing to a small ratio would take a long time for the random walk process to reach $e_{\rm out} > 1$. In order to include such systems for a shorter interval, we introduce the experimental concept of *practical stability* (not yet included in NBODY6). Accordingly, if $e_{\rm out} > 0.99$, we assign the number of outer periods by [cf. Aarseth & Mardling, 2001]

$$n_{\rm out} = 1 + 10\,e_{\rm out}/(1 - e_{\rm out}) \,, \tag{9.17}$$

where the corresponding time interval is used for termination.

Termination of hierarchies is somewhat similar to the case of hard binaries treated by KS in one respect. Based on the predicted apocentre perturbation, $\gamma_{\rm apo} \simeq \gamma[a(1+e)/R]^3$, termination occurs if $\gamma_{\rm apo} > 0.25$ together with $R > a$. The stability condition (9.14) must be checked regularly. This is done conveniently at each *apocentre*. Thus we compare the unperturbed expression $a_{\rm out}(1 - e_{\rm out})$, slightly modified by the empirical factor $1 - 2\gamma$, with $R_{\rm p}^{\rm crit}$ which is now represented by R_0. The simple convention of associating each c.m. particle with a negative identification label, or name, (to be discussed later) allows the variable R_0 to be used for a second purpose. Recall that it is also used as termination distance for regularized soft binaries. Finally, the inner components are again reinitialized as a KS binary in the usual way, whereas the outer component is treated as a single particle or another KS solution, as required.

9.6 Escapers

Star clusters evolve by losing members to the galactic field. It therefore makes sense to remove distant cluster members and concentrate on the bound system. Although the classical tidal radius might be expected to be a good choice for the cluster boundary, simulations show that a significant population is present further out [cf. Terlevich, 1987]. Likewise, stars above the escape energy may linger for a long time when the orbits are calculated in a fixed potential [Fukushige & Heggie, 2000]. Theoretical considerations also suggest that some orbits may reach quite large distances and still return to the cluster [Ross, Mennim & Heggie, 1997]. However, such orbits appear to be of a special type and only a small fraction of escapers in the present models exhibit this behaviour when their motion is investigated. The old criterion of using twice the tidal radius for escaper removal has therefore been retained [Aarseth, 1973], in conformity with other studies [Portegies Zwart *et al.*, 2001].

As is well known [cf. Hayli, 1970; Wielen, 1972], the zero velocity boundary for the symmetrical Lagrange points, L_1 and L_2, is highly flattened in the presence of a galactic field. Likewise, if a particle has a small excess energy, $E_i > C_{\rm crit}$ (cf. (8.17)), it can only escape in the vicinity of the Lagrange points. Consequently, most particles satisfying the energy escape criterion will experience random reflections inside the zero velocity boundary before finding the energy-dependent exit window. Hence kinetic energy in the z-direction is converted to x- or y-motion, and vice versa, by the agency of the core as well as the non-symmetrical tidal torque. This behaviour is analogous to the absence of a third integral for eccentric orbits in certain galactic potentials [Aarseth, 1966b]. For completeness, we mention an extensive analysis of the escape process performed recently [Heggie, 2001].

The central distance of each single and c.m. particle, i, is examined at regular intervals. For open clusters we employ the escape criterion

$$|\mathbf{r}_i - \mathbf{r}_d| > 2\,r_{\rm t}\,, \tag{9.18}$$

where $\mathbf{r}_d$ defines the density centre (to be discussed later) and $r_{\rm t}$ is the scaled tidal radius given by (8.24). At the same time, the energy per unit mass, E_i, is obtained; this includes any external tidal field. Hence escapers from isolated systems may also be treated in a similar way, provided $E_i > 0$ and a nominal value (say $10\,r_{\rm h}$) has been assigned to $r_{\rm t}$. In any case, the quantity $m_i E_i$ is needed in order to correct the total energy, permitting conservation to be maintained.

A given particle is removed from the data structure by compressing all arrays with members $j \geq i$ and reducing N by one. Additionally, all relevant lists must be updated consistently, as well as the total mass and tidal radius. Since the distance between an escaper and other particles is usually fairly large, no force corrections are needed to account for the slight discontinuity. However, care *is* taken if a neighbouring single particle is present that does not satisfy the escape condition, when a delay may be imposed depending on the perturbing force.

In the case $i > N$ which defines a KS solution, we proceed as follows. First the c.m. particle is removed, using the procedure above. The corresponding KS components m_k, m_l of pair $i - N$ are also removed after noting the binding energy $E_{\rm b} = \mu_{kl} h$, with μ_{kl} the reduced mass. Again the latter quantity is subtracted from the total energy and all the list arrays are modified consistently. From time to time even hierarchical systems escape. The treatment is similar to that for binaries, except that additional quantities relating to the inner binary must also be removed. Detailed procedures for escaper removal are discussed in Appendix C.

9.7 Mass loss and tidal interactions

A realistic simulation of stellar systems needs to consider various processes connected with the finite size of stars as well as their change of mass with time. The modelling of synthetic stellar evolution is based on fast look-up algorithms for the radius, luminosity and type as a function of the initial mass and age [Eggleton, Tout & Fitchett, 1989; Tout *et al.*, 1997; Hurley, Pols & Tout, 2000]. Here we are mainly concerned with decision-making aspects for instantaneous mass loss due to stellar winds or supernovae explosions and processes related to tidal interactions.

Each star is assigned an evolutionary time-scale, $t_{\rm ev}$, when the relevant parameters are updated according to some prescription. As well as using the current mass, this scheme requires the initial mass, m_0, as a variable for several purposes. Suitably small time intervals are used for advancing $t_{\rm ev}$ which produces fairly smooth changes in the basic quantities. In the Hermite scheme, all individual values of $t_{\rm ev}$ are examined at small quantized intervals of 100–200 yr coinciding with the end of an integration cycle, or block-step. Each star is also characterized by a stellar type, k^*, in order to distinguish different evolutionary states in the Hertzsprung–Russell (HR) diagram, which facilitates decision-making. For many years a basic scheme was used [cf. Tout *et al.*, 1997] but this has now been replaced by a more general representation [Hurley, Pols & Tout, 2000] comprising some 16 types, ranging from low-mass main-sequence stars to neutron stars and black holes.

The introduction of evolutionary time-scales and types for c.m. particles as well provides greater flexibility of program control. Thus the former specifies the appropriate time for the next mass transfer episode in circularized binaries, whereas the latter is used to define the binary type.

Subsequent to a standard initial state ($k^* = 0$), binaries may undergo chaotic energy exchange ($k^* = -1$) or tidal circularization ($k^* = -2$) until the orbit is circular ($k^* = 10$), followed by Roche-lobe mass transfer ($k^* = 11$, etc.), of which there may be several distinct phases.

The process of tidal circularization has been described in detail elsewhere [cf. Mardling & Aarseth, 2001], together with chaotic motion, and some relevant algorithms will be given in a subsequent section. As far as decision-making is concerned, the orbital elements a, e are updated in a smooth manner either at each pericentre passage of a perturbed binary or, more usually, at the time of rechecking unperturbed motion. Hence, in general, the computational effort here is relatively insignificant, although the recent inclusion of stellar spin (*NBODY4* only) requires integration of some extra equations. Since the regularized coordinates for standard unperturbed binaries by convention specify the first point past apocentre, we need an algorithm for transforming the elements to pericentre. The phase angle is first advanced by half a *physical* period (i.e. $\pi/2$). Using general transformation expressions [Stiefel & Scheifele, 1971], we take $\theta = \pi/2$ and obtain the regularized coordinates and velocity

$$\mathbf{U} = \mathbf{U}_0 \cos\theta + \mathbf{U}' \sin\theta/\nu\,,$$
$$\mathbf{U}' = \mathbf{U}'_0 \cos\theta - \mathbf{U}_0 \sin\theta\,\nu\,, \tag{9.19}$$

where $\nu = |\frac{1}{2}h|^{1/2}$. Finally, if necessary, transformation to exact pericentre employs the analytical solution after solving Kepler's equation. However, an actual integration is performed if the motion is perturbed. Further algorithms are given in Appendix D.4.

An interval of circular motion usually precedes the Roche stage which is triggered by the increase of stellar radii. Since this type of mass transfer is essentially continuous, we need to discretize it in order to fit in with the integration scheme. The process is initiated when one of the stars fills its Roche lobe, with the effective radius [Eggleton, 1983]

$$r_{\rm R} \simeq 0.49q^{2/3}/[0.6q^{2/3} + \ln(1 + q^{1/3})]a\,, \tag{9.20}$$

where $q = m_k/m_l$ is the mass ratio, m_k being the primary. From the known mass transfer rate [cf. Tout *et al.*, 1997], we limit the amount transferred to a small fraction (say 0.5%) of the primary mass. The mass transfer activity is halted temporarily when the corresponding time exceeds the next c.m. force calculation time, whereupon there is a quiescent or coasting stage until $t_{\rm ev}$ is exceeded again. During this process it is important to make sure that the look-up times of the individual components stay ahead of the c.m. value; otherwise mass loss without any transfer would occur. Once an active stage is completed, the relevant type indices and evolution times are updated. More details can be found in section 11.10.

Close binaries consisting of low-mass or degenerate stars are subject to magnetic and gravitational wave braking [cf. Tout *et al.*, 1997]. Now the time-scale might be several Gyr, with correspondingly longer intervals between each orbital modification in the absence of mass loss. For convenience, the angular momentum losses due to magnetic braking and gravitational radiation at small separations ($< 10\,R_\odot$) are combined to yield the final semi-major axis by angular momentum conservation. Finally, the KS variables are scaled at constant eccentricity, followed by standard initialization in the case of perturbed motion (cf. (11.44)).

The actual procedures concerned with mass loss will be described later but it may be instructive here to consider how a supernova event is handled. Let us suppose that a massive component of a hard binary undergoes a sudden transition to a neutron star. According to consensus, such an object will generate a velocity kick because the rapid mass ejection is not symmetrical. Here we mention the main steps for implementing this effect, assuming that the kick occurs at an arbitrary phase (cf. section 15.5). After assigning a random phase shift θ in $[0, \pi]$, equation (9.19) is employed to generate new regularized variables. This is followed by termination of the KS solution, together with corrections of the neighbour forces and total energy resulting from the mass loss. Even if the modified two-body motion is hyperbolic, a new KS regularization is initialized for consistency, subject to the condition $R < R_{\rm cl}$.

9.8 Physical collisions

The interaction of close binaries undergoing mass transfer frequently leads to the stage known as common-envelope evolution [Paczynski, 1976]. A significant fraction of these events results in coalescence of the two stars, accompanied by mass loss and should not be considered as collisions. For practical purposes it is useful to distinguish between hyperbolic encounters and nearly circular motion leading to the common-envelope stage, even though the former may occasionally produce the same outcome.[††] Other types of collision also occur (i.e. induced inner eccentricity) and the possible formation of exotic objects is of particular interest.

The simplest case of collision is between two single stars in a hyperbolic regularized orbit. Although some giant stars are present, estimates indicate that such collisions are quite rare. Thus in a typical open cluster with $N \simeq 10^4$ members, we have $R_{\rm cl}/R_\odot \simeq 5 \times 10^4$. We assume a

[††] Coalescence using the KS method will also be discussed in section 11.10.

provisional collision criterion‡‡ of the type obtained by smoothed particle hydrodynamics or SPH [e.g. Kochanek, 1992],

$$R_{\rm coll} = 1.7 \left(\frac{m_1 + m_2}{2\, m_1} \right)^{1/3} r_1^* \,, \tag{9.21}$$

where r_1^* is the largest stellar radius. Although the decision-making for coalescence is quite different, the actual KS treatment is based on the same principles and need not be commented on specifically here. For clarity, we summarize the main steps of the collision procedure in Algorithm 9.1.

Algorithm 9.1. *Physical collision in KS regularization.*

1	Identify the turning point by $t_0'' t'' < 0$, $R < a$
2	Determine the pericentre distance, $R_{\rm p}$
3	Compare $R_{\rm p}$ with the collision distance, $R_{\rm coll}$
4	Predict back to the exact pericentre
5	Terminate the KS solution in the standard way
6	Form the new c.m. particle with $\mathbf{r}_{\rm cm}$, $\dot{\mathbf{r}}_{\rm cm}$
7	Implement stellar evolution for $m_{\rm cm} = m_k + m_l$
8	Replace body m_l by a massless particle
9	Perform mass-loss corrections of potential energy
10	Update lists to new single body, $m_{\rm cm}$
11	Obtain improved force polynomials for neighbours
12	Initialize force polynomials for $m_{\rm cm}$
13	Correct the total energy, $E_{\rm coll}$ by $\mu_{\rm kl} h$

The first point past pericentre is identified by comparing the sign of the old and new radial velocity, $t'' = R'$, for $R < a$. This algorithm is also used for determining the apocentre of binaries which requires the alternative condition $R > a$. In order to have a regular expression for the distance of closest approach, we introduce the semi-latus rectum

$$p = 4\,(\mathbf{U} \times \mathbf{U}')^2 / m_{\rm b} \,, \tag{9.22}$$

with $m_{\rm b} = m_k + m_l$ for brevity. Since $e^2 = 1 - p/a$ and we have the energy relation $1/a = (2 - 4\mathbf{U}' \cdot \mathbf{U}'/m_{\rm b})/R$, this finally gives

$$R_{\rm p} = p/(1 + e) \,. \tag{9.23}$$

Note that the determination of the eccentricity and semi-major axis from the physical variables of a collision solution is prone to errors and may in fact lead to non-physical values, whereas p contains numerically the factor

‡‡ At present a more general velocity-dependent criterion is still not available.

R which compensates for the division. If the collision condition is satisfied, the actual pericentre is obtained by backwards integration together with a final iteration using the procedure described below. Having replaced two particles by one composite body, we create a so-called 'ghost particle' with zero mass and large distance so that it will be removed as a massless escaper. One possibility not mentioned in the collision algorithm is that step 12 may be replaced by initializing a KS solution of $m_{\rm cm}$ with respect to another close single particle.

The case of collision inside a compact subsystem is much more frequent and quite complicated to deal with. Some of these events are of the common-envelope type which occur if at least one of the components has a dense core and an extended envelope, but technically the treatment is the same (cf. section 12.9). We now consider a situation with more than three chain members since the case $N_{\rm ch} = 3$ leads to immediate termination after some simple procedures, followed by a new KS solution.

The first prerequisite is to have a decision-making algorithm for determining a possible collision. One way to identify the pericentre passage of two bodies in a system of arbitrary membership is to note when the quantity $1/\sum_j R_j^2$ starts to decrease after successive increases. If $\min\{R_j\} < 8 \max\{r_k^*, r_l^*\}$ at the same time, the smallest pericentre distance is determined after the first subsequent function evaluation. In chain regularization, the procedure for obtaining the osculating pericentre distance, $R_{\rm p}$, from (9.23) is given by Algorithm 18.1. The essential steps involved in achieving the objective of identifying tidal interaction or collision candidates are elucidated by Algorithm 9.2.

Algorithm 9.2. *Pericentre determination in chain regularization.*

1	Determine the smallest pericentre distance, $R_{\rm p}$
2	Check distance criterion $R_{\rm p} < 4 \max\{r_k^*, r_l^*\}$
3	Form KS radial velocity from $R' = 2\mathbf{Q}_k \cdot \mathbf{Q}'_k R_k/t'$
4	Evaluate a using non-singular energy expression
5	Obtain pericentre time, $t_{\rm p}$, by Kepler's equation
6	Convert to chain step $\Delta\tau = (t_{\rm p}/a - R'/m_{\rm b})R_k/t'$
7	Choose $\Delta\tau < 0$ if $R' > 0$; otherwise $\Delta\tau > 0$
8	Ensure convergence by $R' < 1 \times 10^{-9}(2m_{\rm b}R_k)^{1/2}$

Here r_k^*, r_l^* denote the stellar radii and m_k, m_l are the masses at each end of the smallest chain separation, R_k. Again the pericentre distance is obtained by (9.22) and (9.23), with $\mathbf{Q}_k$ substituted for $\mathbf{U}$. However, each velocity $\mathbf{Q}'_k$ needs to be modified by the factor R_k/t' (with $t' = 1/L$ here) in order to convert from chain to KS derivatives. Formally the slow-down factor κ should be included but $\kappa = 1$ is imposed here because of possible convergence problems in the solution of Kepler's equation for high eccentricity. Note the extra factor 2 used to initiate the search which is based on osculating elements.

Evaluation of the semi-major axis from the physical variables at small pericentres suffers from numerical uncertainty. Instead we obtain the two-body energy from the regular expression

$$E_{\rm b} = E_{\rm ch} - \mathcal{V}\,, \tag{9.24}$$

where $E_{\rm ch}$ is the energy of the subsystem and $\mathcal{V}$ represents the non-dominant chain and non-chained contributions. From this the semi-major axis is obtained by $a = -\frac{1}{2}m_k m_l/E_{\rm b}$. We remark that collisions represent near-singular conditions and care should therefore be exercised to work in terms of well-defined variables.

An analytical estimate for the pericentre time can be obtained by neglecting the perturbation of any nearby members, whereupon convergence is attained by iteration. The desired expression is derived by the following algorithm [Mikkola, private communication, 1999]. First we establish the unperturbed two-body relation

$$dR'/dt = d(\mathbf{R}\cdot\dot{\mathbf{R}})/dt = m_{\rm b}/R - m_{\rm b}/a\,, \tag{9.25}$$

where $\dot{\mathbf{R}}^2$ has been replaced by $2m_{\rm b}/R - m_{\rm b}/a$. From the definition (4.4) a straightforward integration gives

$$R' - R'_0 = m_{\rm b}\tau - m_{\rm b}t/a\,. \tag{9.26}$$

Setting $R' = 0$ at the pericentre, we obtain the regularized interval[§§]

$$\Delta\tau = t_{\rm p}/a - R'_0/m_{\rm b}\,. \tag{9.27}$$

The pericentre time, $t_{\rm p}$ (< 0 if $R' < 0$) is obtained from Kepler's equation,

$$t_{\rm p} = [\theta - t''_0/(m_{\rm b}a)^{1/2}]t_{\rm K}/2\pi\,, \tag{9.28}$$

where $t_{\rm K}$ is the period. Finally, if relevant, the KS interval (9.27) is converted to chain regularization time units by the scaling factor R_k/t'. Note that $R_k = Q_k$ to high accuracy after reaching a small radial velocity. This is usually achieved in three iterations which cost the same as full integration steps. Since a is obtained by means of the non-singular expression (9.24), the eccentricity is also well determined by $e = 1 - Q_k/a$.

Let us now suppose that the collision condition $R_{\rm p} < R_{\rm coll}$ is satisfied. A few of the procedures are the same as for hyperbolic collision but all are listed by Algorithm 9.3 for completeness.

---

§§ An alternative derivation is given by Stiefel & Scheifele [1971].

Algorithm 9.3. *Implementation of chain collision.*

1	Transform to $\mathbf{r}_j, \dot{\mathbf{r}}_j$ for all chain members
2	Evaluate the two-body energy from $E_\mathrm{b} = E_\mathrm{ch} - \mathcal{V}$
3	Form coordinates and velocity of the c.m. particle
4	Implement stellar evolution for $m_\mathrm{cm} = m_k + m_l$
5	Create a massless particle at location of m_l
6	Add the tidal energy correction $\Delta\Phi$ to E_coll
7	Prescribe new force polynomials for the neighbours
8	Compress the membership list $\{I_j\}$ to remove $j = l$
9	Update the total energy by E_b to ensure conservation
10	Re-initialize the chain with reduced membership, N_ch

The energy E_b (a local quantity) is again obtained by (9.24). In order to maintain energy conservation, the change in potential energy in going from two particles to one is added to the quantity E_coll which forms part of the total energy (defined in the next section). This differential effect, arising from the nearest neighbours, is also included for two-body collisions, where it tends to be quite small. From the negative sign convention, it can be seen that the addition of $\Delta\Phi = \Phi_1 - \Phi_2$ preserves the value of the total energy, $E = T + \Phi$, with Φ_1 due to the interaction between the two colliding bodies and the other chain members, and Φ_2 arising from the c.m. approximation. Since the chain membership $\{I_j\}$ is contained in a sequential list, the massless particle is removed by compressing the array. This allows for re-initialization of the chain regularization directly, unless only two particles are left which implies termination.

In conclusion, we have seen how the membership of a chain regularization may be increased or also decreased by either escape or collision. General algorithms for changing the membership will be outlined in a subsequent section. Although such procedures illustrate the practical usefulness of the chain method, it should be emphasized that the technical treatment requires some complicated programming. However, the advantages of describing near-singular interactions in terms of well-defined variables are tangible.

9.9 Automatic error checks

The star cluster codes, which form the basis for discussion, contain a number of dissipative processes as well as discontinuous dynamical events. In this chapter, we have discussed mass loss, tidal circularization, collisions, escape and the temporary creation of stable hierarchies. Since each event or process is corrected for, it becomes possible to maintain an energy-conserving scheme, and its use as a check of the calculation is included as an option.

The change in total energy is monitored at regular time intervals and is considered to represent the global error, although the main source may well be connected with the integration of some difficult interaction. We compare the relative error, $q_{\rm E} = |\Delta E/E|$, with the specified tolerance, $Q_{\rm E}$, and have implemented the following scheme. If $0.2Q_{\rm E} < q_{\rm E} < Q_{\rm E}$, the calculation is continued normally. On the other hand, if the error is larger there are two courses of action. Thus $Q_{\rm E} < q_{\rm E} < 5Q_{\rm E}$ results in a reduction of the time-step parameters η or η_I and η_R by a factor $(q_{\rm E}/Q_{\rm E})^{1/2}$. However, errors exceeding $5Q_{\rm E}$ are not acceptable and a restart is made from the last save of the common variables, together with a factor of 2 reduction of η or η_I, η_R. A further restart is allowed for, but such cases are usually connected with technical problems which may require attention. Finally, if $q_{\rm E} < 0.2Q_{\rm E}$, the time-step parameters are increased by a small amount unless already at their initial values which should not be exceeded.

Hence by adopting the old procedure of saving all common variables for acceptable solutions [Aarseth, 1966a], we have a practical scheme for controlling the progress of a simulation that can therefore be left to itself on a dedicated machine for long periods of time. Detailed investigations are carried out if the calculation is halted because the error limit has been exceeded. This requires the results to be reproducible, which is usually the case with workstation versions. However, time-sharing on special-purpose HARP or GRAPE computers may sometimes cause problems in this respect, due to small variations in the time ration and reloading of the data which affects the scheduling.

The present data structure allows the total energy to be obtained by adding different contributions which are evaluated separately. Thus the total energy is defined by a sum of ten terms as

$$E = T + U + E_{\rm tide} + E_{\rm bin} + E_{\rm merge} + E_{\rm coll} + E_{\rm mdot} + E_{\rm cdot} + E_{\rm ch} + E_{\rm sub}\,. \quad (9.29)$$

The various quantities are listed in Table 9.1, together with brief definitions. Note that here T and U do not include any internal contributions from binaries or multiple subsystems. This definition arises naturally from the way in which the different classes of objects are integrated. For convenience, we have introduced some new definitions which correspond more closely to code usage, hence $E_{\rm tide} = W$. The collision energy, $E_{\rm coll}$, contains the tidal (or differential) energy corrections associated with the c.m. approximation when combining two particles.¶¶ Moreover, note that the energy budget contains two terms for multiple regularization. Thus $E_{\rm sub}$

¶¶ Differential corrections when starting perturbed two-body motion are also beneficial.

Table 9.1. *Components of the energy budget.*

T	Kinetic energy of single bodies and c.m. particles
U	Potential energy of single and c.m. bodies
$E_{\rm tide}$	Tidal energy due to external perturbations
$E_{\rm bin}$	Binding energy in regularized pairs, $\sum_i \mu_{kl} h_i$
$E_{\rm merge}$	Total internal energy of hierarchical systems
$E_{\rm coll}$	Sum of binding energies released in collisions
$E_{\rm mdot}$	Energy change from mass loss and Roche mass transfer
$E_{\rm cdot}$	Neutron star kicks and common-envelope evolution
$E_{\rm ch}$	Total energy of any existing chain subsystem
$E_{\rm sub}$	Energy in unperturbed triple and quadruple subsystems

remains constant and the original value is subtracted from the total energy at the end of an unperturbed multiple regularization, whereas the internal chain energy, $E_{\rm ch}$, changes with time and cannot be treated in the same way if we want (9.29) to be conserved during the interaction.

Some further comments on the entries in the table are in order. The expression for the potential energy contains the interaction between each KS pair and other pairs, as well as with single particles. For consistency, the energy of each perturbed KS solution is predicted to highest order before being accumulated in $E_{\rm bin}$. As regards hierarchies, contributions from changes in the data structure are also included in the form of differential corrections, both at initialization and termination. Thus we need to take account of the changing nature of the equations of motion, since the resulting c.m. approximation affects the force on nearby particles. Various safety procedures are therefore employed to check the external perturbation on a hierarchical configuration before it is accepted for treatment (cf. Algorithm 11.3).

On conventional computers, the change in potential energy due to mass loss from single stars or binaries is obtained by explicit summation over all members. A separate quantity, $E_{\rm cdot}$, accumulates the kinetic energy contained in the velocity kick when neutron stars are formed. By analogy with the other entries here, one could also introduce a quantity, $E_{\rm esc}$, for the energy carried away by escaping members but, for historical reasons, the relevant correction is subtracted directly. However, information about the different escape processes *is* kept separately for further analysis.

A different strategy for mass-loss correction is possible when using GRAPE. Thus in addition to $\mathbf{F}$ and $\dot{\mathbf{F}}$, the potential Φ is also evaluated by the hardware. For most cases, i.e. significant time-steps and small mass loss, the value at a subsequent time, $t \leq t_0 + \Delta t$, is obtained from

$$\Phi(t) = \Phi(t_0) - \dot{\mathbf{r}}_0 \cdot \mathbf{F}\,. \tag{9.30}$$

Moreover, any external tidal contributions contained in $\mathbf{F}$ are *subtracted* to give the net effect.

Given the general energy expression (9.29) for star cluster simulations, the question of the relative energy error usually quoted in the literature requires special consideration. We first remark that $|E|$ itself may be very large initially when studying primordial binaries. Moreover, the ejection of dynamically formed binaries or fast escapers may result in large changes of opposite sign. Consequently, we adopt a conservative definition of the relative error, based on the energy binding the cluster (discussed later),

$$\alpha = \Delta E/(T + U + E_{\rm tide})\,. \tag{9.31}$$

Again T and U represent the contributions from $N - N_{\rm p}$ single objects, except that the pair-wise potential energies are evaluated more carefully. As a result of mass loss and other dissipative processes, the absolute value of the denominator tends to decrease with time and may in fact become quite small compared with the initial value. However, the energy exchanged, or liberated, in strong interactions may still be large in systems containing significant amounts of so-called 'fossil fuel'. Moreover, the formation of hierarchies also absorbs some of the available interaction energy. Consequently, the accumulated quantity $\sum \Delta E$ for all the energy check intervals provides an alternative measure of the energy error, and in particular any systematic effects are readily evident.

The acceptable error tolerance is a matter of taste and depends on the type of problem being investigated. For general star cluster simulations, a typical value $\alpha \simeq 10^{-5}$ per crossing time can usually be achieved using standard time-step parameters. However, it is desirable to aim for even smaller values above $N \simeq 10^4$ in order to ensure an adequate description of energy generation in the core [Heggie, 1988], particularly because of the longer time-scale required. We defer a general discussion of numerical errors to section 13.2.

To increase the practical usefulness and obtain physically meaningful results, a variety of consistency warnings are included. Termination occurs if any of these are sufficiently serious, i.e. so-called 'danger signals'. This may occur on inappropriate input data or some rare condition that has not been catered for. However, many warnings are only intended for information and their occurrence does not degrade the results.

10
Neighbour schemes

10.1 Introduction

Direct N-body simulations on conventional computers benefit greatly from the use of the Ahmad–Cohen [1973] or AC neighbour scheme. Algorithms for both the divided difference method and Hermite formulation will therefore be discussed in the following sections. We also consider the implementation of the code *NBODY*6++ on a popular type of parallel computer [Spurzem, Baumgardt & Ibold, 2003], since it seems that the future of large-N calculations is evolving in this direction at least for those who do not use the special-purpose HARP or GRAPE machines described previously. The important problem of massive black hole binaries in galactic nuclei is very challenging and appears amenable to direct integration using parallel architecture and neighbour schemes. A direct solution method is described [Milosavljević & Merritt, 2001]. This treats the massive components by two-body regularization, whereas the formation process itself is studied by a tree code. Some of the drawbacks of this method inspired a new formulation where the massive binary is considered as part of a compact subsystem which is advanced by a time-transformed leapfrog method [Mikkola & Aarseth, 2002]. Over the years, the quest for larger particle numbers has also encouraged the construction of partially collisional methods. An early attempt to introduce multipole expansion for the outer cluster regions [Aarseth, 1967] was eventually combined with the code *NBODY*5 and will be considered here since it still forms a viable alternative. Finally, we outline two other hybrid formulations [Quinlan & Hernquist, 1997; Hemsendorf, Sigurdsson & Spurzem, 2002] that combine the self-consistent field method [Hernquist & Ostriker, 1992] with direct integration as well as two-body regularization.

10.2 Basic Ahmad–Cohen method

Having described the theory and initialization procedures for the AC scheme in previous chapters, we now concentrate on some practical algorithms for efficient use. It will be assumed in the following that point-mass calculations are intended since a detailed description of the *NBODY*2 code already exists [Aarseth, 2001b]; otherwise any reference to regularized binary components and centre-of-mass (c.m.) particles can simply be ignored.

In order to focus on the specific tasks, Algorithm 10.1 summarizes the main steps of the integration cycle for the divided difference scheme, together with some procedures for two-body regularization.

Algorithm 10.1. *Integration cycle for the AC scheme.*

1	Select the next particle, i, and define the time by $t = t_i + \Delta t_i$
2	Make a new sorted time-step list, L, if $t > t_{\rm L}$ and adjust $\Delta t_{\rm L}$
3	Advance any regularized solutions up to the current time
4	Decide regular force prediction [case (1)] or summation [case (2)]
5	Search for close encounter if $\Delta t_i < \Delta t_{\rm cl}$ and Δt_i decreasing
6	Predict neighbour coordinates [case (1)] or all particles [case (2)]
7	Combine polynomials for particle i and predict $\mathbf{r}, \dot{\mathbf{r}}$ to order $\mathbf{F}^{(3)}$
8	Obtain the irregular force $\mathbf{F}_{\rm I}^{\rm old}$ and update the times t_k
9	Form new irregular differences and include the term $\mathbf{D}_{\rm I}^4$
10	Initialize any new KS regularization and go back to step 1
11	[Case (1).] Extrapolate $\mathbf{F}_{\rm R}, \mathbf{F}_{\rm R}^{(1)}$ to give $\mathbf{F}_t, \mathbf{F}_t^{(1)}$; go to step 20
12	Evaluate new forces, $\mathbf{F}_{\rm I}^{\rm new}$, $\mathbf{F}_{\rm R}^{\rm new}$, and form new neighbour list
13	Repeat step 12 if $n_i = 0$ and $r_i < 10 r_{\rm h}$; reduce list if $n_i > n_{\rm max}$
14	Adjust the neighbour sphere $R_{\rm s}$ and update the times T_k
15	Construct new regular differences and include the term $\mathbf{D}_{\rm R}^4$
16	Set $\mathbf{F}_t$ from (3.1) and $\mathbf{F}_t^{(1)}$ by combining (2.3) for both types
17	Identify the loss or gain of neighbours and sum the derivatives
18	Update the neighbour list and convert to differences by (2.7)
19	Specify the new regular time-step ΔT_i
20	Assign the new irregular time-step Δt_i
21	Exit on KS termination, new chain or hierarchical merger
22	Check for optional mass loss or updating of stellar radii
23	Continue the cycle at step 1 until termination or output

A regular force calculation is decided on by comparing $t + \Delta t_i$ with $T_0 + \Delta T_i$. If the former exceeds the latter, the regular step is shortened to end at the present time, t. Note that we are comparing the next *estimated* irregular force time with the regular force time. Both the irregular and regular force summations extend over the individual components in case of regularized pairs, unless the c.m. approximation applies. The second regular derivative required at general output times is obtained from a differentiation of (2.1) at $t \neq T_0$ which yields an extra term in $\mathbf{D}^3_{\rm R}$,

$$\mathbf{F}^{(2)}_{\rm R} = \mathbf{D}^3_{\rm R}[(t_0 - T_0) + (t_0 - T_1) + (t_0 - T_2)] + \mathbf{D}^2_{\rm R} \,. \tag{10.1}$$

Likewise for step 11, the regular force and its first derivative are evaluated to highest order (i.e. $\mathbf{D}^3_{\rm R}$) at an intermediate time before the contributions are added to the respective irregular parts. Since a general time is involved, the regular derivative is obtained by the extrapolation

$$\mathbf{F}^{(1)}_{\rm R} = \mathbf{D}^3_{\rm R}(t'_0 t'_1 + t'_0 t'_2 + t'_1 t'_2) + \mathbf{D}^2_{\rm R}(t'_0 + t'_1) + \mathbf{D}^1_{\rm R} \,, \tag{10.2}$$

where $t'_k = t_0 - t_k$ and $t_0 = t$. Substituting $t_k = T_k$, $(k = 0, 1, 2)$ for the regular times then yields the desired value which is added to the irregular force derivative given by

$$\mathbf{F}^{(1)}_{\rm I} = (\mathbf{D}^3_{\rm I} t'_2 + \mathbf{D}^2_{\rm I}) t'_1 + \mathbf{D}^1_{\rm I} \,. \tag{10.3}$$

In the case of a successful close encounter search ($i \leq N$), both particles are predicted to order $\mathbf{F}^{(3)}$ before applying the corrector for particle i. The integration cycle is then terminated and the KS initialization procedure begins (cf. section 8.8). Since the time-steps are not synchronized, the neighbour coordinates are predicted before the c.m. force polynomials are formed.

The new total force evaluation (step 12) follows an irregular step which includes the fourth-order corrector. Some care is required in order to avoid spurious contributions to the regular force differences. Thus in the case of no change of neighbours, the old and new irregular force should be numerically identical to yield the actual first difference (3.5). This is achieved by using the *predicted* coordinates at step 12 instead of the corrected values, and likewise when evaluating derivative corrections.

Appropriate measures are taken in case there are no neighbours. Alternatively, if $n_i \geq n_{\rm max}$, the neighbour sphere radius is reduced by a factor 0.9 and, for particles outside the new boundary, identical contributions are subtracted from the irregular force and added to the regular force. A procedure that adds approaching particles in an outer shell to the neighbour field is beneficial on conventional computers. We also need to make sure that enough neighbours are retained for uneven distributions which invariably occur. In some situations (i.e. for increased accuracy with binaries or shorter CPU time), it may be desirable for the average neighbour number to exceed a specified fraction of the maximum value. An optional procedure is therefore included which increases the predicted membership (3.3) by a suitable factor if $\bar{n} < \frac{1}{2}\, n_{\rm max}$.

The case $i > N$ requires special attention because the neighbour list is used to select the perturbers and the membership may not be sufficient. If $\tilde{R}_s > R_s$, with $\tilde{R}_s = -100\, m_i/h_{i-N}$, the neighbour radius is stabilized on $0.9\, n_{\max}$ by (3.4). Here $\tilde{R}_s$ denotes the maximum perturber distance for equal-mass particles, according to (8.58). Hence for long-lived binaries an attempt is made to increase the neighbour sphere if it is too small for the perturber range $\lambda a(1+e)$, with $\lambda \simeq 100$. This problem only tends to occur for relatively small N, since the inter-particle distance scales as $N^{-1/3}$ and $R_{\rm cl} \propto N^{-1}$. However, systems containing massive binary components require additional consideration because the two-body separation may be fairly large. One simple expedient is to redefine the termination distance by setting $R_0 = 2\, a$ for hard energies ($E_{\rm b} < \epsilon_{\rm hard}$), provided the neighbour radius is sufficiently large (i.e. $a(1+e) < 0.02 R_{\rm s}$).

Because of frequent coordinate predictions, it is convenient to include the c.m. particles in the neighbour list instead of the components, whereupon the latter can be obtained by KS transformations if desired. Hence at every termination the relevant c.m. body needs to be replaced by the two components in their sequential location. Several such list expansions may occur within one regular time-step and some allowance for extra members must therefore be made when allocating the maximum size. Conversely, it may happen that only one component of a future KS solution is included in the neighbour list and regularization occurs before the next update. To circumvent this problem we initialize all the prediction variables to zero such that the current coordinates are used in the force evaluation. Only a small error is incurred here since the separation is comparable to $R_{\rm s}$. The corresponding c.m. particle will then be selected at the next regular step. In the case when primordial binaries are initialized, a recent procedure ensures that both the components are selected together as neighbours, which avoids the difficulty above.

When the force on a single particle is considered, the contribution from any neighbouring c.m. particle is evaluated in the point-mass approximation provided $r_{ij} > \lambda R$, where R is the current two-body separation. In the alternative case, a coordinate transformation is performed and the summation is over both components. However, if the same particle is also a member of the corresponding perturber list by virtue of its larger mass (cf. (8.58)), its effect on the internal motion will be included. This apparent inconsistency can be justified if we note that the orbital energy change, $\dot{h} = \dot{\mathbf{R}} \cdot \mathbf{P}$, cancels to first order when a weak perturbation (here $\gamma \simeq \gamma_{\min}$) is integrated over one period. Hence use of the c.m. approximation according to a distance test avoids additional complications in the evaluation of $\mathbf{F}_{\rm I}$ and $\mathbf{F}_{\rm R}$, which must use the same expression.

A more serious problem with the divided difference formulation is due to large derivative corrections which are difficult to avoid without additional algorithms. Thus one can to a certain extent anticipate some problems by, for instance, selecting a particle undergoing a close non-regularized encounter with an existing neighbour; otherwise large high-order force derivatives may appear. Note that the selection of two such particles instead of one still gives rise to some high derivatives, but the dipole terms cancel. Likewise, a previous neighbour with small time-step is retained out to a distance $2R_{\rm s}$ in order to minimize the derivative corrections ($\mathbf{F}^{(2)} \propto r^{-4}$). Alternatively, such a particle may belong to a soft binary of high eccentricity which becomes regularized at some stage.

Even if the optional treatment to minimize high derivatives is activated, the regular force polynomial may sometimes reveal non-convergent behaviour. Thus the rare case of the second difference being abnormally large compared with the first regular force difference is best handled by neglecting the corrector $\mathbf{D}^4_{\rm R}$ altogether; i.e. if $|\mathbf{D}^2_{\rm R}|(T_0 - T_1) > |\mathbf{D}^1_{\rm R}|$. This behaviour may be due to force derivative corrections for a particle near the boundary that is repeatedly included and excluded when the neighbour radius is modified according to (3.4).

Finally, we consider a special algorithm to alleviate possible problems connected with high velocities. Superfast particles are often ejected from a cluster following strong interactions inside compact subsystems of three or more members, or may be due to a neutron star velocity kick. Unless some precaution is exercised, such a particle may penetrate deeply into the neighbour sphere before the next regular force update, and hence produce an unacceptable change in the irregular force. An optional procedure has therefore been added to create a list of high-velocity particles for possible inclusion in the neighbour field already at a distance of $2R_{\rm s}$, provided the impact parameter is less than $R_{\rm s}$. A high-velocity particle is defined by $v_j^2 > 8v_\infty^2$ together with $\Delta t_j > \Delta t_{\rm cl}$, where v_∞ is the estimated central escape velocity of twice the current rms value. In practice, the limiting square velocity is derived from the expression $16|h_{\rm hard}|$ (cf. (4.1)). The condition $\Delta T_j > 20\,\Delta t_{\rm cl}$ for the regular step ensures that only genuinely freely moving particles are selected. In addition, we may also employ the subsidiary condition $|\mathbf{F}_i|^2 < 4N$ which excludes an ongoing close encounter. Members of the high-velocity list are checked and removed at frequent intervals if $r_j > 3r_{\rm h}$ or if the velocity has decreased below the specified limit.

In conclusion, we see that regularization is beneficial in placing a lower limit on existing time-steps and also assists in suppressing high-frequency force fluctuations that have no significant secular effects. Hence the combination of the AC neighbour scheme with regularization has proved itself, although a considerable programming effort has been invested towards the achievement of this goal.

10.3 Hermite implementation

The Hermite version of the AC scheme (HACS) gives rise to the code *NBODY6* and has many features in common with the standard formulation discussed above. On the whole, the block-step structure allows for a simpler treatment and the main differences are sufficiently distinct to warrant a separate section. The special scheduling has been described in section 9.2 and we therefore move on, assuming that a list of particles due for updating is given.

Since there are usually a number of particles, $N_{\rm block}$, to be advanced at the same time, we employ a different prediction strategy. Thus for small block memberships (say $N_{\rm block} \leq 10$ for $N \simeq 1000$) and no regular force update, the neighbour lists are combined by an efficient sorting procedure for prediction; otherwise the coordinates and velocities of all particles are predicted for simplicity.* Unless already done, any c.m. particles are predicted next, followed by iteration of the regularized time for prediction of $\mathbf{U}$, $\mathbf{U}'$ and KS transformations to obtain the state vector of the components. Most of the procedures concerning the irregular and regular force polynomials are now essentially similar to the previous case, including the use of predicted coordinates and velocities for step 12 of Algorithm 10.1. However, high-order derivative corrections are simplified and the corrector is of different form (cf. (2.23)), containing less terms.

The irregular time-step is determined by the original criterion of chapter 2. Following an earlier suggestion [Makino & Aarseth, 1992], we modify the expression slightly to

$$\Delta t_i = \left[\frac{\eta_I (|\mathbf{F}||\mathbf{F}_{\rm I}^{(2)}| + |\mathbf{F}_{\rm I}^{(1)}|^2)}{|\mathbf{F}_{\rm I}^{(1)}||\mathbf{F}_{\rm I}^{(3)}| + |\mathbf{F}_{\rm I}^{(2)}|^2} \right]^{1/2} . \tag{10.4}$$

Thus the *total* force is used instead of the irregular force, in case the latter happens to be small. This leads to somewhat larger irregular steps since the error of the neighbour force is now a constant fraction of the total force. The same expression would also be beneficial in *NBODY5*.

The basic regular time-step takes the same form as (2.13), with the corresponding regular quantities substituted. However, there are situations where the value of ΔT_i may be increased because the fractional change in the regular force is below the tolerance. We therefore evaluate the predicted change due to an increased trial value $\Delta \tilde{T}_i = 2\Delta T_i$ from

$$\Delta \mathbf{F}_{\rm R} = [(\tfrac{1}{6}\mathbf{F}_{\rm R}^{(3)} \Delta \tilde{T}_i + \tfrac{1}{2}\mathbf{F}_{\rm R}^{(2)})\Delta \tilde{T}_i + \mathbf{F}_{\rm R}^{(1)}]\Delta \tilde{T}_i \, . \tag{10.5}$$

The trial value is chosen on a successful outcome of the convergence test

$$|\Delta \mathbf{F}_{\rm R}| < \eta_R \min\{|\mathbf{F}_{\rm R}|, |\mathbf{F}_{\rm I}|\} \tag{10.6}$$

* Note that regular time-steps are rarely commensurate with the time for small $N_{\rm block}$.

for isolated systems, with a suitable modification of the first force term on the right-hand side for an external tidal field. The final values of Δt_i and ΔT_i are chosen according to the block-step synchronization rules discussed in section 2.6. In order to satisfy decision-making requirements, the irregular time-step is not allowed to exceed the regular value.

In Hermite integration, the second and third force derivatives are obtained at the end of the time-step. Hence it is not necessary to perform the relatively expensive initialization of these quantities by explicit summation, provided a satisfactory time-step can be prescribed. For this purpose we have again adopted an expression of the type (8.56). This form is also used to initialize regular time-steps in *NBODY6*.

Early use of the neighbour scheme was mainly confined to relatively homogeneous systems. However, systems that have experienced core collapse produce a wide range in density. Although the computational cost is dominated by the central region, the original limit on the predicted membership (cf. (3.3)) may be relaxed. Somewhat belatedly we now introduce a modified lower limit outside the half-mass radius,

$$n_{\rm min} = 0.2\, n_{\rm max}(r_{\rm h}/r)\,. \tag{10.7}$$

This condition leads to a gradually decreasing neighbour membership for particles moving into the far halo.

The case of zero neighbours requires special treatment. A similar algorithm exists for standard AC but here we need both the force and its derivative. Very distant particles, say $r_i > 10\, r_{\rm h}$, do not usually have any natural neighbours and are therefore allocated a nominal mass of $0.01\,\bar{m}$ at the *coordinate* centre in order to facilitate the integration. The corresponding first derivative is evaluated explicitly and any optional external perturbations are added to give a well-defined irregular time-step which is usually fairly large. In order to reach the state of zero neighbour number, the radius $R_{\rm s}$ is *reduced* gradually outside a specified central distance according to

$$\tilde{R}_{\rm s} = \max\left\{0.75 R_{\rm s},\, 0.01 r_{\rm h}\right\}, \tag{10.8}$$

which achieves the desired objective unless escaper removal occurs. Note that, in the standard scheme, this modification of the neighbour radius for distant particles may lead to some reduction of the time-steps because of the force derivative corrections.

The HACS scheme provides for a simplified treatment of high-order derivative corrections compared with the standard ACS. As has been discussed elsewhere [Makino & Aarseth, 1992], the high-order derivatives are not actually used for the integration of single particles. However, for some purposes (i.e. diagnostics) it may be useful to predict coordinates and velocities to highest order, and there is also the new feature of regularization to consider. On the other hand, all particles are synchronized at the longest time-step (typically 0.5), whereas output intervals are usually a few time units. Hence the derivative corrections are only performed if the relevant option is activated, except for the case $i > N$ and significant perturbation. The latter is useful in connection with regularization terminations which do occur at arbitrary times.

The higher force derivative corrections on a c.m. particle are not based on the mass-weighted contributions because the component values of $\mathbf{F}$ and $\dot{\mathbf{F}}$ required to construct $\mathbf{F}^{(2)}$ and $\mathbf{F}^{(3)}$ are not available. Consequently, use of the c.m. approximation is only consistent outside the perturber distance, λR, and, for this reason, it is desirable to ensure a sufficiently large neighbour radius or, failing that, place an upper limit on the size of the two-body separation, R.

The Hermite integration cycle is essentially similar to Algorithm 10.1. One difference is that any new KS regularization is initialized at step 3, whereupon the cycle begins again since the sequential data structure may have changed. Moreover, following a successful close encounter search at step 8, the relevant particle indices are saved for the next cycle. The end of each irregular or regular corrector step is characterized by initializing the basic integration variables, $\mathbf{r}_0$, $\mathbf{v}_0$. In HACS, and also for parallel implementations, it is preferable to delay setting the final values of $\mathbf{r}$, $\mathbf{v}$ until the end of the block-step, copied from $\mathbf{r}_0$, $\mathbf{v}_0$. In other words, the *predicted* values are used throughout the current cycle for consistency.

Since memory is no longer a concern, two different regular force derivatives are defined for convenience in HACS. Thus we distinguish between the regular force derivative based on the new neighbours and the derivative (3.9) with respect to the old neighbours. The former is only used to construct the prediction variables (3.6) and (3.7). Note that the latter, which preserves the sum of the derivative corrections at the end of a regular step, may not be updated to include the change of neighbours for single particles. Likewise, the irregular derivatives are treated analogously.

With the block-step scheme, the candidates for KS regularization need to be synchronized before initialization so that both components are advanced in the same block. The time-steps are usually equal but, if necessary, this is achieved by enforcing a step reduction for the particle being advanced. Unless other complications prevail, the new KS solution is initialized at the start of the next cycle by re-instating the previous value of the time. On the other hand, procedures for KS termination, new multiple regularizations or hierarchical mergers are activated at the *end* of the integration cycle. We note that the total number of such initialization procedures is very small compared with the number of block-steps (by a factor of about 10^4) so that coincidence conflicts are extremely rare and not harmful; i.e. this causes a possible delay by one small block-step.

Star cluster simulations often exhibit examples of superfast particles, especially if hard binaries are present. As in the code *NBODY5*, this may result in an approaching particle moving well inside the neighbour sphere before it is identified at the next regular force calculation. The check for possible penetrations is done differently as follows. Since the ejection of a fast particle is usually associated with strong interactions involving KS or chain regularization, it is natural to initiate a special search at terminations that are defined by the control index in Table 7.2. Again the conditions for acceptance described previously are employed. Given a list of high-velocity particles, any contribution to the irregular force is considered together with other optional procedures when evaluating the total force.

A subsequent search for neighbour sphere intersections is made at moderately large commensurate intervals; i.e. $\Delta t_{\rm nb} = 1/32$ (scaled by $(1000/N)^{1/3}$ for $N > 1000$). This provides an opportunity for the regular time-step to be reduced. Hence we treat the case of close and distant neighbours separately. The search itself only considers particles j with regular time-steps $\Delta T_j \geq \Delta t_{\rm nb}$, because these are the most vulnerable. Provided the radial velocity is negative, we form the time until minimum approach by

$$\Delta\tilde{t}_{\rm min} = \min\left\{T_0 + \Delta T_j - t,\ -\mathbf{D}\cdot\mathbf{V}/\mathbf{V}^2\right\}, \tag{10.9}$$

where $\mathbf{D}$ and $\mathbf{V}$ denote the relative distance and velocity. The minimum impact parameter can be estimated if we assume straight-line motion,

$$D_{\rm min} = |\mathbf{D} + \mathbf{V}\Delta\tilde{t}_{\rm min}|\,. \tag{10.10}$$

If the corresponding intruder force is significant, i.e.

$$m_i/D^2_{\rm min} > 0.1|\mathbf{F}_j|\,, \tag{10.11}$$

the regular time-step is reduced by a factor 2, subject to the synchronization condition $T_0 + \frac{1}{2}\Delta T_j > t$, and further reductions are performed if relevant. The above algorithm is also implemented in *NBODY4* for the time-step Δt_i.

A check for removal of high-velocity particles is carried out at intervals of $\Delta t_{\rm nb}$. Such particles are removed from the special list outside a central distance of $3r_{\rm h}$ and also if their velocity falls below the critical value, given by $v^2 = 16|h_{\rm hard}|$ (cf. (4.1)). Moreover, the list is updated to take account of any changes in the particle sequence.

Comparison of the two versions of the AC scheme depends on the hardware even if we consider only workstations or laptops. In fact, on a RISC-based workstation, the total CPU time for the *same* number of steps of both types favoured HACS slightly for a test problem with $N = 1000$. Thus the extra effort to obtain $\dot{\mathbf{F}}$ with HACS is compensated by less predictions and a simpler corrector. The original investigation [cf. Makino & Aarseth, 1992] showed that nearly twice as many steps were needed by the latter method for the same accuracy when softening is used. However, recent laptop comparisons of *NBODY*5 and *NBODY*6 with $N \leq 1000$ yield more comparable energy errors for standard parameters, the latter still being faster per step. The Hermite block-step version is also considerably easier to implement and exhibits greater numerical stability [cf. Makino, 1991a]. Finally, we emphasize that, although of fourth order, the general Hermite scheme has the attractive feature of being self-starting.

10.4 Parallel adaptations

As a first case study, we discuss the adaptation of the parallel code *NBODY*6 to the CRAY T3E [Spurzem, Baumgardt & Ibold, 2003]. This development also aims to provide a portable code for LINUX cluster configurations by a comprehensive code-building algorithm which ideally should reduce to *NBODY*6 for single CPUs. The implementation has necessitated several important changes, both as regards general strategy and exploitation of the different architecture. This code has therefore been given the new name *NBODY*6++ to distinguish it from the original version. The main structural changes facilitate parallelization of three independent procedures as follows:

- Advance all KS solutions up to the end of the current block-step
- Evaluate new irregular forces and apply the corrector
- Obtain regular forces, neighbour lists and perform corrections

The second and third stages have already been fully parallelized. Moreover, the prediction of neighbours has also been included in the second part. In order to ensure that the results do not depend on the order in which the particles in a given block are treated, the final corrector initialization $\mathbf{r} = \mathbf{r}_0, \mathbf{v} = \mathbf{v}_0$ for all the members is delayed until the end of the block-step, as is also done in *NBODY*6.

The parallelization of KS solutions presents a challenge for efficient use of multi-processor machines like the T3E. The non-linearity of the time transformation $dt = R\,d\tau$ does not by itself prevent such procedures because one can always define a set of approximate hierarchical levels. Consider a number of perturbed binaries that are due to be updated during the next block-step. Depending on their location in the cluster, some may have joint perturbers of single particles or even other binaries. It follows that a random assignment to different processors may cause problems of inconsistency in the latter case since predictions are involved. One possible way is to take account of the spatial distribution and construct a binary tree structure in order to facilitate the decision-making. An alternative solution method for the immediate future is to subdivide the block-step by factors of 2 into sufficiently small subintervals to prevent predictions outside the range of validity. All such members in one sub-block can then be distributed to different processors which would achieve a significant speed-up. By analogy with the parallel procedures for the irregular and regular integration steps, it can be anticipated that the number of simultaneous KS solutions will be large enough to justify this approach. Provisional experimentation along the latter lines has been carried out and appears to be promising [Spurzem, private communication, 2001].

In considering this problem, one should envisage studying a large population of primordial binaries (say $N_{\rm b} \simeq 5 \times 10^4$). Such a distribution would contain a significant number of unperturbed binaries, where integration of the corresponding c.m. motions proceeds as for single particles. This leaves the checking for perturbers that occurs on time-scales from one binary orbit up to the c.m. time-step, depending on the local density and velocity distribution. Fortunately, this procedure does not create any conflicts and all the necessary checks can be carried out in parallel, provided the operation is performed as an additional step. Alternatively, by making no distinction, the latter task may be carried out together with the advancement of KS solutions, since the numerical effort to check unperturbed motion is less.

Again the hierarchical time-step algorithm is employed for both the irregular and regular time-steps, which must be individually commensurate. Since the distribution of block-steps is quite wide, with less members at small values in realistic systems, there is some loss of efficiency and serial integration may in fact be preferable. As N increases, so does the number of particles, $N_{\rm gr}$, due to be advanced at the same time. The theoretical prediction $N_{\rm gr} \propto N^{2/3}$ [Makino & Hut, 1988] appears to be in reasonable agreement with present simulations of inhomogeneous systems. Hence for large N, the average number of group members starts to exceed the number of available processors, $n_{\rm proc}$. The typical amount of communication after each block-step is $kN_{\rm gr}/n_{\rm proc}$ double-precision words, where $k = 19$ and $41 + n_{\rm max}$, respectively, for the irregular and regular force calculation. Consequently, it is beneficial to limit the maximum neighbour number as much as possible. Notwithstanding the theoretical prediction $n_{\rm p} \simeq (N/1.8)^{3/4}$ [Makino & Hut, 1988], a constant value, $64 \leq n_{\rm max} \leq 128$, has been found satisfactory for a wide range of particle numbers. We may distinguish between the theoretical speed-up and actual performance. Thus when communication times are neglected, the efficiency does improve up to the maximum processor number. However, at present the computing time only scales linearly with N up to about 512 processors (for $N = 10^4$) owing to memory and communication limits.

The parallel algorithm considered above does not assign specific particles to any processor and every node contains an identical copy of the whole dataset. So-called 'domain decomposition' *is* employed in the recent cosmological tree code *GADGET* [Springel, Yoshida & White, 2001] (and other such codes) which takes advantage of the slow mixing when using comoving coordinates. Accordingly, at the end of each block-step the new data must be broadcast to all the other processors. The net result is that maximum efficiency is achieved at some intermediate number of processors that depends on N [cf. Spurzem *et al.*, 2003]. Thus we have an example where further software development is desirable in order to exploit an existing hardware configuration that is potentially very powerful. Judging from the history of conventional computers and given the incentive, there is bound to be progress in this young subject.†

A separate version of *NBODY*6++ has been prepared for the HARP-3 special-purpose computer [Spurzem *et al.*, 2003]. Such an attempt faces the problem that only the force calculations can be executed in parallel. Moreover, since the irregular force is now evaluated on the host, the regular force is obtained by subtraction after setting the neighbour masses to zero in a second full N summation on HARP. In order to utilize many pipelines, the neighbour lists of all the particles due for regular force updates are first combined, with a subsequent modification of the affected individual components on the host. Hence the additional effort is less for relatively small neighbour numbers which in any case must be below the hardware limit. However, this somewhat laborious procedure suffers from numerical problems because of the different accuracy employed by HARP, which results in reduction of the regular time-steps produced by spurious force derivatives. In view of this exploratory investigation, we may anticipate that future hardware developments will take advantage of the potential gain from using a neighbour scheme.

10.5 Black hole binaries in galactic nuclei

The problem of formation and dynamical evolution of a black hole (BH) binary with massive components is of considerable topical interest. Several past efforts employed direct integration methods to elucidate the

† So-called 'systolic and hyper-systolic algorithms' are now available for this purpose [Dorband, Hemsendorf & Merritt, 2003]. Alternatively, the processors may be organized in a 2D network which yields reduced communication costs [Makino, 2002].

behaviour of such systems but applications to galactic nuclei pose severe limitations with regard to the particle number that can be studied. The formation is usually envisaged as the end product of two separate unequal galactic nuclei spiralling together by dynamical friction, but there are other scenarios. In view of the different numerical requirements for the early and late stages, it is natural to consider this problem in two parts and take advantage of direct integration where it matters most.

A recent investigation [Milosavljević & Merritt, 2001] employed the parallel version of the *GADGET* code [Springel *et al.*, 2001] to study the formation process. This code maintains individual and adaptive timesteps and maps the particles on to an octal tree structure. For initial conditions, two spherical stellar systems containing a central BH mass-point were placed in a moderately eccentric orbit with semi-major axis $a_{\rm G} \simeq 4r_{\rm h}$. The relatively small softening length used, $\epsilon = 0.001$, was well below the BH hard binary separation, $a_{\rm hard} = (m_1 + m_2)/8\sigma^2 \simeq 0.0025$. A mass ratio $m_1/M = 0.01$ and total particle number $N = 256\,{\rm K}$ was adopted. This model is close to current computational limits, although still well short of realistic requirements. Another important aspect here is the presence of a steep stellar density cusp, $\rho \propto r^{-2}$, surrounding each BH, whereas previous studies tended to use shallower King models.

Following the early phase of dynamical friction acting between the two BHs together with their bound subsystems, the collisionless simulation was continued until $a \simeq \epsilon$. However, new datasets were generated by random sampling of the whole population already at an earlier stage when $a \simeq 30a_{\rm hard}$, with effective memberships up to 32 K and enhanced masses. These reduced datasets were then integrated by *NBODY*6++ in order to ascertain the N-dependence of the subsequent evolution. A nearly constant hardening rate, da^{-1}/dt, was observed, with a modest eccentricity increase during shrinkage by a factor of 20. Contrary to previous findings, the hardening rate appeared to be independent of N; this result was ascribed to a larger reservoir of central stars. It is also significant that a simulation with initial supermassive black holes inside steep density cusps produced nuclei with shallow cusps and $\rho \propto r^{-1}$.

The presence of a supermassive binary poses several new technical challenges for the numerical treatment. Since the regularization scheme employed by *NBODY*6 was developed for conventional star clusters containing a realistic IMF, it is essential to modify the strategy in order to prevent inefficient usage. One characteristic feature that requires fresh consideration is the occurrence of persistent orbits of short periods around one of the BHs. For sufficiently small length scales, the effect of gravitational radiation should be included and this may eventually induce coalescence. In any case, a few strongly bound orbits are of little dynamical significance here and such stellar companions can therefore be combined with their BH. Since the dominant binary is surrounded by a large number of stars in bound orbits, the standard criteria for chain regularization and even the KS regularization scheme require modification. In general, the code scales well with the number of processors for a moderate spread in time-steps. However, the scaling was poor when a few particles had small time-steps and so a switch was made to serial integration.

If the decay of the binary continues sufficiently far, energy loss by gravitational radiation will eventually play an important role. The time-scale for this stage to be reached depends on the Brownian motions generated by recoil effects due to slingshot interactions [Saslaw, Valtonen & Aarseth, 1974]. Since the amplitude of the binary wandering is much larger in an N-body simulation than in a real galaxy, the question of the loss-cone replenishment cannot be settled by direct integration alone. However, the ejection of stars by the BH binary tends to lower the central density, permitting the binary an increased amplitude for further interactions. On the other hand, for some density profiles, BH binaries may essentially complete their evolution before gravitational radiation takes over.

The simulation of BH binary evolution is an exciting problem that demonstrates well the power of direct integration and holds great promise. Given the undoubted advantage of treating the BH binary as a perturbed system, it might be worthwhile trying the wheel-spoke regularization discussed in section 5.4. The two BH components would then be permanent members, with a small number of the most critical orbits considered for inclusion in the chain. Although this method has yet to be tried in a serious application, the development is based on the same principles as chain regularization. If it works, such a formulation would in fact lead to much simplification elsewhere because most of the other regularization procedures would become redundant. However, complications are also likely to arise due to the need for a generalized branching chain structure to be developed [cf. Mikkola & Aarseth, 1990]. Hence this idea remains a project for the future. Another promising avenue of dealing with this problem was opened up recently [Mikkola & Aarseth, 2002]. Because of similarities with chain regularization, algorithms for the associated new code NBODY7 will be discussed in a subsequent chapter.

10.6 Hybrid formulations

Many stellar systems are characterized by a core–halo structure where the stars in low-density regions move in essentially collisionless orbits. Algorithms that include the star-by-star treatment in the inner regions and represent the outer parts in an approximate but consistent way can therefore still be considered to be collisional. Although the integration steps increase significantly with central distance in such systems, the force calculation for the inner particles can be speeded up on conventional machines by a fast evaluation of the distant interactions. This point was not considered in an analysis of computational cost [Makino & Hut, 1988].

The use of multipole expansion, discussed in section 3.4, can be formulated as a hybrid method while the collisional aspects are retained. In the following we describe a scheme for combining the direct force summation over particles in the inner regions with external contributions from an expansion in spherical shells [Aarseth, 1967]. The so-called 'shell method' was implemented during the early 1980s in a separate code, *NBODYS*, based on *NBODY*5 but was never used to obtain any published results. Some of the most relevant steps are displayed in Algorithm 10.2, which is an abbreviated form of Algorithm 10.1.

Algorithm 10.2. *Integration cycle for the shell method.*

1	Divide the cluster into radial zones of equal membership
2	Evaluate all the moments at regular intervals, $\Delta t_{\rm sh}$
3	Combine sum of external and internal moments for each shell
4	Select the next particle, i, to be treated (KS, single or c.m.)
5	Predict the coordinates of neighbours or all particles
6	Advance the solution for the irregular time-step, Δt_i
7	Repeat steps 4, 5 and 6 until $t + \Delta t_i > T_{\rm i} + \Delta T_{\rm i}$
8	Perform regular force summation and determine neighbour list
9	Add external and internal contributions from the shell force
10	Form new regular force differences and include the corrector
11	Update irregular and regular differences and set new time-steps
12	Return to step 2 ($t \geq t_{\rm sh}$) or step 4 ($t < t_{\rm sh}$)

The main idea is to divide the system into a relatively small number of zones (e.g. 7) of comparable size to twice the largest neighbour radii. Hence we extend the AC force expression to a sum of three terms by

$$\mathbf{F} = \mathbf{F}_{\rm I} + \mathbf{F}_{\rm R} + \mathbf{F}_{\rm S}\,, \tag{10.12}$$

where contributions from all non-overlapping shells are included in the smooth component, $\mathbf{F}_{\rm S}$. The moments are updated at times $t_{\rm sh}$ with intervals $\Delta t_{\rm sh} = 0.004 t_{\rm cr}$ at a cost of $O(N)$. For each shell we then form separate sums of any internal and external terms of different order (cf. (3.24)), as well as a sequential list of the relevant particles. Accordingly, the main difference with the standard treatment of *NBODY*5 is in the total force evaluation. Given a particle at radial position $\tilde{r}_i = |\mathbf{r}_i - \mathbf{r}_{\rm d}|$ with respect to the density centre, $\mathbf{r}_{\rm d}$, and neighbour radius $R_{\rm s}$, the particles involved in the direct summation are those that fall inside shells in the distance range $[\tilde{r}_i - 2R_{\rm s},\, \tilde{r}_i + 2R_{\rm s}]$. After the determination of the corresponding shell indices, the new regular and irregular force are obtained by a sum over the respective list members. Contributions from the smooth component are then combined with the standard regular force.

The present treatment avoids the problem of boundary crossings if neighbour selection (including high-velocity stars) falls inside a distance of $2R_s$. Moreover, there is no need for interpolation within the associated shell as in the original formulation of section 3.4. We also note that for particles inside the innermost shell, which tend to have the smallest time-steps, only the external contributions are required. Hence the shell moment method appears to be an attractive alternative for allowing increased particle numbers.

Before returning to the problem of black hole binaries, we mention a simpler approach where the AC method for softened potentials was combined with the SCF code to model the dynamical evolution of elliptical galaxies containing a central singularity [Merritt & Quinlan, 1998]. Following the initial approach to equilibrium, a single massive body was introduced. The evolution of the central region was towards nearly spherical shape while the outer parts became axisymmetric, with the rate of change depending on the final BH mass ratio.

The behaviour of black hole binaries has been investigated by a hybrid method with some ingredients of a neighbour scheme. An attempt to cover more realistic mass ratios [Quinlan & Hernquist, 1997] combined the self-consistent field (SCF) method [Hernquist & Ostriker, 1992] with direct integration. Since the basic version of the SCF code employs one expansion centre, the existence of a relatively wide BH binary inside the same system was assumed at the outset. However, the lack of a consistent starting model has certain implications for the results [cf. Milosavljević & Merritt, 2001]. Nevertheless, we summarize this work below as an interesting example of a hybrid method spanning three different techniques.

The dominant two-body motion was included in the point-mass form, whereas attractions from the other stars were added by direct summation using a softened potential. Moreover, the stellar contributions to each BH were split into two parts according to the AC scheme of *NBODY2*. Likewise, a softened potential was employed for the star–BH interactions, with the star–star attractions evaluated according to the SCF method. After some time, the BH binary components became close enough for their relative motion to be regularized. A KS treatment was used, based on the standard Hermite formulation of section 4.6 as implemented in the original version of *NBODY6*. Special efforts were also made to circumvent problems connected with the large mass ratio. Thus stars within a distance of $50a$ of the binary were included in the perturber list regardless of the size of their perturbation. The parameter $\gamma_{\rm min}$ controlling unperturbed two-body motion was also reduced to 10^{-7}.

Galaxy models with density cusps and some 10^5 particles with the masses decreasing towards the centre for improved resolution were selected for study. The softening was chosen as $\epsilon = m_1/v_\epsilon^2$, where the softening velocity, v_ϵ, was 2–3 times the BH orbital velocity for $a_{\rm BH} = a_{\rm hard}$. The two BHs with masses $m_1 = m_2 = 0.01M$ were initialized on nearly circular orbits at a distance $0.5r_{\rm h}$ and spiralled inwards to the centre in about six crossing times. From then on, regularization was employed and subsequent shrinkage by a factor of 12 beyond the hard binary limit (defined above) was observed. After a while, some stars were captured into bound orbits with small time-steps around the BHs and were absorbed by their massive component, subject to the condition $a < 2\epsilon$. In the event, the corresponding growth of the BH masses was at most 0.2%.

The basic integrator starts by generating the first three force derivatives which are not available for the coefficients of the potential expansion. In such situations it is adequate to begin with smaller time-steps and set the higher derivatives to zero, whereupon the next few time-steps can be increased gradually while the proper differences are formed. In this connection, we remark that although the Hermite method requires only the first derivative, it needs to be evaluated precisely for the corrector. It is also reassuring to note that the evolution of the hard binary was independent of the integrator used for the two-body motion, i.e. *NBODY*1, *NBODY*2 or *NBODY*6. However, the accuracy parameters for direct integration need to be chosen conservatively in order for the well-known systematic errors to be acceptable. Although this study did not provide any definite model for the later stages of BH binary evolution and the initial conditions were somewhat unrealistic, it highlighted many essential aspects of a difficult problem and pointed the way forward.

An even more recent development combined the SCF method with the point-mass code *NBODY*6++ [Hemsendorf *et al.*, 2002]. In order to use Hermite integration, the SCF code was upgraded to include evaluation of the force derivative. The initial application did not emphasize the importance of density cusps and used a Plummer model instead, with two massive components $m_{\rm BH} = 0.01M$ placed on the x-axis at $\pm 0.64\, r_{\rm h}$ with small y-velocity of opposite sign. In scaled units, the BH binary formed at $t \simeq 10$ and hardened linearly with time to $a_{\rm BH} \simeq 2 \times 10^{-3}$. The full hybrid code used 64 K and 128 K particles, whereas comparison tests with the basic *NBODY*6++ were restricted to $N = 16\,$K. This new implementation demonstrated that large-N systems can be studied by direct means. However, some numerical difficulties were also reported, suggesting that this is a challenging problem that requires further attention.

11
Two-body algorithms

11.1 Introduction

A large number of algorithms are connected with regularization. Many of these concern the KS treatment which plays a key role in the N-body simulation codes. In this chapter, we derive some expressions relating to the conversion of regularized time, followed by other considerations of a practical nature. A separate section provides essential details of the Stumpff KS method as employed in an N-body code. This is followed by an algorithmic discussion of KS termination. Next we describe decision-making procedures for unperturbed two-body motion which speed up the calculation by a large factor. Another important feature with the same objective is the so-called 'slow-down device', where the principle of adiabatic invariance is exploited. The theory was given previously in connection with chain regularization and here we discuss the KS implementation. Special treatments of stable hierarchies also contribute significantly to enhanced efficiency while retaining the essential dynamics. Finally, the last sections deal with several processes relating to tidal interactions in close binaries that are connected through an evolutionary sequence. We discuss tidal circularization and two-body capture, as well as Roche-lobe mass transfer which all contribute to making star cluster modelling such an exciting and challenging project.

11.2 General KS considerations

We first discuss various general features that are applicable to all the KS methods and also include some aspects of the divided difference scheme, while the next section deals specifically with the Stumpff version.

In order to advance the whole N-body system consistently, we need to integrate the equation of motion for the time (4.39). The conversion from regularized time to physical time is most conveniently carried out by a Taylor series expansion to order n,

$$\Delta t = \sum_{k=1}^{n} \frac{1}{k!} t_0^{(k)} \Delta\tau^k . \tag{11.1}$$

Provided $n \leq 6$, all the necessary derivatives are already known in the high-order integration schemes, and hence the Hermite method applies directly. For the divided differences, $n = 5$ was chosen as sufficient. However, one more order, i.e. $\mathbf{U}^{(5)}$, is available for (11.1) and has been implemented in the Hermite codes for increased accuracy when combined with the Stumpff functions. The high-order coefficients are formed by successive differentiation of (4.39), with the first two terms given by

$$\begin{aligned} t_0^{(2)} &= 2\mathbf{U}' \cdot \mathbf{U} , \\ t_0^{(3)} &= 2\mathbf{U}'' \cdot \mathbf{U} + 2\mathbf{U}' \cdot \mathbf{U}' . \end{aligned} \tag{11.2}$$

Substitution of $\mathbf{U}''$ from (4.37) and $\mathbf{U}' \cdot \mathbf{U}'$ from (4.24) yields

$$t_0^{(3)} = 2hR + m_{\rm b} + R\,\mathbf{U} \cdot \mathcal{L}^{\rm T}\,\mathbf{F} . \tag{11.3}$$

By definition $R'' = t_0^{(3)}$; hence this equation may also be used to obtain the *integrated* value of R as a numerical check of the basic relation (4.29). Although of harmonic oscillator type, it is not clear whether its use would present any practical advantage. Thus R is connected with $\mathbf{U}$ by a summation constraint and the regularized coordinates are already evaluated from linear equations of motion.

An inverse relation is required for interpolation within the interval $\Delta\tau$ in order to determine physical coordinates at a general time. This occurs when the force on other particles due to the KS pair is evaluated and the centre-of-mass (c.m.) approximation does not apply. Let $\delta t = t - t_0$ denote the subinterval since the last KS treatment. We can either obtain a solution of (11.1) by iteration or from the inverse expansion

$$\delta\tau = \sum_{k=1}^{n} \frac{1}{k!} \tau_0^{(k)} \delta t^k , \tag{11.4}$$

in which $n = 3$ usually suffices. From the definition (4.39), $\dot{\tau}_0 = 1/R$ and the next two terms are obtained by differentiation which gives

$$\begin{aligned} \tau_0^{(2)} &= -t_0^{(2)}/R^3 , \\ \tau_0^{(3)} &= (3t_0''^2/R - t_0^{(3)})/R^4 . \end{aligned} \tag{11.5}$$

The division by small values of R is not harmful because it does not affect the integration of the relative motion. Moreover, the c.m. approximation is used if the distance ratio, r_{ij}/R, exceeds some prescribed value (i.e. the parameter λ). Thus the inversion (11.4) is usually called for when the solution of (4.42) is obtained after predicting a nearby c.m. particle in connection with force evaluations. Note that all the quantities on the right-hand side of (11.5) are known for each KS solution.

The prediction of the corresponding values of $\mathbf{U}$ (and $\mathbf{U}'$ for Hermite) is carried out to highest order if required. Alternatively, if the distance ratio exceeds an appropriate value (e.g. 25), it is sufficient to expand the regularized interval to second order only. In this case, less accuracy is required and the coordinate prediction is performed to order $\mathbf{F}_{\rm U}$ (and $\mathbf{U}'$ to order $\mathbf{F}'_{\rm U}$).

Although the program structure for the two KS alternatives discussed here is formally similar, different prediction strategies were employed originally. Thus in the difference scheme it is natural to predict $\mathbf{U}$ and $\mathbf{U}'$ to highest order before evaluating the perturbation and then apply the corrector to all the derivatives according to (2.3). Since the Hermite corrector takes a simpler form, with two terms added to the low-order prediction, it is convenient to treat these contributions separately and this is also in keeping with the spirit of the standard formulation. However, a more careful analysis [Funato *et al.*, 1996; Mikkola & Aarseth, 1998] showed that it is advantageous to predict the regularized variables more accurately. The new procedure differs sufficiently from the basic approach to merit an outlined here.

We first perform a Taylor series prediction of the regularized coordinates and velocity to second or third order in the derivatives of $\mathbf{F}_U$, with the energy stabilization (4.43) included if the perturbation is suitably small (say $\gamma \simeq 0.01$–0.001). Using the transformed physical coordinates and velocity, the perturbation $(\mathbf{P}, \dot{\mathbf{P}})$ is obtained in the usual way. After obtaining the relevant expressions for the corrector, we form the higher Taylor series derivatives where all the terms are evaluated at the beginning of the step. Note that the stabilization factor may be saved at the prediction stage since it must be combined with $\mathbf{F}_{\rm U}$ again for the corrector.* This enables the complete solution for $\mathbf{U}$ and $\mathbf{U}'$ to be written in nested form, instead of combining the predicted values with the corrector. According to the usual rationale of the Hermite formulation, the standard correction procedure may still be used for the energy integration, provided h is predicted to order h'' only, otherwise the above scheme may readily be adopted. Recent experimentation shows that there is little gain in an extension to order $\mathbf{F}_{\rm U}^{(3)}$ in the prediction (with energy stabilization), whereas the addition of one more order to the integration is beneficial in return for about 20 extra operations. When combined with

* This point was not appreciated when using low-order prediction.

the slow-down algorithm, the Hermite KS method is quite effective and deserves attention because of its simplicity.

Let us next consider some aspects connected with the external tidal field. In the standard case (cf. (8.14)), the Coriolis terms are omitted from the perturbation because the contributions to h' cancel identically, as can be seen by taking the scalar product $\dot{\mathbf{R}} \cdot \ddot{\mathbf{R}}$ which implies $\dot{\mathbf{R}} \cdot (\mathbf{\Omega_z} \times \dot{\mathbf{R}}) = 0$. However, the angular momentum would be affected to a small extent. The rate of change of the specific angular momentum is given by $\mathbf{R} \times \ddot{\mathbf{R}}$. The contribution from the Coriolis force is then $-2\Omega_z(X\dot{X} + Y\dot{Y})$, with X, Y the relative coordinates in the plane. For constant semi-major axis this averages to zero over an orbit. Hence we neglect the effect of the rotating coordinate system on the longitude of the periapse since the orientation is of a random nature when it comes to encounters with other particles.

According to the equations of motion (8.14), the linear tidal field gives rise to the contributions

$$\dot{h}_{\rm tide} = \tfrac{1}{2}T_1 \frac{d}{dt}(x_k - x_l)^2 + \tfrac{1}{2}T_3 \frac{d}{dt}(z_k - z_l)^2 \,, \tag{11.6}$$

where $\mathbf{R} = \mathbf{r}_k - \mathbf{r}_l$. Assuming harmonic motion by $\mathbf{R} = \mathbf{A}\cos\omega t$, we obtain $-A_j^2 \cos\omega t \sin\omega t$ from each term. Here $\omega = 2\pi/t_{\rm K}$, with $t_{\rm K}$ the period. Ignoring the different signs, we integrate the change over an orbit for each component j which yields

$$\Delta h_{\rm tide} = A_j^2 T_j \omega \int_0^{t_{\rm K}} \sin\omega t \cos\omega t dt = 0 \,. \tag{11.7}$$

From the definition of the tidal radius (8.24), the relative galactic tidal perturbation in the x-direction for equal masses is

$$\gamma_{\rm tide} \simeq \tfrac{1}{2}N(R/r_{\rm t})^3 \,. \tag{11.8}$$

This is negligible for most KS binaries in large clusters but is included at present for general validity in case large separations should occur. Thus in the approximation (8.58), the magnitude is equivalent to a perturber distance of about $(2/N)^{1/3} r_{\rm t}$ which is usually sufficiently large for unperturbed motion. Note that the assumption of constant period leading to $\Delta h_{\rm tide} = 0$ does not apply in general; however, the effect of the perturbers will in any case dominate when the period is changing.

In the divided difference scheme we employ the standard expressions developed for direct integration, except that now we are dealing with the nine equations (4.37) and (4.38). Note that since the latter gives h' explicitly, we form the higher divided differences of this quantity. Again the respective polynomials are extended to fourth order and the corrector is included in the usual way. For this purpose a nominal reference time, τ_0, is defined and updated by the step $\Delta\tau$ every time. However, the decision-making for advancing KS solutions is based on the corresponding expression $t_i + \Delta t_i$, which is part of the scheduling, unlike in the standard Hermite case where all the physical time-steps are quantized and regularized solutions must be selected by a separate procedure.

Perturbers are determined from the general expression (8.58) for hard binaries, with the substitution $R = a(1+e)$ to allow for possible initialization at smaller separations (e.g. after chain regularization or tidal dissipation). For soft binaries (i.e. $E_b > \epsilon_{\rm hard}$), R is replaced by $\max\{f_\epsilon R_{\rm cl}, R\}$ with $f_\epsilon = 1 - |(E_b - \epsilon_{\rm hard})/\epsilon_{\rm hard}|$. This modification is intended to anticipate any significant expansion during the next orbit. At initialization, the actual apocentre distance is used, irrespective of energy, since the starting value of R may be relatively small.

The perturber selection is formally carried out at the end of the step following the apocentre passage, defined by $t_0'' t'' < 0$ and $R > a$. However, the old perturbers may be retained if the next estimated apocentre time occurs before the corresponding c.m. update, i.e.

$$t_K < t_0 + \Delta t_{\rm cm} - t\,. \tag{11.9}$$

Hence some unnecessary operations are saved for large values of $\Delta t_{\rm cm}/t_K$.

By analogy with the energy stabilization procedure in the difference formulation (cf. (4.43)), we have introduced rectification of the regularized elements at every apocentre subject to (11.9). This is achieved by the following algorithm. Given the correct value of h obtained from integrating h', we seek to modify $\mathbf{U}, \mathbf{U}'$ by the coefficients C_1, C_2 such that $\tilde{h} = h$, where $\tilde{h}$ is the explicit energy expression (4.24). Accordingly, we first write the energy relation in the form

$$h = (2\mathbf{U}'^2 C_2^2 - m_b)/\mathbf{U}^2 C_1^2\,. \tag{11.10}$$

By virtue of angular momentum conservation, $C_1 C_2 = 1$, from which we obtain

$$C_2 = \left\{\tfrac{1}{4} m_b/\mathbf{U}'^2 \pm [\tfrac{1}{2} hR/\mathbf{U}'^2 + (\tfrac{1}{4} m_b/\mathbf{U}'^2)^2]^{1/2}\right\}^{1/2}\,. \tag{11.11}$$

Here the square root sign is chosen according to the eccentric anomaly, with positive value if $\frac{1}{4} m_b/\mathbf{U}'^2 < 1$, or $R < a$. The resulting modifications then yield the corrected values

$$\begin{aligned} \tilde{\mathbf{U}} &= C_1 \mathbf{U}\,, \\ \tilde{\mathbf{U}}' &= C_2 \mathbf{U}'\,, \end{aligned} \tag{11.12}$$

which are used to initialize $R, \mathbf{U}_0$ and t_0''. Since these adjustments are usually very small, there are no further complications. We note that the additional effort amounting to some 13 additions, 22 multiplications, two divisions and two square roots is insignificant compared with the cost of an entire orbit integration. For completeness, an alternative procedure is included for any rare case, which might be a circularized orbit, if the inner square root argument is negative. In that case we choose

$$C_1 = m_{\rm b}/(-8hR\mathbf{U}'^2)^{1/2}\,, \tag{11.13}$$

together with $C_2 = 1$. This corresponds to $C_1 = 1$ for circular orbits or $C_1 = 1/(1-e^2)^{1/2}$ at pericentre or apocentre.

Finally, if $\Delta t_{\rm cm} < \Delta t_{\rm cl}$, a search is made for a possible chain regularization or hierarchical configuration at every apocentre consistent with (11.9) and also at the end of updates of unperturbed motion. The former check is in fact the last procedure of the KS integration cycle.

11.3 Stumpff Hermite version

Since the Stumpff version of KS regularization is now the method of choice, it is useful to concentrate on some aspects of implementation in a separate section. This method differs significantly in construction from the alternatives discussed in chapter 4, particularly when it comes to improving the solution by iteration.

Using the notation of section 4.7, we begin by writing the basic equations of motion at the start of an integration step in Hermite form as

$$\begin{aligned}\mathbf{U}_0^{(2)} &= -\Omega_0\mathbf{U}_0 + \mathbf{f}_0^{(2)}\,,\\ \mathbf{U}_0^{(3)} &= -\Omega_0\mathbf{U}'_0 + \mathbf{f}_0^{(3)}\,,\end{aligned} \tag{11.14}$$

where $\Omega = -\frac{1}{2}h$ and $\mathbf{f}_0^{(2)} = \frac{1}{2}R\mathbf{Q}$ with $\mathbf{Q} = \mathcal{L}(\mathbf{U})\mathbf{P}$ as the perturbed force function, evaluated at the end of the previous corrector cycle. Likewise, the last term of the second equation (11.14) is obtained by comparing with the differentiated form of (4.69) and setting $\Omega = \Omega_0$,

$$\mathbf{f}_0^{(3)} = \tfrac{1}{2}R\mathbf{Q}' + \tfrac{1}{2}t''_0\mathbf{Q} - \Omega'_0\mathbf{U}\,. \tag{11.15}$$

The integration cycle begins by standard prediction of the coordinates and velocities of all perturbers and the c.m. particle. The prediction of $\mathbf{U}$, R and $\mathbf{U}'$ is carried out to highest order which includes the pairwise Stumpff [1962] functions (4.59), $\tilde{c}_4, \tilde{c}_5$ and $\tilde{c}_3, \tilde{c}_4$, respectively, as factors in the two derivatives $\mathbf{U}^{(4)}$ and $\mathbf{U}^{(5)}$. After transformation to global coordinates and velocity, the perturbations $\mathbf{P}, \mathbf{P}'$ are obtained in the usual way and the slow-down factor κ (to be described later) is included if relevant.

The corrector cycle first evaluates the new perturbative functions

$$\begin{aligned} \mathbf{f}^{(2)} &= (\Omega_0 - \Omega)\mathbf{U} + \tfrac{1}{2}R\mathbf{Q}\,, \\ \mathbf{f}^{(3)} &= (\Omega_0 - \Omega)\mathbf{U}' - \Omega'\mathbf{U} + \tfrac{1}{2}R'\mathbf{Q} + \tfrac{1}{2}R\mathbf{Q}'\,. \end{aligned} \tag{11.16}$$

It can be seen that the first of these equations takes its form from (4.69), where the analogy with an Encke-type formulation is evident. This enables construction of the higher derivatives by the Hermite rule (2.20), which yields the expressions

$$\begin{aligned} \mathbf{U}_0^{(4)} &= -\Omega_0\mathbf{U}_0^{(2)} + \mathbf{f}_0^{(4)}\,, \\ \mathbf{U}_0^{(5)} &= -\Omega_0\mathbf{U}_0^{(3)} + \mathbf{f}_0^{(5)}\,. \end{aligned} \tag{11.17}$$

Note that $\mathbf{Q}'$ is obtained from $\mathbf{P}' = R\dot{\mathbf{P}}$ together with $\mathcal{L}(\mathbf{U}')$. The provisional solution of the first equation (4.67) and its derivative, $\mathbf{U}'$, is improved by one full iteration of (11.16) and (11.17) without recalculating the perturbations. Hence the nested solution is evaluated twice to full order, including the pairwise Stumpff functions. As compensation for the extra effort of the iteration, high accuracy is achieved for a somewhat larger integration step.

According to the second equation (4.56), the energy integration

$$\Omega' = -\mathbf{U}' \cdot \mathbf{Q} \tag{11.18}$$

remains the same as for the standard Hermite case. Substitution for $\mathbf{U}^{(2)}$ gives rise to the second derivative

$$\Omega^{(2)} = \Omega_0\mathbf{U} \cdot \mathbf{Q} - \mathbf{f}^{(2)} \cdot \mathbf{Q} - \mathbf{U}' \cdot \mathbf{Q}'\,. \tag{11.19}$$

The two Hermite corrector terms formed from Ω' and $\Omega^{(2)}$ are added to the predicted value to yield an improved solution for Ω at the start of an iteration or at an endpoint.

The corrector cycle ends by saving all the derivatives $\mathbf{U}^{(n)}$, $n = 2, 3, 4, 5$, as well as the perturbative derivatives (11.16). For this purpose it is advantageous to employ the final values of R and R' since the re-evaluation of (11.16) is fast and these expressions also enter $\mathbf{U}^{(2)}$ and $\mathbf{U}^{(3)}$. Finally, it is beneficial to improve the penultimate derivatives by addition of the next order,

$$\begin{aligned} \mathbf{U}^{(4)} &= \mathbf{U}_0^{(4)} + \mathbf{U}_0^{(5)}\Delta\tau\,, \\ \Omega^{(3)} &= \Omega_0^{(3)} + \Omega_0^{(4)}\Delta\tau\,. \end{aligned} \tag{11.20}$$

Unless the step $\Delta\tau$ is constant, new coefficients $\tilde{c}_n$ must be re-evaluated every time. The summation to twelfth order does represent a modest extra cost, estimated as a few per cent of one KS step; however, the improvement in accuracy is substantial [cf. Mikkola & Aarseth, 1998]. The coefficients $\tilde{c}_n$ for $n = 3, 4, 5$ are generated by recursion from $n = 12$. The argument $4z$ then yields $\tilde{c}_5(4z)$ and $\tilde{c}_6(4z)$ by the same method; hence orders $n < 5$ need not be evaluated.[†] This task is most conveniently carried out just after the determination of the regularized step, (8.59), since the next physical time-step (11.1) must be modified by including the coefficients $\tilde{c}_5(4z)$, $\tilde{c}_6(4z)$ in the last two terms (cf. (4.67)). At the same time, the reference energy Ω_0 must be updated and saved. Accordingly, this scheme requires 16 additional variables for each KS pair, including seven Stumpff functions, compared with the standard Hermite KS formulation.

Finally, we note some similarity of this scheme to the recalculation of the dominant term in the time-symmetric method [Kokubo, Yoshinaga & Makino, 1998], discussed in section 2.7. Thus, for small perturbations, the term $\frac{1}{2}h\mathbf{U}$ and its derivative are analogous to the role of the dominant solar interaction. Moreover, the regularized step is also fairly constant if the perturbation is small. Note that a constant value of $\Delta\tau$ does not avoid a re-evaluation of the Stumpff functions because the argument $\Omega\Delta\tau^2$ would still be changing slightly.

11.4 KS termination

The actual termination of two-body regularization requires a number of special procedures to be carried out and is best described in a separate section. In the following we assume the Stumpff version, although many algorithms are fairly similar for the alternative formulations. The main steps are summarized in Algorithm 11.1.

Unless collision occurs, it is advantageous to advance the KS solution to the end of the block-step. In the case of large perturbations, this is achieved by delaying the termination procedure until the physical time-step is smaller than the remaining interval. Alternatively, one or more standard KS steps are carried out *in situ* if required. The residual interval $\delta\tau$ is then obtained by one Newton–Raphson iteration of (11.1) to third order, whereupon the final integration step is performed without any further complications. Since the Stumpff functions depend on $\delta\tau$ by (4.59), the coefficients also need to be recalculated. Consequently, the global time is now defined to be at the end of the block-step.

Any dominant perturbers are predicted to high order before the potential energy interaction between the components and all the perturbers is obtained. Following rectification of $\mathbf{U}$ and $\mathbf{U}'$ to ensure consistency, the

[†] A general algorithm for large arguments is given in the original paper. However, this is not needed here since $z = 0.25\,\eta_U^2$ for unperturbed motion and therefore small.

Algorithm 11.1. *Termination of KS regularization.*

1	Advance the KS solution for pair i until $t_{\rm block} - t_{2i-1} < \Delta t_{2i-1}$
2	Take the final step with $\delta\tau$ obtained by third-order iteration
3	Predict coordinates and velocity for any dominant perturber
4	Evaluate potential energy of binary components and perturbers
5	Rectify the orbit to yield $\mathbf{U}, \mathbf{U}'$ consistent with the value of h
6	Determine current coordinates and velocities of KS components
7	Save basic KS variables in case of a stable hierarchical merger
8	Obtain improved coordinates and velocities of the components
9	Include differential energy correction following rectification
10	Copy modified c.m. neighbour list and radius for the components
11	Move common arrays of components unless most recent KS pair
12	Reduce the pair number, $N_{\rm p}$, by one and redefine $I_{\rm s} = 2N_{\rm p} + 1$
13	Update all relevant lists to correspond with the new sequence
14	Initialize force polynomials and time-steps for the components

differential energy correction is obtained and added to the term $E_{\rm coll}$ for conservation purposes.

If the terminating KS pair is not the last in the sequence (i.e. $i < N_{\rm p}$), all the relevant common arrays are moved down to the first single particle locations, $2N_{\rm p} + 1$, $2N_{\rm p} + 2$, while the subsequent quantities are compressed, and likewise for the corresponding KS and c.m. arrays. It is also necessary to update all common array lists since global locations are referred to. This entails replacement of the c.m. by its components and reduction of any subsequent members, $j > N + i$, by one. Although this is an $O(N)$ procedure, only a few operations per particle are needed, and the overheads for the sequential data structure are less than the cost of polynomial initialization.[‡] In any case, the number of KS terminations is usually quite modest. Finally, in addition to step 14, the force polynomial for any dominant perturber that triggers termination is also initialized in order to be consistent with the current configuration.

Time quantization in the Hermite scheme has certain implications for the time-steps relating to close two-body encounters. In the first place, KS solutions are inevitably initialized at small values of Δt_i, which usually means few block-step members. Consequently, both the irregular and regular steps for the associated c.m. body are severely restricted by the small block-step even if the natural step is large. The inefficiency on termination is less significant, since such procedures are performed at arbitrary block times, where at least the regular time-steps may be assigned more typical values. To compensate, most KS solutions are connected with

[‡] This probably also remains true when obtaining $\mathbf{F}$ and $\dot{\mathbf{F}}$ on the HARP.

binaries which tend to have long life-times and the initial c.m. steps also increase quickly if conditions are favourable. The question remains, however, whether improved strategies can be devised.

11.5 Unperturbed two-body motion

The perturber search for hard binaries frequently results in zero membership for the adopted criterion (8.58). This situation is inevitable, given the N-dependence of the mean particle separation and the typical perturber distance $\lambda R_{\rm cl}$, particularly since many primordial binaries are superhard. If no perturbers are identified following a new search at apocentre, the two-body motion is defined to be unperturbed during the next orbit, with the time-step given by the period, $\Delta t = t_{\rm K}$. Consequently, only the centre of mass needs to be advanced, with the further simplification that the force evaluation is identical to that of a single particle.

The status of an unperturbed binary needs to be rechecked one period later. However, by a more careful analysis of the associated neighbour velocity field, it is possible to estimate the time for approaching particles to come within a certain distance and hence extend the interval of unperturbed motion. The main steps of this procedure are set out below.

From the corresponding list[§] of c.m. neighbours, we determine the particle, j, giving the maximum force as well as the smallest *inverse* travel time, $\beta_s = \mathbf{r}_s \cdot \dot{\mathbf{r}}_s / r_s^2$ for particle index s. In the following all distances and velocities are expressed with respect to the c.m. Although there may not be any approaching particles, we still use this quantity which defines a radial velocity $\dot{r}_s = r_s |\beta_s|$. Let us introduce the perturber boundary for any particle mass $\tilde{m}$ by

$$r_\gamma = R[2\tilde{m}/(m_{\rm b}\gamma_{\rm min})]^{1/3}\,, \tag{11.21}$$

with $m_{\rm b}$ the mass of the binary. The time to reach this boundary with inward motion is then

$$\Delta t_{\rm in} = (r_s - r_\gamma)/|\dot{r}_s|\,, \tag{11.22}$$

with m_s substituted for $\tilde{m}$ in (11.21).

We now evaluate a second travel time for the same particle, based on the acceleration, by

$$\Delta t_{\rm a} = [2\Delta t_{\rm in}|\dot{r}_s| r_s^2/(m_{\rm b} + m_s)]^{1/2}\,. \tag{11.23}$$

Similarly, if $j \neq s$ the dominant body, which may be near a turning point, would have a return time

$$\Delta t_j = [2(r_j - r_\gamma) r_j^2/(m_{\rm b} + m_j)]^{1/2}\,, \tag{11.24}$$

[§] With *NBODY4*, the list is obtained on HARP using an appropriate argument.

with m_j substituted in (11.21). The next unperturbed time interval is then chosen by

$$\Delta t_\gamma = \min\{\Delta t_{\rm in}, \Delta t_{\rm a}, \Delta t_j\}\,. \qquad (11.25)$$

In case the background force dominates, it is also prudent to include the safety condition $\Delta t_\gamma = \min\{\Delta t_\gamma, 2\Delta t_{\rm cm}\}$.

An additional check is performed if the final value of the time interval exceeds the period. Thus if we have $\Delta t_\gamma < 2 \times 10^9 t_{\rm K}$, the number of unperturbed periods is chosen conservatively by

$$K = 1 + \tfrac{1}{2}\Delta t_\gamma / t_{\rm K}\,. \qquad (11.26)$$

Finally, the corresponding time-step itself is taken to be

$$\Delta t_k = K \min\{t_{\rm K}, \Delta t_{\rm cm}\}\,, \qquad (11.27)$$

where $k = 2(i - N) - 1$ denotes the first KS index of the c.m. particle i. The latter precaution is included if the period exceeds $\Delta t_{\rm cm}$. In the case of an extremely short period, with $K > 10^9$, it is sufficient to use the c.m. step itself.

Given the neighbour list, the above procedure is relatively inexpensive and may be speeded up further by restricting the initial search distance. The use of two time-scales provides for the possibility of a more distant perturber approaching first. If there is no such particle and the dominant body is also moving outwards, the above expressions are conservative. Hence the main purpose of the algorithm is to prevent the unperturbed condition from being violated during the next interval, which is chosen as large as possible subject to kinematical considerations.

If it turns out that $\Delta t_k < t_{\rm K}$, a new perturber search is performed in the usual way. In the unlikely event that no perturber is selected, the motion is assumed to be unperturbed during the next period. Alternatively, an update of the physical coordinates and velocity takes place, followed by initialization of the KS polynomials using a slow-down factor $\kappa = 1$.

For completeness, we remark that a counter accumulates the number of elapsed unperturbed orbits, given by

$$n_{\rm K} = (t - t_0)/t_{\rm K}\,, \qquad (11.28)$$

where t_0 refers to the previous epoch. If this counter exceeds the integer limit 2×10^9, it is set to zero again and a second counter is advanced by one. This situation occurs quite frequently in large simulations and demonstrates the necessity of introducing unperturbed two-body motion. An algorithm for partial unperturbed motion, no longer used in the codes, is given by Appendix D.5.

11.6 Slow-down in KS

The theory of the adiabatic invariance for weakly perturbed binaries in chain regularization is given in chapter 5, together with the algorithm. The same principle can be used in KS regularization, where weak perturbations are a characteristic feature to be exploited. Since explicit information is available for the two-body elements, it is possible to change the strategy and only re-evaluate the slow-down factor at the apocentre. In the following we describe the implementation in the Stumpff KS formulation [Mikkola & Aarseth, 1998], but all the procedures have also been adopted in the older codes, *NBODY*3 and *NBODY*5.

Given the basic equations of motion (4.37) and (4.38), the slow-down concept is introduced by scaling the perturbing force and its derivative according to

$$\begin{aligned} \widetilde{\mathbf{P}} &= \kappa \mathbf{P}\,, \\ \widetilde{\mathbf{P}}' &= \kappa \mathbf{P}'\,, \end{aligned} \tag{11.29}$$

where the dimensionless factor κ plays the same role as in chain regularization. In addition, (4.39) should be replaced by

$$t' = \kappa\, \mathbf{U} \cdot \mathbf{U}. \tag{11.30}$$

Hence one regularized period now represents κ actual periods, with the regularized time-step chosen as before. Now the required modifications are simple and only represent minor complications. However, in order to achieve an optimized scheme the determination of κ provides a new challenge. This is desirable since a typical N-body simulation with primordial binaries usually involves large numbers of perturbed periods, with a significant proportion having small perturbations. It is these weakly perturbed binaries that can be studied advantageously with the slow-down procedure.

The value of κ is most conveniently determined at an apocentre point, where the magnitude (but not necessarily the effect) of the perturbation tends to be largest. Not every apocentre passage needs be considered. Thus we only perform this check at the same time as the redetermination of new perturbers, defined as the time for which $\kappa\, t_{\rm K} > t_{\rm cm} + \Delta t_{\rm cm} - t$, where $t_{\rm K}$ denotes the binary period and $t_{\rm cm}$ is the time of the last c.m. integration endpoint.

In order to implement the slow-down procedure, we define κ to be an integer and introduce a hierarchical slow-down vector $I_{\rm sl} = 2^{k-1}$, where k takes the values $1, 2, \ldots$ up to some maximum (say 10). Given the relative perturbation γ, the steps for obtaining the new κ from the old value κ_0 are set out in Algorithm 11.2, originally given in Mikkola & Aarseth [1996].

Algorithm 11.2. *Steps for changing the slow-down factor.*

---

1 Estimate the slow-down factor by $\tilde{\kappa} = \gamma_0/\gamma$, where γ_0 has a suitably small value (say 5×10^{-5})
2 Adopt the standard procedure with $\kappa = 1$ if $\tilde{\kappa} \le 1$; continue the integration if $\kappa_0 = 1$, otherwise go to step 10
3 Determine the largest provisional new index K by the condition $2^{K-1} < \tilde{\kappa}$, with a maximum permitted increase of two levels
4 Obtain the time interval Δt for the perturbation to reach γ_0 by analysing the relative motion of the perturbers
5 Reduce the level K if the provisional factor would yield too large interval, such that $t_{\rm K} I_{\rm sl}(K) < \Delta t$, and update κ
6 Evaluate the eccentric anomaly and obtain the time interval $\delta t_{\rm ap}$ since apocentre passage by use of the eccentric anomaly
7 Find the corresponding regularized time interval $\delta\tau$ by Newton–Raphson iteration from $\delta t_{\rm ap} = \frac{1}{6} t^{(3)} \delta\tau^3 + \frac{1}{2} t'' \delta\tau^2 + R\delta\tau$
8 Integrate $\mathbf{U}, \mathbf{U}'$ and h a step $\delta\tau$ (< 0) back to the apocentre [include new Stumpff coefficients if relevant]
9 Predict the current coordinates and velocities for the perturbers, c.m. and KS components
10 Initialize the KS solution for the standard difference formulation or the Hermite schemes with the new value of κ

Some comments on this algorithm may be helpful. The interval Δt at step 5 is derived by the procedure of the previous section, substituting γ_0 for $\gamma_{\rm min}$ in the relevant expressions. In practice, $\gamma_0 = 5 \times 10^{-5}$ is a good choice for this parameter if $\gamma_{\rm min} = 1 \times 10^{-6}$. The eccentric anomaly, θ, is most conveniently obtained by solving the companion equations

$$
\begin{aligned}
e \cos\theta &= 1 - R/a \,, \\
e \sin\theta &= t_0''/(m_{\rm b} a)^{1/2} \,.
\end{aligned}
\tag{11.31}
$$

From Kepler's equation, the pericentre passage time is then

$$
\delta t_{\rm p} = [\theta - t_0''/(m_{\rm b} a)^{1/2}] t_{\rm K}/2\pi \,, \tag{11.32}
$$

which gives the apocentre time $\delta t_{\rm ap} = \frac{1}{2} t_{\rm K} + \delta t_{\rm p}$. The integration back to the apocentre with an interval $\delta t = -\delta t_{\rm ap}$ before κ has been changed by a discrete factor ensures that the same value is employed during the whole orbit. This is an essential difference from chain regularization, where κ changes slowly after each step when using a self-starting integrator. Another difference is that the effect of changing κ enters directly through the equations of motion and no energy correction is needed. The number of such changes is in any case a small fraction of the number of regularized steps. Thus the block-step scheme that changes κ by a factor of 2 (or 4) is also working well here. Finally, the initialization of KS polynomials is made consistent with the new value of κ by the scaling (11.29).

In principle, the slow-down treatment can be extended to arbitrarily large values of κ. However, it is convenient to employ unperturbed two-body motion if the relative perturbation at apocentre falls below some specified value, traditionally taken as $\gamma_{\rm min} = 10^{-6}$. This facilitates decision-making in the force calculation, since the c.m. approximation is used if there are no perturbers. Otherwise a distance test is required. The relevant procedures for unperturbed motion are given in the previous section.

11.7 Hierarchical mergers

The identification of stable hierarchies uses some of the procedures employed for the chain regularization search. At first sight this may appear incompatible since the latter deals with highly unstable systems. However, as remarked earlier, both subsystem types are usually comparable in size to the close encounter distance and, consequently, any participating binary will inevitably have a small c.m. time-step. Again we recall that, for historical reasons, the terminology *merger* applies to the temporary formation of a stable hierarchical system where the inner binary components are combined into one composite body which is initialized on termination.

The condition $\Delta t_{\rm cm} < \Delta t_{\rm cl}$ at an apocentre passage indicates a possible case for treatment. A general situation is illustrated in Fig. 11.1. Thus in the first instance we need to distinguish between hierarchical structure and strong interactions, but it is desirable to exclude unsuitable configurations at an early stage. This is done by the simple evaluations and checks of Algorithm 11.3.

Some of the above conditions rule out chain regularization which will be discussed in a subsequent chapter. The strategy is based on determining the most likely merger candidate together with the strongest perturber. Note that in the AC codes the neighbour list is used if there are less than three perturbers. After excluding positive radial velocity with respect to the binary, the pericentre separation, $R_{\rm p}$, essentially distinguishes the type of interaction, although stable subsystems require $a_{\rm out} > 0$ as well. In either case the dominant perturber may be another binary, and for cross-section purposes the two semi-major axes are simply added. The vectorial perturbation, $\mathbf{P}$, with respect to $m_i + m_j$ is obtained at step 5. Rare configurations of parabolic type may also be accepted for temporary merger in the pericentre region (step 8), since the associated time-steps

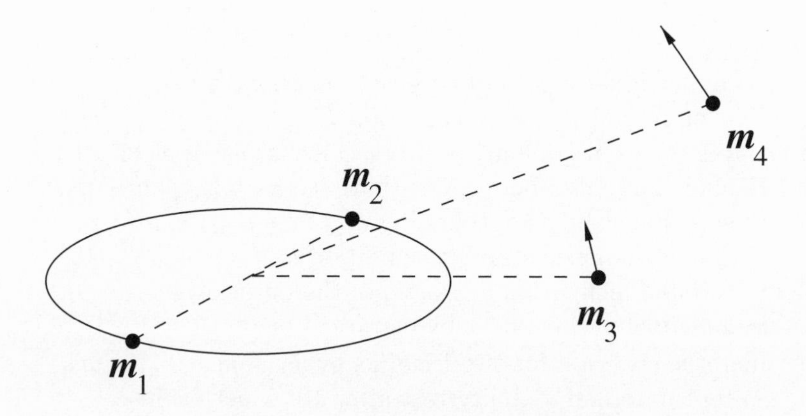

Fig. 11.1. Binary with two perturbers.

Algorithm 11.3. *Hierarchical system search.*

1	Identify the dominant and second perturber, j and p, from m/r^3
2	Form $\mathbf{r} \cdot \dot{\mathbf{r}}$ for m_j and $\gamma_3 = 2m_p/m_{\rm b}(R/r_p)^3$ with respect to m_i
3	Exit on positive radial velocity or $\gamma_3 > 10^4 \gamma_{\rm min}$
4	Evaluate the minimum distance, $R_{\rm p} = a_{\rm out}(1 - e_{\rm out})$
5	Obtain actual perturbing force, $\mathbf{P}$, and $\tilde{\gamma} = \lvert\mathbf{P}\rvert r_j^2/(m_i + m_j)$
6	Define perturbation due to m_p by $\gamma_4 = 2(r_j/r_p)^3 m_p/(m_i + m_j)$
7	Combine semi-major axes if second binary is present; i.e. $j > N$
8	Check stability if $e_{\rm out} > 0.99$, $R_{\rm p} > 10R$, $\max\{\tilde{\gamma}, \gamma_4\} < \gamma_{\rm max}$
9	Continue checking if $r_j > \max\{3R_{\rm grav}, R_{\rm cl}\}$, $R_{\rm p} > 2a_{\rm in}(1 + e_{\rm in})$
10	Consider stability of two binaries if $j > N$ and $R_{\rm p} > a_{\rm in} + a_2$

can become extremely small without significant energy exchange. Subsystems with $r_j > \max\{3R_{\rm grav}, R_{\rm cl}\}$ or large $R_{\rm p}$ are excluded from chain regularization, where $R_{\rm grav}$ is the characteristic gravitational radius defined by (9.10). Finally, in the case of two binaries (step 10), such a system is considered for merging instead if $R_{\rm p} > a_{\rm in} + a_2$, where a_2 is the semi-major axis of the second binary. However, there is usually an intermediate region where neither treatment is suitable; i.e. stability is not satisfied and chain regularization would be inefficient.

Having established a possible hierarchical configuration, we now need to examine various supplementary criteria in addition to the basic one given by (9.14). Some of the main points for consideration are listed by Algorithm 11.4.

As can be seen, the environment is also considered in order to ensure a significant life-time. Since configurations outside the core are likely to survive longer, the energy condition may be reduced for larger distances.

Algorithm 11.4. *Hierarchical stability checks.*

1	Delay for inner region of eccentric orbit: $r_j < a_{\rm out}$, $e_{\rm out} > 0.95$
2	Exit on modified energy condition $E_{\rm out} > \frac{1}{4}\epsilon_{\rm hard}/(1 + r_i/r_{\rm h})$
3	Check for tidal circularization or collision using $a_{\rm in}(1 - e_{\rm in})$
4	Exclude large two-body perturbations: $\gamma_4 > \gamma_{\rm max}$ or $\tilde{\gamma} > 0.05$
5	Ensure sufficient perturbers by $\lambda a_{\rm out}(1 + e_{\rm out}) < 2R_{\rm s}$
6	Evaluate the basic stability condition $a_{\rm out}(1 - e_{\rm out}) > R_{\rm p}^{\rm out}$
7	Determine inclination angle, ψ, of the outer orbit
8	Include inclination effect by empirical factor $f_0 = 1 - 0.3\psi/\pi$
9	Increase criterion for two binaries by $\tilde{f}_0 = f_0 + 0.1(a_2/a_{\rm in})$
10	Accept marginal stability test after 10^4 inner orbits
11	Perform Zare exchange stability test for small inclinations

Accepting $E_{\rm out}$ as one quarter of the hard binding energy, $\epsilon_{\rm hard}$, is perhaps on the generous side, but hierarchies tend to be massive and therefore more robust. In any case, the requirement of sufficient perturbers may be a limiting factor for neighbour scheme codes, particularly with modest values of N. We relax the standard condition slightly by a factor of 2, because the neighbour radius is increased gradually after a new initialization if necessary.

It is well known that inclined systems are more stable for the same value of the outer pericentre distance [e.g. Harrington, 1972]. Consequently, we modify the basic stability criterion (9.14) by an empirical factor (step 8), with the inclination ψ in radians, such that the correction for retrograde motion represents 30%. We evaluate the inclination angle from the scalar product of the two relevant angular momenta, which gives $\cos\psi$. The case of two binaries in a bound orbit requires further attention. Based on experience with the binary–tides problem for two extended stars [Mardling, 1991], the appropriate correction factor is taken as 10% of the smallest semi-major axis ratio. At the same time, the definition of primary and secondary binary is reversed if the latter gives a larger value of $R_{\rm p}^{\rm out}$. With regard to step 8 of Algorithm 11.3, the first five steps of Algorithm 11.4 are omitted here in order to increase the chance of acceptance. Moreover, if $e_{\rm out} > 0.96$, the factor f_0 is reduced further by $10(e_{\rm out} - 0.96)$ to account for the increased practical stability.

On rare occasions, hierarchies may persist over long times even if the stability tests fail. Such configurations often have very large outer eccentricity (say $e_{\rm out} > 0.99$), which results in short times for random energy exchanges in the pericentre region. From chaos theory it is well known that a so-called 'fuzzy boundary' exists; this provides conditions for relatively long life-times. Since we are concerned with *practical* rather than absolute stability here, it may be justified to employ the following heuristic test. Thus if $d_{\rm min}$ falls within the boundary region $f_0\, R_{\rm p}^{\rm out}$ and $0.6\, R_{\rm p}^{\rm out}$, we define *marginal* stability after a large number of attempts (say 10^4–10^5 inner orbits).

It should be noted that the stability condition discussed above refers to the outer component not being able to escape. In this context the Zare [1977] exchange criterion (9.15) is also relevant. Defining the stability measure in terms of the total angular momentum and energy by

$$S = \mathbf{J}^2 E/(\mathbf{J}^2 E)_{\rm crit}\,, \tag{11.33}$$

it has been shown that no exchange can occur if $S > 1$. However, $S < 1$ does not imply exchange since this condition is necessary but not sufficient. The essential reason is that the angular momentum of the outer orbit has opposite sign for retrograde motion and the degeneracy in $\mathbf{J}$ cannot be distinguished from a case of small angular momentum when both motions are in the same sense, and the system is clearly unstable. The exchange criterion is therefore of limited use, but has nevertheless been included for small inclinations since violation would affect stability. Note also that for coplanar motion, the stability boundary (9.14) lies above the exchange limit for $q_{\rm out} < 5$, which is predominantly the case in realistic simulations [cf. Mardling & Aarseth, 2001].

Once formed, a hierarchy may combine with another particle or binary to produce a higher-order binary system of up to a current limit of six members or two subsystems may become bound with small enough eccentricity to be stable. An example of the former type is illustrated in Fig. 11.2. Direct integration of such structures is sufficiently time-consuming for a generalization of the above scheme to be worthwhile since otherwise the loss of efficiency can be significant. The identification of stable higher-order systems is carried out using all the procedures above where some quantities are also used for possible chain regularization tests. However, the conditions for higher multiplicity become more favourable in the so-called 'post-collapse phase' when the central density has decreased, particularly in simulations with mass loss from evolving stars. Hence there is ample hierarchical space for more complex configurations to exist, although their formation rate may be low. Since there is no distinction between different levels of hierarchies (e.g. restricting the dominant perturber m_j to be a binary), two such systems can in principle be combined and, although unlikely, a few examples are seen.

Up to now we have considered algorithms for accepting stable hierarchies of variable membership. Once this task has been carried out, there are still several technical problems to be considered before the solution can be treated as one KS system. Some of the essential stages are presented in Algorithm 11.5.

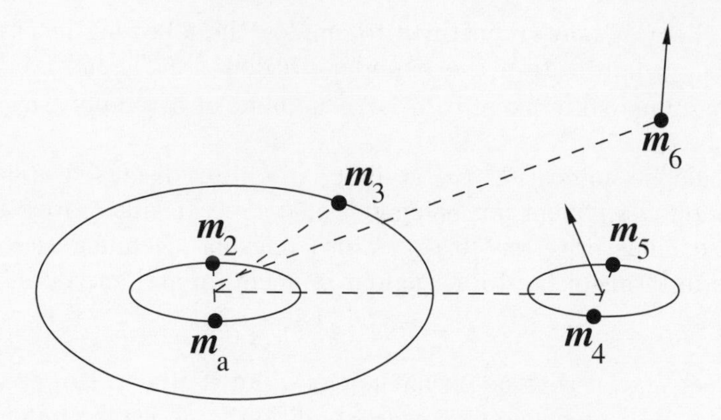

Fig. 11.2. Hierarchical configuration with subsystems.

Algorithm 11.5. *Initialization of merger.*

1	Choose the primary binary to be widest [case $j > N$ only]
2	Save the component masses and neighbour or perturber list
3	Terminate KS regularization of primary and save elements
4	Obtain potential energy of neighbours and primary components
5	Record final values $\mathbf{R}$ and $\dot{\mathbf{R}}$ of the primary components
6	Set initial conditions for the c.m. body by $m_{\rm cm} = m_i + m_j$
7	Create ghost particle $m_l = 0$ with large values of $\mathbf{r}_0, \mathbf{r}$ and t_0
8	Form tidal potential energy and define ghost binary [case $j > N$]
9	Evaluate potential energy of neighbours and inner body, $m_{\rm cm}$
10	Increase the neighbour list if $\lambda a_{\rm out}(1 + e_{\rm out}) > 2R_{\rm s}$
11	Initialize KS polynomials for new binary $m_{\rm cm}$ and m_j
12	Maintain the ghost name for identification at termination
13	Specify negative c.m. name for the decision-making
14	Update $E_{\rm merge}$ by the old binding energy and tidal correction

This sequence is also similar for all the standard N-body codes, except that the AC versions use the neighbour list instead of the perturber list. In the case of two binaries, we select the widest one as primary. Thus for stability checking purposes, a quadruple of unequal binary size may be considered as a degenerate triple and this also gives an advantage at termination (see below). The merger procedure introduces a small discontinuity in the equations of motion since the local force field is simplified to one binary. Hence in order to maintain energy conservation, we add differential corrections due to the potential energy change with respect to the neighbours or perturbers, both for the primary and any secondary.

The new c.m. of the inner components is formed in the usual way. A novel feature here is the introduction of so-called 'ghost particles' in order to maintain relevant quantities of the original second KS component, which is now replaced by m_j. This is achieved by setting $m_l = 0$ after step 6, together with large values of coordinates and the reference time, t_0. Hence special care is needed to avoid removing distant members of zero mass as escapers. Moreover, all the relevant prediction variables, including $\mathbf{D}^3$ or $\mathbf{F}^{(3)}$, are set to zero in order to avoid large numerical terms during predictions. Such a particle will therefore not be advanced, nor will it be part of any neighbour list, once removed. In the case $j > N$, the corresponding binary becomes a ghost. This entails specifying zero component masses and perturber number after carrying out the tidal energy correction. At this stage, the new KS solution can be initialized and all relevant lists made consistent with the new sequence. According to the data structure, all ghost arrays are then exchanged with the second KS component but the precise location is not recorded because it can be recovered later by identifying the relevant name. To facilitate the decision-making, the new c.m. is given a negative name, $\mathcal{N}_{\rm cm} = -\mathcal{N}_k$, related to the name of the first KS component with original mass m_k and the new name is saved. Finally, the dormant binding energy (or energies) is added to the merger energy, $E_{\rm merge}$, together with the relevant tidal corrections. Hence the correction procedures enable an energy-conserving scheme to be maintained throughout the calculation.

It remains to consider termination of the hierarchical motion which is the reverse procedure of initialization. Again we list the essential steps in Algorithm 11.6 for clarity.

Algorithm 11.6. *Termination of merger.*

1 Locate the current index in the merger table
2 Save the neighbour or perturber list for corrections
3 Predict the c.m. motion to highest order
4 Terminate the KS regularization in the standard way
5 Add the old outer component to the perturber list
6 Obtain the potential energy of neighbours and primary c.m.
7 Identify the location, j, of the corresponding ghost particle
8 Restore the basic quantities of the primary components
9 Form new force polynomials for the old outer component
10 Copy the saved KS variables and initialize the primary binary
11 Set unperturbed polynomials for second binary [case $j > N$]
12 Subtract new energy μh from $E_{\rm merge}$ and add tidal correction
13 Compress the merger table, including any escaped hierarchies

Since an arbitrary number of hierarchies may exist at the same time, the procedure begins by locating the desired entry corresponding to the c.m. name. Likewise, the relevant ghost particle is identified by comparing the names of all particles with the separately saved name. The individual masses, coordinates and velocities of the primary components are reconstructed from the saved quantities (cf. Appendix D.2), whereupon the force acting on the old outer component is initialized as for a single particle even if $j > N$, since the perturbation tends to be smaller. After restoring the basic KS quantities, the primary binary is initialized as a perturbed two-body solution, followed by the same procedure for any second binary. The relevant binding energies are subtracted from the merger energy, $E_{\rm merge}$, together with the tidal corrections. Note that this quantity does not revert to zero after one episode since the duration may be significant, during which the environment is subject to change. Finally, the merger table is updated and the entries of any escaped hierarchies (which leave behind a remnant merger energy) are removed for convenience.

As noted above, higher-order systems are also catered for. Although the identification and termination procedures are essentially identical, the rest of the treatment differs in certain respects. To avoid repetitions, some of the relevant algorithms are therefore given in Appendix C.

11.8 Tidal circularization

The process of circularization in close binaries involves several aspects of stellar structure that are not well understood. In particular, a better treatment of the damping time-scale for different stellar types is needed. However, the qualitative features are sufficiently well known to warrant a detailed model which increases the realism of star cluster simulations. Since such binaries are invariably treated by KS or chain regularization, the orbital adjustments are carried out in terms of well-behaved two-body elements. In the following we distinguish between *continuous* and *sequential tidal circularization.* The former process is characterized by a fully consistent treatment of stellar structure [cf. Mardling & Aarseth, 2001] and may be preceded by a stage of chaotic evolution, whereas the latter involves a step-wise adjustment of the orbital parameters [Portegies Zwart *et al.*, 1997].

We first consider the simpler case of sequential circularization which is also the standard implementation in the code *NBODY6* and optional in *NBODY4*. The basic condition for modifying the binary elements is

$$R_{\rm p} < 4r_1^*\,, \tag{11.34}$$

where $R_{\rm p}$ is the predicted pericentre distance[¶] for perturbed or

[¶] Obtained by the regular expression (9.23).

unperturbed motion and $r_1^* = \max\{r_k^*, r_l^*\}$ is the largest stellar radius. The orbit is then integrated or predicted back to the actual pericentre before proceeding. From the eccentricity, $e = 1 - R_p/a$, and angular momentum conservation, the semi-major axis resulting from circularization would be

$$a_{\rm circ} = a(1 - e^2)/(1 - e_{\rm min}^2)\,, \tag{11.35}$$

where a small minimum value, $e_{\rm min} = 0.002$, is used. If $a_{\rm circ} > 4r_1^*$, instantaneous circularization is replaced by a sequential procedure. The new eccentricity, e_1, is then obtained from angular momentum conservation combined with $a_1(1 - e_1) = 4r_1^*$ which gives

$$e_1 = \tfrac{1}{4}a(1 - e^2)/r_1^* - 1\,, \tag{11.36}$$

subject to $e_1 = \max\{e_1, 0.9e\}$. Hence the new semi-major axis becomes

$$a_1 = a(1 - e^2)/(1 - e_1^2)\,. \tag{11.37}$$

Modifications of the binary parameters require corrections of the KS variables, with the new values

$$\begin{aligned} \tilde{\mathbf{U}} &= C_1\mathbf{U}\,, \\ \tilde{\mathbf{U}}' &= C_2\mathbf{U}'\,, \end{aligned} \tag{11.38}$$

where

$$C_1 = [a_1(1 - e_1)/R_p]^{1/2} \tag{11.39}$$

and $C_2 = 1/C_1$ by angular momentum conservation in the absence of stellar spins. The corresponding energy correction is given by

$$\Delta h = \tfrac{1}{2}m_{\rm b}\left(\frac{1}{a} - \frac{1}{a_1}\right)\,, \tag{11.40}$$

with $m_{\rm b} = m_k + m_l$. Finally, the total energy loss, $E_{\rm coll}$, is updated by an amount $\mu_{kl}\Delta h$ and the KS solution is initialized if the motion is perturbed. Since stellar radii tend to increase with time, it may be necessary to repeat the above sequence until the orbit is circularized, taking care to delay if the condition (11.34) is barely satisfied.

One possible problem may arise if a circularized orbit increases its eccentricity owing to the influence of an outer hierarchical component (i.e. the Kozai cycle, to be discussed later). Unless some action is taken, this would allow repeated episodes of circularization until external effects intervene or the hierarchy becomes stable and a merger takes place. Since there is only one pericentre check for each orbit when using two-body or chain regularization, the eccentricity may in fact increase significantly before the next pericentre passage. This could lead to a considerably reduced value of R_p and hence also of $a_{\rm circ}$. One acceptable solution is to replace the factor 4 above by a much larger value (say 50) when e is large (say $e > 0.95$). This produces a more gradual adjustment.

The simulation of primordial binaries combines the subtle interplay of many diverse astrophysical processes. It is therefore natural to take advantage of the synthetic stellar evolution scheme in order to improve the modelling of tidal effects. The basic idea is to represent the stars as polytropes, with an effective polytropic index for evolving stars. Instead of using a factor of 4 in (11.34), the circularization time, $t_{\rm circ}$, is evaluated inside $R_{\rm p} = 10 \max\{r_k^*, r_l^*\}$, assuming constant masses and radii [cf. Mardling & Aarseth, 2001]. Normal tidal evolution is then initiated provided $t_{\rm circ} < 2 \times 10^9$ yr, although the actual time interval is usually much reduced because of the steep dependence on stellar radii which tend to grow. However, normal tidal evolution may begin only if the chaos boundary has been crossed (see next section). In the first instance, we neglect stellar spin and model the evolution of the orbit due to dissipation of the equilibrium tide [Hut, 1981].

Using angular momentum conservation, the new eccentricity is obtained from a fitting function based on appropriate damping constants for different stellar types and the elapsed time interval, $t - t_{\rm p}$ [cf. Mardling & Aarseth, 2001]. For unperturbed two-body motion, the corresponding pericentre distance is

$$R_{\rm p} = R_{\rm p}^0(1 + e_{\rm p})/(1 + e_1)\,, \tag{11.41}$$

where the parameters $t_{\rm p}, R_{\rm p}^0, e_{\rm p}$ are revised only for perturbed motion or stellar evolution updates.‖ The orbital corrections now proceed as above, with the binding energy change for unperturbed motion given by the equivalent expression

$$\Delta h = \tfrac{1}{2} m_{\rm b}(e_1^2 - e^2)/a(1 - e^2)\,. \tag{11.42}$$

In the case of perturbed motion, we superimpose the predicted change in eccentricity, $e_1 - e_{\rm p}$, since the previous pericentre as a differential correction and write

$$\Delta h = m_{\rm b} e(e_1 - e_{\rm p})/a(1 - e^2)\,. \tag{11.43}$$

Alternatively, if the time interval is large (i.e. a dormant hierarchy), the new energy is calculated explicitly from the corrected elements in order to ensure a consistent pericentre.

Binaries undergoing circularization also experience exchange of angular momentum between the orbital and rotational motion. Since the orbital angular momentum is now coupled to the spins, the treatment above must

‖ For perturbed motion $R_{\rm p}^0$ should be replaced by the actual distance.

be replaced by direct integration of all the relevant equations of motion, as will be described in section 15.8.

11.9 Chaotic motions

Highly eccentric binaries may experience a chaotic exchange of energy between the orbit and the tides before reaching a state of normal circularization. This process is also relevant for binary formation by two-body capture and provides an alternative channel for the production of exotic objects. The question of whether the tidal energies that build up can be absorbed by the stars without serious structural changes, or are radiated away, is still open [Kumar & Goodman, 1996; Podsiadlowski, 1996]. However, some of these interactions have parameters that lead to relatively mild chaos and it is therefore worthwhile to explore the consequences of this process in order to have a uniform treatment.

The adopted form of the chaos boundary is based on the generalized Press–Teukolsky [1977] theory which also includes the eccentricity in the similarity variable [Mardling, 1995a]. The corresponding energy dissipation due to two f-modes is obtained from fitting functions for different polytropes [Portegies Zwart & Meinen, 1993]. Detailed expressions for the energy transfer at each pericentre passage are given elsewhere [cf. Mardling & Aarseth, 2001]; here we concentrate on procedures affecting the N-body KS treatment. The chaos boundary itself is a function of eccentricity in terms of the mass ratio and polytropic index. We define chaos to begin if the critical point on the curve for constant angular momentum, $e_{\rm c}$, satisfies a disruption criterion, $e_{\rm c} > e_{\rm dis}$ [Mardling, 1995b], with the reverse condition resulting in enforced collision. Likely candidates for the chaos treatment are selected in the same way as for tidal circularization, which is chosen instead if $e < e_{\rm c}$.

During the chaotic motion we calculate the energy transferred to the stellar oscillations and accumulate the linear and non-linear damped contributions. Knowledge of the total energy lost from the system, together with the loss of angular momentum ascribed to oscillations, allows determination of the new orbital energy and angular momentum which yield the semi-major axis and eccentricity. The corresponding KS solution must now be modified in a consistent way, with the first scaling factor in (11.38) given by (11.39). Since the *orbital* angular momentum is no longer conserved, we use the explicit KS energy relation (4.24) to form the velocity correction factor

$$C_2 = \left(\frac{m_{\rm b} + hR_{\rm p}}{m_{\rm b} + h_0 R} \right)^{1/2} , \tag{11.44}$$

where h_0 is the uncorrected binding energy and R the corresponding pericentre distance. After performing the corrections (11.38), the KS polynomials are re-initialized in the standard way.**

Unless collision occurs during the irregular eccentricity changes, the chaos treatment is terminated when the system has lost enough energy to reach the critical point $e_{\rm c}$ on the chaos boundary [cf. Mardling & Aarseth, 2001]. At this stage transition to normal circularization is initiated in the same way as discussed in section 11.8.

The chaos treatment outlined above also extends readily to the case of initial hyperbolic motion. However, most hyperbolic encounters do not suffer enough kinetic energy loss to become bound even for quite small impact parameters. To reduce the amount of excessive calculations, we consider only periastron separations satisfying

$$R_{\rm p} < R_{\rm p}^{\rm max}\, r_1^*\,, \tag{11.45}$$

where the dimensionless quantity $R_{\rm p}^{\rm max}$ is given by a fitting formula that depends on the mass ratio and polytropic index, as well as the relative velocity at infinity [Mardling, 1996b; Mardling & Aarseth, 2001]. Since this is only a necessary condition for capture, not all such encounters form bound orbits on the first pericentre passage. However, those that do are examined by the standard chaos procedures without any further complications. Note that capture candidates invariably form a KS pair during the approach and, if capture occurs, the pair may still be terminated temporarily as for soft binaries. Hence the possibility exists that an eccentric capture binary may subsequently be recognized as a standard chaotic orbit.

Since the onset of chaotic motion usually involves high eccentricities, such events are likely to occur in strong interactions studied by chain regularization. Although the principle of the treatment is the same, this calls for several new algorithms which will be described in a subsequent section.

11.10 Roche-lobe mass transfer

The stage of Roche lobe overflow follows naturally from tidal circularization and is full of astrophysical complications. All the relevant processes have been discussed in considerable detail elsewhere [Tout *et al.*, 1997; Hurley, 2000; Hurley, Tout & Pols, 2002], and we therefore concentrate on some dynamically-related aspects here. The procedure is initiated upon completion of circularization†† and one of the stars filling its Roche radius, $r_{\rm R}$, defined by (9.20). We distinguish between the primary and secondary

** Unperturbed motion is suppressed during the chaotic stage which is usually short.

†† Defined by a nominal small eccentricity 0.002 for technical reasons.

star, otherwise referred to as the donor and accretor. The former is determined by the binary component having the largest ratio $r^*/r_{\rm R}$, rather than the largest radius.

When combining orbit integration with mass transfer, the question arises as to how this can be done in an efficient and consistent manner. A step-wise approach has been devised, in which a small fraction of the primary mass (at most $\frac{1}{2}\%$) is transferred or lost, subject to the elapsed time not exceeding the next c.m. integration step. This leads to intervals of active mass transfer, interspersed by quiescent coasting until the standard Roche condition is satisfied again. At the end of each cycle, the c.m. look-up time is advanced until the next active episode while the components are assigned larger values to prevent individual evolution. Where relevant, the processes of magnetic or gravitational braking are also implemented in a similar way, with the c.m. look-up time advanced according to a small fractional change in the semi-major axis.

Depending on the masses and mass ratio, four basic Roche stages are possible for each star, associated with main sequence, giant, asymptotic giant branch and white dwarf evolution. During the last stage, a doubly degenerate binary of white dwarf components may start spiralling inwards by gravitational radiation if the period is sufficiently short. This effect is included inside $a = 10\,R_\odot$ and, if enough time is available, the condition for mass transfer may be reached. The ultimate fate is either coalescence for unstable mass transfer or slow orbital expansion.

Some of the basic N-body procedures are connected with the change in mass. First the binary orbit is expanded at constant eccentricity, using the expressions (11.12). Hence the correction coefficient for the KS coordinates is given in terms of the ratio of semi-major axes by

$$C_1 = (a/a_0)^{1/2}\,. \qquad (11.46)$$

Since we perform a homologous expansion at an arbitrary phase, by (4.24) the KS velocity scale factor takes the form

$$C_2 = \left[\tfrac{1}{2}(m_{\rm b} + hRa/a_0)/U_0'^2\right]^{1/2}\,, \qquad (11.47)$$

with U_0' denoting the old regularized velocity. Note that stages of magnetic or gravitational braking that do not result in mass loss are also subject to the same corrections. Following the KS modifications, new global coordinates and velocities are predicted, whereupon standard force and potential energy corrections are carried out. Likewise, the total energy loss, $E_{\rm mdot}$, is updated by the amount

$$\Delta E = \mu_0 h_0 - \mu h\,, \qquad (11.48)$$

due to the change of reduced mass and semi-major axis. The current c.m. coordinates and velocity are also updated to be consistent with the new masses, followed by polynomial initialization of the neighbours.

One complication that arises relatively frequently concerns common-envelope evolution. This occurs when the mass transfer from a giant enters a run-away phase and may lead to coalescence. Alternatively, a superhard binary may remain if there are two cores with enough binding energy to expel the whole envelope [cf. Tout *et al.*, 1997]. A discussion of the relevant procedures is delayed until section 12.9 (cf. Algorithm 12.10) since the treatment is more complicated in chain regularization.

Technically, the case of Roche-type coalescence is essentially similar to the collision that has been discussed in section 9.2 for both KS and chain regularization. However, when using the KS description, we may combine a close neighbour with the composite c.m. body and initialize a new two-body solution directly. This is done, provided the nearest neighbour is a single particle within the close encounter distance, otherwise the new remnant is considered by itself.

Roche cycles are normally completed following the specified mass transfer or orbital shrinkage due to braking without mass loss. However, eventually these processes are terminated. This entails a change to an inactive phase, common-envelope evolution or the formation of a contact binary when both stars fill their Roche lobes.[‡‡] In the former case, the time interval until the next overflow stage is determined in a consistent manner and the c.m. index (k^*) is advanced. This results in a quiescent phase which may be long. The subsequent time intervals are rechecked at appropriate times, using special functions to evaluate the stellar radius. For ongoing mass transfer during the coasting phase ($k^* = 11, 13, \ldots$), the next look-up time is determined from the $\frac{1}{2}\%$ mass transfer rate, subject to limits for single star evolution of the primary.

In general, the scheme outlined above has proved itself and is rather inexpensive, although some cases involving massive main-sequence or Hertzsprung-gap primaries may require several hundred cycles. By its nature, an active Roche condition implies a small semi-major axis in most systems; hence such binaries tend to be unperturbed. However, the treatment does not assume this and there is provision for termination, followed by tidal circularization, if the eccentricity increases as may occur in wide hierarchies. Very occasionally, the treatment of a relatively compact hierarchy may lead to some inefficiency because of the need to terminate a formally stable configuration, although most of the mass-transferring stage can be readily handled *in situ* if desired. Among the rare events noted, we mention the formation of Thorne–Żytkow [1977] objects during the common-envelope stage between a neutron star and a giant.

[‡‡] At present contact binaries are treated by the coalescence procedure.

12
Chain procedures

12.1 Introduction

The basic theory of chain regularization [Mikkola & Aarseth, 1990, 1993] is described in chapter 5, while algorithms that deal with different treatments of physical collisions are detailed in chapter 9. Here we are concerned with a number of additional features that deal with aspects relating to what might be termed the N-body interface, namely how to combine two different solution methods in a consistent way. Strong interactions in compact subsystems are usually of short duration, with the ejection of energetic particles a characteristic feature. First we give some algorithms for unperturbed triple and quadruple systems while the more extensive treatment of perturbed chain regularization is discussed in the subsequent sections. As far as the internal chain subsystem is concerned, this requires extra procedures that add to the program complexity and cost. Having selected a suitable subsystem for special treatment, we also need to consider the change of membership and possible astrophysical processes. Moreover, the question of when to terminate a given configuration requires suitable decision-making for the switch to alternative methods, as well as identification of hierarchical stability in order to prevent inefficiency. Finally, since the implementation of the time-transformed leapfrog scheme has many similarities with chain regularization, we include some relevant algorithms here.

12.2 Compact subsystems

The treatment of strong interactions in subsystems can be divided into two categories. We discuss the case of unperturbed three- or four-body regularization, while perturbed chain regularization is dealt with in the next section. The essential difference in the selection procedure is characterized by the external perturbation, which needs to be smaller in the former case. As far as the present N-body codes are concerned, the Aarseth & Zare [1974] unperturbed three-body and Mikkola & Aarseth [1990] four-body chain regularizations* are usually employed only when a perturbed chain configuration already exists. However, the former method may readily be extended to external perturbations (cf. section 5.3), in which case some of the conditions would be relaxed. Although neither the original three-body nor four-body regularization methods employ the chain concept, the subsequent development can be seen to be a generalization and hence it is natural to consider these methods together.

The selection procedure follows closely the first seven steps of Algorithm 11.3. If a perturbed chain configuration already exists, the stricter condition $\max\{\tilde{\gamma}, \gamma_4\} < 100\gamma_{\min}$ is imposed. The maximum subsystem size permitting unperturbed motion may be estimated by

$$d_{\max} = (100\gamma_{\min} m_{\rm cm}/P_{\max})^{1/3}\,, \tag{12.1}$$

where $P_{\max} = 2\max\{m_p/r_p^3\}$ is the largest perturbation. Hence a perturbation of $100\gamma_{\min}$ on the most distant internal member is acceptable for a short time before termination is enforced. In addition, the conditions of a strong interaction must also be accepted, whereas the opposite applies to the selection of stable hierarchies. There are essentially two such requirements, $R_{\rm p} < 2a_{\rm in}(1 + e_{\rm in})$ and also $R_{\rm p} < a_{\rm in} + a_2$ if there are two binaries (cf. steps 9 and 10). Here the latter condition covers the case of two unequal binaries. Finally, the subsidiary condition $r_j < R_{\rm cl}$ ensures a compact subsystem.

If all the above conditions have been accepted, we have a triple ($n = 3$) or quadruple ($n = 4$) subsystem to be initialized for the integration scheme. The latter case is distinguished by the index of the intruding particle, i.e. $j > N$. Initialization procedures for both systems are listed in Algorithm 12.1.

All these algorithms, except for step 2, also apply to subsystems with four members. Most of these points are self-explanatory and do not require comment. Here $n - 1$ ghost particles are created in the same way as in Algorithm 11.5. In order to conserve energy, the differential correction at step 12 is converted into a velocity change of the c.m. motion as an alternative to a correction of the energy budget itself. In any case, such corrections are relatively small because of the stricter perturbation limit and so is the neglected energy exchange because of the short duration. Finally, in the case of four members the initial separation vectors must be prescribed, with the middle distance usually connecting the two binaries.

---

* The unperturbed four-body regularization of Heggie [1974] as reformulated by Mikkola [1985a] was first used for this purpose [cf. Aarseth, 1985a].

Algorithm 12.1. *Initialization of compact subsystem.*

1. Terminate primary KS solution and any second binary $[j > N]$
2. Allocate dominant binary component as reference body $[n = 3]$
3. Set coordinates and velocities in the local reference frame
4. Evaluate the internal energy and update the subsystem energy
5. Save global (i.e. N-body) indices and attributes of the members
6. Form perturber list inside $\lambda R_{\rm cl}$ from neighbours or full search
7. Obtain the potential energy, Φ_1, of members and perturbers
8. Create $n - 1$ ghost particles and define composite body, $m_{\rm cm}$
9. Initialize force polynomials and time-steps for $m_{\rm cm}$
10. Determine maximum system size for unperturbed motion
11. Calculate potential energy, Φ_2, of $m_{\rm cm}$ and the perturbers
12. Modify the velocity $\dot{\mathbf{r}}_{\rm cm}$ by a factor $[1 + 2|\Phi_2 - \Phi_1|/m_{\rm cm}\dot{\mathbf{r}}^2_{\rm cm}]^{1/2}$
13. Remove all ghost particles from perturber [and neighbour] lists
14. Transform to regularized variables and define initial quantities
15. Introduce distances R_1 and R_2 or select chain vectors $[j > N]$
16. Define the gravitational radius by $R_{\rm grav} = \sum m_i m_j/|E|$

We now turn our attention to some key aspects connected with advancing the internal solutions, and first deal with the three-body case. After transformations to regularized variables in the local rest frame, the integration proceeds according to Algorithm 12.2.

Algorithm 12.2. *Integration of unperturbed triple.*

1. Set initial step $\Delta\tau = [d^3/m_{\rm b}]^{1/2}\delta^{1/10}/t'$, with $d = \min\{R_1, R_2\}$
2. Advance the solution by one Bulirsch–Stoer step
3. Check for tidal dissipation or physical collision
4. Switch reference body if $|\mathbf{R}_1 - \mathbf{R}_2| < \min\{R_1, R_2\}$
5. Perform hierarchical stability test after any tidal dissipation
6. Make temporary exit at end of the c.m. time-step
7. Terminate on escape, collision or $|\mathbf{R}_1 - \mathbf{R}_2| > d_{\rm max}$
8. Continue the cycle at step 2 until termination

The initial step is controlled by the dominant two-body motion, with δ being the absolute tolerance for the Bulirsch–Stoer [1966] integrator (usually 10^{-10} or 10^{-12}). The internal solutions are advanced until the elapsed time exceeds the corresponding c.m. time-step or termination is indicated. In order to facilitate scheduling, we introduce a sub-block of $n_{\rm sub}$ members and determine the minimum value (if any) by

$$\tilde{t}_{\rm sub} = \min\{t_{\rm sub} + \Delta t_{\rm sub}\}, \tag{12.2}$$

where $t_{\rm sub}$ denotes the epoch of initialization and $\Delta t_{\rm sub}$ is the elapsed time interval. Provided $\tilde{t}_{\rm sub} < t_{\rm block}$, the pointer $i_{\rm sub}$ then defines the solution type to be treated, namely unperturbed three-body, four-body or chain regularization. Information about any current multiple regularizations is also useful to exclude new two-body or hierarchical system candidates. If $i_{\rm sub} \le 2$, the composite body of mass $M_{\rm sub}$ is considered as a single particle, consistent with the unperturbed approximation, and no reference to its internal composition is made. The scheme allows for possible extensions to perturbed versions should the need arise.

The case of tidal dissipation will be discussed in a subsequent section. As for physical collisions, the iteration to pericentre follows essentially the procedure of section 9.8. We note that tidal dissipation may lead to significant shrinkage of the inner binary; hence a stability test based on (9.14) is required in order to terminate the treatment. A switch of reference body occurs if $|\mathbf{R}_1 - \mathbf{R}_2|$ becomes the shortest distance. This entails a transformation to local physical variables, followed by a relabelling and introduction of new regularized coordinates and momenta. In addition to the distance test (12.1), we also employ the Standish [1971] radial velocity escape criterion, $\dot{d}^2 > v_{\rm crit}^2$, with

$$v_{\rm crit}^2 = 2M_{\rm sub}\left[\frac{1}{d} + \frac{m_3 m_{3-i}}{d - R_{\rm grav}}\left(\frac{R_{\rm grav}}{m_{\rm b} d}\right)^2\right]. \qquad (12.3)$$

This criterion applies to an escaping body m_i (with $i = 1$ or 2) if we have $d > R_{\rm grav}$ and $\dot{d} > 0$. For completeness, we mention that sharper criteria are available, albeit in more complicated form [Griffith & North, 1974; Marchal, Yoshida & Sun, 1984]. In practice the escape test is delayed until a somewhat larger distance is exceeded, i.e. $(R_{\rm grav} d_{\rm max})^{1/2}$. This ensures a smaller initial perturbation for the new KS solution. Also note that, in the case of escape, termination may be delayed slightly if the orbital phase is unfavourable for the evaluation of new elements (i.e. $R < a$). According to (9.13), we may in fact have $a \ll R_{\rm grav}$ for large values of $v_{\rm crit}$, especially if all the particles are strongly bound. However, in general N-body simulations the subsystem under consideration usually contains at least one energetic binary. Hence this strong inequality may not be reached for typical ejections.

The treatment of unperturbed four-body motion differs slightly from the above algorithm. First, the particle labels are assigned as for standard chain regularization. Moreover, the escape criterion (12.3) does not apply. Instead we compare the maximum size of *any* two-body separation with $d_{\rm max}$; this provides a conservative measure for an escape configuration since the relation (12.1) is a measure of the distance to the subsystem centre.

In the following we suppose that some condition for termination has been satisfied. This requires the procedures listed in Algorithm 12.3 to be carried out.

Algorithm 12.3. *Termination of compact subsystem.*

1	Transform to physical variables, $\tilde{\mathbf{r}}_j, \dot{\tilde{\mathbf{r}}}_j$, in the local frame
2	Identify global indices of ghost particles and note $m_{\rm cm}$
3	Define new quantized time [Hermite] or take $t = t_{\rm sub} + \Delta t_{\rm sub}$
4	Predict current coordinates and velocity of $m_{\rm cm}$ to order $\mathbf{F}^{(3)}$
5	Redetermine the external perturber list in case of changes
6	Evaluate the potential energy, Φ_1, of body $m_{\rm cm}$ and perturbers
7	Restore masses and introduce global coordinates, $\mathbf{r}_j = \tilde{\mathbf{r}}_j + \mathbf{r}_{\rm cm}$
8	Obtain potential energy, Φ_2, of the subsystem and perturbers
9	Adjust velocity $\dot{\mathbf{r}}_{\rm cm}$ by tidal factor $[1 + 2\lvert\Phi_2 - \Phi_1\rvert/m_{\rm cm}\dot{\mathbf{r}}_{\rm cm}^2]^{1/2}$
10	Form global velocities and predict the perturbers to order $\dot{\mathbf{F}}$
11	Update the subsystem tables and the internal binding energy
12	Replace composite particle by the members in all relevant lists
13	Set force polynomials for non-dominant third [and fourth] body
14	Initialize new KS solution for the dominant two-body motion

Again all the above steps are applicable to the four-body case. The only difference is that the new KS solution is identified from the smallest ordered distance. The algorithm for time quantization (i.e. truncation to a commensurate value) will be given in a later section. In the alternative case of the difference formulation, the current time is the sum of the initial epoch and the elapsed interval. We note that now the differential energy correction is performed in reverse order, so that the net effect tends to cancel for a stationary perturber field. The subsystem tables contain entries for each of the three types, with masses, particle names, initial epoch, elapsed interval, maximum size, current value of $\Delta t_{\rm cm}$ and stellar evolution indicator. Chain regularization is also treated on the same footing as far as the internal decision-making is concerned. Finally, we see that, following termination, the integration is continued in the usual way because the new solutions are already in place.

12.3 Selection and initialization

The decision-making involved in the selection of a subsystem for the chain treatment has already been discussed in sections 9.4 and 11.7. Ideally such configurations should be identified before the perturber (a single particle or another binary) approaches so close that the KS solution is terminated. On the other hand, multiple regularization methods are less efficient for large distance ratios; hence the choice of the appropriate moment can be quite delicate, since the dominant binary may be highly eccentric. Consideration of the initial size is also of special relevance for the chain method since this affects the number of perturbers to be included in the equations of motion (cf. section 5.7). Although the search for possible chain candidates, carried out at each apocentre passage during strong interactions (i.e. small c.m. steps), is a good compromise, there are situations when this strategy may be improved to include more frequent checks (cf. section 9.4). Finally, it should be emphasized that only relatively compact subsystems are suitable for treatment. Thus the total number of such events might be about 1000 in a typical star cluster simulation with $N \simeq 10^4$ and some 2000 primordial binaries.

Initialization takes place after a suitable configuration has been selected for treatment. The main steps of this procedure are presented in Algorithm 12.4. If the perturber is another binary ($j > N$), an initial membership of four is equally acceptable.

Algorithm 12.4. *Initialization of chain subsystem.*

1	Terminate primary KS solution and any second binary [$j > N$]
2	Set coordinates and velocities in the local reference frame
3	Evaluate the internal energy and specify global energy, $E_{\rm ch}$
4	Save global (i.e. N-body) indices and attributes of the members
5	Form perturber list from neighbours or full search
6	Create $n-1$ ghost particles and define composite body, $m_{\rm cm}$
7	Initialize force polynomials and time-steps for $m_{\rm cm}$
8	Remove all ghost particles from perturber [and neighbour] lists
9	Include differential corrections of $\mathbf{F}_{\rm cm}, \dot{\mathbf{F}}_{\rm cm}$ due to perturbers
10	Determine minimum value of $t_j + \Delta t_j$ for perturbers and c.m.
11	Select two chain vectors $\mathbf{R}_k$ [or three if $j > N$]
12	Transform to regularized variables and define initial quantities
13	Specify the gravitational radius by $R_{\rm grav} = \sum m_i m_j / \vert E \vert$

Algorithm 12.4 does not contain some of the steps of the corresponding Algorithm 12.1 because the differential potential energy correction is not needed in a perturbed formulation. One extra complication here is that the c.m. force and first derivative are modified by the tidal effect of the perturbers after the standard initialization which assumes a single particle. In order to distinguish the c.m. particle during the subsequent integration, the corresponding name is taken to be zero which is a unique identification since there is only one perturbed chain solution. In the data structure, the components of two KS binaries are placed sequentially at the first single particle location, as is normal for termination. However, in

the case of a triple, or if a fifth member is added later, the single particle is assigned zero mass at its current location. For convenience, the first member is chosen as the c.m. reference body which is therefore assigned the total mass. Note that upon the creation of a new KS solution in the data structure or escape from the chain, the reference body may no longer be found at the original location but it can always be recovered.

Strictly speaking, Algorithm 12.4 as well as Algorithm 12.1 should contain one more procedure; namely to advance the internal solution up to the end of the c.m. step, or first perturber step for the chain, if shorter. This is done for technical reasons in order to distinguish between the very first step and the continuation steps which are treated differently. However, the description of the integration itself is delayed until the next algorithm to avoid repetition.

12.4 Time stepping

The equations of motion for the internal chain are advanced consistently with the rest of the N-body system. Moreover, the associated c.m. body is treated on the same footing as any other particle, except that the expressions for the force and first derivative are modified by the differential effect of the perturbers. Likewise, $\mathbf{F}$ and $\dot{\mathbf{F}}$ for other particles which are members of the chain perturber list are modified in a similar manner, with no distinction made between c.m. or single particles since the additional tidal effect on the former would be second order. The essential steps to advance the internal chain members are given by Algorithm 12.5.

Algorithm 12.5. *Integration of chain subsystem.*

1	Determine $t^* = \min_j \{t_j + \Delta t_j\}$ for c.m. and any perturbers
2	Derive the maximum regularized step from inverting $t^* - t$
3	Advance the solution by one Bulirsch–Stoer step
4	Update the slow-down factor κ for weakly perturbed binary
5	Obtain physical coordinates and velocities in the local frame
6	Renew list and predict perturbers before every c.m. step
7	Check for tidal dissipation or physical collision
8	Switch to a new configuration if the chain is deformed
9	Examine the system for gain or loss of membership
10	Perform hierarchical stability test for large distance ratios
11	Make temporary exit when exceeding the c.m. time-step
12	Terminate on escape or if $\max\{R_k\} > R_{\rm cl}$ $[N_{\rm ch} = 3]$
13	Continue the cycle at step 2 until $t > t^*$ or termination

Some comments on this algorithm may be instructive. The first step, which may well exceed the current block-step, defines the endpoint for advancing the chain solution. There is no simple way to ensure that the physical value of the internal chain step will not exceed the temporary endpoint defined by t^* since the former is known only after the integration step has been completed. If this occurs in the presence of perturbers, the coordinate prediction may be outside the permitted range (i.e. the physical step). A simple solution based on estimating the next chain step from the first-order relation $\Delta\tau \simeq L\Delta t$, with L the Lagrangian may be tried, albeit at the expense of additional integration steps compared with a similar isolated system. However, if L is varying rapidly (i.e. small pericentre distances) this procedure may give rise to excessive values.

An improved algorithm has now been implemented [Mikkola, private communication, 1999]. We first assume that the chain contains one dominant two-body motion and write the relation for the *maximum* regularized time-step as

$$\Delta\tau \simeq \int_0^{\Delta t} [(L - L_{\rm b}/\kappa) + L_{\rm b}/\kappa] dt \,. \tag{12.4}$$

By analogy with the slow-down expression (5.84), the first term excludes the largest contribution and may therefore be considered as slowly varying compared with the second which is treated more carefully. Having split the integral into two parts, we introduce the Lagrangian $L_{\rm b} = E_{\rm b} + 2m_k m_l/r$, with $E_{\rm b}$ the binding energy and r the corresponding separation for the dominant two-body motion, which gives

$$\Delta\tau \simeq [L - (E_{\rm b} + 2m_k m_l/r)/\kappa]\Delta t + \frac{1}{\kappa}\int_0^{\Delta t} (E_{\rm b} + 2m_k m_l/r) dt \,. \tag{12.5}$$

Hence the solution reduces to

$$\Delta\tau \simeq (L - 2m_k m_l/\kappa r)\Delta t + 2m_k m_l Y/\kappa \,. \tag{12.6}$$

The integral $Y = \int r^{-1} dt$ can be solved by iteration, using the c-functions of section 4.7. For the slow-down motion we have $Y = \int r^{-1} dt = \kappa y$, where $y = \int r^{-1} dt$ is evaluated over the reduced interval $[0, \Delta t/\kappa]$. Defining r_0 as the initial value, the Stumpff *hauptgleichung* takes the form

$$\Delta t = r_0 y + \eta y^2 c_2 + \zeta y^3 c_3 \,, \tag{12.7}$$

with $\eta = \mathbf{r} \cdot \dot{\mathbf{r}}$ and $\zeta = m_{\rm b}(1 - r/a)$, where $m_{\rm b}$ now denotes the binary mass. From the property of c-functions [Danby, 1992, p. 174] we have

$$\frac{d}{dy}[y^n c_n(\beta y^2)] = y^{n-1} c_{n-1}(\beta y^2) \,, \tag{12.8}$$

with $\beta = -2h$. In the case of small pericentre distances (say $r/a < 0.1$), a more reliable value of the semi-major axis may be determined by (9.24).

The solution for y can now be obtained by standard iteration, which gives the desired result after we combine the two terms of (12.6). To avoid possible convergence problems for eccentric binaries, it is advantageous to improve the initial guess using a fast bisection procedure. For completeness, the derivatives employed in the Newton–Raphson iteration procedure are given by

$$y' = -(r_0 + \eta c_1 y + \zeta c_2 y^2)\,,$$
$$y'' = -\eta(1 - zc_2) - \zeta c_1 y\,, \qquad (12.9)$$

where $z = \beta y^2$ is the argument of the c-functions which are constructed in analogy with section 4.7. Also note that the slow-down factor κ cancels in the last term of (12.6). The actual choice of the next regularized step is made by comparing $\Delta\tau$ with the current value based on convergence of the integrator. The above procedure is not activated during pericentre iterations when the next step is determined separately after the first derivative evaluation.

For more than three particles, it is occasionally useful to extend the above scheme to include a second two-body term. This may be associated with a close hyperbolic encounter or another (possibly eccentric) binary; hence there is no assumption of a hard binary. The generalization is based on the identification of the two shortest distances in the chain and the estimate of the relative perturbation due to the nearest particle. For this purpose it is sufficient to use an expression for the tidal approximation because the distances are available. If the perturbation is fairly large (say $\gamma > 0.01$), we take $\Delta\tau = L\Delta t$ as the result, otherwise the two acceptable terms may be treated consecutively by a sum of the respective contributions independently. For simplicity the present implementation includes at most one binary with slow-down factor different from unity; however, the algorithm works equally well in the absence of slow-down.

We note that the high-order integrator makes a large number of function calls during each step. This requires that the physical chain coordinates (but not velocities) are obtained by the KS transformation at every stage, together with a consistent prediction of the c.m. and any perturbers. However, the former quantities are also needed to evaluate other terms in the equations of motion and the extra effort is not unduly large for modest perturber numbers. In any case, the external effect (and the calculation of E') can be neglected if we have a small nominal perturbation, say $\gamma_{\rm crit} < 10^{-5}$, with

$$\gamma_{\rm crit} = 2\,\max\{m_p/r_p^3\}\left(\sum R_k\right)^3/m_{\rm cm} \qquad (12.10)$$

denoting the largest contribution. This expression actually provides an over-estimate when the sum of the chain vectors is used since the distance

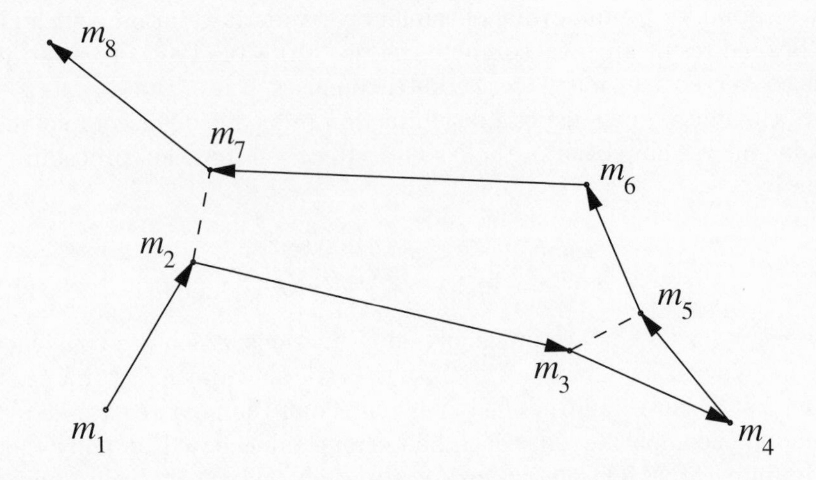

Fig. 12.1. Deformed chain. Dotted lines indicate possible switching.

to the local c.m. would be more appropriate. The slow-down procedure of section 5.8 has proved very beneficial to deal with superhard binaries. This allows arbitrarily small periods to be studied consistently and replaces the inert approximation that was used in the case of binary–binary interactions before the slow-down method became available.

Algorithms for physical collisions can be found in section 9.8 and tidal dissipation will be considered in section 12.9. The next point concerns relabelling of the chain vectors following significant deformation. This involves a check of whether any non-chained distance is the shortest in each triangle formed by two consecutive chain vectors and one non-chained vector, as in the example of Fig. 12.1. A switch also takes place if any other non-chained vector is shorter than the smallest of the chained vectors that are in contact with either end of the vector under consideration. To speed up the operation, distances larger than twice the average length, $\sum R_k/N_{\rm ch}$, are not checked. In order to avoid loss of accuracy, it is beneficial to transform directly from the old chain vectors, $\mathbf{R}_k$, to the new ones. However, round-off problems are less pronounced for the momentum transformations. Thus velocity contrasts are typically proportional to the square root of the corresponding inverse distances. Since the chain deformation is usually modest, there is no need to rescale the next integration step. Finally, it should be emphasized that in the present treatment of chain regularization, all the members are advanced with the same time-step. This is inefficient for large distance ratios but is to some extent alleviated by the slow-down procedure and small membership.

The question of membership change is considered in a later section, followed by hierarchical stability tests and termination procedures.

12.5 Slow-down implementation

The theory of slow-down and its application to chain regularization contained in section 5.8 still leaves several practical aspects to be considered more carefully. Let us first assume that the semi-major axis that enters the expression (5.95) for the energy change is known.

For decision-making it is convenient to employ the sum of tidal forces. With the maximum apocentre, $2a$, as an over-estimate, the relative perturbation is evaluated by

$$\gamma = \frac{8a^3}{m_{\rm b}} \sum_j \frac{m_j}{r_{ij}^3}\,, \tag{12.11}$$

where the summation is over all the other bodies and the distances are computed from the nearest binary component, i, rather than the c.m. In practice, unless there is a large mass dispersion, it is usually sufficient to include neighbouring contributions and use the existing distances, R_k, since a is small during slow-down. In situations with another hard binary as the nearest neighbour, this simple procedure ignores the additional effect of its companion. Alternatively, all the interactions may be added after constructing the chain coordinates, $\mathbf{q}_k$. This permits vectorial contributions to be obtained. Note that the typical membership of a chain configuration rarely exceeds four, in which case there are at most two contributions to the above sum. With more than three members, the use of chain distances for evaluating the perturbation (12.11) may produce an occasional discontinuity if a switch occurs. However, the corresponding energy correction (5.95) does not assume small changes and hence there is no need to employ vectorial quantities even though this would only add slightly to the computational effort.

The expression (12.11) has the desired property of changing slowly, provided the binary is relatively isolated. Although γ is re-evaluated after every integration step, the additional cost is modest because of the large number of function calls with the high-order Bulirsch–Stoer [1966] integrator. The actual choice of κ-values depends to some extent on the overall accuracy requirement as well as the type of problem being investigated. Thus at present an expression of the type

$$\kappa = (\gamma_0/\gamma)^{1/2}\,, \tag{12.12}$$

with $\gamma_0 \simeq 5 \times 10^{-5}$ has been found adequate for small systems. If there are more than four members and several eccentric hard binaries, a value $\gamma_0 \simeq 1 \times 10^{-5}$ is safer in order to avoid undesirable fluctuations. We note that the expression (12.11) contains the square of the binary period, $a^3/m_{\rm b}$, whereas the terms in the summation represent an inverse time-scale squared. This suggests that the square root relation for κ is appropriate, and a suitably small value of γ_0 ensures a regime in which the adiabatic invariance conditions apply. In the case of circularized binaries, the actual semi-major axis may be used in (12.11) instead of the conservative choice $2a$ if suitable precautions are taken to avoid confusion.

We now turn to the evaluation of the semi-major axis which is a delicate matter. Experience shows that even a moderately high eccentricity may pose occasional numerical problems if a is obtained from the physical variables and the binding energy is significant (i.e. a circularizing binary). Instead, we base the determination on the regular expression (9.24) by means of the following algorithmic steps. If slow-down is inactive, we first identify the shortest two-body distance and the corresponding chain index. This permits the sum of m_j/r_{bj}^3 due to any neighbouring members to be evaluated. Substitution of r for $2a$ in (12.11) gives rise to a provisional perturbation, $\gamma^\star$, and no further action need be taken if $\gamma^\star > \gamma_0$. Note that so far only the masses and known chain distances have been used.

If $\gamma^\star < \gamma_0$, the next stage involves transforming the current vectors $\mathbf{Q}, \mathbf{P}$ to chain coordinates and momenta, which are used to derive the individual physical momenta, $\mathbf{p}_j$. From the latter we obtain the kinetic energy, T, and corresponding sum of momenta, $\mathbf{S}$, excluding the dominant two-body contributions due to m_k and m_l. Replacing the associated particle names by indices, we write

$$
\begin{aligned}
T &= \sum_{j \neq k,l} \mathbf{p}_j^2/m_j\,, \\
\mathbf{S} &= \sum_{j \neq k,l} \mathbf{p}_j\,.
\end{aligned}
\tag{12.13}
$$

Hence for three particles there is only one non-singular term. We also obtain the potential energy, Φ_1, from the non-dominant chain, followed by the non-chained part, Φ_2. This gives rise to the perturbing function

$$
\mathcal{V} = \tfrac{1}{2}(T + \mathbf{S}^2/m_\mathrm{b}) + \Phi_1 + \Phi_2\,. \tag{12.14}
$$

Note that the kinetic energy contribution of the binary c.m. follows by virtue of the condition $\sum_k \mathbf{p}_k = 0$. The binding energy relation (9.24) now yields the regular value of a, whereupon the slow-down correction (5.95) is carried out. This can be achieved by a rectification of the accumulated value by setting it equal to $H - E_\mathrm{ch}$ at certain times, or initializing to zero on transition to the non-modified state $\kappa = 1$. Finally, the matrix elements T_{kk} and mass products M_k are modified according to the prescription of section 5.8.

Occasionally, it is desirable to switch off the slow-down procedure even if the adiabatic condition is satisfied. This occurs for example in connection with the determination of a small pericentre which requires careful iteration. Consequently, $\kappa = 1$ is enforced for the relevant two-body interaction, followed by an update of the corresponding mass factors.

12.6 Change of membership

Chain regularization is not restricted to the initial membership of three or four and this introduces additional complications. Consider an existing subsystem being perturbed by another particle. If the latter approaches closely and the new system satisfies the conditions of strong interaction (cf. Algorithm 11.3 and section 12.2), it is more natural to include it with the internal members. Such a search is carried out only if the nominal perturbation (12.10) is significant (> 0.05). The main steps are summarized in Algorithm 12.6 for convenience. The process of absorbing or emitting particles reminds us of molecules recombining or breaking up, according to the binding energy. Note that the absorbed particle may even be an existing KS binary, in which case $N_{\rm ch}$ is increased by two. Because of practical complications, the present treatment is limited to six members but more than four is relatively rare.

Algorithm 12.6. *Addition of perturber to the chain.*

1	Identify the dominant perturber, j, from m_j/r_j^3
2	Obtain the distance, d, and radial velocity, $\dot{d}$
3	Abandon checking if $d > 3\sum_k R_k$ or $\gamma < 0.05$ and $\dot{d} > 0$
4	Specify the outer pericentre distance $R_{\rm p} = a_{\rm out}(1 - e_{\rm out})$
5	Quit further tests if $R_{\rm p} > \sum_k R_k$ and $\gamma < 0.4$
6	Accept the intruder if $\sum_k R_k + d < R_{\rm cl}$
7	Absorb if $d < \sum_k R_k$ and $\dot{d}^2 > \frac{1}{2}(M_{\rm sub} + m_j)/d$
8	Check for inert binary [$j > N$] and switch off slow-down
9	Quantize the current time and predict old c.m. and intruder
10	Redefine chain coordinates and velocities in new c.m. frame
11	Create ghost particle(s) and remove from the relevant lists
12	Re-initialize c.m. polynomials and add tidal force correction
13	Re-evaluate the total energy and the gravitational radius
14	Select chain vectors and transform to regularized variables

Provided the pericentre distance is small, acceptance by any of the criteria above ensures a strong interaction. In principle, the accuracy will also be higher (with somewhat greater cost) by an inclusion of such a particle in the chain structure. The subsequent steps are mostly self-explanatory, except for the time quantization which will be discussed in a later section that deals with termination. In case the membership is increased by two, the intruding binary is terminated in the usual way before the components are expressed in the new c.m. frame. On rare occasions, the new chain may consist of two superhard binaries, defined by the small size $0.01\sum_k R_k$. Since the slow-down procedure presently only applies to one binary, any such intruder is treated temporarily in the unperturbed approximation as an inert particle which is restored on termination. The last few steps are similar to those of Algorithm 12.4. Hence the chain integration can be continued after an update of relevant quantities and selection of chain vectors.

We now turn to the opposite process, namely reduction of membership without termination, since termination will be considered in a separate section. Typically this involves a four-body system which decays to a triple by escape or, alternatively, experiences physical collision. Some aspects of decision-making for these processes were already discussed in sections 9.4 and 9.8, respectively. Chain reduction is only considered if we have $d > 3R_{\rm grav}$ together with $\dot{d} > 0$, where d is the distance to the new local c.m. We distinguish between hyperbolic and nominal escape. Thus if the simplified two-body energy (9.12) is positive, the candidate is accepted provided that also $\max\{R_k\} > R_{\rm cl}$, where the relevant distance must be associated with the first or last chain member. Some additional conditions are described in section 9.4. In the alternative case of sub-parabolic motion, the candidate is accepted if the maximum excursion (when $\dot{d} = 0$ in (9.12)) exceeds $2R_{\rm cl}$. The essential steps associated with particle removal are outlined in Algorithm 12.7.

Algorithm 12.7. *Removal of particle from the chain.*

1	Quantize the current time and predict old c.m. and perturbers
2	Redefine chain coordinates and velocities in new c.m. frame
3	Identify global index of the escaper and check name of c.m.
4	Switch to another c.m. body [case of escaper with zero name]
5	Determine the global index, j, of any exchanged reference body
6	Restore the ghost in relevant lists containing particle j
7	Exchange name of reference body and initialize new c.m. name
8	Update the mass and set new c.m. coordinates and velocity
9	Specify global coordinates and velocity of the escaper
10	Remove the escaper from the relevant subsystem tables
11	Re-initialize c.m. polynomials and associated perturber list
12	Set chain coordinates and velocities of new configuration
13	Form polynomials for the escaper [single particle or KS]

A new point here is that the c.m. reference body may be associated with the escaper. In the beginning it is not known which particle may become an escaper and the first member is therefore selected to represent the new c.m. In that case the identification at step 3 would fail, because the particle name is defined as zero, and the reference body must be re-assigned another chain member, together with redetermination of the corresponding global index (steps 4 and 5). This also entails a restoration of the ghost in the perturber (and neighbour) list, as well as saving the global c.m. name before it is set to zero for identification purposes (steps 6 and 7). Removal of a binary follows similar procedures, where each member is treated in turn before the binary is initialized as a KS solution at step 13. Again the remaining chain configuration is re-initialized according to Algorithm 12.4, starting from step 11 together with re-evaluation of the chain energy (which is also done at step 3 for standard initialization).

12.7 Hierarchical stability

There are several ways in which a chain subsystem may satisfy the condition of hierarchical stability and hence be suitable for the merger treatment described in section 11.5. In the first place, a triple can become stable by decreasing the outer eccentricity due to strong external perturbations. A stable triple may form during a binary–binary interaction in which the least energetic binary is terminated as a KS solution, thereby giving rise to a chain regularization of just three members with the fourth more distant component carrying away the excess energy and angular momentum. However, the ejection of one member from an existing quadruple system is also seen frequently. If the semi-major axes differ significantly, this represents a temporary exchange in which the small binary plays the role of a single particle. Last, we may have five-body or even six-body systems consisting of very hard binaries which can be decomposed into stable subsystems. When relevant, each of these cases needs to be considered in order to avoid loss of efficiency by continuing the chain integration.

In the case of three remaining chain members, the stability check is carried out if $\sum R_k > 4 \min\{R_k\}$. This implies a length ratio in excess of 3 which is near the boundary of the stability relation (9.14) with zero outer eccentricity and equal masses. If $N_{\rm ch} = 4$, we only select systems for further consideration if $\min\{R_k\} < 0.1 \sum R_k$. We distinguish between two binaries and a degenerate triple containing a small binary, as shown in Fig. 12.2(a). The former system is considered if $\max\{R_k\} > 0.7 \sum R_k$, which corresponds to a maximum two-body distance ratio of 4.6 for two equal-size binaries, while the latter case is examined if $\min\{R_k\} < 0.01 \sum R_k$. Unless the middle distance is small, a chain consisting of two binaries is assessed by first evaluating the respective semi-major axes denoted by $a_{\rm in}$ and a_2, where a_2 is the smallest. In addition, the two-body elements $a_{\rm out}, e_{\rm out}$ of the outer binary motion are determined. All these quantities can be obtained using the sorted particle

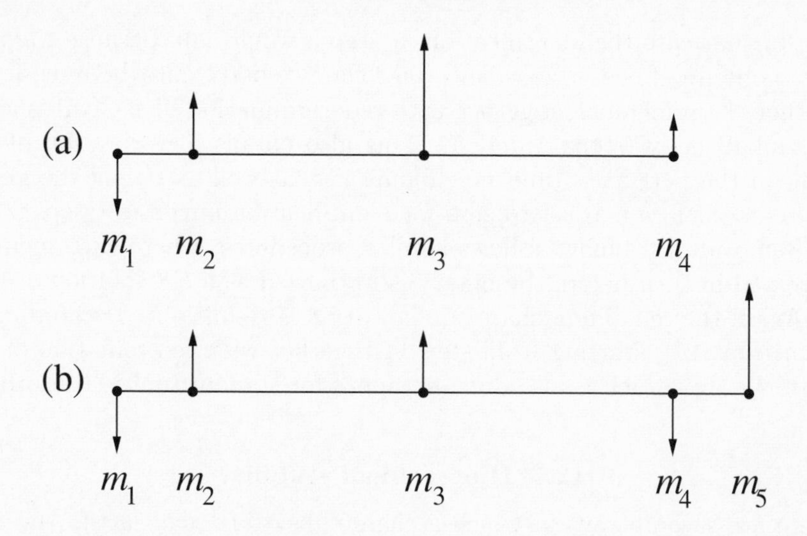

Fig. 12.2. Degenerate triple systems. Schematic illustration of chain configuration with (a) $N = 4$ and (b) $N = 5$.

separations to identify the configuration type. The stability expression (9.14) can be checked after the inclination effect is included, as well as the ratio $a_2/a_{\rm in}$ according to steps 8 and 9 of Algorithm 11.4.

In the alternative situation of a degenerate triple in a four-body system, the small binary is described by its c.m. motion, whereupon the usual stability relations are introduced with no further modification. Note that the binary may occupy any one of the three possible positions in the chain, corresponding to inner, middle or outer membership. Hence such configurations define a higher-order system (discussed in Appendix C), as opposed to a quadruple consisting of two well separated binaries.

Five-body systems constitute another special case for consideration. First we exclude configurations where the ratio of successive chain separations falls in the range 0.25–4, which would be unlikely to satisfy the stability condition. We also ensure that there is a close binary with separation $R_j < 0.04 \sum R_k$ at either end of the chain before continuing. With the sorted indices, I_n, for square separations of the five mass-points, the inner and outer binary masses are defined by $m_{I_1} + m_{I_2}$ and $m_{I_3} + m_{I_4}$, respectively. We now form the three respective distances required for the analysis of Fig. 12.2(b). This enables the inner binary to be selected from the smallest of these distances. The appropriate orbital elements, including the inclination of the outermost member with respect to the inner binary, are evaluated in the usual way. Finally, the stability test (9.14) is carried out, including steps 8 and 9 of Algorithm 11.4.

Chain integration is terminated upon a successful test of the stability condition, or if the subsystem ceases to be compact due to ejection of one or more members. If the innermost binary is highly eccentric, it is desirable to prolong the chain integration slightly in order to avoid initialization at small pericentre separations (i.e. $\min\{R_k\} > a_{\rm in}$). In the following section we discuss the relevant procedures which are also employed when a subsystem decays into two parts due to escape.

12.8 Termination

The termination of a chain integration requires a variety of procedures to deal with the different configurations that may arise. Thus the final membership may be three, four or five, with one or two hard binaries in the second case and usually two in the last. The main steps are summarized by Algorithm 12.8.

Algorithm 12.8. *Termination of chain subsystem.*

1	Transform to physical variables, $\tilde{\mathbf{r}}_j, \dot{\tilde{\mathbf{r}}}_j$, in the local frame
2	Identify global indices of the ghost particles and note $m_{\rm cm}$
3	Modify identification list for special cases $N_{\rm ch} = 2$ or 3
4	Choose the components of the first KS solution as I_1 and I_2
5	Define new quantized time [Hermite] or take $t = t_{\rm sub} + \Delta t_{\rm sub}$
6	Predict current coordinates and velocity of $m_{\rm cm}$ to order $\mathbf{F}^{(3)}$
7	Redetermine the external perturber list in case of changes
8	Restore masses and introduce global coordinates, $\mathbf{r}_j = \tilde{\mathbf{r}}_j + \mathbf{r}_{\rm cm}$
9	Specify global velocities and predict the perturbers to order $\mathbf{F}^{(1)}$
10	Update the subsystem tables and set internal energy $E_{\rm ch} = 0$
11	Replace composite particle by the members in all relevant lists
12	See whether a second KS pair is present [$N_{\rm ch} > 3$; $d_{34} < R_{\rm cl}$]
13	Set force polynomials for non-dominant third [and fourth] body
14	Initialize KS solution for the first and any second dominant pair

In order to identify the global indices of all the chain members at step 2, the name of the reference body (i.e. zero) is first restored. By definition, it is also the only chain member with non-zero global mass and this property is used to ensure a unique determination. The closest particle pair (or dominant term) is selected for KS solution and will usually be a hard binary, but this is not assumed.

Quantization of the physical time is required at the termination of every type of multiple regularization, which employs a non-linear time transformation, in order to assign new commensurate time-steps.[†] Until

[†] A similar procedure is applied for collisions and coalescence when using KS.

recently, a general expression was used for all quantizations of the time (including collision in KS), with the new value given by

$$t_{\rm new} = t_{\rm prev} + [(t_{\rm ch} - t_{\rm prev} + \delta t)/\delta\tilde{t}]\delta\tilde{t}\,. \tag{12.15}$$

Here $t_{\rm prev}$ is the *previous* block time and $\delta t = 0.1\Delta t_{\rm cm}$ a suitably small interval. The new initialization epoch was restricted to the current block time; hence a full N search to determine the next value was also needed a second time after each such initialization because of the way in which the block-step is determined.

Since (12.15) is slightly arbitrary, it has been replaced by the definition

$$t_{\rm new} = t_{\rm prev} + [(t_{\rm ch} - t_{\rm block})/\delta\tilde{t}]\delta\tilde{t}\,, \tag{12.16}$$

where the term in square brackets represents the nearest integer and $\delta\tilde{t} = (t_{\rm ch} - t_{\rm block})/8$ is a small interval, truncated to the next lower hierarchical level by (2.25). The previous block time is employed here because no particles have yet been advanced to $t_{\rm block}$. This yields an appropriate discrete time which is taken to be the new epoch of initialization. Note that for collision, which involves iteration back to pericentre, the new time increment may be negative and even exceed the block-step in magnitude. Likewise, the increment may be negative after change of membership that requires one or more initializations. Since both the c.m. and perturbers are predicted to the new time, this means we ignore a small phase error and the choice of the interval $\delta\tilde{t}$ represents a compromise; a smaller value would also reduce the block-steps significantly. With the unrestricted expression (12.16), however, it is now necessary to perform a *third* search of all values $t_j + \Delta t_j$ in order to ensure the correct particle sequence, bearing in mind that the new time-step may differ from the latest block-step and the sequence has also changed.‡

In the general case of termination without collision, all the steps of Algorithm 12.8 are carried out. If there are more than three members, a second KS pair may be chosen for initialization provided it is well separated from the smallest pair; i.e. $d_{34} < \frac{1}{2}\min\{d_{13}, d_{24}\}$ with the indices denoting the sorted particle labels. The polynomial initialization now proceeds according to type and membership. Thus in the case $N_{\rm ch} = 4$ and one dominant pair, the new force polynomials are first obtained for the two single particles as described in section 8.7, followed by KS initialization according to Algorithm 8.1.

This concludes the standard procedures connected with the chain interface to the N-body system. As can be seen, many delicate aspects are involved and particular attention is paid to the decision-making which

‡ This procedure is only needed in connection with chain and TTL integration.

is quite complicated. Even after several years of experience with the algorithms there are still a number of heuristic procedures which would benefit from improvement.

12.9 Tidal interactions

In this section, we include some technical details of astrophysical processes that are implemented for multiple regularization. Since physical collisions have already been treated elsewhere (cf. section 9.8), we concentrate on aspects connected with tidal dissipation and common-envelope evolution which may also occur inside compact subsystems. In either case, an iteration procedure is needed to determine the smallest two-body pericentre, as described by Algorithm 9.2. Although multiple regularization methods are ideally suited to the study of strong interactions, explicit expressions are not available for the semi-major axis and eccentricity of the dominant motion; hence special algorithms must be developed. In the following we refer specifically to chain regularization but similar treatments are included in the alternative methods.

Since the time-scale for tidal circularization is usually long compared with the duration of chain interactions, we suppress the standard treatment of section 11.8 after initiation but continue to follow any chaotic phase until termination or normal tidal motion is reached. The alternative formulation of sequential circularization is also included [Portegies Zwart *et al.*, 1997], with the main KS algorithms developed in section 11.8. Again only one adjustment is likely to be needed because the stellar radii are not updated during the short interval, although the pericentre distance may change in response to point-mass perturbations.

Let us first consider the less extreme situation of a small pericentre distance, $R_{\rm p} = a(1 - e)$, which falls outside the collision criterion (9.21), where the semi-major axis is obtained by Algorithm 9.2 and $e = 1 - R_{\rm p}/a$. Here we should distinguish between hyperbolic and elliptic motion, since the former may lead to capture. The rare capture process is discussed together with the corresponding KS treatment in section 11.9. It suffices to say that the case of hyperbolic motion is included in the general treatment below which also allows for the dissociation of bound orbits. Once the pericentre iteration has converged, the circularization time is evaluated. If $t_{\rm circ} < 2 \times 10^9\,{\rm yr}$ and the chaos boundary has been crossed, no further action is taken after initialization of the relevant parameters and the standard update follows on termination. More common is the case of a sudden increase in eccentricity, triggering the onset of chaos. In the following we outline the general treatment where the orbital changes may refer to any one of the three formulations, except that the correction procedure differs slightly. The main steps of the orbital modification for the particle pair m_k, m_l are given by Algorithm 12.9.

Algorithm 12.9. *Tidal dissipation in subsystems.*

1	Evaluate the two-body energy $E_{\rm b}$ from non-singular expressions
2	Form osculating elements $a = -\frac{1}{2} m_k m_l / E_{\rm b}$ and $e = 1 - R_{\rm p}/a$
3	Obtain tidal energy and angular momentum change, ΔE, ΔJ
4	Determine the new semi-major axis, $\tilde{a}$, and eccentricity, $\tilde{e}$
5	Derive scale factors for KS coordinates and *physical* momenta
6	Modify the KS coordinates $\mathbf{U}$ and both the momenta, $\mathbf{p}_k, \mathbf{p}_l$
7	Construct the physical chain momenta by the recursion (5.68)
8	Transform to regularized momenta using basic KS relations
9	Update the internal energy by ΔE and differential effect $\Delta \Phi$
10	Carry out stability test according to section 12.7

Again the energy of the dominant two-body motion is derived by the non-singular expression (9.24) and the eccentricity is given by the pericentre relation since the radial velocity is negligible after a successful iteration. Moreover, the tidal energy dissipation and angular momentum change are evaluated in essentially the same way as in section 11.8. Upon obtaining the new elements, the KS coordinate scaling factor C_1 takes the form of relation (11.39), whereupon the new values of the relevant coordinates $\mathbf{Q}_k$ are introduced by

$$\tilde{\mathbf{Q}}_k = C_1 \mathbf{Q}_k \,. \tag{12.17}$$

To determine the second scaling factor, let us consider the case of chaotic motion first. Now we need to scale the physical velocity by the ratio of the pericentre values taken from the basic energy equation (4.8), which yields the explicit form

$$C_2 = \left\{ [\tilde{h} + m_{\rm b}/\tilde{a}(1-\tilde{e})]/[h + m_{\rm b}/a(1-e)] \right\}^{1/2} . \tag{12.18}$$

Since $\tilde{h} = -m_{\rm b}/2\tilde{a}$, this can also be written as

$$C_2 = \left[\frac{a(1-e)}{\tilde{a}(1-\tilde{e})} \right]^{1/2} \left[\frac{1+\tilde{e}}{1+e} \right]^{1/2} . \tag{12.19}$$

Hence the first term on the right-hand side is the inverse of C_1. Note that these coefficients relate to regularized and physical quantities, respectively, and the relation $C_1\, C_2 = 1$ therefore no longer applies. Given C_2, the relative velocity is changed by an amount

$$\Delta \mathbf{v}_{\rm cm} = C_2 (\mathbf{p}_k/m_k - \mathbf{p}_l/m_l) \,. \tag{12.20}$$

This gives rise to the modified physical momenta

$$\begin{aligned} \mathbf{p}_k &= m_k \mathbf{v}_{\rm cm} + \mu_{kl} \Delta \mathbf{v}_{\rm cm} \,, \\ \mathbf{p}_l &= m_l \mathbf{v}_{\rm cm} - \mu_{kl} \Delta \mathbf{v}_{\rm cm} \,, \end{aligned} \tag{12.21}$$

with μ_{kl} the reduced mass and $\mathbf{v}_{\rm cm}$ the local c.m. velocity,

$$\mathbf{v}_{\rm cm} = (\mathbf{p}_k + \mathbf{p}_l)/m_{\rm b} \,. \tag{12.22}$$

For sequential circularization, the effect of stellar spin is not included and angular momentum is conserved. This implies that $C_2 = 1/C_1^2$ since the physical velocity is modified instead of the regularized velocity, $\mathbf{U}'$. The new eccentricity and semi-major axis are then given by (11.36) and (11.37), with the corresponding energy change

$$\Delta E = \tfrac{1}{2} \mu_{kl} (1/a - 1/\tilde{a}) \,. \tag{12.23}$$

We convert to physical chain momenta, $\mathbf{W}_k$, by (5.68) (or the equivalent 5.5) and perform a KS transformation, whereupon the internal energy is updated by ΔE. Since the coefficient C_1 may deviate considerably from unity, we include a potential energy correction due to changes in the non-chained distances as a result of the pericentre adjustment. A stability test is also performed (cf. section 12.7) in case the orbital shrinkage is significant. Unless stability is indicated, the modified system can now be advanced by another integration step according to Algorithm 12.5.

Since the local data structure does not contain the usual c.m. indicators, a special procedure is needed to denote any change in the chaotic motion or tidal circularization index ($k^* = -1$ or -2). This is achieved by use of the sum of the relevant component names for one or two binaries; hence any exchange from a previously tidally active binary will be noted and the old c.m. name removed during subsequent table updates. On termination, the corresponding c.m. name is entered in the relevant table such that the standard KS procedures can be continued. Note that, if required, the pericentre parameters $R_{\rm p}, e_{\rm p}, t_{\rm p}$ can be assigned appropriate values when increasing the global chaos index. To summarize, the present treatment allows one or two tidally active binaries to interact strongly and emerge with the correct status, including possible exchange, and any new dissipation event is also recorded.

Common-envelope evolution may occur in chain configurations if the collision criterion (9.21) is satisfied and at least one of the stars is a giant. The outcome still depends on the energy budget connected with the associated mass loss [Tout *et al.*, 1997], because coalescence is defined if the two cores would spiral inwards so much that Roche overflow occurs before the common envelope is expelled. If coalescence does not occur, we assume the formation of a circularized binary of short period (i.e. typically a few days at most), where the semi-major axis is derived from energy considerations.§ This enables new initial conditions for the two binary components to be specified, albeit with physical variables instead of the KS description of section 11.10.

Once the conditions for coalescence have been satisfied, the treatment is essentially similar to that for a normal collision (cf. Algorithm 9.3), except for the mass loss, Δm. Two new points need to be considered here. To maintain a uniform treatment with the KS method, a list of neighbours inside $(1+\Delta m/M_\odot)r_h/N$ is first formed. This is followed by mass-loss corrections of potential and kinetic energy, together with polynomial initialization for the neighbours. The remaining collision procedures are outlined in section 9.8. Unless termination is indicated (i.e. $N_{ch} = 2$), the chain integration continues in the normal way.

In the case of common-envelope evolution without coalescence, the orbital shrinkage is usually considerable. This is an important process for the formation of very close binaries which is of observational interest. Moreover, such configurations frequently lead to stable hierarchical systems that can be identified as described in section 12.7, together with the stability condition (9.14). Thus with just three members, the probability of this occurring is high since the main requirement is for the outer component to be bound with respect to the inner c.m. Algorithm 12.10 lists the main points in the transition to a superhard binary.

Algorithm 12.10. *Common-envelope evolution without coalescence.*

1	Obtain the two-body energy, E_b, from non-singular expressions
2	Specify $\mathbf{R}_0$, $\dot{\mathbf{R}}_0$ and define $\mathbf{r}_{cm}$, $\dot{\mathbf{r}}_{cm}$ for the close binary
3	Evaluate the semi-major axis, a_0, and energy, h_0, from E_b
4	Perform common-envelope evolution which yields a and Δm
5	Add total energy correction $\mu_0 h_0$; also kinetic energy $\frac{1}{2}\Delta m \dot{\mathbf{r}}_{cm}^2$
6	Construct new relative motion $\mathbf{R}$, $\dot{\mathbf{R}}$ for modified values a and e
7	Form global coordinates and velocities using the old c.m. values
8	Update masses and correct energy by $-\mu h$, with $h = -\frac{1}{2}m_b/a$
9	Include mass-loss correction of potential energy and tidal field
10	Re-initialize chain regularization with current initial conditions

Some of these steps have been discussed before. For consistency with the two-body treatment, we specify a small final eccentricity, e_{min}. If the new binary is initialized at apocentre, the relative coordinates and velocity

§ This has recently been replaced by a sequential procedure that limits the mass loss at each pericentre passage, resulting in a reducing eccentricity.

become

$$\mathbf{R} = \mathbf{R}_0\, a(1 + e_{\rm min})/R_0\,,$$
$$\dot{\mathbf{R}} = \dot{\mathbf{R}}_0 \left(\frac{m_{\rm b}}{a v_0^2}\right)^{1/2} \left(\frac{1 - e_{\rm min}}{1 + e_{\rm min}}\right)^{1/2}, \qquad (12.24)$$

where v_0 is the old orbital velocity. The mass-loss corrections are included in the energy component $E_{\rm cdot}$ (cf. section 9.9) because of the association with a discontinuous process. After the relevant modifications have been carried out, new c.m. polynomials are formed. Since the chain membership is unchanged, standard initialization can now be performed, starting at step 2 of Algorithm 12.4.

12.10 Black hole binary treatment

The possible presence of black hole [BH] binaries in galactic nuclei and even globular clusters is currently receiving much attention [Magorrian & Tremaine, 1999; Miller & Hamilton, 2002]. Given a high-density environment, exotic processes such as tidal disruption of stars outside the Schwarzschild radius and coalescence due to gravitational radiation may occur. As discussed in chapter 10 and a subsequent chapter, several studies have been made of the dynamical behaviour of N-body systems containing a massive binary, and there are still open questions regarding the final stages. Hence, in view of the potential importance, this subject merits further examination.

In the following we describe an implementation of the time-transformed leapfrog method [Mikkola & Aarseth, 2002; hereafter TTL] which has many similarities with chain regularization, especially concerning aspects relating to the N-body interface. We do not address the question of suitable initial conditions, which necessitates astrophysical considerations, but take a simple model with two identical stellar distributions containing a massive central body. The two clusters are placed in an elliptic orbit with overlapping pericentre distance which leads to the rapid formation of a composite system. As a compromise, we choose $m_{\rm BH} = N^{1/2}\bar{m}$ in the case of two equally heavy masses. The initial distribution is taken to be cusp-like with a stellar density profile

$$\rho(r) \propto \frac{1}{r^{1/2}(1 + r^{5/2})} \qquad (12.25)$$

and corresponding 1D velocity dispersion [Zhao, 1996]

$$\sigma^2(r) = \frac{1}{\rho(r)} \int_r^\infty \frac{\rho(r)}{r^2}[m(r) + m_{\rm BH}]dr\,. \qquad (12.26)$$

In view of the different parameters and extra decision-making, a new code called *NBODY7* has been developed [Aarseth, 2003a], with a workstation version analogous to *NBODY6* as well as a subsequent GRAPE-6 implementation. The early evolution of binary hardening that follows the mass segregation stage can be readily studied by the standard KS and chain methods [cf. Quinlan & Hernquist, 1997; Milosavljević & Merritt, 2001]. Eventually, conditions become favourable for a switch to the TTL scheme which is both more expensive and accurate, with $a_{\rm BH} \simeq R_{\rm cl}$ representing a natural choice.

Once the massive binary has been accepted for special treatment, the KS description is replaced by conventional two-body variables, as discussed in section 5.9. In order to facilitate decision-making, the corresponding global particle names¶ $\mathcal{N}_k$, $\mathcal{N}_l$ are assigned negative values and the second component is defined as a ghost particle. The c.m. motion is then initialized in the usual way with a small time-step for the relative motion. At subsequent intervals, suitable subsystem members are selected from the nearest perturbers. Provisionally, we use a distance criterion $d = \lambda_{\rm BH}\, a_{\rm BH}$, with $\lambda_{\rm BH} \simeq 25$ for subsystem members and 50 for external perturbers, with further enlargement on increased membership. Hence the adopted component masses would include relative perturbations out to at least $\gamma_{\rm BH} \simeq 3 \times 10^{-7}$ for $N \simeq 10^5$ and high eccentricity. Special procedures are required to include a regularized binary as two single particles and the inert approximation may be appropriate for superhard energies in the early stages. Likewise, two strongly bound ejected particles are initialized as a KS pair.

The method that advances the internal evolution is essentially similar to the chain treatment, with each interaction evaluated explicitly in the equations of motion that are integrated by the Bulirsch–Stoer method. We take special care to avoid losing precision for large distance ratios by a transformation of coordinates and velocities to the BH binary c.m., otherwise chain vectors are used. Since there are many force evaluations for each binary period, the additional cost of $N_{\rm pert}$ perturbers $\propto N_{\rm BH} N_{\rm pert}$ per step. In order to obtain the net perturbing force on $N_{\rm BH}$ subsystem members of total mass $M_{\rm BH}$, the c.m. contribution

$$\mathbf{F}_{\rm cm} = \frac{1}{M_{\rm BH}} \sum_{i=1}^{N_{\rm BH}} m_i \sum_{j=1}^{N_{\rm pert}} \frac{m_j(\mathbf{r}_j - \mathbf{r}_i)}{r_{ij}^3} \qquad (12.27)$$

is subtracted from the corresponding individual expressions. However, in view of the large mass ratio and small binary size involved, the number of

¶ In the general case of N_1 and N_2 primordial binaries in two subsystems with N_0 members in the first, we would have $\mathcal{N}_k = 1 + N_1 + N_2$, $\mathcal{N}_l = 1 + N_0 + N_1 + N_2$.

perturbers tends to be modest. Moreover, the extra computational effort goes as the inverse period, $\propto a_{\rm BH}^{-3/2}$, whereas $N_{\rm pert}$ diminishes with $a_{\rm BH}^3$ modulated by the central density, which tends to compensate.

Frequent checks are made of the subsystem and neighbour membership. Any changes necessitate the creation and initialization of ghost particles. Some of the relevant procedures are similar to the chain treatment discussed above, with one important exception related to the energy budget. Let $E_{\rm BH}$ denote the subsystem energy at the previous updating, with separate contributions, $E_{\rm pert}$ and $E_{\rm GR}$, due to external perturbers and gravitational radiation, respectively. Accordingly, the error of the subsystem energy is given by

$$\Delta E_{\rm BH} = E_{\rm BH} + E_{\rm GR} + E_{\rm pert} - E_{\rm exp}\,, \tag{12.28}$$

where the last term represents the current energy and is evaluated explicitly for checking purposes only. In order to monitor energy conservation, any contributions from the relativistic terms (denoted GR) and external perturbers are kept separate. Since $E_{\rm exp}$ contains the effect of the external perturbation, the term $E_{\rm exp} - E_{\rm GR}$ is added to the quantity (9.29) and yields the desired constant of the motion. However, the actual errors defined by (12.28) are usually well below the tolerance for so-called 'rectification', whereby $E_{\rm BH}$ would be assigned the explicit value.

The subsystem energy is updated by an amount $\Delta E_{\rm corr}$ after each change of membership according to the expression

$$\Delta E_{\rm corr} = \Delta T_1 + \Delta\Phi - \Delta T_2\,. \tag{12.29}$$

If a new particle is added, ΔT_1 represents the combined kinetic energy of the c.m. and the intruder, and ΔT_2 denotes the final c.m. kinetic energy. These terms have the reverse meaning for ejection. Likewise, $\Delta\Phi$ is the potential energy of the relevant particle with respect to the subsystem, with negative and positive sign, respectively. Moreover, in the case of a binary entering or leaving the subsystem, the internal binding energy is included and appropriate measures taken for the KS treatment. As before, a check is made on the possible reduction of time-steps for high-velocity particles that experience the slingshot effect. The present implementation has the advantage that particles ejected from the subsystem may readily be included in the relevant list at the earliest opportunity.

Following each updating, the energy change due to external perturbers is added to the subsystem energy for convenience and initialized to zero. The corresponding effect on the perturbers is contained in the external budget (9.29). Once a given particle is identified from the perturber list, the corrected force is obtained by a subtraction of the c.m. approximation term, followed by direct force summation over the internal components.

Such differential force corrections are also carried out in connection with polynomial initializations resulting from membership changes.

The consistent advancement of two systems that employ different times requires special care. As far as the subsystem solutions are concerned, each step is completed when $t > t_{\rm block}$. During this interval, the perturbers are predicted at all substeps and it is desirable for the corresponding predictor steps to be convergent. Since the total number of block-steps is comparable to the number of Bulirsch–Stoer steps, this does not appear to present a problem; i.e. the smallest perturber time-steps evolve in a self-similar manner together with the block-step. Following this brief outline, it remains to consider the strategy for adding GR effects.

In the case of an ultra-hard binary containing black hole components, the $(\rm post)^{5/2}$-Newtonian approximation may be included for the most critical two-body interaction. The full expression is given by [Soffel, 1989]

$$\mathbf{F} = \mathbf{F}_0 + c^{-2}\,\mathbf{F}_2 + c^{-4}\,\mathbf{F}_4 + c^{-5}\,\mathbf{F}_5\,, \qquad (12.30)$$

where $\mathbf{F}_0$ denotes the Newtonian force per unit mass and c is the speed of light. The additional terms describe the post-Newtonian acceleration, relativistic precession and gravitational radiation, respectively. Optionally, the latter may be considered without contributions from the more time-consuming precession, which then gives rise to the well-known decay of the semi-major axis and eccentricity. For clarity, some of the main points discussed above are summarized in Algorithm 12.11.

Algorithm 12.11. *Binary black hole procedures.*

1	Identify a hard massive binary for special treatment
2	Define new variables and initialize polynomials
3	Predict coordinates and velocities of perturbers and c.m.
4	Advance the equations of motion up to end of block-step
5	Check gain or loss of subsystem members and update energy
6	Select new perturbers from neighbour list or GRAPE-6
7	Activate GR options for short evolution time-scales
8	Include implementation of BH coalescence or accretion
9	Allow for the case of unperturbed non-GR two-body motion

We distinguish between the situation when the osculating pericentre of the BH binary falls below a specified small value and orbits of field stars making close approaches to one massive component. In view of the time-consuming nature of the additional terms in (12.30), we tentatively consider three different stages of increasing importance which are also reflected in the computational cost. For this purpose we write the classical gravitational radiation time-scale $a/\dot{a}$ [Peters, 1964] in the scaled form

$$\tau_{\rm GR} \simeq 1.3 \times 10^{18} \frac{a_{\rm BH}^4}{\chi(1+\chi)m_{\rm BH}^3} \frac{(1-e^2)^{7/2}}{1 + 73e^2/24 + 37e^4/96}\,, \qquad (12.31)$$

where $\chi = m_2/m_1$ and $m_{\rm BH}$ represents the primary mass. This yields the time in yr if the semi-major axis is expressed in au and the mass in $M_\odot$. As an example, $a_{\rm BH} = 20\,$au and $m_{\rm BH} = 500\,M_\odot$ would require $e \simeq 0.99$ for a time-scale $\tau_{\rm GR} \simeq 10^7\,$yr.

Depending on the problem under investigation, the respective additional perturbations are included at three suitably chosen limiting values of $\tau_{\rm GR}$, with the first effect due to the radiation term. Accordingly, we need to define the astrophysical length and mass units that are already specified in a realistic simulation. Integration of $\dot{a}$ from the inverse (12.31) and the companion equation for $\dot{e}$ from specified initial conditions gives the time to reach a smaller eccentricity. This can be used as a check for the code implementation in the absence of perturbations. Note that the actual time interval tends to be slightly longer than the linear prediction, with the eccentricity decreasing slowly during the early stage.

The expression (12.31) is evaluated after each integration step, using Newtonian expressions for the two-body elements which are usually well defined near the boundaries. An alternative approach, tried at first, was to consider the velocity at the osculating pericentre but this is somewhat arbitrary. In the absence of perturbations, GR coalescence takes place if $\tau_{\rm GR}$ is sufficiently small. Writing the coalescence condition as

$$R_{\rm GR} = f_{\rm GR} G(m_k + m_l)/c^2\,, \qquad (12.32)$$

we take $f_{\rm GR} = 6$ or three Schwarzschild radii if the two massive components are involved. Unless there are more than two black holes, termination usually takes place at this stage or even slightly earlier following significant GR energy loss. It is also of interest to consider the possible accretion of field stars. This can be readily studied by a suitable relabelling of one reference body, together with appropriate decision-making within the integration itself. So far, favourable experience of the GR regime has been gained and the scheme promises to become a valuable tool in future investigations. Some provisional results using GRAPE-6 with up to $N = 2.4 \times 10^5$ particles have already been reported in which GR coalescence was achieved [Aarseth, 2003b].

13
Accuracy and performance

13.1 Introduction

The question of numerical accuracy has a long history and is a difficult one. We are mainly concerned with the practical matter of employing convergent Taylor series in order to obtain statistically viable results. At the simplest level, the basic integration schemes can be tested for the two-body problem, whereas trajectories in larger systems exhibit error growth on short time-scales. However, it *is* possible to achieve solutions of high accuracy for certain small systems when using regularization methods. There are no generally agreed test problems at present but we suggest some desirable objectives, including comparison with Monte Carlo methods. Since large simulations inevitably require the maximum available resources, due attention must be paid to the formulation of efficient procedures. The availability of different types of hardware adds another dimension to programming design, which therefore becomes very specialized. Aspects of optimization and alternative hardware are also discussed, together with some performance comparisons.

13.2 Error analysis

It is a fact of computer applications that an error is made every time two arbitrary real numbers are added. Hence the task is to control the propagation of numerical errors and if possible keep them below an acceptable level. Since the N-body problem constitutes a system of non-linear differential equations the error growth tends to be exponential, as was demonstrated right at the outset of such investigations [Miller, 1964] and emphasized in a subsequent study [Miller, 1974]. Numerical experiments with up to 32 particles were made by measuring the deviation of neighbouring solutions, which were found to have a short time-scale. This implies that such integrations are likely to be invalid after surprisingly short time intervals if we are concerned with nearly exact solutions. In fact, this property poses a challenge to any justification for doing dynamical simulations by direct methods.

The numerical measurement of error growth was carried out only on a few subsequent occasions. Thus the effect of using a softened potential in 25-body systems reduced the rate of divergence of solutions in configuration and velocity space for $\epsilon \simeq 0.08r_{\rm h}$, whereas the behaviour was similar to the point-mass case for $\epsilon \simeq 0.016r_{\rm h}$ [Standish, 1968b]. More extensive simulations for a range of particle numbers up to 340 indicated a mean e-folding time, $t_{\rm e}$, of one crossing time, albeit with a small tendency to decrease with N [Kandrup & Smith, 1991]. It was also shown that only a relatively large value of the softening size affects the result. A similar investigation with 32–512 particles concluded that $t_{\rm e} \simeq t_{\rm cr}/8$ for all the N-values considered [Goodman, Heggie & Hut, 1993]. Finally, the question of N-dependence was tested more fully in a recent systematic study that extended the particle number to 6×10^4 [Hemsendorf & Merritt, 2002]. For the first time a clear trend of the instability growth rate emerged, with $t_{\rm e} \sim 1/\ln N$. However, the practical implications of this surprising result remains to be understood.

On the theoretical side, the question of the appropriate time-scale for error growth in large systems has also been considered. Arguments in favour of a short time-scale have been presented using the estimated closest approach [Heggie, 1988]. More detailed analysis confirmed that the characteristic growth time-scale is indeed shorter than one crossing time for large N [Goodman *et al.*, 1993]. To counteract this, ideas based on the concept of shadow orbits have been presented [Quinlan & Tremaine, 1992] which did much to restore the confidence in direct N-body solutions. Moreover, the statistical properties of global quantities are well behaved [Giersz & Heggie, 1994a,b, 1996, 1997]. Hence it is recognized that other methods need to verify agreement with direct integration where possible in order for the results to be accepted [Spurzem, 1999].

We now turn from general considerations to some practical topics of error measurements. As is well known, even the integration of a binary orbit by direct means is subject to systematic errors that affect both the semi-major axis and eccentricity.* Although such solutions are actually obtained by the more accurate two-body regularization in the present context, it is useful to evaluate the different schemes that are employed in general integrations. Thus, in some sense, a typical cluster orbit has a characteristic eccentric shape that tends to become smoother with

* Here we are not concerned with errors in the longitude of the periapse.

increasing N-values and, being perturbed, therefore requires somewhat more integration steps per orbit than a corresponding isolated binary.

First we compare different methods for two-body orbits with low and high eccentricity. Table 13.1 contains the relative and absolute errors, $\Delta a/a$ and Δe, respectively, together with the number of integration steps for an initial eccentricity $e = 0.1$ and 0.90, as well as $e = 0.99$ in the case of direct integration. The average values per orbit are quoted for an integration over 1000 periods in order to improve the measurement of the secular errors. Results for the standard divided differences are shown first, followed by the Hermite method. These results are obtained using the actual codes NBODY1 and NBODY1H for $N = 2$, where the latter was developed for another purpose and does not use block-steps† [Makino & Aarseth, 1992]. Hence, for *pure* two-body motion, the systematic errors per step are approximately comparable for similar fourth-order implementations. The increased number of time-steps for large eccentricity is due to using a relative force criterion (cf. (2.13)) which is analogous to the Kepler relation $\Delta t_i \propto R^{3/2}$ (or equivalently, $\Delta t_i = (\eta R/|\mathbf{F}|)^{1/2}$ for $N \leq 2$).

It is also of interest to examine the basic schemes implemented for a test particle. Based on the Kepler time-step expression, Hermite integration is somewhat more accurate than divided differences for large eccentricity [cf. Makino, 1991a]. Thus we obtain $\Delta a/a \simeq -1.1 \times 10^{-7}$ and -1.7×10^{-7} for $e = 0.1$ and 105 steps per orbit, whereas $\Delta a/a \simeq -2.0 \times 10^{-6}$ and -4.1×10^{-7}, respectively, for $e = 0.9$ and 340 steps. In the latter case the relative error for the difference method is reduced to -1.5×10^{-7} with the same step number when using a relative criterion based on the force and its second derivative (cf. (2.12)) instead of the Kepler relation. This suggests that the general time-step criterion (2.13) is efficient in minimizing the two-body integration error. In particular, the time-steps are slightly smaller in the pericentre region for the same total step number. However, the corresponding Hermite formulation is in fact slightly worse, yielding $\Delta a/a \simeq -4.9 \times 10^{-7}$. Note that although these methods are formally of the same order, the difference formulation benefits from prediction to order $\mathbf{F}^{(3)}$ for a test particle integration (i.e. $N = 1$), whereas in the general case all the other particles are predicted to low order.

A comparison of three different regularization methods is also included in the table. We illustrate the Hermite KS method, rather than the Stumpff formulation, since the latter does not produce any errors for unperturbed motion. The standard value $\eta_U = 0.1$ is used for all the KS solutions (cf. (8.59)). Without taking the corresponding merits of the numerical effort into account, it can be seen that two-body regularization is significantly better than direct integration for *any* eccentricity.

† The block-step code NBODY1B is less accurate for the same step number here.

Table 13.1. Integration errors for two-body motion.

Method	e_0	$\Delta a/a$	Δe	Steps
Differences	0.1	6.2×10^{-7}	-3.8×10^{-7}	105
	0.9	-1.0×10^{-6}	2.2×10^{-7}	340
	0.99	-4.2×10^{-6}	2.5×10^{-9}	550
Hermite	0.1	-3.7×10^{-7}	8.0×10^{-7}	105
	0.9	-6.5×10^{-7}	-7.8×10^{-8}	341
	0.99	-6.2×10^{-6}	-6.4×10^{-8}	550
Hermite KS	0.1	5.9×10^{-11}	-4.0×10^{-11}	63
	0.9	2.9×10^{-11}	-1.9×10^{-11}	63
Standard KS	0.1	-3.3×10^{-8}	1.6×10^{-8}	63
	0.9	-1.9×10^{-8}	1.8×10^{-9}	63
Stabilized KS	0.1	-1.0×10^{-12}	-5.6×10^{-11}	63
	0.9	-1.0×10^{-12}	-3.1×10^{-10}	63

An earlier comparison of the standard Hermite and time-symmetric KS schemes was rather unfavourable [Funato *et al.*, 1996]. Although no details are given about the algorithm used in the first method, the *estimated* error $\Delta a/a \simeq 1 \times 10^{-6}$ per orbit is consistent with low-order prediction for *half* the usual number of steps per orbit for $e \simeq 0.9$. This reduces to $\Delta a/a \simeq 4 \times 10^{-9}$ when predicting the KS variables to highest order and again using $\eta_U = 0.2$, or 31 steps per orbit (see Table 13.1 for results with $\eta_U = 0.1$). Alternatively, even low-order prediction with $\eta_U = 0.1$ yields a corresponding error $\Delta a/a \simeq -4.4 \times 10^{-8}$. Finally, we emphasize that the time-symmetric method does in fact benefit from high-order prediction, with at least one iteration, as is also the case in the equally accurate Stumpff KS method [cf. Mikkola & Aarseth, 1998].

Table 13.1 also shows the effect of including energy stabilization in two-body regularization (cf. (4.43)). Note that this can be readily done for the difference scheme as well, but only in situations with dominant two-body motion (cf. (3.33)). We now obtain much improved conservation of the semi-major axis at the expense of some deterioration in eccentricity, as has also been found independently [cf. Funato *et al.*, 1996]. It should be emphasized that stabilization is no longer effective in cases of significant perturbation (say $\gamma > 0.001$) since the osculating binary elements are not sufficiently regular. For completeness, we remark that although the time-symmetric scheme is considerably more accurate than Hermite KS for two-body motion, it has only proved itself in three-body scattering experiments and planetesimal N-body simulations [cf. McMillan & Hut, 1996; Funato *et al.*, 1996; Kokubo, Yoshinaga & Makino, 1998].

The results above were obtained using standard codes and hence give an idea of typical integration errors. Note that high eccentricities are often seen, particularly in connection with strongly interacting binaries or Kozai cycles in hierarchical triples, when values of 0.99 and even 0.999 may be reached. Thus, unless some special features are included, such as unperturbed motion in the neighbourhood of small pericentres, direct integration requires considerably more steps and is less accurate than KS. Even so, this strategy has been advocated on the grounds of simplicity [Makino & Taiji, 1998; Portegies Zwart *et al.*, 2001], especially in connection with GRAPE-6.

So far we have been concerned with systematic errors in idealized systems. Since the accuracy is improved for smaller time-steps that take longer to perform, it follows that a compromise must be made when deciding on the value of the time-step parameters η or η_U. Fortunately the high-order methods produce considerable accuracy gains for modest increases in effort and hence there is a tendency to be conservative. It is an interesting question to what extent reliable results may be obtained by choosing rather large time-steps. Thus a three-body scattering experiment appeared to show no noticeable dependence on η for several global quantities when 800 outcomes were combined [Valtonen, 1974]. More general investigations of small systems [Smith, 1974, 1977] also showed that there are no significant differences in the overall characteristics during the first few relaxation times. Moreover, even relatively low accuracy appears to give reliable results. Note, however, that this does not mean that time-step parameters $\eta \simeq 0.05$ (cf. (2.13)) can be recommended since this would lead to rapid disruption of an isolated binary.‡

The emphasis on relative energy errors may be reassuring but is no guarantee of accuracy even though there is a general correlation between $\Delta E/E$ and η. On the other hand, time reversal provides a stricter test of most methods. This aspect was examined at an early stage using NBODY3, where some escapers in small systems (i.e. $N \simeq 25$) could be traced back to the initial conditions after several crossing times. A similar conclusion was reached for the generation of high-velocity escapers in Trapezium-type systems [Allen & Poveda, 1972], where the emphasis was on actual reproducibility.

The time-scale for reproducibility depends sensitively on the history of interactions, as can more readily be ascertained using the regularized three-body code TRIPLE. Thus in the so-called 'Pythagorean Problem' [Szebehely & Peters, 1967], which terminates in escape after about 16 crossing times, time-reversed rms errors in coordinates and velocities of 5×10^{-3} and 2×10^{-3} are obtained with a tolerance of 10^{-10}. This reduces

‡ NBODY1 gives $\Delta a/a \simeq -4\times10^{-5}$ per orbit for $e = 0.9$ with $\eta = 0.02$ or 170 steps.

to 2×10^{-5} and 8×10^{-6}, respectively, if the tolerance is 10^{-12}. On the other hand, the correct outcome of the final binary energy and high eccentricity is also reproduced with a tolerance of 10^{-8}. This is mainly connected with the delay of the time reversal until max $\{R_1, R_2\}/a \simeq 70$, rather than soon after escape, so that phase errors due to the binary accumulate. Hence performing the time reversal instead at a corresponding distance ratio of 9 with tolerance 10^{-9} gave a satisfactory outcome.

Another way of looking at the reproducibility question is to consider the sensitivity of solutions to small changes in the initial conditions. Thus neighbouring trajectories in phase space exhibit exponential growth of their differences, pointing to the chaotic nature of the evolution. Such behaviour is most readily demonstrated in the general three-body problem, where individual orbits can be studied more confidently by regularization methods. However, even with the best methods, some three-body orbits are too complicated to be reproduced by time reversal. One investigation based on global three-body regularization [Dejonghe & Hut, 1986] demonstrated large amplification factors due to small changes in initial conditions; in particular, resonant scattering exhibited great sensitivity. Thus we may adopt the point of view [Mikkola & Hietarinta, 1989] that the numerical solutions of the Hamiltonian system

$$\begin{aligned}\dot{p} &= -\frac{\partial H}{\partial q} + \xi\,,\\ \dot{q} &= \frac{\partial H}{\partial p} + \zeta\,,\end{aligned} \tag{13.1}$$

are explored and the unknown additional functions ξ and ζ include the integration errors. It is further surmised that numerical noise adds to the randomness which may introduce new properties. On the other hand, the concept of shadow orbits enhances the confidence in N-body simulations [Quinlan & Tremaine, 1992].

The choice of time transformation may also affect the accuracy. As discussed previously, the secondary expression (5.57) improves the numerical behaviour during critical triple encounters. However, following subsequent suggestions [Zare & Szebehely, 1975; Alexander, 1986], the alternative $t' = 1/L$ is preferable. Thus in the former case we still have that $t' \propto R^{3/2}$ for an equilateral system, whereas $t' \propto R$ when choosing the Lagrangian. Moreover, it can be shown that if Γ_0 denotes a small deviation from the regularized Hamiltonian, the corresponding equation of motion implies an acceleration correct to second order [Mikkola, private communication, 2001],

$$\ddot{\mathbf{r}} = (1 - \Gamma_0^2)\mathbf{F}\,, \tag{13.2}$$

where $\mathbf{F}$ represents the true physical force per unit mass.

When it comes to errors in large simulations, there is little alternative but to fall back on the energy check. However, much can be gained by combining several independent calculations in order to improve the statistics [cf. Giersz & Heggie, 1994a]. This study showed that the averaged results of a suitable number of statistically independent N-body models converges to predictions obtained by the Fokker–Planck approach, thereby demonstrating the correctness of the models. Thus the ensemble of these integrated models is within the range of physically allowed models or trajectories of the system. Further comparisons between Fokker–Planck and N-body models for two-component systems also showed reasonable agreement [Spurzem & Takahashi, 1995], given the restriction of the former method to isotropic velocities, which was later overcome [cf. Takahashi, 1995].

It is reassuring to know that any significant problem invariably shows up in the calculation of the total energy, which essentially represents the difference between two large numbers but, even so, the loss of precision is usually not too serious.[§] Moreover, the choice of integration parameters is guided by examining smaller systems, as was done above. Now a variety of effects are included, some of which are dissipative. It is therefore very encouraging that energy conservation can be maintained to a satisfactory level, which is usually better than 1×10^{-5} per crossing time for $N \simeq 1000$ and may be considerably less for larger systems.

Since N-body simulations cover a wide range of densities and time-scales, as well as different modes of energy exchange or transport, it is prudent to take a fresh look at the underlying levels of approximations. In general, the largest energy exchanges are connected with close encounters involving one or two binaries, some of which can be studied in confidence by regularization methods. Provided the pre-encounter impact parameters are unbiased, the outcome of strong interactions should also be statistically reliable. Hence it is desirable to direct further efforts towards an analysis of less energetic interactions which are much more numerous.

On the practical side, the relative energy errors produced by the Hermite method appear to be randomly distributed such that the accumulated value is often remarkably small, provided that dominant binaries are treated by regularization. It should also be emphasized that making adjustments to the energy budget does not affect the calculations themselves. Still, the quality of the results depends a great deal on careful development and checking having taken place. However, in the final analysis, it is desirable to make comparisons with independent work, if at all possible. This is done in a subsequent section for the core collapse time which is a characteristic feature of cluster evolution.

[§] The energy associated with close encounters is added as a well-defined quantity.

13.3 Time-step selection

In order to derive the maximum benefit from a high-order scheme, it is essential to employ a sensitive time-step criterion. Since the largest error contributions are usually due to close encounters, it is also important that the time-step is reduced by an appropriate amount already during the approach. Regularization may then be applied below some suitable value (cf. (9.4)). To achieve this objective, we make use of all the available force derivatives in the form (2.13) which has proved itself. This expression is quite general and has several desirable properties. First, it gives a well-defined value in the case of starting from rest when the odd force derivatives are zero, which yields the simple relation (2.12). Moreover, if the force is close to zero, which may occur with an external tidal field, the time-step can still be relatively large because the orbit is smooth.

On a historical note, in the late 1970s it was found that the choice (2.12) was not satisfactory for collapsing systems with softened potential, and the more general expression (2.13) was therefore introduced after some experimentation. Here the basic philosophy is that all the force derivatives should play a role in order to ensure convergence of the Taylor series. Consequently, the actual value of the time-step is not significantly affected by any particular term being unduly small. There are several good reasons for choosing a relative force criterion. This has to do with minimizing individual error contributions as well as the property that the time-steps of two unequal-mass particles will tend to be similar during a dominant two-body encounter. The latter feature is especially useful for identification of candidates selected for regularization.

The property of the time-step criterion (2.13) may be illustrated by an application to the KS formulation. We restrict our attention to unperturbed two-body motion without lack of generality and define the regularized force in the usual way by

$$\mathbf{F}_{\mathrm{U}} = \tfrac{1}{2} h \mathbf{U} \,. \tag{13.3}$$

Applying successive differentiation with respect to the fictitious time and using absolute values on dimensional grounds, we put $U'^2 = \frac{1}{4} m_{\mathrm{b}}$ for a circular orbit (cf. (4.24)) and obtain after some simplification

$$\Delta\tau = \eta_U \left(\frac{|h|U^2 + \frac{1}{2} m_{\mathrm{b}}}{\frac{1}{4}|h| m_{\mathrm{b}} + \frac{1}{2} h^2 U^2} \right)^{1/2} . \tag{13.4}$$

We substitute $U^2 = a$ and $|h| = \frac{1}{2} m_{\mathrm{b}}/a$ so that $\Delta\tau = \eta_U (2/|h|)^{1/2}$, which is the standard form involving the frequency (cf. (8.59)). Hence this exercise yields the desired result without using any factorials.

We now carry out a direct test integration of an isolated binary with NBODY1 in order to illustrate the error behaviour for different time-step

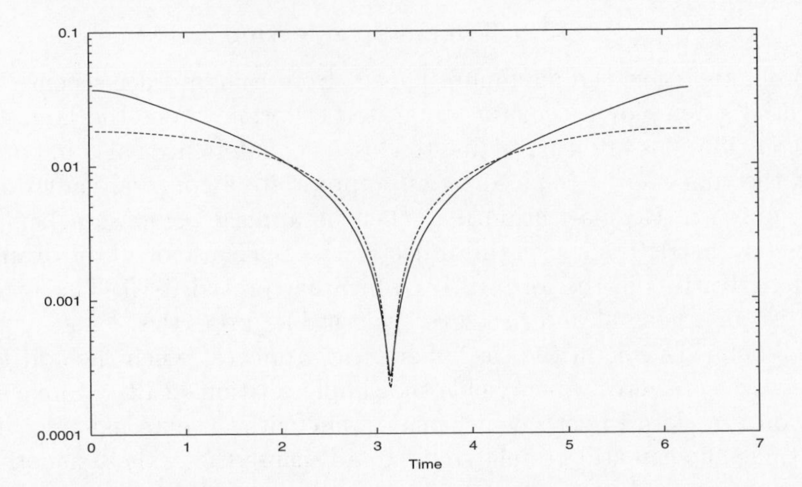

Fig. 13.1. Time-steps for eccentric two-body orbit. Solid line shows the standard relation (2.13) and dotted line is for Kepler's Law.

criteria. We adopt an expression based on Kepler's Law by $\Delta t \propto R^{3/2}$ and compare the errors over 100 periods for low and high eccentricity with the standard criterion. An error sampling after ten periods shows the expected linear growth in the relative error, giving $\Delta E/E \simeq -1 \times 10^{-9}$ per orbit for $e = 0.5$ and both time-step choices adjusted to 570 steps. For $e = 0.9$, the Kepler choice is marginally worse with mean relative errors of -3.8×10^{-9} and -1.8×10^{-9}, respectively, for about 1200 steps per particle per orbit. Thus the general time-step expression (2.13) appears to perform rather well during a close two-body encounter if it can be assumed that the comparison is optimal. As indicated above, for pure two-body motion the simpler criterion (2.12) is slightly better. However, the general form is more sensitive for detecting close encounters.

As shown in Fig. 13.1, the general expression spans a wider range and yields smaller values in the pericentre region, except for the innermost part. It should also be remarked that even soft binaries are chosen for regularization if the eccentricity is sufficiently large, although such binaries are not usually persistent.

13.4 Test problems

Since there are no general analytical solutions beyond the two-body problem, it is difficult to devise suitable tests for large simulation codes even in idealized form. Historically, the so-called 'IAU Comparison Problem' for $N = 25$ [Lecar, 1968] created some confusion, especially since the initial conditions gave rise to collapse and violent relaxation. The alternative approach of comparing the evolution rate with approximate methods was more successful [Aarseth, Hénon & Wielen, 1974]. Subsequently, the agreement with the Monte Carlo method improved further after taking into account the non-dominant terms neglected in the standard relaxation theory [Hénon, 1975]. A more thorough recent effort concentrated on the question of scaling to different values of N and hence bridging the gap with continuum methods [Heggie *et al.*, 1998].

The classical theory of dynamical friction has been tested by combining the gaseous model for single stars with a Monte Carlo formulation for superelastic binary scattering [Spurzem & Giersz, 1996]. This work studied the evolution of hard binaries interacting with the single stars, assuming initial equipartition. The agreement obtained with theoretical expectations also serves as a test of the Monte Carlo treatment of relaxation. Comparison between Fokker–Planck and N-body results again supports the correctness of the theory. Although these attempts are instructive since they establish a connection with theory, they do not provide strict tests of direct N-body codes.

The evolution of idealized cluster models is characterized by core collapse [cf. Spurzem & Aarseth, 1996]. It is therefore natural to formulate a test problem in terms of determining the time of core collapse, $t_{\rm cc}$, for different values of N. This has to some extent been done for the Monte Carlo, Fokker–Planck¶ and gaseous methods. It is instructive to compare these time-scales with some results from direct N-body simulations. The collection of data is presented in Table 13.2 with the scaling $\tau_{\rm cc} = t_{\rm cc}/t_{\rm rh}$.

For the Monte Carlo methods, Column 3 specifies the typical number of so-called 'superstars'. The algorithm of Spitzer & Hart is denoted 'explicit' since it is based on integration of all the superstars, whereas the 'implicit' scheme originated by Hénon [1972, 1975] selects superstars by random sampling. Moreover, recent developments permit a large number of superstars (currently 1–2 $\times\, 10^6$) to be treated so that individual stars can now be represented [Joshi, Rasio & Portegies Zwart, 2000; Freitag & Benz, 2001; Giersz, 2001a,b]. The essential difference between these two formulations is that the time-steps in the former are related to the crossing time, whereas the latter employs a fraction of the relaxation time. Consequently, the implicit method is much faster where its use can be justified.‖ Recently, a new Monte Carlo code has been developed for studying dense galactic nuclei, including tidal disruptions of stars by a central black hole [Freitag & Benz, 2002].

The comparison of numerical methods cannot be undertaken without

¶ The Fokker-Planck equation was first used by A. Fokker [1914] and M. Planck [1917] to describe Brownian motion of particles [cf. Risken, 1984].

‖ See Hénon [1972] or Freitag & Benz [2001] for implicit Monte Carlo algorithms.

Table 13.2. Determinations of core collapse times.

Method	Algorithm	N	τ_{cc}	Reference
Monte Carlo	Explicit	1000	8.0	Spitzer & Hart [1971b]
Monte Carlo	Explicit	1000	18.0	Spitzer & Thuan [1972]
Monte Carlo	Implicit	1000	17.5	Hénon [1975]
Monte Carlo	Implicit	1200	17.0	Stodółkiewicz [1982]
Monte Carlo	Implicit	2000	17.5	Giersz [1996]
Monte Carlo	Implicit	100K	15.2	Joshi *et al.* [2000]
Monte Carlo	Implicit	512K	17.8	Freitag & Benz [2001]
Fokker–Planck	1D	∞	15.0	Cohn [1980]
Fokker–Planck	2D	∞	17.6	Takahashi [1995]
Fokker–Planck	2D	∞	11.8	Einsel & Spurzem [1999]
Gaseous	Anisotropy	∞	15.0	Louis & Spurzem [1991]
Tree	*NBODYX*	1000	22.0*	McMillan & Aarseth [1993]
Direct	*NBODY5*	3000	3.0	Inagaki [1986]
Direct	Parallel	1000	20.0	Giersz & Heggie [1994a]
Direct	*NBODY5*	10^4	20.7*	Spurzem & Aarseth [1996]
Direct	*NBODY6++*	8192	16.4	Baumgardt *et al.* [2002]

* This reduces to 17.2 when using $\gamma = 0.1$ in (1.11).

due allowance for differences in initial conditions, as well as the way in which the measurements are reported. Moreover, the low value of τ_{cc} obtained by Spitzer & Hart was increased by using better diffusion coefficients [cf. Spitzer & Thuan, 1972], whereas $\tau_{cc} \simeq 15$ for a subsequent Plummer model [Spitzer & Shull, 1975a]. On the other hand, a sequence of King models gave $\tau_{cc} = 15.0, 13.0, 11.0$ for $W_0 = 3, 5, 7$, respectively [Joshi *et al.*, 2000]. There is also some indication that the dispersion is quite small, with about 1% spread for 10^5 superstars [cf. Freitag & Benz, 2001]. Note that the gaseous model includes a density-dependent heat source to account for the hardening of binaries formed by the three-body process [cf. Bettwieser & Sugimoto, 1984]. In the latter investigation, $\tau_{cc} \simeq 15.4$ based on $\gamma = 0.4$.

Only one of the table entries [i.e. Inagaki, 1986] relates to a mass spectrum, with power-law exponent $\alpha = 2.5$ and mass ratio 10. As expected on general grounds, this leads to an accelerated evolution. The much shorter time-scale is in qualitative agreement with early Monte Carlo simulations for three discrete mass groups [Spitzer & Shull, 1975b], and also with Fokker–Planck models which produced values of τ_{cc} in the range 3–15 by varying the IMF [Inagaki & Saslaw, 1985]. A similar result of $\tau_{cc} \simeq 3.1$ has been obtained for a King model ($W_0 = 5$), continuous mass function, simplified stellar evolution and tidal field [Giersz, 2001b]. Also note that rotation reduces the evolution time [Einsel & Spurzem, 1999].

Since the N-body results are expressed in terms of the half-mass relaxation time, $t_{\rm rh}$ (cf. (1.11)), the theoretical scaling factor, $N/\ln(\gamma N)$, has been absorbed and the small internal dispersion is reassuring, considering that the results of Giersz & Heggie are for a King model instead of the Plummer model. Likewise, the core collapse times for the continuum models are mostly in reasonable agreement with the N-body results. Thus it appears that core collapse times for direct integrations are mutually consistent, as has also been emphasized previously** [cf. Spurzem, 1999]. Finally, we note that the traditional value of the Coulomb factor for equal masses, $\gamma = 0.4$, is too large. The smaller value $\gamma = 0.11$ has been determined by N-body simulations [cf. Giersz & Heggie, 1994a], whereas Freitag & Benz [2001] used $\gamma = 0.14$.

Most large-N simulations are not of the idealized type considered above. However, we need to be confident that the basic process of relaxation is modelled correctly before introducing complications, and this appears to be the case. In this spirit, we may go on to include additional dynamical effects due to an IMF with or without primordial binaries, as well as rotation. Now the parameter space becomes too big for specific comparisons and the results must be taken on trust to a large extent. This makes it important to be aware of the possible dispersion so that a given calculation can be interpreted properly [cf. Giersz & Heggie, 1994a,b]. Although there is no good theory of fluctuations to guide us, it is fortunate that global quantities tend to become better defined with increasing N or the use of ensemble averages. This leaves the difficult question of scaling the results to larger N which is yet to be resolved, even though attempts have been made [cf. Heggie *et al.*, 1998; Baumgardt, 2001].

N-body simulations provide many opportunities for comparison with theory. Although analytical work is invariably based on approximations, the theory of dynamical friction is sufficiently well established to permit detailed comparison with numerical experiments [Bonnell & Davies, 1998]. The phase of mass segregation is of particular interest because it precedes core collapse and can therefore be compared more readily with approximate methods. Moreover, the presence of a mass spectrum gives rise to an early evolution that may be compared with theoretical estimates [cf. Spitzer, 1987]. In particular, a two-component system is more amenable to analysis and therefore constitutes a possible test problem. The outcome is also of astrophysical interest since massive stars are usually found in the inner cluster region [Raboud & Mermilliod, 1998b]. However, the question of a possible initial segregation still remains open.

** A recent case $N = 64\,\mathrm{K}$ gave $\tau_{\rm cc} = 17.8$ [Baumgardt, private communication, 2002].

13.5 Special-purpose hardware

Ideally, a FORTRAN N-body code should work on any computer. However, the situation has become more complicated with the construction of both parallel and special-purpose machines. In the former case, the main new feature is message-passing between the different processors in order to advance the integration consistently. Such procedures take the form of a replication of all common variables across the processor network in order to keep them synchronous. Moreover, the standard data structure needs to be modified slightly [cf. Spurzem *et al.*, 2003].

The development of parallel computers is undoubtedly favourable for many tasks in comparison with the older vector supercomputers that seem to have gone out of fashion as far as direct N-body simulations are concerned. It provides opportunities for attacking large computations in new ways that inevitably bring rewards to those who make the effort. However, the cost-effectiveness of such expensive installations should be questioned, since what we now call medium-sized simulations (i.e. $N \simeq 1000$) can be made on laptops in a few hours without any overheads.

An exciting new technology for special-purpose computers has emerged during the last decade. It started with the GRAPE family [Sugimoto *et al.*, 1990], where the force evaluations are performed by hard-wired operations that supply the result to a host machine. Later, the more accurate HARP-1 was designed for collisional calculations [Makino, Kokubo & Taiji, 1993] and in 1994 the Cambridge HARP-2 with eight pipelines was acquired, yielding an actual performance of 1.9 Gflops for $N = 10^4$. Subsequently, an innovative project led to the more powerful GRAPE-4 [Makino *et al.*, 1997] with peak performance of 1 Tflop [see also Makino, 1996a]. In addition to significantly improved precision, these machines provide the explicit derivative of the force, thereby facilitating a fourth-order integration scheme. Since HARP is also designed for the Hermite method, we continue to use the original name to distinguish it from the low-precision GRAPE-3 and GRAPE-5 versions which are mainly employed in cosmological or collisionless simulations.

In some sense, this new technology is a backwards step because it relies on the brute force method of solution. However, the large gain in speed outweighs the loss of efficiency and makes this solution method highly cost-effective. Several smaller versions have been used for dedicated studies of star clusters and planetary dynamics, and the experience is favourable. HARP-2 calculates $\mathbf{F}$ and $\dot{\mathbf{F}}$ for up to eight particles at the same time after predicting all N coordinates and velocities on the hardware. A more powerful version called HARP-3 with up to 88 simultaneous force calculations has also been in use, with an actual performance of about 20 Gflops for $N = 4 \times 10^4$, near the chip memory limit 42 K.

Table 13.3. Pipe-line efficiency.

Machine	Pipes	N	Levels	Cycles	Efficiency
HARP-2	8	10 543	16	5.1×10^5	0.962
HARP-3	88	10 543	16	7.3×10^4	0.519
HARP-3	88	20 461	18	3.2×10^5	0.644
GRAPE-6	48	50 705	20	2.6×10^6	0.826
GRAPE-6	48	50 081	22	6.7×10^6	0.731
GRAPE-6	48	100 145	22	1.1×10^7	0.841

The code *NBODY4* has been specially adapted for HARP, as well as the new GRAPE-6 [Makino, 2003; Makino, Fukushige & Namura, 2003], and many features are discussed elsewhere in this book. Here we comment on some aspects relating to the performance and first consider the simplified case of single particles. Although the number of force evaluation pipes is relatively small compared with the number of processors on a CRAY T3E supercomputer, there are many occasions when only a few particles are advanced at the same time. Such small time-steps are invariably connected with terminations of KS or chain regularization and the subsequent initialization. Hence it becomes a question of the pipeline efficiency due to the hierarchical time-step distribution. Some examples are illustrated in Table 13.3 for three different types of hardware.

We have included the two older Cambridge machines as well as the newly acquired GRAPE-6 with theoretical peak of 1 Tflop and 600 Gflops actual performance for $N = 256\,\mathrm{K}$ with P4 2GHz host.[††] Although the latter has 32 chips, there are 48 virtual pipelines for parallel force calculations. Each entry is for an appropriate value of N, together with a typical number of hierarchical levels. In general the number of levels increases slightly with N and depends on the core–halo density contrast. Moreover, the shape of the time-step distribution is a reflection of the density.

We assume a bell-shaped curve with relative membership distribution $P(l) = 2^{l-1}+l$ for level l up to $l^* = l_{\max}/2$ and $P(l) = P(l-1)/2$ for $l > l^*$, where $l_{\max}$ is the maximum number of time-step levels. The symmetry is preserved by setting $P(l^* + 1) = P(l^*)$ and using even values of l^*. This distribution is unfavourable for a parallel treatment but representative of realistic cluster models. The efficiency is obtained by considering the number of force evaluations during an interval equal to the maximum time-step, and adding the total number for each block-step according to the weights $P(l)$. Column 3 shows the corresponding particle number

[††] A more powerful version with 1024 chips and four hosts has reached 11.5 Tflops in a binary BH simulation with $N \simeq 1 \times 10^6$ particles [Makino & Fukushige, 2002].

and the last column displays the efficiency. The actual hardware cost is reflected by the total number of cycles given by Column 5 because the CPU time is independent of the block-size.

A comparison of HARP-2 and HARP-3 for a Plummer model with $N = 10^4$ equal-mass particles yielded a factor of 4.8 in the ratio of CPU times when the initialization cost was excluded. Given a substantially faster host on HARP-2, this relative performance agrees quite well with the last two entries of Table 13.3. As for GRAPE-6, the efficiency for $N > 5 \times 10^4$ should be judged in relation to its performance. In any case, the host is a bottleneck for realistic simulations with stellar evolution and primordial binaries.‡‡ However, the amount of slowing-down cannot be quantified without reference to a specific model.

Working with the special-purpose hardware gives several opportunities for optimization. Thus there is a useful feature that some tasks can be carried out on the host while HARP or GRAPE-6 is busy evaluating the forces. The main time-saving procedures consist of predicting the next sequence of particles on the block (if any) and correcting the previous block members. Moreover, the facility for creating a list of neighbours inside a specified distance is invaluable since this is needed for many different purposes. Among some of the main requirements are KS perturber selection, finding the closest neighbours for the chain and merger procedures, checking for unperturbed motion and performing force corrections due to stellar mass loss. This entails making a suitable initial guess of the search distance, taking into account the location in the cluster and, where relevant, the semi-major axis, as well as the perturbation itself since the latter is a good indicator of the nearest neighbour.

Some of the most frequent uses of the neighbour list are concerned with the redetermination of perturbers. It is necessary to ensure a sufficiently large membership for including more distant massive particles (cf. (8.58)). On the other hand, when only a few close neighbours are needed, the following strategy has been adopted. Given the distance to the density centre, r_i, the core radius, r_c, and membership, N_c (to be defined later), we choose a basic search distance

$$d = (r_c^2 + r_i^2)^{1/2} / N_c^{1/3} \tag{13.5}$$

if $r < 2r_h$, otherwise $d = r_i - 1.9r_h$ subject to a suitable maximum. When primordial binaries are present, the neighbour list is used heavily for checks of unperturbed motion. However, we may take advantage of a faster hardware facility that returns only the closest particle index.

‡‡ Notwithstanding early globular cluster simulations of primordial binaries [Spitzer & Mathieu, 1980], a cost analysis of direct integration in the large-N limit did not consider this aspect and concluded that the decreasing formation rate would lead to reduced computational requirements for the binaries [Makino & Hut, 1990].

Since HARP is used together with a workstation host, the actual performance depends on many factors which affect the load balance. One important aspect concerns the treatment of perturbed binaries which is described in Appendix E. In addition to obtaining differential corrections of $\mathbf{F}$ and $\dot{\mathbf{F}}$ for the c.m. as well as the associated perturbers, the relative motion must be integrated on the host. From the experience so far, the performance is degraded when a large binary population is present, especially for binding energies near the hard/soft boundary that usually require a significant perturber membership. Hence if an effective algorithm can be developed for parallelizing such binaries on a multi-processor machine or so-called 'Beowulf configuration', this may well prove a better way to carry out such demanding simulations.

At present some exciting hardware developments are emerging that will undoubtedly become part of future configurations. Thus Field Programmable Gate Arrays (FPGAs) may replace the host for carrying out certain time-consuming tasks [Kuberka *et al.*, 1999]. For example, in the limit of large N, it appears feasible to evolve stars or treat collisions by SPH in real time. One project has already made significant progress using a workstation host combined with an FPGA board and the GRAPE [Spurzem *et al.*, 2002]. Thus the FPGA architecture facilitates the evaluation of neighbour interactions in an analogous manner to the AC scheme. Thus a successful implementation of SPH has been achieved with considerable gain in performance compared with a standard GRAPE configuration. Moreover, such a hybrid architecture has the potential to overcome bottlenecks on the host for a variety of applications. We also emphasize the advantage of a short development time for new software as well as rapid reconfiguration of existing algorithms in much the same way as current libraries are used.

As far as the GRAPE hardware is concerned, it is highly desirable to improve the design by including a general neighbour scheme in order to achieve higher performance. At the same time, the rapid advances being made in computer design will surely continue to present new opportunities, especially as regards parallel developments. In addition to the new breed of parallel supercomputers [Tamaki *et al.*, 1999; Brehm *et al.*, 2000], we mention the recent trend of constructing Beowulf PC clusters that involves combining a large number of fast processors. Thus one such system consisting of 20 Pentium 4 processors has recently been installed and used for the code *NBODY6++* described in section 10 [Spurzem, private communication, 2002]. In conclusion, judging from the substantial progress on many fronts since the early days, it can be anticipated that further software innovation and technological improvements will open up many new exciting avenues.

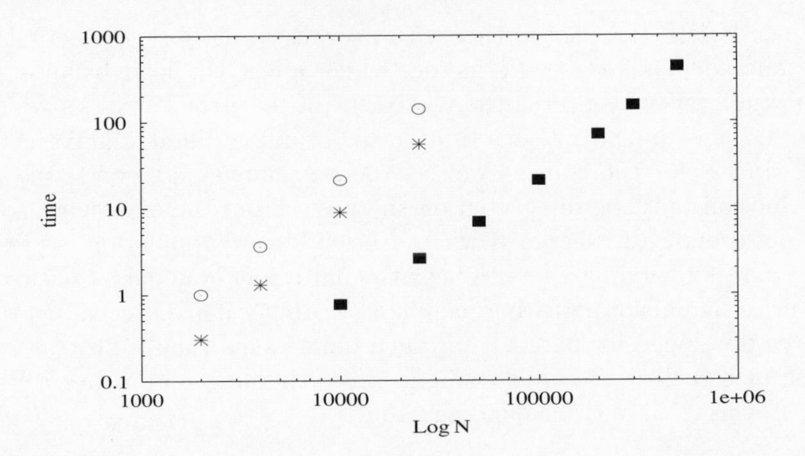

Fig. 13.2. CPU time comparisons. The symbols are: asterisks for the workstation tests, open circles for HARP-2 and filled squares for GRAPE-6.

13.6 Timing comparisons

We conclude this chapter with a brief comparison of performance in order to illustrate the power of special-purpose hardware and also demonstrate what can be achieved on a standard workstation or fast laptop. The actual performance of a code depends on many factors. As an illustration of the capability of the AC method, the Hermite version with small softening was used to obtain some characteristic timings, although *NBODY6* was found to be be similar. Further calculations were made with the special-purpose machines HARP-2 and GRAPE-6.

Figure 13.2 displays the CPU times for a few values of the particle number and an equal-mass Plummer model in virial equilibrium. Standard accuracy parameters $\eta_I = 0.02$, $\eta_R = 0.03$ were used with softening $\epsilon = 4r_h/N$, while $\eta = 0.02$ on the special-purpose computers that do not use softening. A short integration to $t_{\rm f} = 2$ was chosen, during which there were only a few close hyperbolic encounters. It was found experimentally that increasing the predicted neighbour number (3.3) by a factor of 2 produced the minimum CPU times. Note that the original expression was based on tests with $N \leq 200$, influenced by memory considerations that are no longer relevant. To a good approximation, the total CPU time in minutes per scaled time unit can be expressed as $T_{\rm comp} \simeq A(N/10\,000)^{2.1}$, where $A \simeq 4.5$ for a 500 Mflops (2 GHz) workstation. Likewise, we obtain $T_{\rm comp} \simeq 6.7(N/50\,000)^{1.7}$ for GRAPE-6 and N in $[5\times10^4, 3\times10^5]$. Hence the AC exponent is somewhat larger than the original value 1.6 for small N [Aarseth, 1985a], also found by Ahmad & Cohen [1974], but in full agreement with Makino & Hut [1988] for $N \leq 1000$.

In the case of special-purpose hardware, the additional time spent on

Table 13.4. Comparison of integration steps.

N	Irregular	Regular	$\bar{n}$	time
2500	9.83×10^5	1.03×10^4	41	2.1
2500	9.83×10^5	1.05×10^4	57	2.1
10 000	5.21×10^6	4.32×10^5	82	44.8
10 000	5.21×10^6	4.05×10^5	154	44.6
25 000	1.56×10^7	1.10×10^6	136	288.6
25 000	1.56×10^7	9.88×10^5	304	287.0

initialization is not included here since the host computer is performing the most expensive tasks. Comparing a 500 Mflops workstation with HARP-2 itself for $N = 10\,000$, the respective times are 9.0 and 21 minutes. Thus 5.28×10^6 integration steps on the latter and 57 equivalent floating-point operations for each interaction yields 2.4 Gflops with the enhanced host. These timings are consistent with the theoretically predicted speed-up factor of $(N/2)^{1/4}$ for the AC method [Makino & Aarseth, 1992]. Note also that the number of time-steps are remarkably similar.

A second purpose of the timing tests was to compare the present AC code with the proposal for a different form of the predicted neighbour number based on theoretical considerations [Makino & Hut, 1988]. In the event, the suggested choice $n_{\rm p} = (N/10.8)^{3/4}$ gave a slightly smaller CPU time than the bracketing factors $N/10$ and $N/12$ with the chosen parameters for $N \leq 10^4$. The results are summarized in Table 13.4 for a 100 Mflops workstation. The corresponding entries are for the modified original method, followed by the Makino–Hut proposal. Note the similar number of irregular time-steps, while the final average neighbour numbers are significantly different. In fact, constant small neighbour numbers provide better parallelization and load balancing in parallel execution and would also be an advantage for the design of the next generation of special-purpose hardware [cf. Spurzem *et al.*, 2003]. In this connection, we mention a simulation with $N = 10^4$ and $n_{\rm max} = 100$ which took 60 minutes and yielded an acceptably small energy error. The corresponding number of integration steps was 5.20×10^6 and 7.11×10^5, with $\bar{n} = 15$. Hence it *is* possible to employ small neighbour numbers with some loss of efficiency which is not a major concern on parallel computers.

In conclusion, we note that the workstation tests with $N = 25\,000$ represent a factor exceeding 10^3 with respect to the initial investigation of von Hoerner [1960], although the latter study covered somewhat longer time intervals. Moreover, GRAPE-6 is already a factor of 20 faster than the corresponding host in spite of using the brute-force method.

14
Practical aspects

14.1 Introduction

In the preceding chapters, we have described a variety of numerical methods and their implementation. Given these tools and algorithms, it should be possible in principle to construct direct N-body codes for dynamical studies. However, the combination of methods introduces considerable complications which have taken many years to master and it is therefore much easier to take over one of the large codes.* On the other hand, it would be good programming practice to implement a stand-alone code based on direct summation with softened potential or three-body regularization, where decision-making is considerably simplified.

In the following we provide some hints to facilitate the use of one of the main codes, as well as guidelines for implementing movies. These range from practical comments on getting started to producing a meaningful description of the results. Special counters that record events of interest provide further information. Graphics packages are generally avoided in the codes for reasons of compatibility. However, two versions of stand-alone regularization codes are available with software for movie making which is a good way to study dynamical interactions. We also discuss some diagnostic aspects that may assist in the dissemination of the results and outline various strategies for identifying numerical problems.

14.2 Getting started

Each computational project invariably has certain objectives which are defined by choosing specific initial conditions. This may also entail minor code changes to deal with particular requirements. A general understanding of the data structure then becomes desirable because any modification

* Most of the codes are freely available on `http://www.ast.cam.ac.uk/~sverre`.

may have unforeseen repercussions. Ideally, a user manual should be available, describing relevant machine-related aspects, together with input templates. However, this is a big task and to avoid duplication it is hoped that this book will provide most of the information needed. The manual will therefore mostly be limited to some practical issues. In addition, a list of all the routines and their main purpose has been prepared.

To start with, an appropriate choice of the maximum array sizes for the common blocks (cf. Table 7.6) must be made. Memory is usually not an issue on modern computers but unduly large array declarations produce unwieldy common data saves, and in any case recompilations are very fast. Before compiling, it is also necessary to select the most suitable FORTRAN compiler for the chosen hardware.

The main input parameters and options are listed in Tables 7.3 and 7.4, respectively. One useful feature of the codes concerns the verification of acceptable values of the main input data, otherwise termination occurs. Since the network of options is rather complicated, it is difficult to detect any mutual inconsistency even after consulting the definitions in the table. We distinguish between idealized and realistic cluster simulations, where the latter require several additional options related to stellar evolution, primordial binaries and finite-size effects. To assist users, input templates have been constructed for both types of system. As far as input data are concerned, many are dimensionless and appropriate to a wide range of particle numbers, whereas most of the others have dynamical significance. Although the number of input lines is quite small for a standard calculation, additional options call for some insertions which can be confusing for the unwary.

Before making changes, any existing test results should be consulted and compared, but such results are often both machine- and compiler-dependent. Hence a satisfactory outcome, defined by the energy check (9.31), is perhaps the best one can hope for in a complicated code when the luxury of an alternative reliable solution method is not available. Hopefully, there should be no difficulties when making a long standard calculation without modifications. More stringent tests are possible with the three-body or chain regularization codes in the form of time reversed solutions, which can often be reproduced to high accuracy.

As for making changes, a distinction should be made between including more data analysis and other implementations affecting the integration scheme. The former should be straightforward, whereas any genuine modifications require a good understanding of the code design and data structure. In particular, great care must be taken when introducing additional variables, and this also holds for data analysis. Still, the data structure is sufficiently versatile to permit manipulation of complicated configurations, as can be seen from the treatment of ghost particles.

14.3 Main results

Every practitioner has different ideas on how to display the results. As a compromise, summaries of interesting quantities are available at regular time intervals. Some of this information is not fully documented but headings assist in the interpretation and inspection of the relevant part of the code should lead to clarification. Among the various optional data available are Lagrangian radii for different mass percentiles, relevant core parameters and distributions of binary properties, as well as integration steps. However, there is no attempt to determine the density profile since the Lagrangian radii are usually preferred.

An optional general data bank is also produced with prescribed frequency in terms of the standard interval for main results. This contains the basic variables $m_j, \mathbf{r}_j, \dot{\mathbf{r}}_j$ for all single particles and c.m. bodies, as well as other useful data such as particle identity, potential energy, local density contrast and stellar type. Thus, if desired, all data analysis can be performed afterwards using this information which also contains some relevant global quantities readily identified by code inspection. Moreover, it is a simple matter to include additional data summaries of any desired information, which can always be constructed from the basic variables. Note that the data bank contains only the *current* cluster members when using the optional escape procedures, and hence is much reduced in size at later times. In any case, it is not appropriate to retain distant particles in the orbit integration, at least when using a linearized tidal field.

Since a given state is likely to contain hierarchical systems, it is desirable to reconstruct the properties of the associated ghost particles (cf. section 11.7). This is achieved by means of identifying the corresponding merger index that enables the relevant quantities to be obtained, as described in detail by Appendix D.2. When it comes to the analysis of hierarchical binaries, the energy of a dormant KS pair may be recovered from the merger table by combining the reduced mass with the corresponding saved value of the binding energy (cf. Appendix D.2).

The rare case of an active chain regularization occurring at the time of data analysis also needs to be allowed for since ghost particles play a similar role here. This is much simpler than for hierarchical systems and consists of the following two steps. First the global location of each member is determined by comparing the names $\mathcal{N}_j$ with the chain table names, taking care to include the case of zero name for the reference body which may not always come first in the sequence because of possible ejections. This permits the combination of the global coordinates and velocity of the reference body with the individual local values, as is done for perturbation calculations. Hence the basic data arrays may be modified accordingly.

Further procedures are invoked by employing a second time interval, $\Delta t_{\rm adj}$, which is usually smaller for convenience but this is not a requirement. Among these are the energy check, escaper removal and updating of the density centre as well as of the integration parameters (cf. section 9.9). In the Hermite versions, and especially *NBODY*6, it is slightly advantageous to make use of time intervals that are commensurate with unity or the first few fractions $1/2^n$ [cf. Makino & Aarseth, 1992], when all of the most sensitive solutions are known to full accuracy.

The structure of the codes relating to the generation of data files is relatively simple. After all the coordinates and velocities have been predicted to highest Order and the current density centre evaluated, any special procedure for data analysis may be implemented. Since the evolution rate of a cluster tends to slow down with time, it may be desirable to sample the data less frequently. This is achieved by increasing both time intervals by a factor of 2, subject to the checking interval being less than the current crossing time (which increases significantly) and the cluster binding energy sufficiently small in absolute value.

14.4 Event counters

One good way to describe the history of a simulation is to make use of special counters, since this type of information cannot be reconstructed from the general data bank which only provides occasional snapshots. We distinguish between various counters monitoring the integration and others that refer to events of one kind or another. It may be instructive to illustrate some of the main counters from a typical simulation with 1800 single stars and 200 primordial binaries, as displayed in Table 14.1. These results were produced by a workstation using *NBODY*6, hence some astrophysical processes discussed in this book are not included.

The results were obtained using the input data specified by Table 7.3 which is also part of the standard template when primordial binaries and stellar evolution are included (except for $n_{\rm max} = 140$). The ratio of irregular and regular time-steps gives a measure of code efficiency, where the separation of time-scales increases slowly with N. Given that relatively smooth motion still requires about 100 regular time-steps for a typical cluster crossing, the actual ratio is decided by the amount of close encounters and other irregularities. Moreover, the extent of any core–halo structure is also relevant. As for regularization activity in general, this depends on any primordial binaries being present. In the present example, the amount of chain regularization is fairly modest as measured by the *internal* step counter, *NSTEPC*, whereas the external c.m. time-steps

Table 14.1. Characteristic counters.

Name	Definition	Counts
NSTEPI	Irregular time-steps	8.0×10^7
NSTEPR	Regular time-steps	2.1×10^7
NBLOCK	Block steps	5.1×10^6
NKSTRY	Regularization attempts	2.4×10^6
NKSREG	KS regularizations	2.2×10^3
NKSHYP	Hyperbolic regularizations	800
NKSMOD	KS slow-down modifications	6.6×10^4
NKSPER	Unperturbed two-body orbits	4.6×10^{11}
NMERGE	Hierarchical mergers	359
NEWHI	Independent new hierarchies	25
NCHAIN	Chain regularizations	86
NSTEPU	Regularized time-steps	4.2×10^7
NSTEPC	Chain integration steps	1.1×10^5
NMDOT	Stellar evolution look-ups	5.0×10^4
NSN	Supernova events	5
NWD	White dwarfs	90
NCOLL	Stellar collisions	4
NBS	Blue stragglers	2
NSYNC	Circularized binaries	9
NSESC	Single escapers	1833
NBESC	Binary escapers	176
NMESC	Hierarchical escapers	2

tend to be larger. An assessment of the relative numerical effort of these procedures is difficult in view of the different solution methods. However, when using HARP, an occasional long-lived chain configuration may slow the host computer noticeably and therefore affect the load balance.

Unperturbed two-body motion is employed extensively in the simulation. This is reflected in the choice of the initial period distribution which has a lower cutoff of 5 days, whereas the hard binary limit is near $a_{\rm hard} \simeq 200\,{\rm au}$. On the other hand, modifications of the slow-down factor κ for KS solutions (cf. Algorithm 11.2) occurred surprisingly few times. This may be taken to mean that a given value is applicable over many orbits before any adjustment by a factor of 2 either way and also that unperturbed motions are predominant.

The last part of the table refers to aspects connected with stellar evolution. Given the actual disruption time of about 2.0 Gyr, the total number of look-up procedures is quite modest. This is essentially due to the small fraction of stars that evolves substantially during this time. The number of collisions is also small, as might be expected. In addition to the two blue stragglers and two other collisions, there were two so-called 'contact binaries' following coalescence in circularized orbits. Note that the counters give the accumulated as opposed to the current number. Finally, as a bonus, one hyperbolic capture was recorded. It also appears that significantly larger cluster ages are needed to produce the maximum observable number of blue stragglers at any one time [Hurley *et al.*, 2001a], which can therefore only be reached for larger cluster models.

Escape of binaries is a common process in this type of simulation. This includes a significant percentage of low-mass binaries which retain their pre-escape values of a and e, and are therefore ejected as point-mass objects. Out of the 176 binaries that escaped before termination at $N_{\rm f} = 2$, eight did not consist of the original components. The escape of hierarchies generally takes place after the half-life stage, when the central potential is weaker. It has been noted that such systems tend to be centrally concentrated, as is also the case for binaries [Aarseth, 2001a; Aarseth & Mardling, 2001]. In the present small simulation, there are typically one or two stable systems at one time but this increases to about ten for $N \simeq 10^4$ and even 50 for $N \simeq 3 \times 10^4$ with 50% binaries. From the counter of independent hierarchies, the formation rate is quite small but, in compensation, the life-times are often long.

The Hermite integration scheme gives rise to highly satisfactory energy conservation, bearing in mind the many dissipative events. Thus the sum of all relative energy errors, $\sum \Delta E/E$, amounts to only about -1×10^{-4} for the whole calculation. Likewise, the alternative measure of adding the actual deviations, $\sum \Delta E$, represents less than 0.03% of the standard initial energy, E_0. Finally, we remark that the test calculation presented here involves many energetic interactions and is therefore quite challenging.

14.5 Graphics

The introduction of graphics in a code invariably raises questions of compatibility. Hence it may seem preferable to produce separate versions that can only be used with commonly available software. Examples of the alternative approach are contained in the *NEMO* package [Teuben, 1995] and also the *STARLAB* facility [McMillan & Hut, 1996].[†] Here the user can generate a variety of initial conditions and perform some integrations illustrated by movies. However, most of these procedures are based on the C and C++ programming languages which may deter some.

The philosophy adopted here is to include movie making facilities in the special FORTRAN codes that employ multiple regularization, *TRIPLE* and *CHAIN*, and also the simplest N-body code, *NBODY1*. There is a

[†] See `http://www.astro.umd.edu/nemo` and `http://www.sns.ias.edu/~starlab`.

choice between the $X11$ graphics that is generally available on UNIX systems and the popular PGPLOT package, but not all permutations have been implemented.‡ This provides good opportunities for further programming developments and is therefore suitable for classroom exercises. There is considerable educational benefit in visual demonstrations of even quite simple dynamical interactions.

One new feature when making computer movies in real time of few-body systems with the regularization codes is that the calculations are usually too fast and need to be slowed down artificially. In any case, it is desirable to have control over the viewing time. The amount of delay depends on the hardware which may have widely different cycle times and experimentation is needed. Moreover, in some cases the standard integration steps are too big for the visualization to appear continuous and may need to be reduced by a specified factor. The reason is that the high-order Bulirsch–Stoer [1966] integrator is used, where smaller tolerances tend to increase the number of function calls rather than decrease the time-steps. In practice it is convenient to plot the motion after each integration step, instead of using constant intervals of physical time, but this may not suit every purpose and can easily be changed.

Other types of graphics may readily be incorporated in the codes. We mention simultaneous displays of global quantities of interest, such as the density and velocity profile or the synthetic HR diagram. Relevant data for the latter purpose are available as an optional feature. However, the requirements will generally depend on the nature of the project and any implementations are therefore best left to the practitioner.

It is clear from these modest beginnings that graphics representation may take diverse forms. This tool can be readily used to good advantage in order to gain insight into complicated processes, such as dynamical interactions of compact subsystems or, indeed, scattering experiments. It can also be employed for diagnostical purposes when other means fail. A particular mode of graphics not mentioned here concerns 3D displays which are particularly suited for visual illustrations of N-body dynamics. Hence we may anticipate considerable creative efforts in this area, ranging from personal studies to *IMAX* displays of large simulations.

14.6 Diagnostics

Simulation codes tend to produce much special information that is only consulted during the development phase or if there are technical problems. Given that some calculations may require months of dedicated effort, this is good programming practice, and in any case the data may not appear

‡ Some movie versions are available on `http://www.ast.cam.ac.uk/~sverre`.

in hardcopy form. In this section, we are mainly concerned with the type of information needed for disseminating results, whereas the question of error diagnostics is discussed in the final section. In the first place, it is useful to distinguish between general data banks and information about different kinds of dynamical configurations or unusual objects.

The structure of the basic data bank has already been discussed above. Another systematic table of data is concerned with binaries and is sampled with the same frequency. It consists of the following twelve entries for each binary in standard notation: $E_{\rm b}$, e, $E_{\rm cm}$, $\mathbf{r}_{\rm cm} - \mathbf{r}_{\rm d}$, m_k, m_l, $P_{\rm b}$, $\mathcal{N}_k$, $\mathcal{N}_l$, k_k^*, k_l^*, $k_{\rm cm}^*$. Here $E_{\rm cm}$ is the specific binding energy of the c.m. body and $P_{\rm b}$ is the period in days. This optional table facilitates tracing the history of a given binary that may experience several exchanges or memberships of hierarchies. We exploit the program data structure and include only binaries treated by two-body regularization for convenience, whereas softer binaries may be identified from the basic data bank if desired, albeit at some cost. The case of a hierarchical system being present is denoted by a special value (i.e. -10) of the c.m. index, $k_{\rm cm}^*$, and the entries for $\mathcal{N}_l$ and k_l^* relevant to the outer component are given instead. However, an additional table provides more specific information on hierarchical systems [cf. Aarseth, 2001a].

The question of new binary formation is of considerable interest. A binary is still defined as primordial even if it exchanges both its components, since the binding energy is more relevant dynamically than the identity of the components. Also note that both the mass and absolute value of the binding energy tend to increase as a result of exchange [Heggie, 1975]. The identification of any newly formed binary poses an interesting challenge. For this purpose, we have adopted the following algorithm. A special list containing the names of the 20 most recently terminated KS components is maintained and checked for membership when the next two-body solution is initialized. The second condition that must be satisfied is that the two particle names should not differ by one if the first is a primordial component. However, any hierarchical binary is recognized as being new, and likewise for a KS pair containing one or even two independent primordial components if not identified in the termination list.

Given that long-lived hierarchies are an important feature of star cluster models, the question of their origin and evolution deserves attention. The merger algorithm constitutes one useful tool for this purpose. Such systems are only accepted for treatment if the *outer* binding energy exceeds $\frac{1}{4}\epsilon_{\rm hard}$, softened by the central distance factor $1 + r/r_{\rm h}$. The actual formation process itself is often elusive and not well defined. However, we may obtain some interesting data on earlier stages by recording failed mergers, taking care to avoid too many successive attempts [cf. Aarseth, 2001a]. Recall that this search is performed at each apocentre of a KS binary and also after the check for unperturbed motion, in either case provided that the c.m. time-step is suitably small.

One type of hierarchical formation is associated with binary–binary collisions where the fourth body carries away the excess energy and angular momentum [Mikkola, 1984a,b]. If chain regularization is used, the outer triple component is likely to be strongly bound with respect to the inner c.m. Such interactions are readily identified if the stability condition (9.14) is satisfied since this leads to enforced termination. The fascinating question of quadruple system formation remains, although some tentative clues are beginning to emerge [Aarseth & Mardling, 2001].

Further comments on the energy budget may be instructive. According to Table 9.1, the sum $E_{\rm bin} + E_{\rm merge}$ represents the current energy of two-body motions, with the second term containing the internal energy of any inactive hierarchical components. During chain regularization, however, the quantity $E_{\rm ch}$ defined in Table 9.1 is usually dominated by one or more binaries and should therefore be combined with the above sum to give a more representative value of the total energy available in fossil fuel. For some purposes, we may refer to $E_{\rm merge}$ as *latent energy* [cf. Aarseth & Mardling, 2001]. Since the escape of hierarchical triples or quadruples affects the merger energy, the relevant contributions are subtracted from $E_{\rm merge}$ to yield a consistent value after an escape, likewise for any higher-order systems, where each subsystem is treated separately.

It is useful to distinguish between the total energy (9.29) and the part that binds the cluster, i.e. $E_{\rm bind} = T + U + E_{\rm tide}$. Assuming $E_{\rm bind} < 0$, this gives the current equilibrium crossing time

$$t_{\rm cr} = \left(\sum m_i\right)^{5/2} / (-2E_{\rm bind})^{3/2} \,. \tag{14.1}$$

Because of energy generation due to binary processes and mass loss from evolving stars, $|E_{\rm bind}|$ decreases significantly with time, thereby increasing $t_{\rm cr}$ even in tidally truncated systems. This can be seen qualitatively by writing $E_{\rm bind} \simeq -M^2/4r_{\rm h}$ at equilibrium which yields $t_{\rm cr} \simeq 2(2r_{\rm h}^3/M)^{1/2}$. By analogy with the adiabatic invariance relation $am_{\rm b} =$ const for a binary undergoing slow mass loss, it follows that both terms contribute to the increasing crossing time. Finally, we emphasize again that the particular form of the total energy expression is completely regular since the additional terms of (9.29) are obtained from well-defined quantities.¶

Another table of interest provides information about distant escaping particles that are usually removed from the calculation since the linearized tidal field would not be appropriate any longer. One line for each escaper (single particle, binary or composite c.m.) with non-zero mass

¶ The *initial* value of $E_{\rm ch}$ is well defined and so is the integrated change.

contains the quantities $t_{\rm phys}$, $m_i/M_\odot$, $v_\infty^2/v_{\rm rms}^2$ and v_∞, where $t_{\rm phys} = T^* t$ is in Myr and v_∞ is in $\rm km\,s^{-1}$. The third term measures the specific escape energy scaled by the current rms velocity, which decreases with time. The distribution of terminal escape velocities displays a long tail of superfast particles due to strong interactions involving binaries. Thus for rich open cluster models with $N \simeq 10^4$ and $r_{\rm h} \simeq 3\,{\rm pc}$, examples in excess of $100\,{\rm km\,s^{-1}}$have been noted, compared with a typical initial value $v_{\rm rms} \simeq 2\,{\rm km\,s^{-1}}$. The summary of results also gives additional information for each escaping binary or any inner hierarchical components, which may be extracted and used for analysis (cf. Appendix C.4).

Let us now turn briefly to some astrophysical aspects. The construction of sequential synthetic HR diagrams illustrates well the opportunities for observational comparisons [see e.g. Hurley *et al.*, 2001b]. The cluster simulation codes contain optional procedures for creating such data banks. One instructive feature is that binaries undergoing different evolutionary stages are displayed with special symbols to facilitate identification. Many of these objects are monitored using event counters as discussed above. In addition, other files contain histories of exotic objects formed by collision or coalescence via the common-envelope process.

As the modelling improves, further processes involving mass transfer will be included. In particular, there is considerable interest in the different kinds of objects that may be produced, especially since some of them have short life-times and hence may not be observable in practice. Since the scheme of combining dynamics with stellar evolution is relatively new and quite complicated, it is worthwhile to compare many of these results with population synthesis procedures, where the astrophysics is based on the same algorithms [cf. Hurley, Tout & Pols, 2002]. This is particularly relevant for primordial binaries that have experienced small perturbations to their orbital elements. In conclusion, the provision of a consistent treatment enriches the possible outcomes and is therefore certain to improve our understanding of star cluster evolution and furnish a comprehensive description of exotic objects.

14.7 Error diagnosis

In the previous section, we commented on a variety of useful data banks that are produced by the simulation codes and also discussed some related algorithmic aspects. Naturally, the availability of such results presupposes that there are no technical problems. The question then arises of what to do if this is not the case. Below we offer some simple suggestions that are connected with the topic of diagnostics.

Several main types of behaviour may be distinguished when discussing numerical problems in N-body simulations. In the first category we have abrupt termination caused by a non-standard condition, such as division by zero or another illegal operation. It is often possible to pinpoint the exact location in the code, using special software. However, identifying the nature of a problem and where it goes wrong is only half the story, and some understanding of the relevant algorithm will usually be needed in order to find the right cure. It may be possible to perform a temporary modification unless something is fundamentally wrong.

The case of infinite looping is another common occurrence that is usually more difficult to isolate, and a few hints on how to proceed are given here. Since the process of numerical integration consists of advancing the solutions by many discrete time intervals, the first task is to ascertain the precise moment when things begin to go wrong. One way is to restart the calculation from a recent common save and record some information at the start of every new cycle or block-step, taking care to stop on any unreasonably small value of the current step due to convergence problems. In order to determine the onset of any deterioration, it is also useful to make a note of the time-step counters and the given particle identity. Without any further clues, different processes can be ruled out by elimination.

In the alternative case of termination due to large energy errors it is desirable to monitor the solution in as much detail as possible, paying special attention to the high-order derivatives which are sensitive to any discontinuities. One tell-tale sign of a force irregularity in the Hermite scheme is that the *natural* time-step decreases abruptly, whereas the quantized value may be reduced by at most a factor of 4. Hence a time-step reduction trap may facilitate identification at an early stage, instead of waiting for any extreme condition to develop.

Several other strategies are available if the problem is not connected with standard integration. For instance, options relating to mergers, two-body regularization or multiple regularization may be activated to provide further details about these features. Should the problem be due to the chain method, two deeper levels of diagnostics are available with information about each integration step. Likewise, the value of the basic KS parameters R, h, γ may be obtained after each cycle following the last reliable initialization or termination. Moreover, if an unperturbed KS pair is being updated and the c.m. step is small, there could still be problems in procedures relating to tidal dissipation, or the search for stable hierarchies may just fail. Also note that the calculation may slow down significantly in the presence of a long-lived hierarchy that falls outside the stability limit and this may give rise to errors. If all this does not reveal anything unusual, it may also be a good idea to look carefully at any treatment of stellar evolution. A picture of the general state should eventually emerge after examining all the available clues. However, it is important to recognize that there are other processes on short time-scales which are intrinsically difficult to handle. We also mention the troublesome case of eccentric Roche lobe overflow for an inner hierarchical binary due to inclination effects, which requires repeated circularization.

Finally, we return again to the question of what to do if the energy check is not satisfactory. This situation is often related to the existence of some borderline configuration treated by a less efficient algorithm, thereby giving rise to systematic errors. For example, in relatively small systems studied by the AC method, a massive wide binary may not be accepted for KS regularization because the c.m. neighbour list is too small (e.g. $\lambda R > R_s$). This problem is usually remedied by assigning a relatively large maximum neighbour number, rather than the standard value of $2N^{1/2}$. Other difficulties may be associated with long-lived hierarchical systems, treated by direct integration or chain regularization, which fail narrowly to satisfy the stability criterion. In any case, a general understanding of the current state of the system is desirable, and further information may be obtained by activating various optional procedures.

One useful strategy is to sub-divide the checking interval in order to ascertain whether the error accumulation is gradual or caused by a sudden glitch. Such behaviour might point to the presence of a massive binary that is too wide to be accepted for KS treatment, or a hierarchical system just outside the stability limit. The latter may be formed in binary–binary collisions which are a common feature of realistic cluster simulations. Note that, depending on the nature of the problem, the results may not always be reproducible in some codes with complex decision-making, because a few variables are not part of the saved data. Moreover, any extra prediction of coordinates and velocities to high orders for neighbour schemes could also affect the outcome.[‖]

Since the structure of N-body codes is invariably complicated, it is not always easy to diagnose the initial cause of a problem. Hence trying to follow a certain suspect sequence may yield many false trails which must be investigated. Once a troublesome procedure has been located by a process of elimination, the final search can be narrowed down by energy checks using the bisecting principle. In particular, the last good value of the relevant time-step counter may be employed in order to limit the amount of diagnostic information before identifying the underlying cause.

After determining the exact cause of a problem, the task of improvement is by no means simple and may necessitate significant modification. This is especially the case if some new condition has been encountered that falls outside the range of the specified parameters. In conclusion, we may therefore say that a complicated code is never fully tested and there are always new challenges to be faced.

‖ For simplicity, present code versions do not employ separate variables for this purpose.

15
Star clusters

15.1 Introduction

Simulating star clusters by direct N-body integrations is the equivalent of scaling mountains the hard way. At any time the maximum particle number depends on hardware and is therefore limited by technology. Some of the methods that have been described in this book are ideally suited to studying the classical point-mass problem. In addition, a wide variety of astrophysical processes can be included for realistic modelling of actual clusters. Recently the simulations have been devoted to systems with up to $N \simeq 10^4$ particles which includes rich open clusters. However, with the construction of the GRAPE-6 special-purpose computer we are now able to investigate small globular clusters as observed in the Large Magellanic Cloud (LMC) [Elson *et al.*, 1998].

In the following we concentrate on various aspects of star cluster simulations not covered in earlier chapters. We first describe algorithms for determining the core radius and density centre which are useful tools for data analysis. For historical reasons, idealized models (i.e. isolated systems) are also considered, particularly because of their relevance for more approximate methods. After further discussions of the IMF, we return to the subject of assigning primordial binaries and illustrate their importance by some general results. External effects due to the tidal field and interstellar clouds form an important ingredient in star cluster modelling even though the latter are rarely studied. Algorithms for combining stellar evolution with the dynamical treatment have been outlined previously. Here we review procedures for mass loss, tidal interactions as well as collisions and highlight the implications for cluster simulations. In addition, processes connected with stellar rotation are also included. We conclude by addressing aspects relating to globular cluster simulations, especially as regards the projected usefulness of the present scheme.

15.2 Core radius and density centre

Most star clusters tend to have regular shapes with a pronounced density increase towards the centre. However, the symmetry centre used by observers often deviates significantly from the c.m. in a simulation. This is mainly due to recoil when fast particles are ejected from the core following strong interactions. Already the early N-body discussions [Aarseth, 1963b; von Hoerner, 1963] recognized the need for measuring global quantities with respect to the maximum particle concentration.

The concept of density centre [von Hoerner, 1963] is based on assigning individual densities, $\rho_j = (3/4\pi)3.5m/r_{j3}^3$, defined by the third nearest neighbour distance, r_{j3}, and forming the expression

$$\mathbf{r}_{\rm d} = \sum_{j=1}^{N} \rho_j \mathbf{r}_j / \sum_{j=1}^{N} \rho_j \,. \tag{15.1}$$

Likewise, the core radius is introduced empirically as a weighted sum over the individual central distances by

$$r_{\rm c} = \sum_{j=1}^{N} \rho_j |\mathbf{r}_j - \mathbf{r}_{\rm d}| / \sum_{j=1}^{N} \rho_j \,. \tag{15.2}$$

These definitions served their purpose well for a long time. Subsequently, a systematic investigation of this problem [Casertano & Hut, 1985] proposed the sixth nearest neighbour as the best criterion. Although based on equal-mass particles, this formulation has been used by several N-body practitioners after generalization to a mass spectrum.

The present codes employ a slightly different procedure. It is based on the principle of convergence, namely that the density centre determination should give essentially the same result by excluding the outer parts. Since this is an N^2 process, we restrict the sample to the innermost $N/5$ members, based on the provisional guess,* $\max\{3r_{\rm c}, r_{\rm h}\}$. A sorted list of the six shortest distances is formed, based on the over-estimated value $2(6/n_j)^{1/3}R_{\rm s}$, with n_j the neighbour number and $R_{\rm s}$ the neighbour radius. In the case of the HARP code, the (square) central distances are first sorted, whereupon local neighbour lists are created using an r-dependent initial guess which increases gradually outwards. Individual densities are now assigned according to the generalized expression

$$\rho_j = \sum_{k=1}^{5} m_k / r_6^3 \,, \tag{15.3}$$

* More than half the particles are usually located outside the half-mass radius.

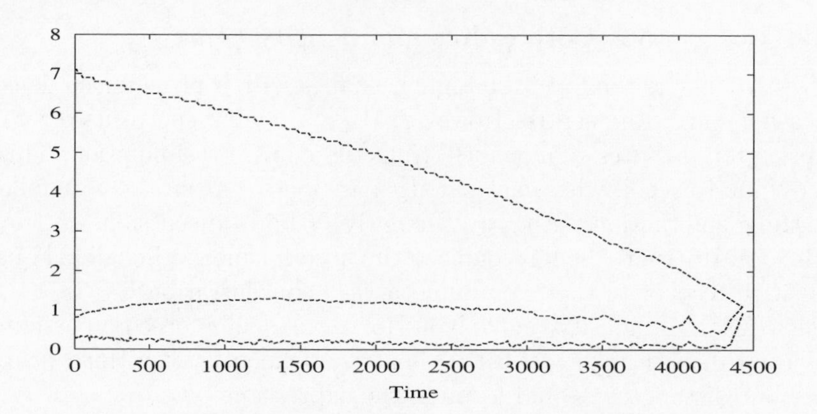

Fig. 15.1. Core radius, half-mass radius and tidal radius as functions of time in Myr. The depicted model has $N_s = 8000$, $N_{\rm b} = 2000$ and $r_{\rm h} = 3\,{\rm pc}$.

where r_6 is the sixth shortest distance. Note that Casertano & Hut showed that an unbiased estimator for equal masses is provided by defining the density $\rho_k = (3/4\pi)(k-1)m/r_k^3$, thus omitting the kth neighbour, and this has been adopted. With this definition, the density centre is again given by (15.1), where the summation is over the reduced membership, N^*, with all single KS components replaced by the corresponding c.m. body. However, the core radius used by *NBODY*5 since 1986 is taken to be

$$r_{\rm c} = \left(\sum_{j=1}^{N^*} \rho_j^2 |\mathbf{r}_j - \mathbf{r}_{\rm d}|^2 / \sum_{j=1}^{N^*} \rho_j^2 \right)^{1/2} . \tag{15.4}$$

Provided $N^* \geq N/5$, this expression converges for different central samples and agrees quite well with the more expensive form (15.2), which does not have this property. The evolution of three characteristic radii are shown in Fig. 15.1 for a typical star cluster model.

Although the operational definition differs from the classical expression [cf. King, 1962], it describes the characteristic size of the core well and may also be used for some observational applications.† For completeness, we define $N_{\rm c}$ as the corresponding membership inside $r_{\rm c}$. Likewise, we may define the density-weighted average of the density by [Casertano & Hut, 1985]

$$\rho_{\rm d} = \sum_{j=1}^{N^*} \rho_j^2 / \sum_{j=1}^{N^*} \rho_j \,. \tag{15.5}$$

† Recall that the observational core radius employs the e-folding density. In the following, the more common phrase 'core radius' will be used instead of 'density radius'.

The maximum density, $\rho_{\rm max} = \max\{\rho_j\}$, is also recorded after scaling by the mean density, $\frac{1}{2}M/r_{\rm h}^3$. Up to now it has been used for the determination of the close encounter distance (9.3) but this may well be too conservative and has recently been replaced by $\rho_{\rm d}$. In practice it is sufficient to update the density centre and core radius at the same time as performing the total energy check and escaper removal. Perhaps the procedure most affected by the discontinuous re-evaluation of the density centre concerns the codes *NBODY*5 and *NBODY*6 which impose an optional shortening of the regular time-step inside compact cores according to

$$\Delta\tilde{T}_i = \tfrac{1}{2}(1 + |\mathbf{r}_i - \mathbf{r}_{\rm d}|^2/r_{\rm c}^2)\,\Delta T_i\,. \tag{15.6}$$

Finally, we note that the results may not be reproducible in detail for different checking intervals when using the algorithms above. In particular, the optional updating of regularization parameters that depend on the core density is carried out after each total energy calculation.

15.3 Idealized models

Broadly speaking, N-body simulations have at least three main objectives. At the simplest level, equilibrium models with equal masses are studied in order to make comparisons with more approximate methods, such as the Monte Carlo, gaseous or Fokker–Planck description [Heggie & Aarseth, 1992; Makino, 1996a; Spurzem, 1999]. At the other end of the spectrum, the effort is devoted to the modelling of astrophysical processes [Terlevich, 1987; Portegies Zwart *et al.*, 1997; Aarseth & Heggie, 1998; Hurley *et al.*, 2001a,b]. The third approach is taken much less frequently and consists of investigating detailed dynamical processes such as escape [van Albada, 1968], binary formation [Aarseth, 1972a] or the formation of hierarchical systems [Kiseleva *et al.*, 1996; Aarseth, 2001a]. This type of investigation may readily be undertaken for idealized models, where the addition of a mass spectrum broadens the scope.

Now that powerful methods are available, it can be anticipated that the study of classical processes will be rewarding. Hence both the two-body and chain descriptions are ideally suited for investigating strong interactions involving one or more binaries. The former may be used to examine the formation of binaries or different types of hierarchical configurations, whereas the latter often gives rise to very energetic escapers and highly eccentric binaries. Moreover, the fascinating question of stable triples forming in binary–binary collisions‡ [Mikkola, 1983] also needs to be addressed further [cf. Aarseth, 2001a]. We emphasize that the use of

‡ Here we adopt the physicist's definition in the sense of a cross section.

regularization techniques enables the determination of well-defined orbital elements in the neighbourhood of numerical singularities.

The subject of core collapse has a long history but much of it is based on an obsession with the singular density solution which turned out to be a mirage. It required the direct numerical approach to resolve this dilemma finally, even though the crucial point that a central source absorbs the excess energy was first inspired by analytical work [Hénon, 1961]. This remarkable study, which was ahead of its time, laid the foundations for a proper understanding of globular cluster evolution. From the so-called 'homology model', the formation of a central singularity is relatively fast, i.e. about 7×10^9 yr for a $10^5\, M_\odot$ cluster. Moreover, in order to resolve the predicted infinite density, it was proposed that negative energy accumulating at the centre is absorbed by binaries and multiple stars, as hinted by the earliest N-body simulation [cf. von Hoerner, 1960].

The homology model was confirmed by an early continuum simulation using the Fokker–Planck method [Cohn, 1979], which resulted in the density relation $\rho \propto r^{-2.0}$ in the region between the shrinking core and the halo just prior to core collapse. Following technical refinements, the non-isothermal self-similar structure led to an asymptotic solution $\rho \propto r^{-2.23}$ [Cohn, 1980]. In a pioneering theoretical investigation, Lynden-Bell & Eggleton [1980] determined the power-law index –2.21 from fundamental principles. Subsequently, such similarity solutions have also been obtained for gaseous models in several investigations [Heggie & Ramamani, 1989; Louis & Spurzem, 1991]. However, the central energy source proposed by Hénon sets the scale for the maximum density and was later implemented in the Monte Carlo method [Hénon, 1975].

N-body simulations of core collapse with equal-mass particles were hampered by small-N limitations for a long time. However, comparison of averaged solutions for different particle numbers in the range 250–1000 [Giersz & Heggie, 1994a] shows excellent agreement during the pre-collapse phase when the time is scaled by the standard relaxation time factor $N/\ln(\gamma N)$. Some of these simulations employed a modified version of the code NBODY1 which included a special two-body regularization method based on time smoothing [Heggie, 1972a]. Likewise, anisotropic gaseous and isotropic Fokker–Planck models are in good agreement with ensemble-averaged N-body simulations both in the pre- and post-collapse phase after including an N-dependent energy generation by three-body binaries [Giersz & Heggie, 1994a; Giersz & Spurzem, 1994]. Moreover, the stochastic nature of the energy generation was emphasized. Statistical N-body studies also revealed self-similar solutions in the post-collapse phase in spite of energetic escapers [Giersz & Heggie, 1994b]. More recently, it was demonstrated that the core collapse time is at least halved for equal-mass models with initial rotation [Boily, 2000]. The first such post-collapse simulations have also been reported [Kim *et al.*, 2002].

A good example of core collapse with the direct method for $N = 6000$ and a mass spectrum was obtained after some six months dedicated integration [Aarseth & Heggie, 1993]. Another heroic effort using NBODY5 with $N = 10^4$ and equal-mass particles yielded further insight and demonstrated conclusively that core collapse is halted by the formation of a few central binaries [Spurzem & Aarseth, 1996]. Following these investigations, the time was ripe to explore other initial conditions.

The introduction of primordial binaries for equal-mass systems adds another degree of complexity. One study compared gas-dynamical solutions with direct integrations for $N = 2500$ during the pre-collapse stage and found reasonable agreement in the presence of primordial binaries [Heggie & Aarseth, 1992]. There have also been other notable efforts to explore the effect of primordial binaries in isolated systems [McMillan, Hut & Makino, 1990, 1991]. Equal-mass systems with up to 20% hard binaries were studied, with the result that most binaries were destroyed in binary–binary collisions. However, a system of equal-mass single stars and binary components tends to exhibit early mass segregation, thereby enhancing the collision rate. A similar later investigation [McMillan & Hut, 1994], which included the galactic tidal field, argued that a critical binary fraction above about 10% ensures the retention of sufficient binaries in the later stages. The question of binary survival during the early phase of violent relaxation has also been addressed using an approximate method [Vesperini & Chernoff, 1996]. This simulation considered a population of 5% medium-hard binaries in collapsing clusters of 1000–5000 equal-mass stars. Thus a characteristic N-dependent cutoff energy ensured survival of the initial collapse phase.

Idealized Monte Carlo models with primordial binaries were also explored at an early stage [Spitzer & Mathieu, 1980]. This work employed approximate rate coefficients for binary–binary interactions adapted from analytical and numerical results for single star–binary interactions [cf. Heggie, 1975; Hills, 1975]. Again the use of equal masses for all particles gave rise to a collapsing core dominated by binaries that may nevertheless display some characteristic features of late cluster evolution. Thus over half the energy released went into reaction products which escaped, in qualitative agreement with N-body simulations, albeit for small binary fractions [Heggie & Aarseth, 1992]. Modelling of primordial binary interactions has now been much improved [Giersz & Spurzem, 2000].

Looking beyond core collapse, the occurrence of core oscillations was first demonstrated by both the fast gas-dynamical and Fokker–Planck methods [Bettwieser & Sugimoto, 1984; Cohn, Hut & Wise, 1989]. After several attempts, the crowning achievement in the quest for the equivalent N-body effort was accomplished by Makino [1996a,b] who used $N = 32\,\mathrm{K}$ on the 1 Tflop GRAPE-4 with NBODY4, although its full power was not available. Thus the central density varied by a factor of 1000 and several oscillation cycles were seen. Although of little astrophysical consequence for real star clusters, this calculation provided considerable justification for the validity of alternative methods. In this connection, we remark that three-body effects are not important during the pre-collapse phase of large idealized systems, whereas the modelling of more advanced stages by approximate methods require the addition of energy production by binaries based on semi-analytical considerations.

Early claims to have reached this goal [Makino, Tanekusa & Sugimoto, 1986; Makino & Sugimoto, 1987] have been re-assessed in the light of further work. The apparent gravothermal oscillations exhibited in models with $N = 100$ and 1000 are now interpreted as being due to stochastic binary activity. A linear stability analysis of gravothermal oscillations [Goodman, 1987] showed that such stochastic fluctuations dominate below $N \simeq 7000$. It is generally agreed that two ingredients are needed to establish gravothermal oscillations, namely a temperature inversion in the core, followed by an expansion without energy production due to binary interactions or three-body encounters. These features do occur in the pioneering gaseous models of Bettwieser & Sugimoto [1984] and are taken as signatures of the gravothermal nature. A subsequent simulation with $N = 10^4$ narrowly failed to find convincing evidence for such oscillations [Spurzem & Aarseth, 1996]. On the other hand, slow gravothermal *expansion* was confirmed in an N-body system with slight initial temperature inversion for 3000 particles [Heggie, Inagaki & McMillan, 1994].

The question of scaling N-body models has received further attention. Simulations with NBODY6++ for $N \leq 16\,\mathrm{K}$ [Baumgardt, 2001] showed that the life-time does not scale with the relaxation time when the escapers are removed beyond a cutoff distance, irrespective of whether an external tidal field is present. However, there is no scaling problem if escaping stars are removed promptly on exceeding the critical energy. We also mention an investigation of long-term evolution of isolated clusters [Baumgardt, Hut & Heggie, 2002]. Plummer models with equal masses and particle numbers $N \leq 8\,\mathrm{K}$ were again studied with NBODY6++. The focus was on post-collapse evolution until complete dissolution which involved extremely long time-scales. The final structure was characterized by two parameters, the current particle number and half-mass radius. As might be expected for equal masses, binary activity was relatively unimportant.

Notwithstanding the remarkable recent accomplishments, we note that some of the general features of cluster evolution were already exhibited in the earliest models based on very modest particle numbers [von Hoerner,

1960; Aarseth, 1963b; Wielen, 1967; van Albada, 1968]. Moreover, the subsequent history of the Monte Carlo methods [cf. Hénon, 1972; Spitzer & Hart, 1971a; Stodółkiewicz, 1982] illustrates that new avenues of investigation can yield fruitful results.

15.4 Realistic models

It has been known for a long time that there are binaries in open clusters. However, the realization of their dynamical importance was slow to take hold until the results of N-body calculations became available. On the other hand, some well established examples of suitable cluster members (i.e. not too short period) were actually beginning to emerge during the early epoch of realistic simulations [e.g. Wickes, 1975].[§] In spite of considerable observational efforts, the main properties of open clusters – i.e. the IMF, mass segregation, binary abundance and membership – was only placed on a firm footing relatively recently [Mathieu, 1983; van Leeuwen, 1983]. Likewise, the birth of the HST has provided a wealth of high quality data to inspire simulators of globular clusters.

The first step towards increased realism is to introduce a mass function. This has the general effect of reducing the relaxation time and increasing the escape rate. Although the classical half-mass relaxation time is well defined for an equal-mass system [Spitzer, 1987], this is not the case for a mass spectrum. On the numerical side, we are still without a satisfactory operational definition of the relaxation time even though this was already attempted in the early days [von Hoerner, 1960; Aarseth, 1963b]. This is mainly due to the problem of adding energy changes in such a way that the result does not depend on the time interval. The best one can do is therefore to invoke the time-scale for dynamical friction [Spitzer, 1969] when considering early stages of mass segregation. In fact, there has been much confusion in the literature between the two concepts because they give similar values for $N \simeq 100$ (cf. (1.11) and (1.14)).

Many theories of escape have been proposed but comparisons with numerical models are generally unfavourable [Wielen, 1972, 1975]. Although such comparisons are fraught with uncertainty, i.e. small number statistics and model dependence, the theory of Hénon [1969] appears to show some qualitative agreement. This permits the influence of the most massive members on the escape rate to be calculated explicitly for an isotropic Plummer model without mass segregation. The predicted escape rate per

[§] But see Aitken [1914] for an early reliable determination of HD 30810 with period 16.6 yr, eccentricity 0.445 and inclination 9° in the Hyades.

crossing time for a group of N_i members of mass m_i is given by

$$\frac{\Delta N_i}{t_{\rm cr}} = -\frac{\sqrt{2}N_i}{4M^2}\sum_{j=1}^{k} N_j\, F_{\rm ij}(\nu)\, m_j^2\,, \qquad (15.7)$$

where the interaction coefficient $F_{\rm ij}$ is a known function of the mass ratio $\nu = m_i/m_j$ and there are k discrete groups. Hence the contributions from different masses may be evaluated from the IMF. As an illustration, in a cluster with $N = 500$, $f(m) \propto m^{-2}$ and a mass ratio of 32, the eleven heaviest members provide 50% of the total escape probability when considering the lightest particle. Likewise, only the three heaviest particles contribute 50% to the much smaller escape rate of the fourth body. It can also be seen that, with this mass function, the dependence is even steeper than $\propto m^3$. Combining two particles in a binary therefore enhances their effect considerably, although the general tendency will be somewhat less for relaxation. To contrast with an equal-mass system, the theory predicts just one escaper for 30 crossing times (i.e. $F(1) \simeq 0.076$), compared with 46 actual and 32 predicted escapers in the quoted example [Aarseth, 1974]. On the other hand, some of the seven escapers in the equal-mass model with $N = 250$ may have been due to binaries or a steeper central density which are not accounted for by the theory.

In order to evaluate the importance of the mass spectrum, it is useful to introduce the notion of an effective particle number [Farouki & Salpeter, 1982], defined by

$$N_{\rm eff} = M^2/\sum m_i^2\,. \qquad (15.8)$$

Hence a steeper IMF produces a smaller value of $N_{\rm eff}/N$ which reflects the increased graininess of the system and from this one can also understand qualitatively why the evolution rate speeds up. As an example, the standard HARP model with 8000 single stars and 2000 hard binaries gives $N_{\rm eff} = 3751$, whereas the actual particle number is 12 000. As a cluster evolves, the heaviest stars undergo mass loss, whereas there is some preferential escape of light members, such that the mean mass remains nearly constant for some time. Consequently, $N_{\rm eff}$ tends to increase slightly.

The tendency for only a small depletion rate in simulations with $N = 250$, tidal field and mass loss [Aarseth & Wolf, 1972] contrasts with most theoretical expectations. This result was later confirmed for somewhat larger models ($N \leq 1000$) that included interstellar clouds as well as stellar evolution, and where the absence of core collapse was noted [Terlevich, 1987]. A flattening of the mass function was also observed in HARP-3 simulations with NBODY4 for $N = 4096$ [Vesperini & Heggie, 1997]. The models included point-mass external tidal field, disk shocking and simplified stellar evolution with instant mass loss. An increasing mass fraction in white dwarfs was seen during the later stages (also see Hurley & Shara, 2003). Although disc shocking has no direct differential effect, it still promotes preferential escape via the existing mass segregation. However, these results were obtained without primordial binaries and are based on a somewhat idealized tidal field, as well as a relatively small particle number, and can therefore not be scaled to globular clusters.

Fokker–Planck models also show clear evidence of the evolution rate dependence on the IMF for isolated systems [Inagaki & Saslaw, 1985]. The evolution of the mass function was emphasized in subsequent Fokker–Planck simulations appropriate for relatively small globular clusters [Lee, Fahlman & Ricker, 1991]. This work included the effect of binary heating as well as a tidal field. A gradual flattening of the mass function was found due to the preferential evaporation of low-mass stars.

A large survey of globular cluster evolution was undertaken with the Fokker–Planck method [Chernoff & Weinberg, 1990], which included the effect of stellar mass loss and an external tidal field for circular orbits at different central distances. One important result was that mass loss during the first 5 Gyr appeared to be sufficiently large to disrupt weakly bound clusters (i.e. $W_0 \leq 3$) with a Salpeter [1955] IMF. However, further collisionless calculations with GRAPE-3 yielded some agreement but a factor of 10 longer life-times [Fukushige & Heggie, 1995].

Eventually, direct N-body simulations using GRAPE-4 with $N = 32\,\mathrm{K}$ [Portegies Zwart *et al.*, 1998] confirmed the longer life-times. This disagreement for non-isolated systems is mainly due to the implementation of an instant energy escape criterion in the isotropic Fokker–Planck scheme, whereas a more recent anisotropic formulation admits an apocentre criterion that delays escape [Takahashi, 1997]. Finally, the improved treatment of escape in the anisotropic Fokker–Planck method produced good agreement with the direct N-body approach¶ [Takahashi & Portegies Zwart, 1998].

Several systematic studies of small to medium-size isolated clusters have been made in order to elucidate their general behaviour. A comprehensive investigation for unequal masses [Giersz & Heggie, 1996, 1997] showed that average global quantities are well defined even for modest values of N (i.e. 250–1000). As expected, the core collapse is accelerated when a mass spectrum is included. This work also provides a better idea of the dispersion in the core collapse times and mass percentile radii. Somewhat surprisingly, post-collapse solutions appear to be homologous, with no significant evidence of further mass segregation. Moreover, massive stars escape due to binary activity and even dominate the relative depletion of low-mass stars. However, this trend should become less pronounced in

¶ Also see the so-called 'Collaborative Experiment' [Heggie *et al.*, 1998].

clusters where stellar evolution is important. As expected, the addition of a tidal field speeds up the escape rate significantly, yielding a factor of 20 shorter half-life for a tidally truncated cluster.

The effect of the IMF itself has been evaluated in another systematic study using *NBODY5* [de la Fuente Marcos, 1995, 1996a,b, 1997]. Here the main purpose was to compare the outcomes as a function of particle number in the range $50-500$ for five different choices of the IMF. In particular, it is possible to distinguish between a standard power-law IMF and more recent proposals which invariably contain some structure. This series of papers included various realistic effects, such as mass loss from evolving stars, primordial binaries and a tidal field. The question of the ultimate fate of open clusters has also been examined [de la Fuente Marcos, 1998]. Thus it appears that the stellar types of poorly populated clusters provide important clues about their original membership. Finally, the fate of brown dwarfs in open cluster was considered [de la Fuente Marcos & de la Fuente Marcos, 2000], confirming earlier work of a modest preferential escape of low-mass stars.

The choice of an IMF for star cluster simulations should be guided by observations of young systems. It is becoming increasingly clear that the distribution of low-mass stars is significantly depleted with respect to the classical power-law index $\alpha = 2.3$ [Kroupa, private communication, 1999], and this has been implemented (cf. section 8.2). On the other hand, the maximum mass is rather arbitrary. Hence a conservative choice of $15\,M_\odot$ seems more appropriate for open clusters instead of some value representing the stochastic nature of star formation. In any case, the number of stars above, say $5\,M_\odot$, is relatively small and their evolution times short.

As for the slope of the IMF, tidally truncated clusters with low central concentrations and $\alpha \leq 2.5$ are more likely to be disrupted before reaching core collapse [Chernoff & Weinberg, 1990; Aarseth & Heggie, 1998]. Likewise, for increased concentration, the disruptive effect of mass loss by stellar evolution occurs at smaller values (i.e. $\alpha \leq 1.5$). Hence such work provides important constraints on the IMF of old clusters. Strong tidal fields are also important for compact clusters near the Galactic centre. Recent N-body simulations with 12 000 stars and $r_{\rm h} \simeq 0.2\,$pc led to dissolution in about 50 Myr at 100 pc [Portegies Zwart *et al.*, 2002], whereas similar models obtained with *NBODY6* for $N \leq 3000$ and a flatter IMF gave about 10 Myr [Kim *et al.*, 2000].

An ambitious survey of N-body models has recently been completed with *NBODY4* and a powerful GRAPE-6 configuration [Baumgardt & Makino, 2003]. Different families of multi-mass clusters up to $N = 128\,$K were studied with emphasis on the r-dependence of a logarithmic tidal field. A significant depletion of low-mass stars was found, together with scaling laws for the life-times as a function of the eccentricity. Although neutron stars were retained, the adoption of a shallower initial slope for masses below 0.5 $M_{\odot}$ gives a more realistic description. Also note that the gradient of a logarithmic tidal field is less than for the point-mass case.

The IMF is also very important when including primordial binaries. Here there are additional questions to consider, such as the mass ratio distribution and period range, as well as the binary fraction itself. Much observational data has become available since the earliest exploratory simulations with so-called 'initial binaries' [Aarseth, 1975, 1980]. Even so, the measured binary fractions differ significantly, ranging from possibly as high as 95% for young clusters [Duchêne, 1999] to more modest values of about 48% in the inner region of the Pleiades [Raboud & Mermilliod, 1998a] and $\geq$ 38% in $M67$ [Montgomery, Marschall & Janes, 1993]. The presence of spectroscopic binaries in M67 is another pointer to realistic initial conditions [Mathieu, Latham & Griffin, 1990]. Already there is some evidence for hierarchical triples in open clusters such as the Pleiades [Raboud & Mermilliod, 1998a]. A remarkable quadruple system in NGC 2362 contains a massive O-type close binary which itself has an inner triple component [van Leeuwen & van Genderen, 1997]. This is hardly surprising in view of the numerical experience [cf. Aarseth, 2001a; Aarseth & Mardling, 2001]. Initial conditions for primordial triples have been implemented in the present N-body codes, but not yet explored. Now the question of mass ratios becomes quite delicate and the lack of observational evidence necessitates some experimentation.

The general effect of primordial binaries on cluster evolution soon became apparent [Giannone & Molteni, 1985]. This work was among the first to use NBODY5 for star cluster simulations. Several models with 300 equal-mass objects and a binary fraction $f_{\rm b} = 0.2$ were considered, with constant binding energy for each model in the range 5–50 times the mean kinetic energy. Super-elastic binary–binary encounters accounted for about 80% of the total binary exchange, leading to a spread in the energy distribution. In view of the relatively short time studied ($t \simeq 25 t_{\rm cr}$), the models with the hardest binaries experienced much less energy exchange and consequently had a smaller effect on the core expansion. This early investigation also emphasized that encounters with very hard binaries produce high-velocity escapers, whereas the energy released in interactions with medium hard binaries gives rise to general cluster expansion. Although somewhat limited by present-day standards, this first systematic exploration of primordial binary dynamics formed a template for future work.

The possibility that young high-velocity stars may originate in strong interactions involving superhard binaries was studied at an early stage with NBODY5 [Leonard, 1988; Leonard & Duncan, 1988]. The emphasis was on including a population of massive binaries with short periods. Although the particle number was relatively modest (i.e. $N \simeq 50$), this was a serious test for the code which did not yet employ chain regularization, but it did include unperturbed three- and four-body regularization as well as a more primitive treatment of hierarchies. Subsequently, the simulations were extended to larger systems ($N \simeq 500$) without binaries [Leonard & Duncan, 1990], and it was demonstrated that there would have been enough close encounters to produce energetic escapers by binary–binary collisions. The emphasis of these studies was to associate the origin of runaway stars with the fast escapers of both singles and doubles which are characteristic of such systems. Hence this work did much to demonstrate the energetic importance of primordial binaries.

With more accurate distance and radial velocity observations becoming available, it is now possible to determine the common origin of runaway stars. Thus a recent investigation [Hoogerwerf, de Bruijne & de Zeeuw, 2000, 2001] demonstrated several convincing examples of backwards integrations with intersecting orbits in young stellar groups such as the Trapezium. Since some of these examples contain an eccentric binary, this provides evidence for the dynamical ejection scenario. An interesting question concerns the limiting velocity following strong interactions when due allowance is made for the stellar radii [Leonard, 1991].

Considerable efforts have now been devoted to the study of primordial binaries. Some of this work was mainly concerned with including a binary fraction of up to about 20% for equal-mass systems [McMillan, Hut & Makino, 1990, 1991]. The results indicate that the binary fraction decreases with time so that less fossil fuel will be available to halt core collapse. However, in view of theoretical uncertainties, such results cannot be scaled to globular clusters. The situation may be more favourable if a tidal field is included and the binary fraction exceeds about 10% [McMillan & Hut, 1994]; however, this conclusion is still based on equal masses for the single stars which leads to artificial mass segregation. Subsequently, it was shown that only a small binary fraction is needed to affect the cluster evolution significantly when an IMF and stellar evolution are considered [Aarseth, 1996a]. The reason is that the most massive single stars are depleted first, leaving the binaries with an increased central concentration of up to about 50%. Moreover, it appears that these binaries are able to survive in an environment of reduced central density.

The fraction in hard binaries was later increased from 5% to 20% which may be more representative of young clusters [cf. Aarseth, 2001a]. Note that the KS treatment of binaries at intermediate energies, say $a \simeq a_{\rm hard}$, are relatively expensive when using the HARP special-purpose computer. This is due to the interactions between binaries and nearby particles being evaluated on the host, as described in Appendix E. In addition, the perturbed and unperturbed two-body motions are also advanced on the host, which represents a considerable overhead.

A variety of models with maximum binary fractions of up to 100% have also been investigated with NBODY5 [Kroupa, 1995a,b,c; Kroupa, Petr & McCoughrean, 1999]. An interesting new idea of inverse dynamical population synthesis [Kroupa, 1995a] has important implications. It is proposed that the observed field distribution of multiple systems can be accounted for by considering formation in characteristic aggregates which are then dispersed. An argument is given in favour of 100% binaries in small clusters containing 400 stars, with the initial size as a free parameter. The half-mass radius is chosen to be in the range 0.08–2.5 pc, based on observations of young clusters. Depending on the cluster size, about 50–90% of the binaries are hard initially. Moreover, stellar evolution effects are not included because of the short time-scale associated with small values of $r_{\rm h}$. Following complete disruption of the clusters, a surviving binary population of up to 60% is seen, in qualitative agreement with observations of the Galactic field.

A subsequent paper [Kroupa, 1995b] contains the first detailed models of the Galactic field based on including binaries of short periods. An original concept called *eigenevolution* is proposed, whereby the elements of short-period binaries are modified to take account of the observed correlations between period, eccentricity and mass ratio. The derived two-body elements may then be used as initial conditions for dynamical simulations or population synthesis. This suggests that the closest binaries in a realistic initial distribution should already be circularized during the pre-main-sequence evolution. An investigation of different initial velocity states [Kroupa *et al.*, 1999] found that the observational evidence favours expanding models, following earlier gas expulsion. Moreover, these simulations show the effect of kinematical cooling due to disruption of wide binaries which has not been reported before. In another series of experiments, correlations between binary ejection velocities and periods were studied for different half-mass radii [Kroupa, 1998]. Further analytical considerations were also employed to constrain the most likely initial conditions for the Orion Nebula Cluster [Kroupa, 2000].

A recent study attempted to model an evolutionary sequence of cluster evolution, starting with an equilibrium system which contained two-thirds gas [Kroupa, Aarseth & Hurley, 2001]. The gas contents was reduced smoothly on a short time-scale such that the cluster began to expand. However, a significant part remained bound, with a slowly expanding halo. Interestingly, this model of the Pleiades leads to well-defined predictions that can be tested observationally by future space missions. For completeness, we note that similar ideas for explaining associations have been tried before [Lada, Margulis & Dearborn, 1984; Goodwin, 1997].

It is a notable feature of the standard models with primordial binaries

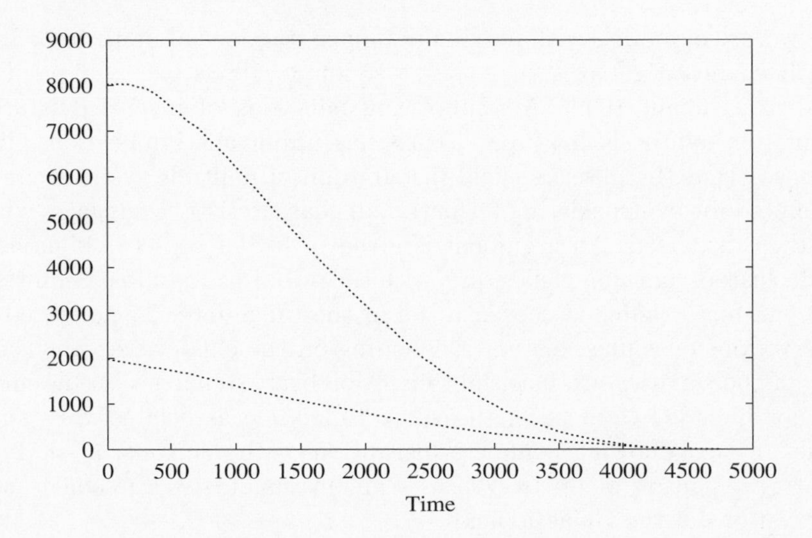

Fig. 15.2. Cluster membership as function of time in Myr. Single stars and binaries are plotted separately.

that a non-negligible fraction of the escapers have high velocities. Thus if we exclude the neutron star kick velocities, typically 0.4% of the terminal velocities exceed $10\,\mathrm{km\,s^{-1}}$, compared with an *initial* rms velocity of $2.0\,\mathrm{km\,s^{-1}}$ that declines by a factor of 2 at the half-life stage. Many of these velocities are due to energetic events connected with chain regularization and can therefore be studied in detail. Such interactions also lead to large binary recoil velocities with subsequent escape. Another characteristic development is the growing number of particles satisfying the energy escape criterion (8.17) without moving outside twice the tidal radius. This resident field star population often exceeds 50% during the later stages and forms an interesting topic for further study [Heggie, 2001; Baumgardt, 2002]. However, the empirical approach of adopting escape beyond $2r_{\rm t}$ seems justified on observational grounds.

To conclude this section, Fig. 15.2 shows the membership of single particles and binaries as a function of time for a typical model containing all the processes that have been discussed. The present model has a half-life of 2×10^9 yr which can account for most old open clusters; however, larger initial particle numbers would give even greater ages. Note that including mass loss due to stellar evolution prolongs the life-times [de la Fuente Marcos & de la Fuente Marcos, 2002]. Hence the general cluster expansion and associated lengthening of the relaxation time more than compensates for the increased escape due to the tidal field. On the other hand, the presence of primordial binaries (treated by stellar evolution) and a tidal field reduces the life-time [Wilkinson *et al.*, 2003].

15.5 Stellar evolution

The implementation of stellar evolution poses many technical complications but is essential for describing real star clusters. It was realized at an early stage that mass loss from evolving stars would be important, and some simple schemes were tried [Wielen, 1968; Aarseth & Wolf, 1972; Aarseth, 1973; Terlevich, 1983, 1987]. All these attempts employed an instantaneous mass loss due to supernova events and it was not until much later that more realistic mass-loss procedures due to stellar winds were included [Aarseth, 1996a; Portegies Zwart *et al.*, 1999]. This new development was based on fast look-up functions for synthetic stellar evolution, originally formulated by Eggleton, Fitchett & Tout [1989] and subsequently refined into a comprehensive network that included binaries [cf. Tout *et al.*, 1997]. The work of Hurley, Pols & Tout [2000] extended and improved the fitting scheme from solar composition to any value of the metallicity Z in the range 0.03–0.0001. For computational convenience, it is assumed that all the stars in a cluster have the same composition, although different initial ages may be assigned if desired. The scheme for rapid binary evolution was recently enlarged to include more complicated systems, as well as the synchronizing effect of tides due to stellar spins [Hurley, Tout & Pols, 2002]. All these algorithms have been implemented in the code NBODY4, whereas NBODY6 only treats mass loss from single stars and collisions at present, but addition of other processes may be anticipated. An alternative scheme for realistic star cluster simulations [Portegies Zwart *et al.*, 2001] is summarized in Appendix F.

Given the initial mass, m_0 in $M_\odot$, and age (defined below), the fitting functions provide information about the stellar radius, r^*, luminosity, l^*, core mass, m_c, and evolution type, k^*, as a function of time. A total of 16 different types are recognized, ranging from low-mass main-sequence stars to black holes and massless remnants. The mass loss itself is specified by a separate algorithm where the main wind loss for evolving stars is given by a Reimers-type expression [Kudritzki & Reimers, 1978],

$$\dot{m} = -2 \times 10^{-13} r^* l^* / m \tag{15.9}$$

in $M_\odot \mathrm{yr}^{-1}$, with r^*, l^* expressed in solar units and averaged over an interval. Several refinements have also been introduced for different categories such as luminous stars and pulsating supergiants [cf. Hurley *et al.*, 2000]. As an indication of the early work, a cluster model containing 12 500 single stars and 2500 binaries with non-solar metallicity has been simulated on HARP-3 by NBODY4 [Elson *et al.*, 1998].

In order to determine the mass loss, Δm, due at the look-up time, $t_{\rm ev}$, we introduce a second time-scale, $t_{\rm ev0}$, which is updated to the current time after each adjustment stage. Another quantity, $\tau_{\rm ev}$, called the epoch

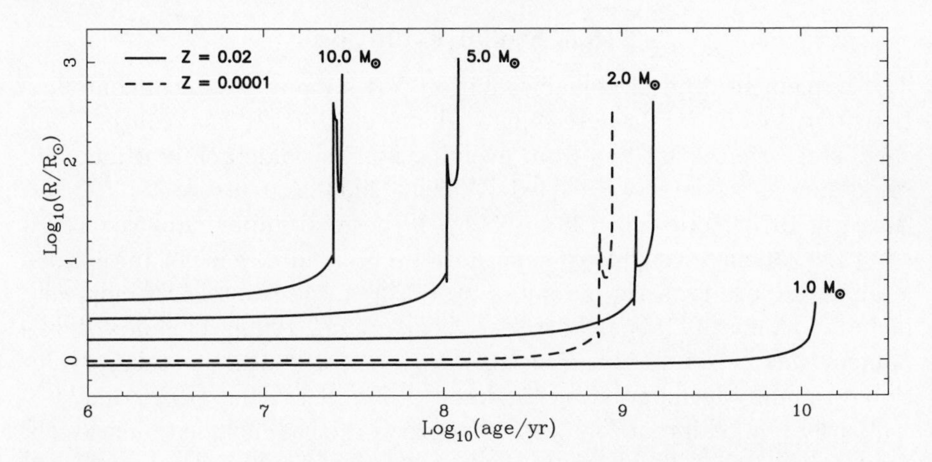

Fig. 15.3. Synthetic stellar evolution. Tracks of radii are shown for different masses and two metallicities as function of time (courtesy J. Hurley, 2003).

is also assigned to each star and used for rejuvenation purposes [cf. Tout *et al.*, 1997]. Thus the apparent age, $t_{\rm ev} - \tau_{\rm ev}$, becomes the time argument for the HR diagram fitting functions, whereas the mass loss is obtained by $\Delta m = \dot{m}(t_{\rm ev} - t_{\rm ev0})$. A new value of the look-up time is prescribed such that the mass loss or change of radius will be suitably small. This is mainly determined by the expression

$$\Delta t_{\rm ev} = \min\{\tau^*/20,\, \Delta t_{\rm rem}\}\,, \tag{15.10}$$

where τ^* is the characteristic evolution time and $\Delta t_{\rm rem}$ is the remaining interval until a type change occurs. However, rapidly evolving supergiants (i.e. $k^* = 5, 6$) are assigned a more conservative value of $\tau^*/50$. Since $\Delta t_{\rm rem}$ may be small, a lower limit of 100 yr is chosen at present. Some evolutionary tracks are shown in Fig. 15.3 for $Z = 0.02$ and 0.0001.

Any mass loss is assumed to be instantaneous because the wind velocity is high compared with the escape velocity. Corrections of the total potential and kinetic energy along the lines of section 12.9 are therefore made in order to ensure a conservative scheme. In the codes *NBODY*5 and *NBODY*6 the former contribution is summed at the same time as the force polynomials are updated, whereas *NBODY*4 only improves the neighbour forces and obtains the potential directly from HARP or GRAPE-6.

Supernova events or black hole formation are also implemented above corresponding initial masses of about 8 $M_\odot$ and 25 $M_\odot$, although the latter limit is highly dependent on the mass-loss rate and remnant mass. We tentatively assign a kick velocity sampled from a Maxwellian distribution with 1D dispersion of 190 km s^{-1}[Hansen & Phinney, 1997]. In the case of single stars, we restrict the value to 10 V^* (cf. (7.4)) which suffices to ensure fast escape; i.e. considerably higher values would affect the energy budget unduly, as well as the ability to deal with superfast particles. Looking to the future, a low-velocity tail seems to be needed in order to account for neutron star retention in globular clusters.‖ A general procedure for obtaining the kick velocity is given in Appendix D.1.

The kick velocity, $v_{\rm kick}$, with randomized directions is added to the pre-supernova value and the corresponding kinetic energy increase subtracted from $E_{\rm cdot}$ (cf. (9.29)). In addition to the potential energy correction and force update due to the mass loss, the force derivative is also modified by the increased velocity in the Hermite method. Again we perform a mass-loss correction, $\frac{1}{2}\Delta m v_0^2$, with v_0 the pre-kick velocity. Finally, after constructing a neighbour list when using HARP, new force polynomials are obtained for the neighbours as well as the remnant itself.

If a binary component is subject to supernova mass loss, the velocity kick algorithm requires special consideration. First a random orbital phase is selected to avoid biasing and the two-body solution is advanced according to (9.19). We estimate the limiting (i.e. parabolic) disruption velocity based on an approximate remnant mass of $1.4\,M_\odot$. The actual kick velocity is then selected by taking $\min\{v_{\rm kick}, v_{\rm f}\}$, with

$$v_{\rm f} = \left(\frac{2(m_{\rm b} - \Delta m)}{R} + 100\,V^{*2}\right)^{1/2}. \tag{15.11}$$

The binary may remain bound for some values of the mass ratio and semi-major axis but the recoil is likely to result in escape from small clusters at least [Aarseth & Mardling, 1997]. Following KS termination, the velocity kick is implemented and energy corrections carried out as for single stars. Finally, the two-body motion is re-initialized as a KS solution if the separation satisfies the standard criterion $R < R_{\rm cl}$ since this is still a close encounter and we may also have $\dot{R} < 0$.

Among other processes that involve stellar astrophysics, we have already discussed Roche lobe overflow in some detail (cf. section 11.10). As far as the energy dissipated during tidal circularization is concerned (cf. section 11.8), the binary components are modelled as polytropes and the relevant parameters are updated following changes in radius or mass. The current implementation of stellar evolution is essentially capable of reproducing all main types of stars and binaries and hence the results can be represented in the form of synthetic HR diagrams, either for solar metallicity [Aarseth, 1996a; Tout *et al.*, 1997; Portegies Zwart *et al.*, 2001] or the more general case required for globular clusters [Hurley, 2000;

‖ A new class of high-mass X-ray binaries has recently been proposed in which mass-transferring systems produce rapidly rotating cores, with smaller asymmetrical recoils of the neutron stars [Pfahl, Rappaport & Podsiadlowski, 2002; Pfahl *et al.*, 2002].

Hurley *et al.*, 2001b]. In view of the modest effort so far, many exotic objects and transient phases have yet to be identified, whereas the diverse observational data have been collected over a long period.

The present implementation of an energy-conserving scheme for different types of mass loss has proved highly successful. It appears that the assumption of instantaneous ejection is quite realistic because essentially no gas is observed in both open and globular clusters after the initial phase. This pleasant state of affairs makes star cluster simulations ideally suited to a gravitational many-body description but, as we have seen, stellar evolution plays a vital role and enriches the results. However, this also increases the complexity of analytical considerations. In this connection we remark that a theory of mass segregation for an evolving IMF is still lacking and would be very beneficial for understanding star cluster evolution. On the observational side, there is some evidence of an evolutionary effect which appears to have a complicated signature [Raboud & Mermilliod, 1998b]. The significant population of white dwarfs in both open and globular clusters also presents interesting constraints on the dynamical models [von Hippel, 1998] (but see Fellhauer *et al.*, 2003).

15.6 Tidal capture and collisions

A number of cluster models that include tidal dissipation have been studied [Aarseth, 1996a,b; Mardling & Aarseth, 2001]. However, technical developments have been going on in parallel and most of these simulations should be considered experimental. The situation regarding tidal capture and collisions is less complicated when it comes to implementation, although there are many astrophysical uncertainties about both processes, especially in relation to the deposition of energy and possible mass loss [Podsiadlowski, 1996].

Capture or collision events involving single stars are relatively rare, at least in open cluster simulations. This is particularly the case for tidal capture, where only one model in about ten yields a permanent bound state [cf. Mardling & Aarseth, 2001]. At present there is some indication that capture is more likely to occur between unbound members in compact subsystems treated by chain regularization. Thus the interaction between a single particle and a binary presents an enhanced cross section with respect to classical two-body encounters. For the future one would like to know the relative proportion involving giants or supergiants since such questions are difficult to estimate theoretically, although there have been attempts [Davies *et al.*, 1998; Bailey & Davies, 1999]. An additional uncertainty facing theoretical estimates is connected with the binary fraction in the core, which may be much higher than expected because stellar evolution promotes mass segregation [cf. Aarseth, 1996a, 1999b].

On the theoretical side, various questions need to be clarified before the capture theory [Mardling, 1995a,b, 1996a,b] can be accepted as a viable process. However, it *is* encouraging that at least some candidate events have been identified in the simulations. If substantiated, further indirect arguments [Mardling, 1996c] attempting to rule out alternative explanations of X-ray binary formation may revive the theory.

When it comes to considering collisions, it is useful to distinguish between classical two-body encounters of single cluster stars and hyperbolic interactions in unstable multiple systems. We also emphasize that some close encounters that satisfy the pericentre criterion (9.21) end up being classified as coalescence following the common-envelope process. Hence for some purposes, like counting collision impacts, these events should also be added. Alternatively, close binary white dwarfs may form, with gravitational radiation time-scales of Gyrs [Shara & Hurley, 2002].

In order to gain some information about the collisions and also near-misses, we examine some relevant data for all hyperbolic encounters inside the tidal capture distance (11.45). Of special interest is the pre-encounter velocity, $v_\infty = (2h)^{1/2} V^*$, or the corresponding eccentricity

$$e = 1 + 2hR_{\rm p}/m_{\rm b}\,, \qquad (15.12)$$

evaluated at pericentre. Based on a small sample, the hyperbolic excess is typically about $1\,{\rm km\,s^{-1}}$ for cluster models with rms velocity $2\,{\rm km\,s^{-1}}$ and somewhat more near the centre. Hence the low end of the velocity distribution provides the dominant contribution to the cross section.[**]

Other quantities, such as masses and stellar types as well as the central distance and density contrast are also of interest. Up to now such data have been scarce and further insight into the collision process must therefore await more systematic investigations. However, the interpretation of some blue stragglers as collision products of main-sequence stars appears to be promising.

At an early stage, Sandage [1953] presented evidence for blue stars in the HR diagram of the globular cluster M3. In the search for explanations, the binary mass-transfer hypothesis was discussed in some detail by McCrea [1964]. A stellar dynamical approach was initiated by Hills & Day [1976], who estimated several hundred collisions in the life-time of a typical globular cluster. As far as simulations are concerned, from binary–binary scattering experiments Hoffer [1983] noted the preponderance of close approaches in strong interactions and surmised that physical collisions and coalescence would be enhanced. Inspired by the dynamical evidence, Leonard [1989] examined the physical collision hypothesis. It was emphasized that the number of events could be significant due to

[**] Recall that the rms relative velocity is augmented by $\sqrt{2}$ in a Maxwellian.

the combined effects of mass segregation and increased cross section from strong binary–binary interactions.

The collision hypothesis was put on a firmer footing by Leonard & Linnell [1992] who concluded that collisions of such stars can only account for about 10% of the observed objects in M67 and NGC 188. Further considerations [Leonard, 1996] supported the view that both the mechanisms of mass transfer and physical collisions are required to explain the observations. Although binaries containing a blue straggler would favour the mass transfer mode of origin, a few such objects may also form from the coalescence of an inner binary of triple systems. However, the presence of up to 10% blue stragglers in main-sequence binaries inside 6 core radii of a low-density globular cluster [Bolte, 1992] suggests that the mass transfer mechanism plays an important role [cf. Ferraro *et al.*, 2003].

Early simulations with $N_s = 5000$ and $f_b = 0.05$ indicated that a small number of blue stragglers are formed in each model [Aarseth, 1996b]. This has been confirmed by an independent study with $N_s = 1024$ and $f_b = 0.5$ [Portegies Zwart *et al.*, 2001]. However, these simulations still contain too few stars to yield a significant number of such objects at any one time and hence the results are not suitable for specific comparisons with observed clusters rich in blue stragglers.

Recently, a more ambitious modelling of $M67$ [Hurley, 2000; Hurley *et al.*, 2001a] demonstrated that the observations may be accounted for by a combination of Roche-type mass transfer and physical collisions. The initial model consisted of 5000 single stars and a binary fraction $f_b = 0.5$, giving $N = 15\,000$. A large number of blue stragglers were produced, with a maximum of 29 co-existing at one time, in good agreement with the observations. On average, about half the blue stragglers formed as the result of dynamical interactions. Hence population synthesis alone is not sufficient to explain the observations [Hurley *et al.*, 2001a]. Moreover, the maximum number of blue stragglers was only reached after some 4–5 Gyr which favours rich clusters. Now that more data are becoming available from ground-based and HST observations [Bolte, Hesser & Stetson, 1993; Grindlay, 1996; Johnson *et al.*, 1999], the theoretical challenge is sharpened. At the same time, there is a renewed effort to improve our understanding of collision remnants [Bailyn, 1995; Lombardi, Rasio & Shapiro, 1996; Sills *et al.*, 1997], following earlier pioneering investigations by several workers [cf. Davies, Benz & Hills, 1991; Benz & Hills, 1992]. Note that, so far, primordial triples have not been studied.

Hence, in conclusion, it can be seen that open cluster simulations provide a fascinating laboratory for studying astrophysical processes without recourse to any scaling. The accumulation of high-precision data from HIPPARCOS and HST will surely do much to inspire further work of increased realism.

15.7 Hierarchical systems

Given the importance of binaries for cluster dynamics, it is hardly surprising that stable hierarchies should form. This feature was anticipated in early code implementations [Aarseth, 1985a] by the merger procedure discussed in section 11.7. It was subsequently put to the test in simulations with superhard primordial binaries [Leonard, 1988; Leonard & Duncan, 1988, 1990] which necessitated differential energy corrections to be introduced. The point here is that the total energy has a small discontinuity when going from a three-body to two-body description, and *vice versa*, and this is sufficiently large to be noticeable in the energy budget of high-precision calculations unless corrections are carried out.

It has been known for some time that stable triples may form as the result of binary–binary collisions in scattering experiments [Mikkola, 1984a,b,c]. However, only a few tentative investigations have examined the frequency and life-times of hierarchies [Kiseleva *et al.*, 1996; de la Fuente Marcos *et al.*, 1997]. A more systematic study is now under way, both as regards the aspects of formation and evolution [cf. Aarseth, 2001a; Aarseth & Mardling, 2001]. First we summarize some of the criteria relating to the stability of triples and quadruples. A more technical discussion of higher-order systems is given in Appendix C.

The question of stability has received much attention and some historical notes are in order. The early numerical studies of Harrington [1972] led to a simple criterion for equal masses. This was soon replaced by a general relation [Harrington, 1975], with the outer pericentre condition

$$R_{\rm p}^{\rm crit} = A\left[1 + B\ln(\tfrac{2}{3} + \tfrac{2}{3}q_{\rm out})\right]\,a_{\rm in}\,, \tag{15.13}$$

where $q_{\rm out} = m_3/(m_1 + m_2)$ denotes the outer mass ratio and $a_{\rm in}$ is the inner semi-major axis. Fitting of the coefficients yielded $A = 3.5$, $B = 0.7$ and $A = 2.75$, $B = 0.64$ for direct and retrograde motion, respectively. A subsequent refinement [Bailyn, 1987] led to the modified expression for direct motion [cf. Valtonen, 1988]

$$A = (2.65 + e_{\rm in})(1 + q_{\rm out})^{1/3}\,, \tag{15.14}$$

with $e_{\rm in}$ the inner eccentricity. This stability criterion was employed in the N-body codes for a number of years.

Eventually, a more systematic investigation was undertaken which resulted in an improved relation [Eggleton & Kiseleva, 1995]. The empirical fitting function for prograde planar motion is given by[††]

$$Y_{\rm crit} = 1 + \frac{3.7}{Q_3} - \frac{2.2}{1 + Q_3} + \left(\frac{1.4}{Q_2}\right)\left(\frac{Q_3 - 1}{Q3 + 1}\right), \tag{15.15}$$

[††] Corrected for a typographical sign error.

with $Q_3 = (m_b/m_3)^{1/3}$ and $Q_2 = (\max\{m_1/m_2, m_2/m_1\})^{1/3}$. The stability check is accepted, provided that

$$R_p^{out} > Y_{crit}\,(1 + e_{in})a_{in}\,. \tag{15.16}$$

Hence Y_{crit} is measured in terms of the inner apocentre distance. Note the important parameter space restrictions to zero outer eccentricity and inclination, whereas retrograde and inclined orbits are known to be more stable [Harrington, 1972, 1975]. The fitting function (15.15) was undoubtedly beneficial, particularly in regard to the mass dependence, and served its purpose well for a time. However, the outer eccentricity which plays an important role does not enter in any of the early derivations.

A more general stability criterion, discussed in section 9.5, has now been in use for several years [cf. Mardling & Aarseth, 1999]. More recently a new criterion was derived from first principles, where the empirical scaling factor C in (9.14) can be determined numerically from fast fitting functions [Mardling, 2001, 2003a]. Thus an efficient algorithm provides the desired result in terms of arbitrary masses and orbital elements (see section 18.5 for a discussion). Compared with (15.15), the semi-analytical relation (9.14) requires larger pericentre ratios for increasing e_{out}. Given the constraint that the *outer* binding energy should essentially exceed $\frac{1}{4}\epsilon_{hard}$ (defined by (4.1)), this limits the maximum value of the outer eccentricity. Even so, relatively large eccentricities (e.g. $e_{out} > 0.98$) are seen during the post-collapse phase. On the other hand, configurations with large e_{out} that do not satisfy the stability criterion often persist for long times because the energy exchange at periastron is very small and of a random nature. An experimental modification that defines practical stability has therefore been introduced (cf. (9.17)).

In most open star cluster models there is enough *hierarchical space* for quadruples as well as higher-order systems to exist, even if the innermost binary is not close to contact. Thus with a half-mass radius of 6×10^5 au for a rich open cluster, nested stable hierarchies could exist having semi-major axes of, say, 60, 6 and 0.6 au, with significant eccentricities and still allow an extra compact inner binary. Various algorithms for dealing with such systems are described more fully in Appendix C.

Even in isolation, some hierarchical systems display complex evolutionary histories. Thus Roche-lobe mass transfer may tend to destabilize the system and eventually lead to a strong interaction [Kiseleva & Eggleton, 1999]. The possibility also exists that the orbital inclination is sufficiently high to induce significant eccentricity growth. This fascinating process is known as Kozai cycles [Kozai, 1962]. According to the assumption underlying (9.14), a_{in} = const for a stable triple. Conservation of angular momentum then leads to a cyclic relation between the inner eccentricity

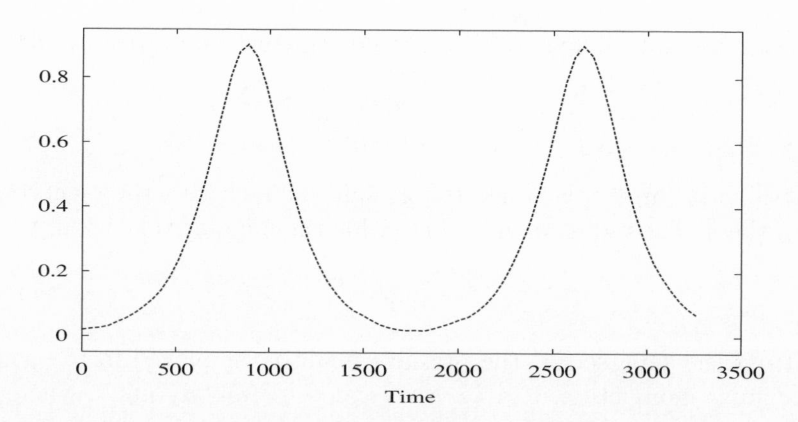

Fig. 15.4. Kozai cycle. Inner eccentricity as function of time.

and orbital inclination, ψ, of the form $\cos^2\psi(1-e^2) = \text{const}$ in the limit of an unchanged outer orbit, or rather as the sum of two constants.

Figure 15.4 illustrates an example of eccentricity modulation for a triple with scaled masses $m_1 = m_2 = 1, m_3 = 0.6$ and period ratio of 10, starting with $e_{\rm out} \simeq 0.1$ and inclination $\psi \simeq 80°$. The maximum value of 0.90 is reached in a cycle time of only about 200 inner periods, with $R_{\rm p}/R_{\rm p}^{\rm crit} \simeq 1.5$ when the empirical inclination effect is included (cf. Algorithm 11.4). For completeness, the time-scale for a Kozai cycle is given by [Heggie, private communication, 1996]

$$T_{\rm Kozai} = \frac{T_{\rm out}^2}{T_{\rm in}} \left(\frac{1+q_{\rm out}}{q_{\rm out}}\right) (1-e_{\rm out}^2)^{3/2} g(e_{\rm in}, \omega_{\rm in}, \psi)\,, \tag{15.17}$$

where the function g is generally of order unity but depends on the argument of periapsis, $\omega_{\rm in}$ [cf. Mardling & Aarseth, 2001].

15.8 Spin–orbit coupling

All stars are endowed with angular momentum which is an imprint of the formation process. Even though the spin vector of single stars is essentially constant during most of the evolution, the rotational velocity may vary considerably. However, for most purposes of cluster simulations, stellar rotation only plays a role in binaries with relatively small pericentres. In the following we discuss some characteristic stages of evolution where significant spin–orbit interactions may be operating. We distinguish between normal and circularizing binaries, as well as hierarchical configurations and Roche lobe overflow, where the latter represents an endpoint.

Initial rotational velocities are assigned for each star by a fit to the observational data [Lang, 1992] for main-sequence stars [Hurley, 2000],

$$v_{\rm rot} = \frac{330m^{3.3}}{15 + m^{3.45}} \,. \tag{15.18}$$

This value in $\rm km\,s^{-1}$ is converted to spin aligned with the orbital axis. Given the stellar radius in $R_\odot$, this yields the angular velocity in $\rm yr^{-1}$,

$$\Omega = 45.35 \frac{v_{\rm rot}}{r^*} \,. \tag{15.19}$$

Omitting the superscript, the primary quantity employed in the code is the angular momentum, $J = kmr^2\Omega$, scaled to N-body units, where k is a structural constant. It is convenient to adopt a representation as the sum of two contributions, due to the envelope and core, by writing

$$J = [k_2(m - m_{\rm c})r^2 + k_3 m_{\rm c} r_{\rm c}^2]\Omega \,, \tag{15.20}$$

where $r_{\rm c}$ here denotes the *stellar* core radius, $k_2 = 0.1$ and $k_3 = 0.21$ [cf. Hurley *et al.*, 2000]. This enables the instantaneous angular velocity to be recovered when required.

The spins (15.20) are updated according to mass loss and/or change of radius, with relevant expressions for the stellar type. We assume that an amount Δm is ejected from a spherical shell at the surface, which gives $\Delta J = \frac{2}{3}\Delta m r^2\Omega$. Magnetic braking for convective stars is also included in the form [Hurley *et al.*, 2002]

$$\dot{J}_{\rm mb} = -5.83 \times 10^{-16} m_{\rm env}(r\Omega)^3/m \,, \tag{15.21}$$

with $m_{\rm env}$ the envelope mass. For a normal binary, the spins are updated independently of mass and radius, according to the above scheme. Finally, the case of the companion accreting some of the mass lost from the primary is also treated [cf. Hurley *et al.*, 2002].

Tidal effects come into play when stellar oscillations are induced during close periastron passages. This condition may be reached after a sudden transition to high eccentricity or new binary formation, and also by secular perturbations (i.e. Kozai cycles) inside a stable hierarchy. The subsequent process of tidal circularization then acts to reduce the eccentricity, with consequent period shortening due to angular momentum conservation. The torque acting on the tidal bulge gives rise to angular momentum transfer, such that the stars tend to become synchronous with the orbital velocity at periastron. Hence the orbit loses angular momentum to the stars as it shrinks and also in general due to increasing radii.

The theory of equilibrium tides provides a framework for consistent description of the evolution of binary parameters. We adopt the simplified model of Hut [1981] in which small deviations are considered and both spin axes are assumed to be normal to the orbital plane. Let us first consider the case of tidal circularization for a hard binary treated by two-body regularization, discussed in section 11.8. Note that for a typical cluster model with $N = 10^4$, most KS binary components would have a very long circularization time since $t_{\rm circ} \propto (a/r)^8$. The basic theory of the spin treatment has been presented elsewhere [Hurley *et al.*, 2002]; hence we only summarize some salient points of the numerical procedures.

The equations of motion for the eccentricity and both stellar spins are integrated by the fourth-order Runge–Kutta method (RK4) using appropriate time intervals determined by the largest derivative as well as the departure from synchronous rotation. Thus a given interval is subdivided to ensure convergence, and care is taken to avoid overshooting in the nearly singular region $a(1-e) \geq 4r_k^*$. The new semi-major axis is obtained from total angular momentum conservation, whereas the angular velocities are converted to spins using relevant values for the dimensionless radius of gyration, $k_{\rm g}$.

An update of the regularized orbital elements is also required. Since circularization is implemented at pericentre, the coefficient for the KS coordinates $\mathbf{U}$ is given by (11.39). Likewise, use of the energy relation, (4.24), yields the corresponding coefficient for $\mathbf{U}'$ by (11.44) after obtaining the new value of h. In the case of unperturbed two-body motion, these corrections are performed after each perturber check. Note that an unperturbed binary may reside inside a wide hierarchy without any Kozai cycles being activated. However, (15.17) implies a cycle time $> 1000T_{\rm out}$ at the standard perturber boundary $a_{\rm out} \simeq 100a_{\rm in}$ for the lower limits $e_{\rm out} = 0$, $e_{\rm in} = 0$, whereas typical eccentricities would make the cycle time considerably longer, assuming unperturbed motion at the outer pericentre. Consequently, external perturbations would modify such an orbit and thereby randomize any secular effects on the inner binary.

Circularization is deemed to have taken place when the eccentricity reaches a small value, currently $e_{\rm min} = 0.002$, which still allows the pericentre algorithm based on t'' (or R') changing sign to be employed. Following circularization, the spin synchronization treatment is continued because the stellar radii or semi-major axis may still change. The equilibrium angular velocity is defined in terms of the instantaneous valve by

$$\Omega_{\rm eq} = f_2(e^2)\Omega_{\rm orb}\left[\frac{1}{f_5(e^2)(1-e^2)^{3/2}}\right], \tag{15.22}$$

where f_2 and f_5 are simple algebraic functions of eccentricity given by Hut [1981]. Since the synchronization time-scale $\propto (a/r)^6$, the secondary component may not reach this state before the orbit has been circularized. However, the associated spin angular momentum also tends to be correspondingly smaller. Although there is observational evidence for synchronized spins in several circularized binaries, an established triple with circular inner orbit contains a rapidly rotating component which poses a puzzle [Saar, Nordström & Andersen, 1990].

Next we consider the effect of spin–orbit coupling in hierarchical systems. In *NBODY4*, the inner eccentricity change is modelled by an averaging method in systems that may become tidally active [cf. Mardling & Aarseth, 2001], and the effect of spin has been included. Given the inner and outer orbital elements, the maximum eccentricity, $e_{\rm max}$, can be calculated analytically in the absence of external perturbations [Heggie, private communication, 1996]. Since favourable conditions may exist at the end of chain regularization, the final values of ψ and $e_{\rm max}$ are recorded for further analysis, and likewise at the start of any hierarchical initialization.

The secular evolution of the inner binary of a hierarchical system is studied by an averaging method for the rate of change of the Runge–Lenz and specific angular momentum vectors. However, only systems with $e_{\rm max} > 0.9$ and $t_{\rm circ}(e_{\rm max}) < 2 \times 10^9$ yr are considered. The inner Runge–Lenz vector is given by

$$\mathbf{e}_{\rm in} = \dot{\mathbf{R}} \times \mathbf{j}/m_{\rm b} - \mathbf{R}/R\,, \tag{15.23}$$

where $\mathbf{j} = \mathbf{R} \times \dot{\mathbf{R}}$. The rates of change of these vectors depend on the accelerations produced by the tidal and spin bulges, tidal dissipation, possible relativistic precession, as well as the third body. The corresponding equations of motion, averaged over one inner orbit, take the form

$$\begin{aligned} \dot{\mathbf{e}}_{\rm av} &= b_1\widehat{\mathbf{e}} + b_2\widehat{\mathbf{q}} + b_3\widehat{\mathbf{j}}\,, \\ (d\mathbf{j}/dt)_{\rm av} &= c_1\widehat{\mathbf{e}} + c_2\widehat{\mathbf{q}} + c_3\widehat{\mathbf{j}}\,, \end{aligned} \tag{15.24}$$

with $\widehat{\mathbf{q}} = \widehat{\mathbf{e}} \times \widehat{\mathbf{j}}$. The contributions to b_i and c_i from the third body perturbation in the quadrupole approximation[‡‡] were derived by Heggie [1996], while the tidal and spin contributions are based on Eggleton, Kiseleva & Hut [1998] for rotation axes normal to the orbital plane. Convenient expressions for the complete formulation are given by Mardling & Lin [2002]. In units of au, $M_\odot$ and yr, the period of the relativistic precession is [Holman, Touma & Tremaine, 1997]

$$T_{GR} = 3.4 \times 10^7 (1 - e^2) T_{\rm in} a_{\rm in}/m_{\rm b}\,. \tag{15.25}$$

Finally, the oblateness effect is modelled using appropriate apsidal motion constants [Schwarzschild, 1958].

The different contributions depend sensitively on the inverse powers of the semi-major axis and $(1 - e^2)$, with $b_1 \propto 1/a^8(1 - e^2)^{13/2}$ and

‡‡ A more general octupole order treatment is available [Ford, Kozinsky & Rasio, 2000]. However, note that the quadrupole terms dominate the evolution for high inclinations.

$b_2 \propto 1/a^{13/2}(1-e^2)^4$ from the tidal dissipation and apsidal motion terms, whereas the relativistic precession part of the latter $\propto 1/a^{5/2}(1-e^2)$. Consequently, the eccentricity growth due to the Kozai cycle may be suppressed by the apsidal motion or relativistic precession. Although this may seem extreme, such values have been noted on many occasions. The reason for this behaviour is that the starting values of eccentricity and inclination may be favourable, which makes for large values of $e_{\rm max}$.

The relevant equations of motion are integrated by the RK4 method, taking special care with the time-step choice for large eccentricities [cf. Aarseth & Mardling, 2001]. This is done either at every outer apocentre passage for perturbed KS solutions or after each check of unperturbed motion. The basic integration step is taken as the harmonic mean

$$\Delta t = \tfrac{1}{2}(T_{\rm in} \min\{T_{\rm Kozai}, T_{\rm q}\})^{1/2}, \tag{15.26}$$

where $T_{\rm q}$ is the growth time-scale due to the quadrupole term, with additional safeguards for large eccentricities. All the elements are redetermined at the end of each integration interval, thereby facilitating standard KS initialization at termination. Some relevant KS procedures are summarized in Appendix D.3.

The question of appropriate termination criteria requires careful consideration. One possibility is to initiate tidal circularization once the corresponding time-scale, $t_{\rm circ}$, falls below a specified value. The rationale for such a decision is that this would lead to eventual circularization since the Kozai cycle would not be operative in the case of unperturbed two-body motion following some shrinkage (cf. (15.17)). However, it is beneficial to include all the effects until the largest value of the eccentricity is reached, which occurs for $\mathbf{e} \cdot \dot{\mathbf{e}} < 0$ subject to $\dot{e}_{\rm av} > 0$. In fact, this condition also ensures that the maximum shrinkage is achieved. Note that the oblateness and relativistic precession are not taken into account during the standard circularization treatment. Hence the outcome depends on the tidal eccentricity damping versus induced growth due to the outer body, combined with the effect of spin.

Hierarchical evolution often leads to large values of the inner eccentricity even though the coalescence condition is usually avoided. Since angular momentum is conserved during circularization, the final semi-major axis may become quite small. This mechanism for orbital shrinkage might therefore play an important role in close binary formation and, in fact, there exists a correlation between the observed period and eccentricity [Mermilliod & Mayor, 1992], albeit with some exceptions [Verbunt & Phinney, 1995]. As for possible applications to observed systems, the quadruple star μ Orionis contains two circular binaries of comparable short period in an eccentric wide outer orbit [Fekel *et al.*, 2002]. The effect of any high inclinations would therefore be subject to strong damping.

Although the averaging procedure is capable of dealing with the complete history of circularization, it is more convenient to initiate the standard treatment upon attaining the maximum eccentricity or a sufficiently short time-scale. Thus hierarchical binaries with relatively short periods and significant perturbation would also approach the circular state more rapidly. If we exclude rare cases when the external effect is able to overcome the damping, there follows an interval of circular motion before the Roche-lobe overflow condition is satisfied. Depending on the remaining interval, the binary components may evolve significantly and hence acquire further angular momentum from the orbit. Likewise, this process also operates during subsequent quiescent phases.

The case of angular momentum exchange during inactive periods is studied in a similar way. Now we simplify the treatment by assuming zero eccentricity, leaving the two spin equations to be integrated. In general, the qualitative effect of including stellar rotation is to shorten the time-scale until the next epoch of mass transfer. Hence all the stages leading to possible mass transfer include spin effects in a uniform and consistent way. Finally, it is encouraging that at least the primaries tend to be well synchronized during circularization.

15.9 Globular clusters

The forthcoming attack on the globular cluster problem represents a grand challenge for N-body simulators. This exciting possibility is now becoming feasible due to the construction of the new special-purpose computer called GRAPE-6 [Makino & Fukushige, 2002; Makino, Fukushige & Namura, 2003]. As an indication of the rapid progress in hardware developments, a measured speed of 29.5 Tflops has recently been achieved [Makino, private communication, 2002]. At the same time, further efforts in software development will be required in order to exploit the increased performance to the full.

An important external effect acting on most globular clusters is due to tidal shocks when the orbit crosses the Galactic plane with high velocity. Theoretical heating rates [Spitzer, 1987; Gnedin & Ostriker, 1997] indicate that the resulting internal heating is significant and should be included in the simulations. Thus the Hermite scheme requires an external potential that can be differentiated explicitly twice in order to yield an analytical expression for the first derivative of the force. It should be emphasized that simplicity is desirable in view of the need to include these perturbations on the host machine. The code *NBODY4* contains prescriptions for implementing a shock by the impulsive approximation. At a given moment, the individual vertical velocities are increased by adding a Maxwellian component with dispersion σ_z. Since this is a discontinuous process, new force polynomials are initialized using HARP or GRAPE-6, but in view of the long Galactic oscillation period this is not too expensive. Finally, the resulting kinetic energy gain is added to the total energy in order to maintain conservation.

So far, only a simplified model has been used to examine the effect of disc shocking [Vesperini & Heggie, 1997] and, in any case, no direct N-body simulation of a realistic globular cluster has yet been performed. Such simulations require appropriate synthetic stellar evolution (i.e. low metallicity) and a suitable potential for eccentric orbit integration, as well as a fairly large particle number (say $N \simeq 10^5$). Now that the GRAPE-6 (32 chip version) is available, the latter condition can essentially be met without primordial binaries. Moreover, a general stellar evolution scheme [cf. Hurley *et al.*, 2000] has already been implemented. The new models discussed in section 8.6 should therefore be very useful for beginning the study of globular cluster evolution in a 3D environment.

Globular clusters provide ideal testing grounds for several exotic processes. Particularly the question of neutron star retention needs to be re-addressed in the light of current observations of low-mass X-ray binaries which imply that a significant population is present [Pfahl, Rappaport & Podsiadlowski, 2002]. Based on the current modelling of velocity kicks, it is difficult to envisage sufficient retention even assuming strong mass correlation in the primordial binary distribution. If corroborated further, this may strengthen the search for alternative modes of neutron star formation. Observational capabilities of detecting the white dwarf sequence [Cool, Piotto & King, 1996] will also place further constraints on the numerical models.

Because of the high central densities, a number of collisions may be expected to occur in globular clusters since the stellar population contains a significant fraction of red giants. On the other hand, there are also many degenerate objects and low-mass stars with small radii which complicates the estimate. An early pioneering investigation [Hills & Day, 1976] suggested that physical collisions are likely, with ~ 300 events predicted for a typical globular cluster. On the basis of current simulations with primordial binaries, this estimate is probably far too low. Collisions may also deplete red giants at the Galactic centre [Davies, 2001]. Moreover, collisions are needed to account for the substantial number of blue stragglers that are now being reported [Grindlay, 1996; Guhathakurta *et al.*, 1998; Shara *et al.*, 1998]. These objects exhibit pronounced central concentration and an individual mass determination [Shara, Saffer & Livio, 1997] is consistent with the collisional scenario. Scaling up from a simulation of $M67$ [Hurley *et al.*, 2001a], it appears that a significant population of such objects would form preferentially in richer clusters, with both physical collisions and Roche-lobe mass transfer contributing.

Close encounters not involved in collisions may produce tidal capture binaries. Indications from an ambitious early Monte Carlo simulation are that this process may be important in globular clusters [Stodólkiewicz, 1985]. Hence calculations with increased particle numbers should provide an opportunity for testing the capture theory [Mardling, 1996a,b,c]. Thus even a small fraction of surviving capture binaries for every collision event will have important repercussions. It can also be anticipated that several types of exotic binaries that are not identified in population synthesis studies originate as the result of dynamical interactions, either by orbital shrinkage or exchange. In particular, the astrophysical implications of the latter process deserves more careful study, and already early scattering experiments demonstrated its efficiency [Hills, 1975].

The binary fraction in globular clusters is also of considerable interest. In view of the small formation rate of hard binaries expected from theoretical estimates [Spitzer, 1987] and numerical scattering experiments [Aarseth & Heggie, 1976], nearly all the observed binaries may be primordial or formed by exchange. In spite of detection difficulties, the evidence has been mounting for a significant binary fraction [Hut *et al.*, 1992; Meylan & Heggie, 1997]. To quote a specific case of a young globular cluster in the LMC [Elson *et al.*, 1998], the binary fraction was determined to be about 35% near the centre, decreasing to 20% further out. In addition, dynamically important non-luminous binaries should be present in considerable numbers. Hence binaries may contain enough fossil fuel to prevent late stages of core collapse from being reached. However, the calculation of the energy transfer to the core will prove a severe test for the direct integration methods. In addition, a study of the escape process itself is of fundamental interest and will provide opportunities for theoretical comparisons. We also mention an observational project to search for high-velocity escapers from globulars [Meusinger, Scholz & Irwin, 2001].

Notwithstanding accumulating observational evidence, the question of possible scenarios for the formation of supermassive objects is also of interest in relation to globular clusters. In this connection we mention a first attempt to account for the growth of compact objects based on direct N-body simulations [Lee, 1993]. This model assumes a dense star cluster where binaries form by tidal capture, with the energy loss due to gravitational radiation. The code *NBODY*5 was employed to study systems with $N = 1000$ equal-mass stars and high velocity dispersion, $\sigma \simeq 2000\,\mathrm{km\,s^{-1}}$. Following binary formation by GR capture, the KS formulation proved ideal to include the dominant gravitational radiation part of (12.30) which can be treated analytically if there are no perturbers. By analogy with coagulation of planetesimals (to be discussed later), the mass spectrum evolved by runaway growth featuring only one dominant object. Thus, after a stage of normal evolution, this process is facilitated by mass segregation on a typical core collapse time-scale.

A scenario for the possible formation of intermediate-mass black holes (IMBHs) in young compact clusters has been described, inspired by new observations [Ebisuzaki *et al.*, 2001]. The main idea is again that the products of runaway collisions between massive stars sink to the cluster centre by dynamical friction before completing their evolution, whereupon they combine to form an IMBH. The rapid growth of one massive star through collisions has been reported in N-body simulations of compact clusters [Portegies Zwart *et al.*, 1999], although there is currently some uncertainty about the retention of newly formed BHs. If some of the clusters survive while spiralling towards the galactic nucleus, the IMBH will reach the centre followed by cluster disruption. The liberated black holes may then form binaries with other IMBHs and undergo coalescence by gravitational radiation. However, a quantitative model for the emergence of the central supermassive black hole (SMBH) is still not available. Hence it remains to be seen whether this sequential process can replace more conventional evolution paths involving gas accretion for the growth of an SMBH.

More quantitative models for runaway growth of IMBHs have now been presented [Portegies Zwart & McMillan, 2002]. Direct N-body simulations with up to 64 K particles were performed on GRAPE-6. Dense clusters with initial half-mass relaxation times below 25 Myr are dominated by stellar collisions involving the same star. Such collisions result in the runaway growth of a supermassive object which may produce an IMBH.

Finally, on the astrophysical side, we mention a recent GRAPE-6 simulation with implications for more massive BHs in the centres of globular clusters [Baumgardt *et al.*, 2003]. Thus a conventional model with $N = 128\,\mathrm{K}$ stars obtained by using *NBODY4* and appropriate stellar evolution was able to reproduce the observed velocity profile in M15 without requiring the presence of a BH mass in excess of about $10^3\,M_\odot$.

Several technical aspects need to be re-assessed when moving to large-N simulations. As far as multiple regularization methods are concerned, it would be advantageous to be able to treat several perturbed interactions at the same time. This can be readily achieved by extending the three-body and global four-body regularizations to include external perturbers. The relevant theoretical formulations have already been given in section 5.2 and Appendix A. It then becomes a question of developing the corresponding interface as described in chapter 12, using similar principles. Note, however, that here the membership is fixed so that collision or escape is simplified by immediate termination. Hence one possible, but extremely rare, problem for the triple treatment would be the approach of a strong perturber with the two other methods being used. In any case, a procedure exists for terminating a less critical configuration of the same type by comparing the respective perimeters. Experience so far suggests that the chain regularization is only employed some 500 times during a typical simulation with $N = 10^4$. Since the interaction times are usually short, this probably represents less than 0.1% of the total time.

Looking ahead to general cluster simulations with $N \simeq 10^5$ as the next objective, it is by no means certain that the present time-step criterion (cf. (2.13)) will prove to be sufficiently sensitive. Thus one might envisage conditions with a large mass dispersion which would make direct integration of close encounters for $R > R_{\rm cl}$ less accurate, unless the time-step parameter η is reduced $\propto N^{-1/6}$ [cf. Makino & Hut, 1988]. Otherwise it may become necessary to introduce subsidiary criteria for the time-step to ensure convergence, such as an evaluation of the predicted force change [cf. Wielen, 1974], use of the fourth force difference or the difference between the predicted and corrected velocity [Makino, 1991a]. The latter feature has already been adopted by taking advantage of the high-order prediction on the new hardware.[§§] A study of larger particle numbers may also necessitate a more careful appraisal of other sources of errors.

Finally, we consider an aspect connected with the increased time-span for larger systems which have longer life-times. Since the time itself is only represented to about 16 figure accuracy and the time-steps tend to decrease $\propto N^{-1/3}$ [Spurzem, Baumgardt & Ibold, 2003], it is desirable to maintain precision (cf. 12.2 and 12.16) and ensure the commensurability condition $\mathrm{mod}\,\{t/\Delta t_i\} = 0$ by a rectification of the scaled time. A second offset time, $t_{\rm off}$, is then used for converting to physical units which are mainly needed for stellar evolution purposes. This procedure entails subtracting the quantities $t_0, T_0, t_{\rm ev}, t_{\rm ev0}, \tau_{\rm ev}$ by the offset interval, $\Delta t_{\rm off}$, whenever the scaled time, t, exceeds this value. In addition, a few global quantities used for decision-making need to be updated. All standard integration procedures are unaffected by the rescaling, so that the value of the scaled integration time remains in the interval $[0, \Delta t_{\rm off}]$. Since the smallest time-steps[¶¶] usually occur in connection with the quantization of time at chain termination, it may also be possible to delay most new initializations until the end of the block-step. This would still allow rare cases of extending the solution beyond the pericentre of eccentric binaries at the time of escape (cf. section 12.2). Hence it can be expected that the present integration scheme will prove itself in large-scale simulations with minor modifications. In conclusion, the shape of things to come is now under discussion [Hut *et al.*, 2003; Heggie & Hut, 2003].

[§§] Recent experience with the GRAPE-6 for $N = 60\,000$ revealed some rare cases of failure by the standard time-step criterion (2.13) which were eliminated after comparing with (2.16) for large values ($\Delta t_i > 0.01(6 \times 10^4/N)^{1/3}$). Notwithstanding the low-order prediction, the subsidiary criterion (2.16) proved adequate with $\epsilon_{\rm v} = 0.001$.

[¶¶] Currently 1.8×10^{-12} at hierarchical level 40, compared with total times $\simeq 5 \times 10^3$.

16

Galaxies

16.1 Introduction

Recent years have seen impressive advances in the simulation of collisionless systems. However, a wide range of problems in galactic dynamics are still amenable to direct N-body integration and will therefore be discussed for completeness. This chapter also serves as a historical review of several topics that blossomed from a primitive beginning. All such problems are characterized by the employment of a small softening of the potential which reduces the effect of close encounters. We begin by describing an application of the grid perturbation method to a ring of interacting molecular clouds in the inner part of the Galaxy [Aarseth, 1988b]. This enables a realistic number of clouds to be considered. The tidal disruption of dwarf spheroidal galaxies orbiting the Milky Way has also been studied by direct means [Oh, Lin & Aarseth, 1995].

More extreme interactions of galaxies often result in the formation of one system, subsequently denoted as a remnant to distinguish this process from mergers used for hierarchical stellar configurations. Studies of galaxy interactions involving black holes have also become topical [Makino & Ebisuzaki, 1996; Makino, 1997]. This problem is particularly challenging because of the difficulty of scaling to realistic conditions. Small galaxy groups and clusters are ideal for N-body simulations if close encounters can be treated using a softened potential. Early simulations [Aarseth, 1963b] were primitive but some basic aspects of dynamical evolution were elucidated. On larger scales, cosmological problems have created much interest over the years, with increased realism both as regards initial conditions and particle number. The early investigations [Gott, Turner & Aarseth, 1979; Dekel & Aarseth, 1984; Itoh, Inagaki & Saslaw, 1990] demonstrated the basic behaviour of expanding systems and will be described as background material for modern cosmological simulations.

16.2 Molecular clouds

The development of the grid perturbation method outlined in section 3.5 was aimed at simulating the ring of giant molecular clouds in the inner part of the Galaxy [Aarseth, 1988b]. Here the main idea is to include the perturbing effect due to the nearest clouds, with the dominant motion arising from the smooth density distribution, thereby providing an analogy with a thin planetary system. The galaxy potential of Miyamoto & Nagai [1975] is convenient for studying motions in the 4–6 kpc ring. In order to consider the effect of spiral arms, we include a perturbing acceleration of the form [Tomisaka, 1986]

$$g_r(r,\theta) = -\frac{V_{\rm rot}^2(r)}{r}(1 + A\sin\phi) + r\Omega_{\rm p}^2\,,$$
$$g_\theta(r,\theta) = -\frac{V_{\rm rot}^2(r)}{r}A\sin\phi\tan i\,, \tag{16.1}$$

in polar coordinates r, θ. Here $V_{\rm rot}$ is the mean circular rotation velocity and A is the amplitude of the spiral perturbation. The corresponding phase is given by

$$\phi = \phi_0 - \ln(x^2 + y^2)/\tan\, i - 2\theta - \Omega_{\rm p}t\,, \tag{16.2}$$

where i is the pitch angle, $\theta = \tan^{-1}(y/x)$ the orbital phase and $\Omega_{\rm p}$ the pattern speed, set to corotation at 15 kpc. Hence this logarithmic trailing spiral gives rise to a radial and tangential force component, scaled by the galactic radial force.

Again the active perturber bins are determined by a mass-dependent expression which is taken as

$$n_i = 1 + (m_i/m_{\rm B})^{1/2}N_{\rm b}\,, \tag{16.3}$$

where $m_{\rm B}$ is the mass of the biggest initial cloud and $N_{\rm b}$ is a parameter; i.e. typically $N_{\rm b} = 4$ for a total of $N_{\rm a} = 100$ perturber bins. Hence $n_i = 1$ for the smallest masses and $n_i \simeq 5$ for clouds at the maximum initial mass, which represents a dimensionless perturbation of $\simeq 10^{-4}$. The adopted individual time-step algorithm is based on the orbital period modified by the perturbation, and the equations of motion are advanced by standard fourth-order differences. Cloud–cloud collisions are modelled by inelastic accretion, with giant clouds exceeding $2 \times 10^6\,M_\odot$ subject to disruption into a number of fragments after 5×10^7 yr. As a refinement, the force due to a small population of giant clouds ($m > m_{\rm B}$) is included by direct summation.

Knowledge of the perturbation facilitates collision determinations. For a separation r_{ij} outside the softening length, the relative perturbation due to a cloud of mass m_j with respect to the centrifugal force, $F_{\rm g} = V_{\rm r}^2/r_{\rm g}$, can be estimated by

$$\gamma_i = m_j r_{\rm g}/r_{ij}^2 V_{\rm r}^2\,. \tag{16.4}$$

A close encounter search is made only if γ_i exceeds an appropriately scaled value obtained at the minimum overlapping separation, $4\,\epsilon_0$. Adopting constant cloud density, the softening size (or half-mass radius) is taken as $\epsilon = \epsilon_0\,(m/m_0)^{1/3}$, with the conservative choice $\epsilon_0 = 5\,{\rm pc}$ for $m_0 = 10^4\,M_\odot$. Hence the interaction potential can be written in the form $\Phi_i = -m_j/(r_{ij}^2 + \epsilon_i^2 + \epsilon_j^2)^{1/2}$ [cf. White, 1976]. The collision search for particles with a suitable small time-step is in the first instance restricted to the current perturber bin, with coordinates predicted to order $\mathbf{F}_j$. As usual in such simulations, the overlapping condition $2\,(\epsilon_i + \epsilon_j)$ defines an inelastic collision, whereupon the new c.m. body is introduced in the standard way.

The astrophysical modelling outlined here has many similarities with two other contemporary investigations [Tomisaka, 1986; Kwan & Valdes, 1987] which produced qualitatively similar results. As for the numerical methods, the former study included all the clouds in the force summation and was therefore quite expensive, whereas the latter only considered interactions inside 400 pc for 10^4 clouds of initial mass $10^5\,M_\odot$. In comparison, the mass-dependent cutoff in the grid perturbation scheme falls in the range 250–1250 pc at 4 kpc for a smaller minimum mass, whereas the overheads of the method are relatively modest. Although the present scheme was an improvement of the computational efficiency, with more careful treatment of the massive clouds than in the cutoff procedure, it has not been fully exploited yet.

Four models with $N = 5000$ clouds distributed at constant surface density inside a 4–6 kpc ring of thickness ± 0.1 kpc were studied. The initial masses were selected from an IMF in the range 10^4–$5 \times 10^5\,M_\odot$ and assigned circular orbits. Consequently, the eccentricity growth due to perturbations gives rise to an increasing velocity dispersion defined by

$$\sigma_{\rm p} = \left[\frac{1}{N}\sum \dot{r}^2 + (v_\theta - v_{\rm c})^2\right]^{1/2}, \tag{16.5}$$

where $v_\theta - v_{\rm c}$ represents the departure from circular motion. In the absence of coalescence and the spiral density wave, $\sigma_{\rm p} \simeq 7\,{\rm km\,s^{-1}}$ already after 1.25 rotation periods. Including coalescence led to $\sigma_{\rm p} \simeq 10\,{\rm km\,s^{-1}}$ with some 6600 collisions at 1.5 rotation periods. In comparison, there were 8000 collisions and $\sigma_{\rm p} \simeq 7\,{\rm km\,s^{-1}}$ when the spiral arm effect was introduced. Finally, these provisional simulations showed some evidence of mass segregation in the form of smaller velocity dispersion and half-thickness of the most massive clouds which formed by coalescence.

16.3 Tidal disruption of dwarf galaxies

A number of dwarf spheroidal galaxies are in orbit around the Galaxy at distances that are sufficiently close to induce strong perturbations. The process of tidal disruption has been investigated by direct N-body simulations, using the equations of motion in section 8.5. A softened potential of the form $\Phi = -m/(r^4+\epsilon^4)^{1/4}$ was used with *NBODY*2 in order to minimize the departure from the Newtonian form [Oh, Lin & Aarseth, 1995]. Since the r-dependence is quite steep and $\epsilon \ll R_s$, it is justified to ignore the softening during the regular force summation.* Possible problems with softening and alternative expressions have been discussed by several authors [Dyer & Ip, 1993; Theis, 1998; Dehnen, 2001]. The question of an optimal choice for ϵ has also been addressed [Merritt, 1996].

In order to estimate the two-body relaxation, two isolated isotropic King [1966] models with $W_0 = 2$ and $N = 1000$ were studied over 20 Gyr. The second model consisted of two populations containing 500 members differing in mass by a factor of 2. Detailed comparison showed a slight decrease of the half-mass radius of the heavy population, whereas r_h expanded by less than 10% for the light members. Moreover, only about 10% of the outermost mass distribution expanded beyond the initial limiting radius. These results indicate that direct N-body simulations with equal masses are appropriate for studying the tidal disruption of dwarf spheroidal galaxies when relaxation plays a minor role.

One set of models employed a logarithmic potential for the Galaxy and a range of cutoff radii, r_k/r_t, with respect to the tidal radius. A direct comparison of two models in circular orbits showed that more than half the mass escaped well beyond the tidal radius after the first period for $r_k/r_t = 1.9$, whereas a dwarf galaxy with $r_k/r_t = 1$ survived intact on a Hubble time.

In the case of eccentric orbits, the tidal radius at perigalacticon was used [King, 1962]. For the point-mass potential and mass ratio α we have

$$r_t = \left[\frac{\alpha}{3+e}\right]^{1/3} (1-e)\,a\,, \tag{16.6}$$

with a somewhat more complicated expression for the logarithmic potential [cf. Oh *et al.*, 1995]. Hence this definition tends to increase the disruption rate. Even so, a model with $r_k/r_t = 1.9$, $a = 100\,\mathrm{kpc}$, $e = 0.5$ and mass $M_d = 2 \times 10^6\,M_\odot$ was only marginally disrupted on a Hubble time. Since $r_t \propto M_d^{1/3}$ and the density distribution is shallow, the central density remained nearly constant throughout the evolution.

Models with a point-mass potential of the Galaxy were also considered. In the case of comparable periods, the tidal force is about twice as strong

* This simplification is not recommended for the sensitive Hermite method.

as for a logarithmic potential. Despite a stronger tidal force during perigalacticon passages, the results obtained are remarkably similar, although the disruption times are somewhat less. Moreover, similar dwarf galaxies on circular orbits tend to be less stable. One possible reason for this behaviour is that these dwarf galaxies only spend a fraction of a period in the innermost part of their orbit, where the velocity is higher.

These simulations were made a long time ago, when computing power was quite modest. Nevertheless, some useful results were still obtained. Finally, we remark that much more ambitious simulations of dwarf galaxies with $N \simeq 10^6$ have been performed by the *SUPERBOX* particle-mesh code which uses moving nested grids [Fellhauer *et al.*, 2000].

16.4 Interacting galaxies

In the preceding section, some justification for modelling collisionless systems by direct integration was given. Such arguments were particularly relevant in the early days when the particle number was limited to a few hundred. In this spirit, a series of N-body simulations of galaxy collisions was carried out using *NBODY2* [Gerhard, 1981, 1982]. The main emphasis of this work was on investigating the role of the angular momentum vectors in the formation of the final system.

The initial galaxy models consisted of a differentially rotating disc of $N_{\rm d} = 124$ equal-mass particles orbiting a small central mass. The disc was stabilized by a live spherical halo with membership $N_{\rm h} = 125$. As in the previous section, possible two-body effects were studied first. Here the relevant time-scales are somewhat shorter, justifying the slightly smaller particle numbers. Several experiments were performed with different choices of impact parameter, initial c.m. velocity and angle between the two disc angular momentum vectors. The results showed that prograde rotation favoured the formation of a composite system, with the strength of the interaction determined by the relative orientation of the spin vectors. A typical outcome was the formation of a remnant after about ten crossing times, with an internal density $\rho \propto r^{-3}$, whereas the mass loss was only a few per cent. Analysis of the final velocity field revealed some interesting features which were ascribed to the oblate structure.

A more systematic exploration of colliding galaxy models was made for small ($N \simeq 30$ each) systems [Roos & Norman, 1979]. In this work, cross sections were determined for a range of parameters that were later used in cosmological simulations as well as for an investigation of rich galaxy clusters [cf. Aarseth & Fall, 1980; Roos & Aarseth, 1982]. For completeness, we also mention some early simulations of interacting galaxies with *NBODY2* and rather larger particle numbers ($N \simeq 250$ each) which set the scene for subsequent work [White, 1978, 1979].

Simulations of interacting galaxies have recently been revived in connection with black hole binaries in galactic nuclei. The availability of the GRAPE-4 special-purpose computer enabled quite large values of N to be considered by direct integration [Makino & Ebisuzaki, 1996; Makino, 1997]. This work employed the Hermite version of *NBODY*1 [Makino & Aarseth, 1992]. The first paper used $N = 16\,\mathrm{K}$ and a softening of $\epsilon = 1/128$. Different BH masses were studied in order to evaluate their effect on the structure of the final remnant. Following the formation of a remnant, a duplicate was constructed as the progenitor for the next simulation and the BH binary was artificially replaced by one particle. The main observationally related result was the formation of a shallow density cusp, $\rho \propto r^{-\alpha}$ with $\alpha \leq 1$ [see also Nakano & Makino, 1999].

The second paper concentrated on the evolution of the central BH binary, starting from similar initial conditions but now with a total number of particles up to 256 K. A softened potential was used for interactions between field particles, whereas all other forces were evaluated by the $1/r^2$ law. Starting with two model galaxies in parabolic orbit at separation $d = 10$ (in standard N-body units) and black hole masses $m_{\rm BH} = 1/32$, the remnant was formed by $t \simeq 30$ and the BH components became bound soon afterwards. Following an early phase of rapid hardening independent of N, the evolution rate tended to the power-law $dE_{\rm b}/dt \propto N^{-1/3}$ for a wide range of N, and by $t \simeq 60$ the final value approached the characteristic cluster energy of -0.2 in scaled units.

Two similar efforts which essentially employed point-mass interactions in the inner region together with two-body regularization have already been discussed in section 10.5 [cf. Quinlan & Hernquist, 1997; Milosavljević & Merritt, 2001]. Again an early hardening of the BH binary was followed by a stage of constant energy gain until the calculations were terminated for technical reasons. Much can also be learnt by studying the three-body problem. A recent investigation [Blaes, Lee & Socrates, 2002] focused on the Kozai mechanism in the presence of GR effects for a hierarchical triple of supermassive black holes. Integration of the averaged equations for the orbital elements showed a significant reduction of the coalescence time. Moreover, the inner binary may now sample interactions with a larger fraction of stars, thereby alleviating the loss-cone effect. This still leaves an interesting competition between the time-scales of the outer orbit shrinkage and the inner binary undergoing GR decay.

The numerical simulations may be confronted with theoretical predictions. Here the process of mass segregation plays an important role in models involving infall which are currently in vogue. Moreover, it appears that the early evolution is independent of N because the loss-cone depletion has not begun. However, the use of a relatively large and constant value of $M_{\rm BH}$ for different N-values does affect the scaling to realistic conditions. Hence the question of the maximum excursion of the BH binary requires further study. Likewise, the related problem of Brownian motions has still not been resolved [cf. Hemsendorf *et al.*, 2002]. The outcome has implications for the possible formation of massive triples before binary coalescence by gravitational radiation can take place. It can therefore be anticipated that greater efforts will be directed towards resolving this fascinating problem, and hopefully direct N-body integrations will play a significant part [cf. Mikkola & Aarseth, 2002; Aarseth, 2003a,b].

16.5 Groups and clusters

Some of the earliest N-body simulations were concerned with small groups and clusters of galaxies [Aarseth, 1963b, 1966a]. This problem is numerically easier than point-mass dynamics and is justified because galaxies have extended mass distributions, permitting their interactions to be modelled by a softening of the Newtonian potential (cf. (2.8)). Direct integrations were performed for clusters with $N = 50$ and 100 over a few crossing times. Since different masses were included for the first time, this was sufficient to demonstrate mass segregation and significant binaries were also noted. A subsequent investigation [Aarseth, 1969] added initial rotation which gave rise to flattening, enhanced mass segregation and shorter relaxation time. Because of the general expansion of isolated systems, the rotational kinetic energy decreases. At the same time, angular momentum is transported outwards, in analogy with accretion discs.

Following the development of the *NBODY*2 code, larger systems could be studied [White, 1976]. Moderately expanding initial conditions with $N = 700$ members of different mass were used to model the Coma galaxy cluster. Several subclusters which formed during the early phase provided conditions for violent relaxation after turn-around, producing a final equilibrium system with core–halo structure. Already at this stage there was significant mass segregation, in qualitative agreement with observations of luminosity segregation. The local infall scenario became popular in later years when attention turned to expanding models.

A more consistent treatment [Roos & Aarseth, 1982] included inelastic coalescence based on determinations of cross sections from N-body simulations [Roos & Norman, 1979]. Again an initially expanding system was used, with a cosmological density parameter $\Omega_{\rm in} = 1.75$ and $N = 700$ particles selected from a standard luminosity function. This gave rise to a maximum expansion factor $R_{\rm max}/R_{\rm in} = (1 - \Omega_{\rm in}^{-1})^{-1} \simeq 2.4$.

The galaxy cluster model was evolved over 1.5×10^{10} yr with a virialized radius of 2.5 Mpc and total mass $1.5 \times 10^{15}\, M_{\odot}$. At the end of this interval, the membership decreased to 240 with an enhanced proportion of heavy particles. Most of the coalescence took place during the early phase when the velocity dispersion was less than the internal value and the radii had not been reduced by tidal stripping. The amount of mass segregation at the end was quite strong, in agreement with the investigation above. Thus it appears that much of the mass segregation that occurs in the subsystems survives the violent relaxation during the overlapping process. It was also suggested that tidal stripping due to collisional mass loss in the core of a rich cluster limits the radii of such galaxies to 20 kpc within 5×10^9 yr. Furthermore, the stripped material may form part of the envelopes of central giant galaxies. On the debit side, it may be remarked that halos are not treated in a self-consistent way since their destruction during tidal interactions is not included and this will lead to an overestimate of the coalescence rate. Nevertheless, the attempt to account for evolutionary effects in galaxy clusters illustrates well the versatility of the direct N-body approach.

The more realistic cluster simulations described above were based on the idea of early expansion, followed by turn-around and recollapse. This description set the stage for cosmological simulations which also had a modest beginning when measured by present-day standards. Finally, we note an early attempt of studying compact galaxy groups dominated by dark matter [Barnes, 1984] where again *NBODY*2 was used with $N \simeq 450$.

16.6 Cosmological models

The idea of studying galaxy clustering on a local scale that includes the brightest 100−1000 members is attractive if it can be assumed that mass is correlated with luminosity. Consequently, most of the early investigations of expanding systems adopted this simplifying hypothesis.

Cosmological N-body simulations began to take shape in the mid-1970s with some innovative explorations of the basic clustering process [Aarseth, Gott & Turner, 1979; Gott, Turner & Aarseth, 1979]. The important question of initial conditions was also addressed in the first paper. Initial density fluctuations of the type

$$\delta\rho/\rho \propto m^{-1/2-n/6} \tag{16.7}$$

were chosen, with m the mass within an arbitrary radius and the index n denoting any departure from a Poisson distribution. Different models with $N = 1000$ and 4000 particles were considered, the latter having a universal mass function. To increase the range of separations and hence amplitude for the two-point correlation function, some point-mass calculations were also made with a precursor of *NBODY*5.

For a given initial mass distribution and cosmological density parameter, $\Omega_{\rm in}$, the velocities were chosen by the expression

$$\mathbf{v}_i = H_{\rm in}\mathbf{r}_i + \mathbf{v}_{\rm p}\,, \tag{16.8}$$

with scaled Hubble's constant $H_{\rm in} = (2M/\Omega_{\rm in})^{1/2}$ and $\mathbf{v}_{\rm p}$ representing possible random motions. The latter decay as the result of expansion and cooling, whereas peculiar velocities are generated by departures from homogeneity. Closed and open models were studied, with $\Omega_{\rm in} = 1$ and 0.77, and corresponding expansion ratios $R_{\rm max}/R_{\rm in} = 10$ and 32. During the expansion, over-dense regions are retarded and eventually form bound clusters which approach virial equilibrium, whereas negative density fluctuations lead to voids of growing size.

The correlation function of the models display a power-law behaviour with index $\alpha \simeq 1.8$. This agrees well with observed values in the range 0.1–10 Mpc. Another test of galaxy clustering simulations was provided by analysis of group catalogues [Turner *et al.*, 1979]. Comparison of simulations with observational group catalogues of bright galaxies favoured open models ($\Omega_0 \simeq 0.1$) and $n = -1$ which produce a wider range of cluster sizes than purely Poisson-type initial conditions. A third paper [Gott, *et al.*, 1979] examined the slope of the correlation function in more detail and concluded that the value of Ω cannot be determined solely from the corresponding observations.

The general idea that bright ellipticals may be the result of coalescence [Toomre, 1977] was first explored in a cosmological simulation using the code *COMOVE* [Aarseth & Fall, 1980]. The capture cross section for different impact parameters and velocities was obtained by combining previous N-body experiments [van Albada & van Gorkom, 1977; White, 1978; Roos & Norman, 1979]. It was assumed that inelastic coalescence occurs instantaneously at the first perigalacticon passage, with the new internal energy of the remnant given by

$$E_{\rm r} = E_1 + E_2 + E_{\rm b}\,, \tag{16.9}$$

where $E_{\rm b}$ is the orbital binding energy. Adopting a final Plummer model without mass loss, this gives the galaxy half-mass radius as

$$r_{\rm g} \simeq -0.2\,m^2/E_{\rm r}\,. \tag{16.10}$$

The corresponding softening size was taken to be $\epsilon \propto m^{1/2}$. In addition, the 1000 particles were assigned randomly oriented spins for the purpose of analysing the dimensionless spin parameter $J|E_{\rm r}|^{1/2}/GM^{5/2}$ based on angular momentum conservation.

Many of the remnants were formed in a hierarchical sequence, with the early stage characterized by the turn-around and infall of weakly bound pairs. The preponderance of nearly radial orbits resulted in a radius–mass relation $r_{\rm h} \propto m^{0.85}$ and a low value for the characteristic remnant spin. Open models with $\Omega_{\rm in} = 0.69$ were studied, and the final value $\Omega_0 = 0.1$ was reached after expansion by a factor of 22. The main conclusion based on these simulations was that accretion plays an important role in the clustering process, at least if massive halos are assumed. Thus a significant fraction of dark matter distributed homogeneously in similar models only reduced the collision rate by a modest amount [Findlay, 1983]. The latter study incorporated the effect of spin generated by tidal torques induced by the neighbours but the final rotation rates were still small.

Cosmological clustering scenarios were also examined for anisotropic initial conditions which produce flattened superclusters [Dekel & Aarseth, 1984]. So-called 'pancake simulations' were made with $N = 10^4$ by first evolving the models kinematically according to the Zel'dovich approximation and subsequently by self-consistent means, using the *COMOVE* code with small softening. As a sign of the efficiency of the comoving code and bearing in mind the modest computing power (i.e. VAX 11-780, < 1 Mflop), the total CPU times were only about 50 hr for an expansion factor of 8. The observed excess of the two-point spatial correlation function on large scales, which does not agree with the standard hierarchical clustering picture, was reproduced in the case where superclusters collapse to form flat structures. It also follows that the latter are young dynamically which has implications for galaxy formation. This investigation emphasized the importance of more realistic initial conditions and their relevance for the shape of the correlation function.

The above simulations also give rise to quite large voids that are nearly empty of galaxies. Analysis of voids based on earlier numerical models showed that the characteristic size distribution does not depend significantly on $\Omega_{\rm in}$ or the initial conditions [Aarseth & Saslaw, 1982]. From mass conservation, the typical sizes of voids must exceed the cluster sizes since the density enhancements can be arbitrarily large. Comparison of the largest voids in the simulations with observations indicate that the latter are larger, in apparent conflict with the standard gravitational clustering picture. However, a similar analysis of simulations with flattened superclusters for somewhat richer memberships might narrow this gap.

Finally, we mention the use of the vectorized version of *COMOVE* [cf. Aarseth & Inagaki, 1986] for comparing gravitational clustering with thermodynamic theory. In a first paper [Itoh, Inagaki & Saslaw, 1988], it was shown that this theory gives a good description of the simulations as well as observations. Generalization to two mass components [Itoh, Inagaki & Saslaw, 1990] also proved possible, using the single-component thermodynamic distribution function applied to each mass group separately. Hence, in conclusion, we see that the early N-body simulations of galaxy clustering provided considerable stimulus towards theoretical developments which led to the present picture of a dominant dark matter distribution.

17
Planetary systems

17.1 Introduction

In the last few years, the subject of dynamical planetary formation has undergone a remarkable transformation. Thus we now have an increasing database of actual observed systems which provides much material for theoretical and numerical work. It therefore seems appropriate to devote a small chapter to direct N-body simulations of planetary systems. The emphasis in the early days was on performing idealized calculations to ascertain whether the eccentricity growth due to weak perturbations could lead to significant accretion by collisions. With the increase of computing power, more realistic modelling has become feasible by direct methods, but powerful tree codes are an attractive alternative. The latter technique has proved effective for studying both planetesimal systems and planetary rings. Following the discovery of many extra-solar systems, the question of stability has become topical. Stability is often addressed using symplectic integrators which are outside the main scope of this book. In the following we distinguish between planetary formation and planetesimal dynamics. This division is somewhat arbitrary but planetesimal simulations are usually concerned with particles distributed in a thin annulus which therefore represents larger systems.

17.2 Planetary formation

Although there are early integrations of planetary dynamics relating to Bode's Law [Hills, 1970], it took another decade for the subject proper to get under way. The first attempt to model an evolving mass distribution [Cox & Lewis, 1980] suffered shortcomings in that the system appeared to terminate its evolution before reaching a realistic final configuration. However, this study employed Keplerian orbits outside the sphere of influence and hence neglected distant encounters. Thus even an initial uniform eccentricity distribution up to $e_{\max} = 0.1$ of 100 planetesimals with 1.6 lunar masses each in the terrestrial region was insufficient to prevent isolation of the feeding zones of the massive bodies.

A full N-body model, also in 2D, was more successful [Aarseth & Lecar, 1984; Lecar & Aarseth, 1986]. Here the grid perturbation method described in section 3.5 was employed. The collision criterion used was based on overlapping impacts of the two radii, assuming realistic density and inelastic coalescence without mass loss.* This model began with $N = 200$ lunar-size planetesimals in circular orbits in the range 0.5–1.5 au with constant surface density $\Sigma = 10.5\,\mathrm{g\,cm^{-2}}$, or mass $m_0 = 6.6 \times 10^{26}\,\mathrm{g}$. Perturbations out to about $N_\mathrm{b} = 5$ azimuthal bins were included on each side (out of a total of 100 bins), corresponding to about 300 times the sphere of influence, $R_\mathrm{s} = (2m_0/M_\odot)^{2/5}a$, at $a = 1\,\mathrm{au}$.

A final membership of six bodies was reached after $\sim 5 \times 10^4\,\mathrm{yr}$. Starting from circular orbits, the mean eccentricity grew to about 0.06 by $t \simeq 10^4\,\mathrm{yr}$ while the membership decreased to about 20. At this stage, all perturbations as well as the indirect force terms were included. Comparison of the three most massive bodies showed no evidence of runaway; the third largest mass increased at a similar rate as the maximum mass. The mass spectrum evolved according to

$$dN \propto m^{-q}\,dm\,, \tag{17.1}$$

where the exponent decreased smoothly from a large value to below unity with $q \propto 1/\bar{e}$. Although the final masses were planet-like, the eccentricities tended to be rather large. In spite of the simplifying assumptions of 2D, equal initial masses and no fragmentation, the direct approach showed promise and pointed the way to future developments.

This work also provided a theoretical framework for predicting the particle number and mean eccentricity in a differentially rotating system, where the competing processes of eccentricity growth by perturbations and damping due to collisions are operating. Including gravitational focusing yielded an evolution equation

$$N_0/N = 1 + t/\tau\,, \tag{17.2}$$

with $\tau \simeq 1300\,\mathrm{yr}$ from the data, which reproduced the characteristic knee in the numerical results. However, taking into account the variation of the collision frequency and separation with the mean mass resulted in a less satisfactory agreement. The prediction of $\bar{e}$ was more successful and

* Note that the simple device of placing the remnant body at the c.m. position with corresponding velocity does not conserve orbital angular momentum strictly in a system with differential rotation since the velocity changes are unequal.

yielded a power-law $\bar{e} \propto t^{\beta}$ with the exponent between $\frac{1}{2}$ and $\frac{2}{3}$, which bracketed the data during the first 10^4 yr.

Considering the final configuration, the spacing between nearby orbits is proportional to $1/N$, whereas the radial excursions are proportional to e. Since the former behaves as t while $\bar{e}$ evolves as $t^{2/3}$ at most, the spacing between adjacent orbits gradually prevents further interactions from taking place and the evolution ceases. An intriguing result was that both the mean eccentricity and the maximum mass varied as the same power of time, in direct contradiction with a classical theoretical prediction [Safronov, 1969] that $\bar{e}$ should vary as some characteristic mass to the $\frac{1}{3}$ power. As a further pointer to the delicate state of the final system, two overlapping orbits appeared to be in a stable configuration. Thus the integration was extended another 6×10^4 yr without collision occurring, suggesting the existence of a commensurability condition. Assuming comparable inclinations in 3D to the final eccentricities, the estimated time-scale would increase by a factor of 4000 [cf. Aarseth & Lecar, 1984].

A subsequent 2D study also based on the grid perturbation method included the important process of fragmentation, as well as some improvements of the long-range treatment of massive bodies [Beaugé & Aarseth, 1990]. Cross sections for collision outcomes were taken from scaled laboratory experiments and known structural parameters which determine the critical velocity for shattering. Three velocity regimes were distinguished.[†] First, there is no cratering below a critical impact velocity given by

$$V_{\rm crit} = 2S/V_{\rm s}\rho\,, \tag{17.3}$$

with S the crushing strength, $V_{\rm s}$ the sound speed and ρ the density. In this case the two colliding bodies separate with the rebound velocity which is reduced by the coefficient of restitution. For intermediate impact energies, the cratering process was modelled by mass loss which depends on kinetic energy and involves the mass excavation coefficient. A proportion of this material is not ejected from the bodies, leading to a net mass transfer after neglecting a small external mass loss. Here the cumulative fraction of ejecta with velocity exceeding v was approximated by

$$f(v) \simeq c_{\rm ej}\, v^{-9/4}\,, \tag{17.4}$$

with the coefficient $c_{\rm ej}$ depending on physical properties. Thus the body shatters for sufficiently large impact energy. From laboratory experiments, this is likely to occur if the impact energy per unit volume, $\mathcal{E}$, exceeds the impact strength. The distribution of fragment masses depends on the

[†] See the earlier work by Cazenave *et al.* [1982] for a similar independent treatment.

biggest fragment, determined by the empirical relation

$$m_{\max} = \tfrac{1}{2}(m_k + m_l)\left(\frac{\mathcal{E}}{S}\right)^{-1.24} . \tag{17.5}$$

As a compromise, each fragmenting body m_k, m_l is represented by four slowly escaping bodies and a remaining core, with appropriate masses sampled from a theoretical distribution. Finally, the ejection velocities were again chosen according to (17.4).

In order to prevent unduly large particle numbers being reached, the minimum fragment mass was taken as $0.01\,m_0$, with the initial value $m_0 = 1.15 \times 10^{26}$ g. The equal-mass bodies were again placed in circular orbits but this time between 0.6 and 1.6 au to allow for some inward drift due to energy dissipation during collisions. Moreover, three different surface densities $\Sigma \propto r^{-\alpha}$ with $\alpha = 0, -\frac{3}{2}$ and $-\frac{1}{2}$ were studied. In two models, the particle number decreased steadily during the entire phase after the early formation of a few planetary embryos[‡] which grew in a hierarchical or runaway manner. As shown by one model, the competition between accretion and fragmentation can sometimes go against the general trend. However, eventually the emerging embryos accreted the lighter bodies. Final configurations of three different models with $N = 200$ planetesimals yielded four principal bodies with moderate eccentricities on a time-scale of $\simeq 5 \times 10^5$ yr. Although the results of this simulation were quite encouraging, we note the restriction to 2D for computational reasons, as well as the absence of gas drag or a significant low-mass population.

This simulation illustrates three characteristic stages of evolution. During the first few thousand years the eccentricities grow slowly, ensuring that all collisions are inelastic; hence N decreases smoothly. Once the average eccentricity becomes significant (i.e. $\bar{e} \simeq 0.04$), fragmentation sets in and the particle number exhibits episodes of temporary increase. The planetesimal population can be divided into two groups, embryos and low-mass bodies, which evolve differently. The former retain low eccentricities and, by virtue of the larger masses, fragmentation is unlikely to occur; hence further growth takes place leading to protoplanets. However, the fragments tend to decrease in size, further enhancing the shattering probability until many reach the minimum mass. The third and final stage begins once the two populations become well separated, with the protoplanets accreting orbit crossing fragments. Some of the protoplanets have unstable motion and experience secular eccentricity increase until they collide with another massive body. Thus the final configuration consists

[‡] Here an embryo is defined as having mass $m > 4\,m_0$.

of a few planets with low eccentricities and marks the transition from chaotic to regular behaviour.

In view of the long evolution time-scale estimated for realistic systems, the simulations described above could not be extended to 3D. However, an interesting new approach was made by restricting attention to a narrow ring [Cazenave, Lago & Dominh, 1982]. The models consisted of 25 bodies distributed within a small and increasing range of semi-major axis, eccentricity and inclination. Direct integrations by the Bulirsch–Stoer [1966] method were performed for times up to $\sim 10^8$ days. The equilibrium result $\bar{i} = \frac{1}{2}\bar{e}$ for the inclination already obtained by statistical methods was confirmed. Accretion with fragmentation was adopted and the final solution was dominated by one body with small eccentricity in spite of some increase of the ring size. Although somewhat idealized, these early studies served as a useful introduction to much later N-body investigations, to be described in the next section.

With more powerful computers and several years of dedicated CPU time, 3D simulations for the terrestrial region are now possible [Chambers & Wetherill, 1998]. A series of models based on $N = 24$–56 isolated and nearly coplanar embryos with semi-major axes distributed in the range 0.55–1.8 au were studied over time intervals of 10^8 yr. The orbit calculations employed a symplectic integrator [Levison & Duncan, 1994] which was modified to deal with close encounters using the Bulirsch–Stoer method inside about 2 Hill radii [Chambers, 1999] and the results are therefore relevant to the present discussion. Combination of the two methods led to a small systematic growth of the total energy which is acceptable, bearing in mind the long duration.

In the approach to the final state, secular oscillations in eccentricity and inclination play an important role. First, neighbouring planets in nearly overlapping orbits contribute short-period cycles. In addition, models containing Jupiter and Saturn display resonances with periods of $10^6 - 10^7$ yr. The secular oscillations modify the perihelion and aphelion distances on time-scales that are short compared with the collision time. Hence, in order to avoid collisions, the protoplanetary orbits must satisfy non-overlapping conditions. Depending on the extent of the latter, further evolution takes place on increasing time-scales unless commensurability relations prevail. Even so, several objects of smaller mass survive until the end. However, the 3D simulations still exhibit relatively large final eccentricities. Consequently, $e_{\rm max}$ is too large to permit more than about two final terrestrial planets to be formed in simulations including Jupiter and Saturn. It may be noted that large mean eccentricities occur at early stages and this leads to enhanced values during secular oscillations. In conclusion, this long-term study illustrates the effect of secular resonances on time-scales of 10^5 yr which are best observed in simulations. However, some evidence of commensurability relations may also be found in 2D simulations [cf. Lecar & Aarseth, 1986].

17.3 Planetesimal dynamics

On theoretical grounds, the expected accumulation time for 3D models is a factor of 10^2–10^3 times longer than for similar 2D models. Since larger particle numbers are also desirable, this calls for a different approach. A first attempt at direct integration in 3D [Aarseth, Lin & Palmer, 1993] provided evidence for the onset of runaway growth, albeit for small systems. Here the particle in box method [Wisdom & Tremaine, 1988] was modified to include self-gravity, as described in section 3.6. This entails additional force summation over the eight neighbouring boxes by employing periodic boundary conditions and is therefore much more expensive. Individual time-steps were used again, together with the scheduling of *NBODY2* (cf. section 9.2).

The basic model contains $N = 100$ particles with mass $m_0 = 8 \times 10^{-11}\, M_\odot$, size $r_0 = 2 \times 10^{-6}$ au and box size $S_b = 0.04$ placed at 1 au. These parameters are based on computational considerations together with a minimum solar nebula of five Earth masses inside 1 au. The box size should not exceed a few per cent for the self-similar approximation to remain valid and should also be much larger than the Hill (or Roche) radius, $r_H = (2m_0/3)^{1/3}a$ for comparable masses. Hence, with these parameters, $S_b \simeq 106\, r_H$. An optional drag force was also tried, with components

$$
\begin{aligned}
F_x &= -\alpha_1 \dot{x}\,, \\
F_y &= -\alpha_2(\dot{y} + \tfrac{3}{2}\Omega x) - \alpha_4\,, \\
F_z &= -\alpha_3 \dot{z}\,,
\end{aligned}
\tag{17.6}
$$

where α_k represents the drag coefficients.

The particle in box method was used to study a variety of models both in 2D and 3D. A theoretical analysis shows that the dynamical evolution is regulated by energy transfer from the Keplerian shear to dispersive motions via gravitational scattering and dissipation due to collisions [cf. Palmer, Lin & Aarseth, 1992]. The state of the system is characterized by the Safronov [1969] number

$$
\Theta = Gm_0/2r_0\sigma^2\,. \tag{17.7}
$$

When the velocity dispersion is low, i.e. $\Theta \gg 1$, scattering is more efficient than dissipative collisions and σ increases until dynamical equilibrium is attained, with $\Theta \simeq 1$. One purpose of using both 2D and 3D models was to verify the prediction that identical equilibrium conditions are reached.

Since the collision time-scale in 3D is much longer, the use of 2D models also saves significant computational effort. Results of 2D simulations were used to confirm earlier analytical work that dynamical equilibrium is established with a velocity dispersion comparable to the surface escape velocity of the dominant mass population. There was some evidence for an early stage of runaway coagulation in three models with different initial velocity dispersion. Moreover, the collision time and asymptotic equilibrium value of σ agreed well with theoretical predictions.

The 3D simulations covered time-scales of 10^2 yr, or about 100 times shorter than in 2D. Now there is clear evidence for energy equipartition and mass segregation towards the midplane, due to dynamical friction. Provided that most of the mass resides in the low-mass field planetesimals, equipartition is sustained and creates favourable conditions for runaway coagulation [cf. Ida & Makino, 1992; Kokubo & Ida, 1996]. However, in view of the modest particle number, the self-similar approximation may not always be applicable to the intervals studied and hence only the onset of runaway can be established. Still, a model with initial σ-values near equilibrium only developed a maximum mass ratio $m/m_0 = 7$ (with $N = 76$) by $t = 1.2 \times 10^4$ yr, whereas the two heaviest bodies already showed strong sedimentation towards the origin of the fundamental $z - v_z$ plane. In conclusion, this work demonstrated that many dynamical processes in planetesimal systems are amenable to theoretical analysis. Moreover, some basic predictions [cf. Palmer *et al.*, 1992] were verified numerically, including conditions for the onset of runaway.

An alternative way of studying large planetesimal systems is to consider a narrow ring of particles [cf. Cazenave *et al.*, 1982]. In one such investigation [Kokubo & Ida, 1995], a swarm of up to 1500 low-mass planetesimals were assigned circular orbits centred on 1 au with small inclinations. This calculation was made possible using the HARP-1 and HARP-2 special-purpose computers with the Hermite integration method. Two nearly circular protoplanets were found to accrete mass in a stable manner, provided their separation exceeded about $5r_{\rm H}$. As the protoplanets grow, their orbital separation becomes smaller when measured by the Hill radius. This leads to a gradual repulsion until the separation reaches a value when the heavy body scattering becomes ineffective. If there are more than two such bodies with small separation, the repulsion may be replaced by eccentricity growth due to mutual scattering with consequent collision or ejection.

The question of runaway growth has also received much theoretical attention. This effect was confirmed numerically by the study of the coagulation equation under the assumption of equipartition of kinetic energy [Wetherill & Stewart, 1989]. However, the statistical description is based on average quantities and hence is not appropriate for the critical stage containing massive bodies. Conclusive evidence of runaway growth was finally obtained by direct integration in 3D [Kokubo & Ida, 1996]. Here $N = 3000$ planetesimals of mass $m_0 = 10^{23}$ g were distributed in a ring of width 0.04 au with Gaussian eccentricities and inclinations having rms dispersions $\bar{e} = 2\bar{i} = 2r_{\rm H}/a$. In any case, the initial distributions of m, e, i relax quickly on a time-scale of about 10^3 yr and lead to producing a continuous power-law mass function $dN/dm \propto m^\alpha$ with $\alpha = -2.5$. The device of using five times larger radii than determined by a density $\rho = 2\,{\rm g\,cm^{-3}}$ was adopted on the grounds that only the accretion time-scale is reduced, whereas going from 2D to 3D changes the growth mode. Still, this argument depends on assumptions about the role of close encounters which is difficult to estimate in differentially rotating systems, hence an independent check is desirable (see below).

As expected [cf. Aarseth *et al.*, 1993], dynamical friction was effective during the later stage. When the largest body in the mass distribution reached $\simeq 10^{24}$ g, it became detached and showed runaway growth, defined as an increase in the ratio $m_{\rm max}/\bar{m}$. This is in qualitative agreement with the theoretical condition for runaway growth, $2\,\Omega\, r_{\rm H} \leq v_{\rm esc}$, where $v_{\rm esc}$ is the surface escape velocity of the massive body. It is also highly significant that a 2D simulation showed a reduction in the maximum mass ratio from 30 to 15 in 2000 yr, whereas the present 3D mass ratio increased from 10 to 140 in 20 000 yr. This behaviour is supported by theoretical considerations comparing growth rates derived from the two-body approximation [cf. Kokubo & Ida, 1996]. Thus in the low-velocity regime ($v_{\rm rel} < v_{\rm esc}$ and $v_{\rm rel} \geq 2\,\Omega\, r_{\rm H}$), $dm/dt \propto r^2$ in 3D and $dm/dt \propto r$ in 2D, where a more favourable gravitational focusing factor has been ignored in the former case. Finally, for completeness, runaway growth is defined by

$$\frac{d}{dt}\left(\frac{m_1}{m_2}\right) = \frac{m_1}{m_2}\left(\frac{1}{m_1}\frac{dm_1}{dt} - \frac{1}{m_2}\frac{dm_2}{dt}\right) > 0\,, \tag{17.8}$$

with $m_1 > m_2 > m_0$.

Further explorations of runaway growth [Kokubo & Ida, 1998, 2000] included some new aspects. In the first study, the late runaway stage was investigated with two protoplanets inside a swarm of low-mass bodies. Now several similar-sized protoplanets grew, keeping their orbital separation larger than $5r_{\rm H}$, while most planetesimals remained small. This process, called *oligarchic* growth, is due to the slow-down of runaway growth after the coalescence of some protoplanets.

The second study, also with the Tokyo HARP-3 and GRAPE-4, included both gas drag and inelastic collisions. This time, realistic radii were used in order to test the earlier results based on enhanced values, whereas $N = 3000$ particles of equal mass 10^{23} g with $\Delta a = 0.021$ au were selected. In order to avoid the inward drift due to gas drag, the surface density of the planetesimal ring was kept constant by employing a scheme for treating boundary crossings, taking care not to introduce artificial heating. At $t = 2 \times 10^5$ yr, the maximum mass reached about $200\,m_0$, with $\bar{m} \simeq 2\,m_0$, hence most of the mass resides in the smallest bodies. Subsequently, the power-law exponent became only slightly less steep ($\alpha \simeq -2.2$), maintaining most mass in small bodies and hence favouring runaway growth. In a second model, an initial mass function with $N = 4000$ particles in the range 2×10^{23}–4×10^{24} g and $\alpha = -2.5$ was chosen, using a ring width $\Delta a = 0.085$ au. The final evolution is characterized by partial runaway growth interrupted by a few collisions between massive bodies (as surmised in Kokubo & Ida [1995]), after which oligarchic growth takes over. The time-scale for reaching protoplanets with masses $\sim 10^{26}$ g is about 5×10^5 yr, which agrees well with some analytical estimates and also with results of the multi-zoned coagulation equation.

In conclusion, the planetesimal simulations of Kokubo & Ida discussed above have provided much insight into the processes that control the important late stage of planetary formation. Although fragmentation and inelastic rebound have not yet been taken into account here, this work illustrates well the versatility of the direct approach.

Tree codes provide another way of studying larger particle numbers. The hierarchical structure of a tree code allows large dynamic range in density at modest extra cost for each force evaluation [Richardson, 1993a,b]. One such code called *PKDGRAV* has been adapted from cosmological applications to the case of planetesimal and ring simulations [Richardson *et al.*, 2000]. The integrator employs the leapfrog method with individual time-steps taken to be quantized by factors of 2 as in the Hermite scheme. Here the basic time-step is obtained from

$$\Delta t = \eta \, (r/|\mathbf{F}|)^{1/2} \,, \tag{17.9}$$

where r is the distance to the Sun or the dominant neighbour. A similar criterion based on the force and first derivative would be more appropriate but the latter quantity is not readily available here. In any case, (17.9) does have the desirable property of yielding a fixed number of time-steps for an unperturbed two-body orbit. However, additional safeguards must be included in order to ensure that collisions are detected.

When a tree code is used, the cost of identifying the n_s nearest neighbours $\propto n_s \log N$.[§] The collision time for a pair of approaching particles with radii r_1 and r_2 is determined by the expression

$$t_{\rm coll} = -\frac{\mathbf{r} \cdot \mathbf{v}}{v^2} \left\{ 1 \pm \left(1 - \left[\frac{r^2 - (r_1 + r_2)^2}{(\mathbf{r} \cdot \mathbf{v})^2} \right] v^2 \right)^{1/2} \right\} , \tag{17.10}$$

[§] In direct N-body simulations, a search is only made for particles with small Δt.

choosing the smallest positive solution. Hence the collision procedure must be performed if a value of $t_{\rm coll}$ is smaller than the corresponding time-step. This expression is applicable here because the leapfrog implementation is second order in time, permitting a quadratic prediction within small time-steps. Coalescence is assumed if the relative velocity is less than the escape velocity, otherwise the particles are made to bounce with energy dissipation due to coefficients of restitution and surface friction [Richardson, 1994, 1995]. A check on the angular velocity prevents the breakup limit being exceeded, in which case a bounce is enforced. The code design is based on providing initial conditions and performing extensive data analysis by separate software, as is also partly the case for the older N-body codes. Moreover, a general data visualization package called *TIPSY* is freely available.

Although the tree code is still undergoing development, results of several experiments have been presented. A comparison with the model for $N = 4000$ discussed above reproduced the behaviour leading to oligarchic growth and runaway. Again five or six protoplanets were separated by 5–10 Hill radii after 20 000 yr, with $m_{\rm max}/\bar{m} \simeq 90$. These two different methods also agreed well on the number of inelastic collisions (2884 *vs* 2847) and small eccentricity of the largest body ($\simeq 0.002$).

The pilot version of *PKDGRAV* is already able to study $N \sim 10^6$ particles over several hundred periods using parallel supercomputers, whereas $N \sim 10^5$ is feasible on clusters of workstations. However, future requirements call for long-term solutions in order to span different evolutionary stages. One suggested way to achieve substantial improvement is to divide the Hamiltonian into a Keplerian component using the classical f and g functions and a perturbed part. This would allow long time-steps to be used for most orbits, which have well separated Hill radii. A successful outcome of such a scheme depends mainly on the ability to distinguish between regimes of weak and strong interactions. Early tests for the outer giant planets show that this is indeed possible over time-scales $\sim 10^6$ yr [cf. Richardson *et al.*, 2000]. Given the potential gains, it can be anticipated that an algorithm for encounter detection will be devised.

Further developments of the planetesimal code have been described elsewhere [Stadel, Wadsley & Richardson, 2002]. Hopefully, a full collision model, including cratering, mass loss and fragmentation will be implemented in due course, together with gas drag. The improved code may be used to obtain better resolution for simulating so-called 'rubble piles' which are a characteristic feature of many Solar System objects [Leinhardt, Richardson & Quinn, 2000]. The effect of the impactor mass ratio has recently been studied with the aim of parameterizing planetesimal growth [Leinhardt & Richardson, 2002]. A review of the properties of such gravitational aggregates is also available [Richardson *et al.*, 2003].

17.4 Planetary rings

In this section, we first consider some applications of the tree code *BOX_TREE* [Richardson, 1993a,b] to problems of planetary rings and rubble piles. The initial development of this code was aimed at studying thin rings of high density, where collisions are frequent. Such systems are ideally suited to a description based on self-similar patches orbiting the planet, where the box size, S_b, is larger than the radial and vertical particle excursions but much smaller than the central distance. More recently the code *PKDGRAV* has in fact proved more suitable for ring simulations as well [Richardson, private communication, 2002]. Thus it is more flexible in allowing a rectangular self-similar patch to be modelled and a larger number of ghost replicas may also be included.

A range of particle masses and radii are required for realistic simulations of planetary rings. In this work, a power law IMF $n(m) \propto m^\alpha$ was used. With $f = N(m)/N$ as the fractional cumulative distribution, individual masses are generated with f-values increasing smoothly from 0 to 1 in steps of $1/N$ by

$$m_i = m_{\rm min} \left\{ 1 + f_i \left[\left(\frac{m_{\rm max}}{m_{\rm min}} \right)^{\alpha+1} - 1 \right] \right\}^{1/(\alpha+1)} . \qquad (17.11)$$

Assuming constant particle density, the size distribution can be obtained by substituting $\tilde{\alpha} = (\alpha - 2)/3$ in (17.11), together with the appropriate scaling factor.

Some models were first compared with the original scheme by excluding self-gravity [Wisdom & Tremaine, 1988], whereas mass and spin effects were also added in later models. The collision frequency is controlled by the optical depth which for non-uniform particle sizes is given by

$$\tau = \Sigma \pi r_i^2 / S_b^2 . \qquad (17.12)$$

Models with a basic box membership $N = 50$ integrated over 30 periods show excellent agreement for small and large values of the optical depth. Of specific interest is the equilibrium values of the vertical velocity dispersion, σ_z, and corresponding particle density, $n(z)$. Likewise, $n(z)$ averaged over the equilibrium interval compares well with the analytical model of Goldreich & Tremaine [1978] for $\tau = 0.2$.

The collision treatment was generalized to include normal and tangential coefficients of restitution, ϵ_n and ϵ_t. In order to treat the collision, vectors $\mathbf{R}_1, \mathbf{R}_2$ connecting the sphere centres to the impact point are introduced. The corresponding angular velocities, $\boldsymbol{\omega}_i$, combined with the linear spin velocities, $\boldsymbol{\sigma}_i = \boldsymbol{\omega}_i \times \mathbf{R}_i$, and the relative velocity, $\mathbf{v} = \mathbf{v}_2 - \mathbf{v}_1$, yield the relative spin velocity at impact as $\mathbf{u} = \mathbf{v} + \boldsymbol{\sigma}_2 - \boldsymbol{\sigma}_1$. The kinetic energy loss in such a collision can be obtained from the new relative spin velocity in terms of the original normal and tangential values,

$$\tilde{\mathbf{u}} = -\epsilon_{\rm n}\mathbf{u}_{\rm n} + \epsilon_{\rm t}\mathbf{u}_{\rm t} \,. \tag{17.13}$$

Post-collision linear and angular velocities were derived, taking into account the impulsive torques suffered by the two colliding spheres and the resulting energy loss, combined with angular momentum conservation. Let I_1, I_2 denote the moments of inertia of the two bodies touching at the point of impact and define $\alpha = R_1^2/I_1 + R_2^2/I_2$. For the case of two colliding uniform spheres, the new linear and angular velocities are then given by [Richardson, 1994]

$$\begin{aligned}
\tilde{\mathbf{v}}_1 &= \mathbf{v}_1 + \frac{m_2}{m_1+m_2}[(1+\epsilon_{\rm n})\mathbf{u}_{\rm n} + \beta(1-\epsilon_{\rm t})\mathbf{u}_{\rm t}]\,,\\
\tilde{\mathbf{v}}_2 &= \mathbf{v}_2 - \frac{m_1}{m_1+m_2}[(1+\epsilon_{\rm n})\mathbf{u}_{\rm n} + \beta(1-\epsilon_{\rm t})\mathbf{u}_{\rm t}]\,,\\
\tilde{\boldsymbol{\omega}}_1 &= \boldsymbol{\omega}_1 + \beta\frac{\mu}{I_1}(1-\epsilon_{\rm t})(\mathbf{R}_1 \times \mathbf{u})\,,\\
\tilde{\boldsymbol{\omega}}_2 &= \boldsymbol{\omega}_2 - \beta\frac{\mu}{I_2}(1-\epsilon_{\rm t})(\mathbf{R}_2 \times \mathbf{u})\,,
\end{aligned} \tag{17.14}$$

with $\beta = 1/(1+\alpha\mu)$ and μ the reduced mass. Realistic simulations were made with central box memberships in the range 200–3200 for up to ten periods in the less time-consuming cases. The typical size range 0.5–5.0 m chosen implies a considerable mass dispersion and hence provides a good opportunity for examining equipartition effects. As expected, the smallest bodies have the largest angular velocities and vertical excursions.

When different sizes are included, multi-layered loosely bound aggregates tend to form. However, some of the large-scale features are comparable in size to the central box. This artifact introduces disruptive perturbations due to the periodic boundary conditions which may produce more transient structures than in real systems. Among other aspects requiring attention are a wider size distribution as well as a significant sticking probability for small particles. Hence future simulations need to be both larger and include more physics. On the technical side, the closely packed aggregates also enhance the gravitational interactions with consequent time-step reduction. Improved modelling of planetary rings therefore needs to overcome several challenging problems before the results reach the impressive quality of observations from space. Notwithstanding such difficulties, the power of the present method was illustrated well by the rubble pile simulation of Comet D/Shoemaker–Levy 9 during the recent collision with Jupiter [Richardson, Asphaug & Benner, 1995]. The general problem of tidal disruption has also been discussed in more detail [Richardson, Bottke & Love, 1998].

Among several other ring simulations not mentioned so far, we finally discuss some recent work carried out on special-purpose computers. One such investigation [Daisaka, Tanaka & Ida, 2001] employed the HARP-2 and GRAPE-4 to study the evolution of a self-gravitating ring based on the particle in box scheme (cf. section 3.6). Given the extra computing power, larger domains may be included. This approach enabled a more careful assessment of the transient wake structures first noted in earlier simulations [Salo, 1992; Richardson, 1993a, 1994]. Different models with up to about 42 000 particles in the central box were examined for a range of parameters in the regime of small total disc mass ($M_{\rm disc}/M_{\rm planet} \simeq 10^{-8}$). The presence of strong wakes which develop when the optical depth becomes significant (i.e. $\tau \geq 0.5$) also leads to an increased number of inelastic collisions. Applications to Saturn's B-ring show an enhancement of the effective viscosity by a factor of about 10 and is considerably larger than theoretical values.

A direct N-body simulation of the Uranian rings has also been performed [Daisaka & Makino, 2003]. In this model the mass fraction was increased artificially from 10^{-9} to 10^{-6}, with $N = 40\,000$ particles assigned small initial eccentricities and inclinations inside a narrow ring of width $\Delta r = 0.02a$. Two shepherding satellites of comparable mass were included, together with the J_2 oblateness term. The self-consistent evolution was followed over time-scales of many thousand mean periods using GRAPE-6. In addition to the force and its derivative, this hardware provides the index of the nearest neighbour which can be used to determine the time-step as well as checking for inelastic collisions. Following the initial growth of the eccentricity, the ring developed an elliptical shape which preserved a uniform precession rate for long times, with characteristic features of variability. This surprising behaviour is not in accord with theoretical expectations and the complex time variation exhibited may be observable by the proposed Cassini mission. Hence, in conclusion, detailed observations of the Solar System offer unrivalled opportunities for testing numerical simulations in the high-density regime.

17.5 Extra-solar planets

The remarkable new observations of extra-solar planets have inspired considerable efforts by theoreticians and simulators alike. Here we discuss briefly a few numerical investigations relating to the vital questions of evolution and stability. However, the excursion into this young and fascinating subject should only be regarded as an introduction. It has become apparent that even systems with two or three planets orbiting the central star display many interesting aspects. Surprisingly, there is little historical tradition for studying such systems. It is also somewhat paradoxical that after more than a generation of N-body simulations with length-scales increasing from pc to hundreds of Mpc, the attention should now be on scales of a few au or even metres in the case of planetary rings. In the following we discuss some characteristic features resulting from numerical studies of two and three interacting planets and conclude by reviewing recent simulations with emphasis on orbital stability.

Planetary systems containing two members form the simplest type for numerical exploration and yet not much is known about this problem. A recent investigation illustrates well the complicated behaviour which may be examined by direct integration [Ford, Havlickova & Rasio, 2001]. Systematic studies were made of two Jupiter-like planets in nearly circular orbits and small inclinations around a solar-type star. The semi-major axis ratio was chosen to lie in a narrow range just outside the stability boundary, $a_1/a_2 = 0.769$ [Gladman, 1993]. Integrations for up to 10^7 yr combined two different methods. Thus strong interactions were treated by the Bulirsch–Stoer [1966] method, whereas a symplectic integrator was used for intermediate interactions. Finally, unperturbed Keplerian orbits were adopted if the outer separation exceeded $100a_1$ and $\dot{r}_2 > 0$. Four possible outcomes were recorded: (i) escape, (ii) planet–planet collision, (iii) planet–star collision, and (iv) integration time exceeding 10^7 yr. These conditions occurred with relative frequency of about 50, 5, < 1 and 45%, respectively, where the termination time represented some 1.6×10^6 inner orbital periods. The integration time limit was sufficiently long for convergent branching ratios to be reached. Only equal-mass planets were investigated here; consequently, the probability of escape would be affected by different mass ratios. Inspection of the final two-planet systems reveals that they are usually locked in a resonant configuration with nearly 3:2 period ratio, and the pericentres remain anti-aligned which minimizes the exchange of energy and angular momentum.

Short-term interactions of extra-solar planets have also been considered [Laughlin & Chambers, 2001]. Careful modelling of one observed two-planet system combined direct integration with an iterative scheme. The self-consistent fitting technique resulted in optimal values of the inclination, thereby reducing greatly the uncertainty due to this troublesome degeneracy. Direct integration based on the derived elements showed the system to be stable over at least 10^8 inner periods. Finally, it was emphasized that mutual planetary perturbations should be included when making fits to the observed reflex velocity curve.

The interpretation of three planets orbiting the star υ Andromedae poses some interesting questions of evolution and stability. Given the shortest period of only 4.6 d, direct N-body integrations of this four-body system over relevant time-scales would be prohibitive. Even so, early explorations were made using the Bulirsch–Stoer method [Laughlin &

Adams, 1999; Rivera & Lissauer, 2000]. A later investigation based on a symplectic integrator and improved data concluded that this system may be stable for at least 100 Myr on the assumption of nearly planar orbits and inclination mass factors ($1/\sin\psi$) up to 4 [Lissauer & Rivera, 2001]. Although the innermost planet is weakly coupled to the motion of the outer planets, its eccentricity exhibits large oscillations for some parameters, indicating that resonance conditions are important. In particular, the alignments of the periastron longitudes seen in the experiments have a direct bearing on the long-term stability. With such a short inner period, other effects are likely to play a role. One comprehensive study [Mardling & Lin, 2003] modelled the additional processes of tidal interaction and spin–orbit coupling between the innermost planet and the host star, as well as relativistic precession. During this complex interplay, the two outer planets evolved towards alignment of their pericentres.

Given modern ideas that most stars are formed in associations or clusters, it would also seem natural to suppose that planetary systems originate in such environments. The question then arises as to how the planetary orbits are modified by stellar perturbations. Some consequences of this cosmogony have been addressed in scattering experiments and N-body simulations. We first discuss the statistical approach of studying the outcome of typical stellar encounters on a single planetary orbit [Laughlin & Adams, 1998]. A star Jupiter system was placed on a circular orbit at 5 au and subjected to perturbations from passing binary stars sampled from a Trapezium-type cluster with typical density $n_0 \simeq 10^3\,\mathrm{pc}^{-3}$ and velocity dispersion $1\,\mathrm{km\,s}^{-1}$. Effective cross sections for different outcomes were obtained for 40 000 scattering experiments generated by a Monte Carlo approach [cf. Hut & Bahcall, 1983]. Direct integrations of the four-body problem were again made using the Bulirsch–Stoer method.¶ The probability of disruption per 100 Myr was found to be 13%, whereas a considerably larger fraction received moderate increases in eccentricity. An estimated 5% of such planets may be ejected in a typical cluster over this interval, whereas only 3% developed high eccentricity. The scattering mechanism can account for some observed eccentric systems as well as a few 16 Cygni-type hierarchies, although extremely short periods are unlikely to be produced in this way.

Self-consistent modelling of giant planet eccentricity evolution have also been carried out. In one such study [de la Fuente Marcos & de la Fuente Marcos, 1997], realistic N-body simulations of medium-sized open clusters ($N \leq 500$) were made with NBODY5. Here 10–50% of the stars were assigned planetary companions in nearly circular orbits with initial

¶ For small N-body systems without regularization, the second-order method is more efficient [cf. Mikkola & Tanikawa, 1999b].

separations of 6–10 au. The results provided evidence for a high survival probability. Hence only about 2% of the planetary systems were disrupted before escaping and nearly 90% escaped from the cluster without any significant changes in the two-body elements. A few stable hierarchies of the 16 Cygni type were noted out of 500 models. Similar disruption rates were obtained in a subsequent investigation that included some models with primordial binaries [de la Fuente Marcos & de la Fuente Marcos, 1999]. It was also pointed out here that high ejection velocities ($v > 25\,\mathrm{km\,s^{-1}}$) may be produced in hierarchical configurations, whereas encounters by single stars are rarely disruptive.

Statistics on the distribution of minimum encounter separations have been obtained by N-body simulations [Scally & Clarke, 2001]. Models of young clusters with half-mass radius $r_{\mathrm{h}} = 1\,\mathrm{pc}$ containing 4000 single stars were studied using NBODY6. Thus by the current age of the Orion Nebula Cluster ($t \simeq 12\,\mathrm{Myr}$), only about 4% of the stars experienced an encounter closer than 100 au. Stellar encounters are therefore unlikely to play a significant role in destroying proto-planetary discs.

The question of free-floating planets in clusters was addressed in another N-body simulation [Hurley & Shara, 2002]. The motivation for this work was to examine the evolution of the disrupted planet population in rich open clusters. Three simulations were made using NBODY4 on GRAPE-6. Each model consisted of $N_{\mathrm{s}} = 18\,000$ single stars and $N_{\mathrm{b}} = 2000$ binaries, with typical velocity dispersion of $2\,\mathrm{km\,s^{-1}}$ and core density $n_0 \simeq 10^3\,\mathrm{pc^{-3}}$. Some 2000 or 3000 planets were placed in nearly circular orbits around randomly chosen single stars with a uniform semi-major axis distribution in 1–50 au or 0.05–50 au. Contrary to recent claims [Bonnell *et al.* 2001, Smith & Bonnell, 2001], this consistent simulation demonstrated that free-floating planets can form a significant population in such clusters. Thus, of the liberated planets, nearly half were retained after 1 Gyr. This fraction decreased to 12% after 4 Gyr when some 10% of the systems had been disrupted. Consequently, the preferential depletion of these light objects occurs on a half-mass relaxation time-scale.

Finally, we mention some attempts to constrain the birth aggregate of the Solar System. Thus both N-body simulations [de la Fuente Marcos & de la Fuente Marcos, 2001] and scattering experiments combined with star formation considerations [Adams & Laughlin, 2001] favour an origin in relatively small clusters. In a recent investigation [Shara, Hurley & Mardling, 2003], 100 solar system models comprising the Earth and Jupiter are placed in a cluster of 20 000 stars containing 10% primordial binaries. Provisional results suggest that significant changes, occur on a time-scale of 100 Myr. Hence this would support the picture of our Solar System being formed in relatively small stellar aggregates that also have shorter life-times.

18
Small-N experiments

18.1 Introduction

In this last chapter, we return to the subject of numerical experiments in the classical sense. Using the computer as a laboratory, we consider a large number of interactions where the initial conditions are selected from a nearly continuous range of parameters. Such calculations are usually referred to as scattering experiments, in direct analogy with atomic physics. Moreover, the results are often described in terms of the physicist's cross sections, with the results approximated by semi-analytical functions. The case of three or four interacting particles is of special interest. However, the parameters space is already so large that considerable simplifications are necessary. In addition to the intrinsic value, applications to stellar systems provide a strong motivation. Naturally, the conceptual simplicity of such problems has also attracted much attention.

In the following we discuss a number of selected investigations with emphasis on those that employ regularization methods. It is convenient to distinguish between simulations and scattering experiments, where in the former case all particles are bound. The aim is to obtain statistical information about average quantities such as escape times and binary elements, and determine their mass dependence. Small bound systems often display complex behaviour and therefore offer ample opportunities for testing numerical methods. On the other hand, scattering experiments are usually characterized by hyperbolic relative velocities, the simplest example being a single particle impacting a binary. The outcome of a wide range of parameters may be studied, using an automated procedure for generating the initial conditions. Such experiments also present good opportunities for comparing analytical cross sections for use in applications. Finally, the last section is devoted to the topics of chaos and stability which represent the ying and yang of stellar dynamics.

18.2 Few-body simulations

Even the general three-body problem hides an amazing variety of solutions which have inspired much numerical work. The intricate orbits displaying successive close encounters provide a serious test of integration methods. On a historical note, the first such numerical integration was carried out by Strömgren [1900, 1909] for a hierarchical triple over part of an outer orbit, with energy conservation to six figures.* Soon there followed a pioneering investigation by Burrau [1913] of a system with comparable separations and masses. This work introduced the concept of halving time-steps which succeeded in dealing with two close encounters and appeared to demonstrate that the orbit was not periodic. These initial conditions also served as a template for other studies.

The full complexity of the Pythagorean Problem was unravelled much later after the implementation of KS regularization [Szebehely & Peters, 1967]. Three comparable mass-points start from rest in a rectangular configuration and perform a celestial dance by repeated close encounters until the lightest body escapes, leaving behind a highly eccentric binary. Although the total angular momentum is zero, inspection of the orbits (or the corresponding movie) shows that the minimum moment of inertia occurs when the lightest body bisects the binary components near their apocentre. Hence the triple encounter that produces the final ejection is relatively mild; i.e. the perimeter, or sum of distances, is a factor of 10^3 greater than the closest two-body separation. Note also that in order to escape when starting from rest, the ejected body must pass between the binary components with retrograde motion to compensate for the angular momentum of the binary. In the context of three-body regularization [cf. Aarseth & Zare, 1974], we remark that only two changes of reference body are needed for the entire evolution. As can be seen from a movie, this implies that the dominant mass controls the evolution.

Although much can be learnt by studying one system, it is necessary to consider an ensemble of initial conditions in order to reach general conclusions. Early explorations based on small samples already yielded relevant information on the distribution of escape times and mass dependence [Worral, 1967; Agekian & Anosova, 1968a,c]. Numerical problems restricted the fraction of examples leading to completion in the former study, which employed the fourth-order Runge–Kutta method (RK4). Completion problems also occurred at the start of the second investigation. However, the introduction of Sundman's [1912] time smoothing enabled close two-body encounters to be treated more accurately. In either case, the process of escape was demonstrated beyond doubt. Eventually,

* Using the *TRIPLE* code on a laptop, the outer body reached $R_1 \simeq 73$ at $t \simeq 215$ in just 0.03 s, with turn-around at $R_1 \simeq 175$ and eventual exchange at $t \simeq 2840$.

two-body regularization became feasible in stellar dynamics [cf. Peters, 1968a,b] and improved the integration accuracy, thereby facilitating more systematic investigations.

In one 2D survey of 800 initial conditions [Standish, 1972], only 38 cases did not result in escape. Of these, 26 belonged to a set containing significant angular momentum which would favour hierarchical stability. For the sample as a whole, the mean escape time was about $29\,t_{\rm cr}$, which confirmed the earlier work. The semi-major axes distribution showed a wide range below the limiting value for escape, $a_c = m_k m_l/2|E|$. An interesting correlation between small initial rotation and large final eccentricity was also noted. As for mass dependence, the heaviest particle escaped in 29 cases out of 480 for a randomized IMF; however, in no case did the escaping mass exceed that of the remaining binary.

A similar investigation [Szebehely, 1972] surveyed 125 examples starting from rest, again based on the Pythagorean triangle, with integer masses 1–5. This integration used a 15th-order recurrent power series, together with the KS method for the two closest bodies, and maintained the relative energy error below 10^{-11} throughout. Here 92 cases led to escape before the limit of 150 time units was reached.† The largest body escaped once only, with another six when the two dominant members were equal. The general features of interplay and ejection (which implies return) were also emphasized, together with a discussion of discontinuous escape times.

In view of the round-off sensitivity of numerical solutions [Dejonghe & Hut, 1986], the upper limit of 150 time units may be rather optimistic. All 14 examples with escape times exceeding $t_{\rm esc} \simeq 100$ were repeated using the code *TRIPLE*, and only six were found to agree. Typical relative energy errors were 10^{-14} and the strict test of time reversal was applied successfully to verify these solutions. As for the other eight cases, most formed binaries but only one passed the time reversal test with rms error in position and velocity below the acceptable limit 1×10^{-3}. Moreover, the reproducible examples showed some evidence of a large minimum perimeter and corresponding less frequent changes of reference body. We may therefore conclude that use of the energy error by itself is questionable for establishing exact integrations without taking the duration and type of orbit into account, or performing time reversal.

When using the statistical approach, high accuracy is not essential to obtain meaningful results, provided the sample is sufficiently large. In one such study [Valtonen, 1974], about 200 experiments were performed in 3D with four different time-step parameters giving rise to relative rms energy errors 5×10^{-4}–3×10^{-2}. The distributions of eccentricity, terminal escape velocity and life-time did not show any clear accuracy dependence.

† Since the total mass was not fixed, $t_{\rm cr}$ for the systems varied from 3.88 to 7.03.

Subsequently, a survey of some 2.5×10^4 cases was carried out [Valtonen, 1976]. The initial conditions consisted of a binary with an incoming third particle mostly bound to the c.m. Again the integration employed a two-body regularization technique for the closest particle pair [Heggie, 1973]. The total angular momentum has a significant effect on the escape velocity, with small values leading to more energetic escape, confirming the enhanced effect of triple encounters. Likewise, the final eccentricity was also higher, as expected from stronger interactions. In earlier discussions of this work, the term *gravitational slingshot* was coined to describe the preferential acceleration of a particle with small pre-encounter impact parameter [Saslaw, Valtonen & Aarseth, 1974].

According to several authors, the expected equilibrium eccentricity in 2D and 3D is given by the probability distributions $f_2 = e(1-e^2)^{-1/2}$ and $f_3 = 2\,e$, respectively. Likewise, the predicted escape dependence on mass in a three-body system goes approximately as $P(m) \propto m^{-2}$ in 2D, in qualitative agreement with several numerical investigations [cf. Standish, 1972; Valtonen, 1974; Anosova, 1986]. Properties of the final state have also been determined by a statistical theory based on phase-space averaging in regions of strong interactions [Monaghan, 1976; Nash & Monaghan, 1978]. This approach led to predictions of eccentricity distributions in 2D and 3D, as well as escape velocity and mass dependence which compared well with existing numerical results for low angular momentum. Finally, concerning the life-times of bound systems, the distribution is quite wide and of exponential form [Valtonen, 1988]. The main parameters affecting the latter are mass dispersion, total angular momentum and energy.

An extensive project of three-body studies was initiated already in 1964 by the Leningrad school [Agekian & Anosova, 1968a,c]. This work is summarized in the following [Anosova, 1986]. Given two particles on the x-axis, initial positions of the third body were sampled uniformly from the positive quadrant of a circular region. Corresponding velocities were chosen with the virial ratio, $Q_{\rm vir}$, in the range 0–0.5. The numerical integrations were carried out by the RK4 method with Sundman's time smoothing, yielding typical final energy errors $\Delta E/E \simeq 10^{-4}$. A total of 3×10^4 experiments were considered. The following characteristic orbital types were distinguished: close triple approach, simple interplay, ejection with return, escape, stable revolution (hierarchical configuration), Lagrangian equilibrium configurations, collision and periodic orbit [cf. Szebehely, 1971]. By analogy with atomic physics, the concept of resonance was introduced later [Heggie, 1972a, 1975].

About 20% of the systems did not terminate in escape as defined rigorously (cf. (12.3)), due to temporary ejection. The notion of conditional escape was therefore introduced for distances exceeding (10–15) $R_{\rm grav}$, with $R_{\rm grav} = \sum m_i m_j/|E|$. In fact, the mean escape time for isolated systems is dominated by a few examples and can be arbitrarily large. Hence it may be preferable to discuss the probability, $P(t)$, that escape occurs after a given time. In the case of actual escape, the half-life was estimated as $t_{1/2} \simeq 80\, t_{\rm cr}$ with some reduction for increased mass dispersion. Mapping of the initial conditions leading to escape after the first triple approach showed a number of islands delineating the same outcome. These regions exhibit complex structure which was examined in some detail, revealing further properties. Thus the boundaries of the subregions correspond to distant ejections due to the first triple approach. The most important parameter for the outcome is given by a geometrical ratio involving m_1 and m_2 with respect to the escaper, m_3.

Subsequent studies with the *TRIPLE* code based on similar selection procedures for initial conditions of equal-mass systems have also been made [Johnstone & Rucinski, 1991; Anosova, Orlov & Aarseth, 1994]. In the first instance, 2850 examples were examined. The time until the final triple encounter leading to escape was determined as $f(t) \propto \exp(-0.69\, t/t_{1/2})$, with $t_{1/2} = 90\, t_{\rm cr}$, and the other parameters were in essential agreement with the original investigations. In the second study, two regular domains with a sample of 2500 experiments were examined. This allowed a comparison between the two regions as well as two methods. Moreover, a distinction between predictable and non-predictable results were made, where a time-reversed solution for the latter could not be obtained within coordinate or velocity rms errors of 10^{-3}. In the event, the similarity of statistical results for various quantities indicate that the basic evolution is independent of the selection method as well as the numerical method.

The role of close triple approaches may be studied by monitoring the moment of inertia. We note that escape is not necessarily associated with the smallest value. However, the last interaction is usually of fly-by type in a hyperbolic passage [Anosova, 1991]. As for the eccentricities, 2D systems with unequal masses tended to have $\bar{e} \simeq 0.81$ for the largest samples and $\bar{e} \simeq 0.71$ in 3D, consistent with theoretical expectations. A small proportion of long-lived hierarchical systems satisfying stability tests [cf. Harrington, 1972] were correlated with significant angular momentum. In an earlier investigation [Anosova, Bertov & Orlov, 1984], systems with non-zero angular momentum showed that 95% decay after a triple encounter and the life-times increase with rotation.

The addition of post-Newtonian terms to the standard equations of motion was considered in section 5.9. Since this approach appears to be rarely used in N-body simulations, we mention a similar treatment for the three-body problem [Valtonen, Mikkola & Pietilä, 1995]. The post-Newtonian approximation of order 2.5 was again adopted, together with the

Bulirsch–Stoer [1966] integrator. In view of the large relativistic precession term, the equations of motion were derived from a time-transformed Hamiltonian,

$$\tilde{H} = [H - E + \sum \mathbf{r}_k \cdot \mathbf{F}_k(t)]/L\,. \tag{18.1}$$

The third term contains the external forces represented by the relativistic corrections and L is the Lagrangian energy. The basic Pythagorean configuration was studied, with length unit of 1 pc and three different mass units of $10^5, 10^6$ and $10^7\,M_\odot$. All these examples resulted in escape of the lightest body at progressively earlier times than in the original problem. However, when the mass unit was increased to $10^8\,M_\odot$, coalescence of the black holes took place slightly before the time of the closest two-body encounter. Hence in this scaling sequence, a large limiting mass separates escape and coalescence.

For completeness, we also mention some relativistic simulations that made use of the *TRIPLE* code to study three-body interactions of $10\,M_\odot$ black holes [Kong & Lee, 1995]. The post-Newtonian approximation (radiation and low-order precession) was included in the regularized equations of motion by modifying the momenta according to (5.29). Close encounters between a binary and single black holes led to dissipation and coalescence in about 10% of the examples for hardness parameters $x \equiv -3E_{\rm b}/m\sigma^2 > 100$, where σ is the 1D velocity dispersion.

Moving to slightly larger classical systems, we note the scarcity of such simulations. If we restrict attention to $N \leq 10$, there are only a few applications to stellar dynamics. Following the pioneering work of von Hoerner [1960, 1963] which has already been discussed, the subsequent exploration of isolated few-body systems revealed general trends applicable to larger memberships (Poveda, Ruiz & Allen, 1967). Thus in a provisional study of 13 systems with $N = 10$ members, every case showed a final binary with binding energy exceeding the total energy [van Albada, 1967]. This investigation was extended to 29 systems which yielded much valuable material [van Albada, 1968]. The formation of energetic binaries consisting of massive components was emphasized. Again the binding energy of the final binary exceeded the total energy in most cases. Moreover, significant initial rotation gave rise to shorter evolution times. Some long-lived triples and even a few quadruples were noted and their formation was correlated with the total angular momentum. In one system, the inner eccentricity grew to exceed $e = 0.99999$ and may well have been the first N-body demonstration of a Kozai [1962] cycle.[‡]

In this work, the fascinating process of escape and the dependence on mass was elucidated to a degree that has not been surpassed. The following classification system for encounters leading to escape was proposed:

[‡] Similar behaviour was noted in later simulations with $N = 250$ [Aarseth, 1972a].

- Multiple encounters: normal binary and triple encounters and all encounters involving more than three stars
- An encounter between a binary and a single star, with the binding energy of the binary increasing
- Ejection from a member of a bound triple system
- Exchange of a binary component and ejection of the original body

In practice, most escapers may be associated with one of these processes which are also known to occur in larger systems. Hence a similar analysis to ascertain their relative importance in star cluster simulations would be highly desirable.

An early exploration of dynamical decay of general four-body systems [Harrington, 1974] considered the possible channels of outcome. The integrations employed a ninth-order Runge–Kutta–Nyström method for the second-order equations, together with a rarely used energy stabilization technique [Nacozy, 1972]. The experiment contained 100 examples each in 2D and 3D. Without regularization, a substantial fraction of the 2D systems were terminated prematurely because of close encounters. With relatively unstable quadruples, nearly 50 binaries and about 10 stable triples resulted from each of the samples. After this first attempted exploration of hierarchical formation in bound few-body systems, the subject seems to have been neglected.

Some experience has been gained with chain regularization applied to small systems [Sterzik & Durisen, 1995, 1998; Kiseleva *et al.*, 1998; Sterzik & Tokovinin, 2002]. Thus for $N \leq 10$, all the particles may be included in the special treatment, with external perturbations added if desired. An attempt to choose the outcome of cloud collapse calculations as initial conditions was made for five-body systems [Sterzik & Durisen, 1995]. The initial conditions started with a mean spacing of 300 au, total mass $3\,M_\odot$ and a small amount of rotational energy. Some 5000 examples were integrated for each of eleven models until the first escape occurred. The corresponding velocity was of primary concern, with more than half the escapers exceeding $3\,\mathrm{km\,s^{-1}}$ which would account for observations of some T Tauri stars associated with molecular clouds. The effect of a mass spectrum for $N \leq 5$ was also examined in a similar subsequent work [Sterzik & Durisen, 1998].

A second investigation focused on life-times and escape velocities of systems with 3–10 particles [Kiseleva *et al.*, 1998]. Nearly 10^4 different initial conditions were studied for each model, characterized by different mass dispersion and membership. With a scaling of 100 au for the initial radius and a mean mass of $1\,M_\odot$, about 1% of the escapers exceeded

$30\,\mathrm{km\,s^{-1}}$. Escape was defined as hyperbolic motion with respect to the c.m. for distances exceeding $10\,R_{\rm grav}$. At the end of long integrations up to $10^3\,t_{\rm cr}$, about 50% terminated in a binary while some 10% remained in hierarchical configurations without any stability tests being applied.

A novel attempt was made recently to compare the relative alignments of inner and outer orbits in observed triples with results from simulations of small groups ($N \leq 10$) [Sterzik & Tokovinin, 2002]. Although the observational dataset contained only 22 systems, there was some overlap with the numerical results of the final triples. Comparison of this sample with the plot of $T_{\rm out}$ against $e_{\rm out}$ inverted from the relation (9.14) yielded a significant percentage outside the stability boundary for planar systems. This excess is consistent with inclined configurations being more stable than coplanar prograde ones [cf. Harrington, 1972]. However, the numerical sample probably also contains long-lived unstable triples which have been classified as stable.

18.3 Three-body scattering

The idea of laboratory experiments in stellar dynamics [Hénon, 1967] takes its purest form in three-body scattering. There were several early attempts at such simulations, albeit with limited usefulness because of incomplete data [Yabushita, 1966; Agekian & Anosova, 1968b]. Nevertheless, some characteristic trends were observed. The subject was given a firm foundation in the first systematic survey involving over 10^4 examples of single star–binary encounters [Hills, 1975]. Three families with different fixed masses and a circular binary were used to construct a sequence of initial conditions by varying the intruder velocity and impact parameter.

Each experiment was characterized by two dimensionless parameters; the energy, $\alpha = (V_{\rm f}/V_{\rm c})^2$, and impact parameter, p, scaled by the semi-major axis, a_0. Here $V_{\rm f}$ is the velocity at infinity of the field star and $V_{\rm c}$ the minimum value for dissociating the binary, given by

$$V_{\rm c} = \left[\frac{m_1 m_2 (m_1 + m_2 + m_3)}{m_3 (m_1 + m_2) a_0}\right]^{1/2} . \qquad (18.2)$$

Only a few discrete values of p were sampled since the results can be fitted to a Gaussian and hence the parameter space is reduced. This work established the so-called 'water-shed effect' whereby the binary orbit tends to shrink for small $V_{\rm f}$ but loses energy if the pre-encounter velocity exceeds a well-defined value which depends on the mass ratios. Thus the cross-over at $\alpha_{\rm c} \simeq 0.57$ was found for equal masses and circular binary orbits. In addition to obtaining collisional cross sections for the energy change, it was shown that the probability of exchange is surprisingly large for equal masses and approaches unity if the field star is dominant. We note that the process of exchange is often associated with resonance, during which there are repeated sequences of temporary binaries with alternating ejections. If long-lived, such configurations present considerable demands on the computational resources as well as accuracy.

A series of further papers was devoted to various aspects of encounters between single stars and hard binaries [Hills & Fullerton, 1980; Fullerton & Hills, 1982; Hills, 1989, 1990, 1991, 1992]. The numerical technique was improved by the use of time smoothing [Heggie, 1972a], together with a high-order integrator. Determination of the average rate at which a hard binary increases its energy by interacting with single stars showed an inverse dependence on V_f but no dependence on a_0. Hence all hard binaries increase their binding energies at the same rate. Since momentum is also conserved, the corresponding recoil kinetic energy can be determined and used to examine the heating of star clusters. It has been shown theoretically [Heggie, 1972b, 1975] that the largest contributions are due to a small number of discrete encounters. Following such a cascade, the associated recoil velocity may eventually become sufficiently large for the binary to escape. Subsequent experiments were made for different mass ratios. The earlier value for the cross-over point between hard and soft binaries was confirmed as $\alpha_c \simeq 0.5$ for $m_3 \leq \frac{1}{2}(m_1 + m_2)$, whereas $\alpha_c \simeq 10$ for $m_3 \geq \frac{3}{2}(m_1 + m_2)$. A discussion of intruder mass and velocity effects [Hills, 1990] emphasized that the energy change cross-over does not in general occur at the hard/soft limit but rather depends on the orbital velocity ratio, V_f/V_{orb}.

In another investigation, encounters between massive intruders and binaries were studied in order to elucidate the process of tidal breakup of the components [Hills, 1991]. Velocity-dependent cross sections were obtained for the processes of exchange and dissociation, as well as for changing the binding energy and eccentricity. A characteristic feature after exchange of the massive intruder is that the new semi-major axis can be quite large, and likewise for the ejection velocity. However, the final binding energy also increases significantly; this ensures survival in future encounters with field stars. Regarding the main motivation for this work, the ratio of cross sections for tidal breakup and exchange was determined as $\sigma_{tidal}/\sigma_{ex} \simeq 2.5\, r^*/a_0$, with r^* the stellar radius. Altogether 8×10^5 such encounters were simulated and yielded a wealth of information. An earlier less ambitious study was concerned with the outcome of similar experiments with low-mass intruders of small velocity [Hills,1983]. The total cross section for increasing the energy was found to be independent of the orbital eccentricity. The probability of temporary capture reached a maximum of 20% when the closest approach was about $2a$.

Extensive scattering experiments with a much wider range of intruder masses were also performed [Hills, 1989, 1992]. A total of over 10^5

experiments were carried out requiring about two years of dedicated workstations as a cost-effective alternative to supercomputers. Again a Monte Carlo sampling of binary orientations was used [Hills, 1983], with typical relative energy errors below 10^{-6} even for long-lived resonances. Cross sections tend to approach simple analytical limits for small and large intruder masses. These simulations show that there is a critical impact parameter beyond which no exchange can occur. The corresponding closest approach distance is given by

$$R_c \simeq 2.1\, a_0 \left[(m_3 + m_\mathrm{b})/m_\mathrm{b}\right]^{1/3} . \tag{18.3}$$

Hence hyperbolic intruder velocities tend to lower the exchange limit with respect to the classical criterion for bound orbits [cf. Zare, 1976]. Results were presented for the dimensionless exchange cross section as a function of intruder mass. It goes to zero below $m_3/m_1 \simeq 0.3$ and reaches a plateau beyond $m_3/m_1 \simeq 10$.

Another interesting feature in this study is the temporary formation of triples with long life-times. At small impact parameters, the probability increases from 0.4 for low-mass intruders to a maximum of 0.95 near $m_3/m_1 = 0.55$ and then declines. Further complicated behaviour is seen when the original binary is first captured by a massive intruder. Either an exchange occurs with the other component being ejected in a slingshot interaction or the binary is disrupted gently, which gives rise to a long-lived triple. In conclusion, this classical series of papers contributed significantly to the understanding of three-body processes.

After describing some studies of interactions between single stars and binaries, we consider the process of binary formation itself. An early project to determine the probability of binary formation by three-body encounters was highly successful [Aarseth & Heggie, 1976]. Initial conditions of three incoming stars with impact parameters ρ_i were selected according to $\rho_i = r_0\, Y^{1/2}$, where r_0 is the maximum permitted value and Y is randomized in [0,1]. The integration started well before the time of maximum interaction, with the additional conditions of unbound two-body motions and negative total energy, whereas termination occurred when one pair separation exceeded the largest initial value. Here the *TRIPLE* code proved efficient in its first application.

Each set of experiments is characterized by a specified value of the dimensionless parameter $X = r_0\, V^2/2m$. By choosing $V^2 = Nm/2r_\mathrm{V}$, the scaled virial theorem yields $X = Nr_0/4r_\mathrm{V}$. An increasing number of initial conditions were considered for X in the range 0.5–1024, with 5×10^6 cases at the upper limit. However, actual integration is not necessary if the smallest initial two-body energy exceeds an appropriately chosen critical value. This rejection technique enabled quite large X-values to be selected without loss of positive outcomes while maintaining small error bars. A least-square solution gave a power-law for the probability of binary formation as

$$P_{\rm bin} \simeq 4\,X^{-2}\,. \tag{18.4}$$

This numerical solution is consistent with the asymptotically largest contribution obtained by a theoretical derivation[§] [cf. Aarseth & Heggie, 1976]. Based on this result, it was shown that the binary formation rate per crossing time is independent of N and that essentially all such binaries are soft. The present work provided quantitative answers to a classical problem and the numerical results guided the theoretical derivation, demonstrating the usefulness of the experimental approach.

The case of equal masses continued to attract attention. With greater coverage of parameter space, improved comparison with analytical approximations becomes feasible. One such ambitious undertaking [Hut & Bahcall, 1983] studied over 10^6 scattering experiments without making assumptions about dependence on orientation, phase angle or impact parameter. The results yielded cross sections for ionization, exchange and resonance scattering as functions of intruder velocity. The total cross section for an event Z can be obtained from the numerical results by

$$\sigma_Z(v) = F n_Z(v)/n_{\rm tot}(v)\,, \tag{18.5}$$

where $n_{\rm tot}$ is the total number of initial conditions for which event Z can occur, n_Z is the number of actual events and F is a factor that includes gravitational focusing.

Although the humble RK4 integrator was used, careful analysis justified the acceptance of relatively large energy errors. The essential point is that the total error of a cross section can be determined in Monte Carlo sampling, taking into account all the sources of uncertainty. Of the nine independent initial parameters that are required for a complete mapping of phase space, the masses and eccentricity were assigned fixed values, whereas the four angular variables were randomized. The impact parameter was sampled uniformly in ρ^2 up to a maximum given by

$$\rho_{\rm max}(v) = (C/v + D)a\,, \tag{18.6}$$

with C and D appropriate constants, reflecting the effect of gravitational focusing. Finally, the intruder velocity was scaled according to (18.2) and chosen in a specified range.

Additional experiments were later made for high intruder velocity [Hut, 1983]. Further analysis of the numerical results combined with analytical approximations yielded an accurate description over a complete range of

§ A similar study [Agekian & Anosova, 1971] for two X-values gave a factor of 9 smaller probability at $X = 5/\sqrt{3}$, although the initialization procedures were identical.

parameters [Hut, 1983, 1993]. Thus differential cross sections were derived for binary–single star scattering at high intruder velocity, together with total cross sections for ionization and exchange which were shown to be independent of binary eccentricity. These results for equal masses agree with analytical calculations [Heggie, 1975]. The water-shed effect for equal masses and a thermal binary eccentricity distribution was redetermined, yielding $\alpha_{\rm c} \simeq 0.48$ [Hut, 1993], in fair agreement with earlier values [Fullerton & Hills, 1982] based on zero impact parameter. Comprehensive fitting formulae for exchange reactions of arbitrary mass ratios have also been obtained [Heggie, Hut & McMillan, 1996]. The semi-analytical fitting functions are accurate to 25% for most mass ratios. In order to facilitate exploration of the three-body problem further, an automated scattering package has been developed and the software is freely available.¶

Special three-body scattering experiments are sometimes motivated by astrophysical considerations. One investigation examined configurations where the intruding star is more massive by a factor of 2, which is relevant to globular cluster cores [Sigurdsson & Phinney, 1993]. These simulations again employed the RK4 integrator. Analysis of 1×10^5 experiments showed that for moderately hard binaries, exchange is the dominant process. Moreover, exchange of a heavy field star actually increases the cross section in future encounters since the binary tends to widen while becoming more strongly bound. It was also found that resonances contribute significantly to physical collisions.

To conclude this brief review of some salient three-body scattering experiments, it is evident that this beautiful problem still contains a wealth of fascinating complexity which is waiting to be unravelled.

18.4 Binary–binary interactions

The addition of a fourth body increases the number of variables considerably. Thus with two initial binaries, we have four more angles, as well as a second mass ratio, semi-major axis and eccentricity. Hence it is not surprising that so few studies have been made in four-body scattering. Moreover, this problem is also harder numerically and some kind of regularization method is beneficial.

The first modest outcomes of binary–binary encounters already showed some features of interest [Thüring, 1958]. The impact of two equal circular binaries was integrated by hand, with the number of differential equations reduced by half due to symmetry. The choice of initial conditions led to escape, with the binaries considerably more strongly bound and

¶ The STARLAB software can be found at `http://www.sns.ias.edu/~starlab`.

having reversed orbital spins.[‖] After examination of the c.m. velocity gain, the modern concept of a hyper-elastic encounter was defined. Because of the analogy with binary–binary scattering, we also mention that further electronic calculations of impacting four-body systems by the Adams–Moulton method were performed subsequently [Thüring, 1968]. This work introduced the additional concepts of elasticity, interaction energy, as well as hard and soft configurations, which are still topical.

The early 1980s saw a pioneering investigation which was not repeated for a long time [Mikkola, 1983, 1984a,b,c], as is often the case with a hard mountaineering ascent. Some 3000 experiments were performed in the first study, which was restricted to equal masses and semi-major axes. The impact parameters were randomized to give an equal number of impacts on equal areas up to a limiting value, i.e. $\psi(\rho) = 2\,\rho/\rho_{\rm max}^2$. This sample proved sufficient for obtaining meaningful results of fly-by interactions as a function of the impact energy. Four types of final motions were delineated for purposes of data analysis:

- Two binaries in a hyperbolic relative orbit
- One escaper and a hierarchical triple system
- One binary and two escapers
- Total disruption into four single stars

If none of these categories were confirmed, the configuration was classified as undecided and the integration continued. The majority of outcomes were of the first type, with the third type well represented. This dataset also provided early evidence of hierarchical triple formation (type two), especially at low impact energy, although stability was not checked.

A further 1800 cases of strong interactions with smaller values of the maximum impact parameters were also investigated. Such interactions are of particular interest and may lead to quite complex behaviour. Thus one star may be ejected in a bound orbit and return after the escape of a second ejected star. The problem is then reduced to the three-body case. Alternatively, the softest binary may become more weakly bound and return for a strong interaction with the escape of only one component. In both these processes the remaining triple can be very wide. Analysis of these configurations based on two values of the critical outer pericentre ratio for an approximate criterion [cf. Harrington, 1972] yielded a high proportion of stable systems, with about 20% at modest impact energies. The question of exchange with two binaries surviving was also addressed. Thus at the smallest impact energy, the probability of exchange was about

[‖] This property depends on initial conditions which may also result in exchange.

50%. Finally, the distribution of energies in the case of two escapers was given. More general initial conditions were studied later [Mikkola, 1984b,c] and a start was made on the important extension to unequal masses [Mikkola, 1985b, 1988]. The provisional result $P \propto (m_i m_j)^3$ was obtained for the probability of the remaining binary masses in both three- and four-body systems.

A special integration method was devised in order to deal with the different configurations discussed above [Mikkola, 1983]. At the beginning of this work, a fourth-order polynomial method for direct integration was used together with the two-body regularization method of section 4.5 [Burdet, 1967; Heggie, 1973]. However, persistent configurations consisting of a close binary and two eccentric elliptical orbits proved too expensive. A code combining three different solution methods was therefore developed. Long-lived hierarchies were treated as perturbed Keplerian orbits using the classical variation of parameters method, and more than one relative motion could be studied at the same time. Direct integration was employed for non-hierarchical systems not benefiting from regularization. Finally, the global regularization method [Heggie, 1974] was modified to a more convenient form for practical use [Mikkola, 1985a]. This three-part implementation was quite efficient and contained sophisticated decision-making. As a unifying thread, all the relevant equations of motion were advanced by the powerful Bulirsch–Stoer [1966] method.

For completeness, we record another contemporary binary–binary scattering experiment [Hoffer, 1983]. A time-smoothing technique [Heggie, 1972a] was used, combined with the c.m. approximation for a hard binary and the equations were integrated by the RK4 method. Because of numerical problems, this study was unable to deal with hard binary interactions and the results are therefore of limited validity. However, the importance of physical collisions in such encounters was highlighted. Over the subsequent years, there have been very few four-body scattering experiments. A preliminary study of scattering cross sections [Hut, 1992] for equal masses emphasized the frequent occurrence of hierarchical triple formation. A distinction was made between systems preserving one of the inner binaries and those undergoing exchange. Given the relatively small effort so far in this fundamental problem, the time is ripe for a renewed attack.

A thorough comparison of regularization methods for binary–single star and binary–binary scattering was very informative [Alexander, 1986]. All the current multiple regularization schemes were examined [Aarseth & Zare, 1974; Zare, 1974; Heggie, 1974], together with different choices for the time transformation. Although the initial conditions of the main comparisons were based on symmetrical configurations starting from rest, some results were also presented for more general scattering experiments which gave rise to hierarchical triples.**

A number of careful tests were made for three-body and four-body systems, including time reversal, and the Poincaré control term was also included [Baumgarte & Stiefel, 1974]. From the general form (5.66) of the equations of motion, we have the relation [cf. Alexander, 1986]

$$\frac{d}{d\tau}\left[\frac{1}{2}(H - E_0)^2\right] = -(H - E_0)^2 \frac{1}{g}\frac{dg}{d\tau}. \tag{18.7}$$

Hence stabilization of the integral $H - E_0 = 0$ is achieved when g increases along the solution path. This condition is satisfied for the time transformation $g = 1/L$ if the Lagrangian is *decreasing*; i.e. after pericentre passage. It may therefore be beneficial to reverse the sign of each control term when g is decreasing. This device was already used without regularization [Zare & Szebehely, 1975]. However, care must be taken near critical situations with regularization since the regularity is lost.

After testing several alternative time transformations, as well as the standard forms, the inverse Lagrangian was favoured on grounds of accuracy and efficiency. Somewhat surprisingly, the wheel-spoke regularization [Zare, 1974] appeared to give smaller residual errors than the global method for $N = 4$. However, it should be noted that only one type of initial conditions were examined, in which close triple encounters may have been less important. Based on these results, we would therefore expect the subsequent chain regularization method [Mikkola & Aarseth, 1993] to be competitive for binary–binary scattering. As described earlier, it has already proved itself in N-body simulations with primordial binaries.

A special code for analysing scattering experiments was also developed [Alexander, 1986]. By analogy with (18.2), the critical velocity in the four-body case is defined as

$$V_{\rm c}^2 = \frac{m_1 + m_2 + m_3 + m_4}{(m_1 + m_2)(m_3 + m_4)}\left[\frac{m_1 m_2}{a_1} + \frac{m_3 m_4}{a_2}\right]. \tag{18.8}$$

This quantity can then be used to define the dimensionless hardness parameter, $x = (V_{\rm f}/V_{\rm c})^2$. Comparison of a three-body scattering experiment with hardness $x = 10^4$ showed agreement in the final values of a and e for the different methods. However, in the case $N = 4$ and $x = 10^2$, the identity of the second escaper was not the same; consequently the final binary elements did not agree. There were many temporary ejections and exchanges and the assumption of unperturbed two-body motion also introduced further uncertainty. The code was subsequently used for

** The adopted stability criterion only refers to exchange; cf. (9.14) regarding escape.

binary–binary scattering experiments [Leonard & Linnell, 1992] in connection with the collision hypothesis for blue stragglers.

Further binary–binary scattering experiments were made using astrophysical parameters, including realistic stellar radii, for modelling collisions [Leonard, 1995]. Thus a cluster potential with rms escape velocity $3.5\,\mathrm{km\,s^{-1}}$was assumed for reference. The binaries were generated with masses 2–$20 M_\odot$ and uniformly distributed periods 1–100 d for a total of 2000 experiments. A considerable fraction of O-type stars formed by collisions were ejected with sufficiently high velocities ($> 30\,\mathrm{km\,s^{-1}}$) to be associated with runaway stars. It was also emphasized that it is difficult to distinguish between runaway stars originating by dynamical slingshot ejection and the supernova mass-loss mechanism in binaries.

The globular cluster M4 contains a unique triple system where one of the inner binary components is a pulsar. A model based on exchange was proposed to account for the existence of the old pulsar inside a soft outer binary inferred from observations [Rasio, McMillan & Hut, 1995]. In this scenario, it was assumed that the inner binary ($a_{\rm in} = 0.8$ au) only suffered a small perturbation during a close binary–binary encounter, consistent with the small eccentricity, $e_{\rm in} = 0.025$. Accordingly, a three-body scattering experiment was performed using the *STARLAB* package, with the exchange probability estimated after excluding a minimum impact parameter of $3a_{\rm in}$ obtained from separate calculations. Even so, the maximum probability of triple formation was found to be about 50% in the most favourable case of a pre-encounter binary with semi-major axis 12 au which would typically form a wide outer binary, $a_{\rm out} \simeq 100$ au.

Strong binary–binary interactions in star cluster simulations often lead to physical collisions with the possible formation of exotic objects [cf. Mardling & Aarseth, 2001]. An extensive series of such scattering experiments was undertaken with a view to clarifying this question [Bacon, Sigurdsson & Davies, 1996]. The selection of initial conditions followed earlier three-body studies [Hut & Bahcall, 1983]. A set of experiments was defined by fixed semi-major axes, masses and a range of intruder velocities, the other elements being sampled by Monte Carlo techniques. A total of 10^5 experiments were performed, divided into 25 sets. Both the RK4 integrator and chain regularization were used. However, the secular drift in total energy became unacceptable with the former when the semi-major axes ratio exceeded 2.

The following five usual types of outcome were examined: fly-by, exchange, breakup, stable triple and unresolved experiment. The latter reached a maximum of 5% in the case of equal semi-major axes, with 2% being typical. Surprisingly, the fraction of exchange cases was quite small and reached at most 4%. A somewhat higher proportion of exchange for equal masses, likewise with two surviving binaries, was seen in an earlier experiment for a small sample [Mikkola, 1983]. Stable triples were defined according to the simplified Harrington [1972] criterion,

$$R_{\rm p}^{\rm crit} = 3\,(1 + e_{\rm in})\,a_{\rm in}\,, \tag{18.9}$$

which does not include the masses (cf. (15.13) and (15.15) for improved expressions). Moreover, in view of (9.14), the numerical factor increases by 2.0 for a characteristic outer eccentricity of 0.9 instead of zero. Bearing in mind that many systems exceed the stability ratio by modest amounts, the adopted definition is somewhat optimistic.

A primary motivation for the above work was to estimate the rate of tidal encounters and physical collisions from binary–binary interactions in globular clusters. This requires the determination of the minimum periastron distance, $r_{\rm min}$, which is about three times the radius of a main-sequence star for significant tidal effects [Press & Teukolsky, 1977]. The derived cumulative cross section may be compared with a power-law fit

$$\tilde{\sigma}(\min_{j \neq i}\{r_{ij}\} \leq r_{\rm min}) = \sigma_0 \left(\frac{r_{\rm min}}{a_{12}}\right)^{\nu}\,, \tag{18.10}$$

where a_{12} is taken to be the geometric mean of the semi-major axes. The resulting graphs are conveniently approximated by piece-wise fitting for two values of the exponent, ν. With this approximation, the number of tidal interactions or physical collisions may be estimated from typical cluster core densities and binary properties. In the latter case, these estimates range from a few to about 500 for core densities in the range 10^2–$10^6\,{\rm pc}^{-3}$ and binary fractions $f_{\rm b} \simeq 0.1$. Note that such estimates are highly uncertain since eccentricity-induced processes in hierarchical systems cannot easily be taken into account [Mardling & Aarseth, 2001].

Recently an under-estimate of the smallest two-body distances was noted which affects the lower end of the cumulative cross section [Rasio, private communication, 2001]. Thus, in the calculations, $r_{\rm min}$ was simply taken as the minimum distance, specified at the end of each integration step. It is a property of regularization methods that the time-steps may be large during close encounters, provided the perturbation is small. Below we give a general algorithm for determining the pericentre distance which is also appropriate for three-body and chain regularization [Mikkola, private communication, 1990].

In this general-purpose algorithm, $\mathbf{U}'$ represents the actual derivative used in the equations of motion, together with the corresponding t', and the relevant two-body terms are identified. A well-defined result for the eccentricity is obtained since the division by R at step 4 cancels; hence the pericentre distance $Q_{\rm p}$ is regular. The same algorithm also applies

Algorithm 18.1. *Pericentre determination.*

1	Form the square of the KS cross product, $C = \|\mathbf{U} \times \mathbf{U}'\|^2$
2	Set physical transformation factor, $D = (2R/t')^2/(m_1 + m_2)$
3	Construct the semi-latus rectum by $p = CD$
4	Obtain the inverse semi-major axis, $a^{-1} = (2 - D\|\mathbf{U}'\|^2)/R$
5	Derive the eccentricity from $e = (1 - p/a)^{1/2}$
6	Evaluate non-singular pericentre distance, $Q_{\rm p} = p/(1 + e)$

to standard KS regularization. Note that this formulation is based on osculating solutions of the two-body problem and should therefore only be used for small relative perturbations.

Finally, we mention another approach where multiple regularization methods are used to investigate scattering outcomes in more detail. Thus Monte Carlo simulations of star clusters containing primordial binaries can be considered as a series of scattering experiments in parallel. First results of an implementation using unperturbed three-body [Aarseth & Zare, 1974] and Heggie [1974] four-body regularization have already been obtained [Giersz & Spurzem, 2003]. Hopefully, the more general method of perturbed chain regularization can also be included in this treatment. Hence the Monte Carlo method allows all the usual processes discussed earlier to be examined in considerable detail.

18.5 Chaos and stability

Given the complexity of the three-body problem, it is perhaps surprising that the case of 1D did not receive much attention in the early days. A series of papers [Mikkola & Hietarinta, 1989, 1990, 1991] describes complicated behaviour of chaotic type. In this section, we first consider a few investigations that emphasize such features and also provide applications of the three-body and chain regularization methods. The 1D formulation has already been elucidated earlier, with the Hamiltonian for equal masses given by (5.14).

Two final states of solutions are possible in 1D problems. Break-up is defined as a binary with energy ratio $E_{\rm b}/E > 1$, together with the third particle outside $100\,a$. The ratio $z = E/E_{\rm b}$ is a convenient parameter for displaying results since it is confined to the interval [0,1]. It turns out that the life-time of the interplay phase divides the interactions into two well-defined regions in both directions of time. In so-called 'zero interplay time', a close triple encounter occurs only once early on and the outcome is a continuous function of the initial conditions. For longer interplay times, the orbits depend sensitively on the initial conditions and hence are chaotic. There is also a small region of infinite interplay time of quasi-periodic motions in the neighbourhood of the well-known Schubart [1956] periodic orbit.

A second similar investigation was made for positive energies. The characteristic types of final motions are either a binary and a single particle or complete ionization. Properties of the orbits are summarized in 2D maps of initial values and the transition from negative to positive energy is illustrated. For completeness, we mention the third paper which is mainly concerned with a survey of the periodic orbit for a grid of masses. The numerical solutions are used extensively to map the regions of stability and compare the results with linear stability analysis. Remarkably, linearly stable orbits transformed to 3D survived over at least several thousand periods when integrated directly by the method of section 5.2.

Recently an analogous study was made of the symmetrical four-body problem in 1D [Sweatman, 2002a]. Application of chain regularization yields a simple Hamiltonian of the type (5.14) with four well-behaved equations of motion. Again the motion can be divided into two main categories, where scattering orbits are characterized by systems starting and finishing as subsystems of binaries and single particles, whereas for bound orbits all the four bodies are confined. Moreover, scattering orbits are subdivided into fast and chaotic types, in which all bodies come together either once or a number of times, respectively. A subregion was chosen to illustrate the chaotic behaviour in greater detail and the elements of a Schubart-type periodic orbit were determined by iteration.

Returning to 2D, the effect of chaos was seen in an investigation which applied perturbed velocities to the initial conditions of the Pythagorean Problem [Aarseth *et al.*, 1994a]. A novel feature in these experiments was to ascertain reproducibility by including time reversal after escape and to measure the final rms errors in position and velocity, σ_x and σ_v. An initial tolerance of 10^{-13} was used with the *TRIPLE* code. Each experiment was repeated with 100 times reduced tolerance if the errors were not acceptable; i.e. $\max\{\sigma_x, \sigma_v\} < 1 \times 10^{-3}$. This procedure, which allows a determination of successful experiments, may be considered a template for accuracy studies, although such stringent criteria are not recommended for scattering experiments.

The final escape process in the vicinity of the standard Pythagorean configuration has also been examined [Aarseth *et al.*, 1994b]. Based on 831 examples, the angle, Ψ, between the velocity vector of the escaper and the line bisecting the binary components favoured $\Psi > 45°$ by a ratio of 2:1. The significance of this conclusion has been criticized [Umehara & Tanikawa, 2000]. The latter work introduced the velocity vector product

$$\Phi_{jk} = v_j v_k \sin\left(|\theta_j| - |\theta_k|\right), \tag{18.11}$$

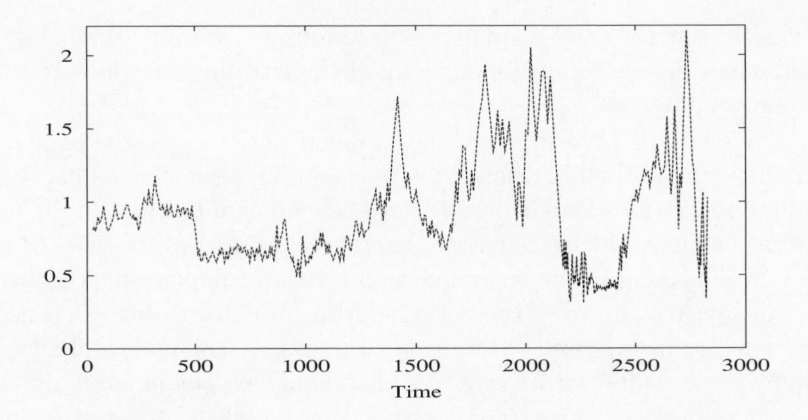

Fig. 18.1. Chaotic behaviour of outer apocentre distance. Time is measured in crossing times of the triple system.

where θ_i is the angle spanned by $\mathbf{r}_{jk}$ and $\mathbf{v}_i$. On the other hand, the corresponding radial velocity is given by

$$U_{jk} = v_k \cos\theta_k - v_j \cos\theta_j \, . \tag{18.12}$$

Although the second condition of positive radial velocity agrees with the region for positive Φ_{ij} and equal masses, the two tests are not equivalent for $v_j \gg v_k$; consequently the former does not always apply.

Actual cluster simulations deal with a wide variety of hierarchical configurations where chaos often plays an important role. It is therefore not surprising that some systems are close to the stability boundary as defined by (9.14). However, the temptation to accept a long-lived subsystem as stable even if this condition is not met must be resisted, albeit at some cost. Here we give an example that illustrates this point, taken from a realistic star cluster simulation ($N_{\rm s} = 800$, $N_{\rm b} = 200$). This episode began with a compact binary–binary interaction in which one of the components acquired escape energy during the first approach. The outer pericentre distance of the new triple was marginally (i.e. 3%) inside the stability value after due allowance for the inclination effect in which $\psi \simeq 130^{\circ}$. This necessitated a lengthy integration by chain regularization until the outer small mass was eventually ejected.

Data from the initial four-body configuration was extracted and used by the stand-alone chain code for further examination with slightly smaller tolerance and the same outcome was reproduced on a comparable time-scale in the absence of external perturbations. Figure 18.1 shows the apocentre distance of the outer body, evaluated near each turning point, over a time interval of about 2800 crossing times with some 500 excursions. After an irregular build-up to larger distances, the reverse process gave rise to considerable shrinkage, followed by another stage of successive energy gains, whereupon ejection took place. Although the time interval is too long to guarantee accurate solutions, the characteristic chaotic behaviour is nevertheless instructive.

The results of scattering experiments may also be used to illustrate the chaotic nature of gravitational interactions instead of obtaining cross sections. As we have seen above for the 1D case, this approach is generally based on producing phase-space maps which reveal chaotic structure. One such study [Boyd & McMillan, 1993] adopted standard selection of initial conditions for three-body scattering and used the *TRIPLE* code. The emphasis was on displaying maps of the entire region over which resonance can occur. Regions where simple ejection, fly-by or exchange take place are separated by so-called 'rivers of resonance scattering'. Enlargements of such a region reveal intricate nested patterns of alternating smooth and irregular bands on decreasing scales. The existence of such subdivided bands is a hallmark of fractal structure. Moreover, in the case of equal masses, the structure within these regions is due to the existence of a constant probability of escape from the interaction region at each close approach.

In order to study long-lived systems, a novel use was made of the total kinetic energy which contains information about the internal dynamics. Thus the case of a hierarchical configuration with non-zero outer eccentricity shows this quantity varying on two time-scales. For equal masses, the two frequencies, ω_1, ω_2, may be recognized by transforming the sum of the two Keplerian energies into a relation of the form

$$(1/2)^{1/3}\omega_{\rm b}^{2/3} + 2\,(1/2)^{1/3}\omega_3^{2/3} = \text{const}\,. \tag{18.13}$$

These frequencies can be determined accurately and the result shows a smooth variation as binding energy is transferred between the inner and outer orbit. However, there are times when the system cannot be decomposed into two separable orbits. The departures from the conservation law can be used to identify epochs when interesting non-hierarchical interactions or transitions are taking place.

The question of practical unpredictability has been addressed by examining the phase-space structure of chaotic three-body scattering [Mikkola, 1994]. The complexity of the problem was reduced by choosing initial values along curves in phase space and analysing the final binary properties. Numerical investigations with the *TRIPLE* code showed that phase space is divided into separate zones, isolated by hyper-surfaces of parabolic disruption. Thus there are regular regions characterized by short interplay times, whereas chaotic behaviour is usually associated with long interplay times and ejections without escape. Since the discontinuous singular surfaces are very dense, a small numerical error can move the trajectory across a surface, making the outcome different and hence unpredictable. This behaviour is connected with an earlier idea that escape times are discontinuous [Szebehely, 1972].

Although the Pythagorean configurations undoubtedly display chaotic behaviour, certain regular patterns may be seen from movies. Thus in the original system with masses 3, 4 and 5, there are nearly 30 independent two-body encounters, and yet the two lightest bodies are never the closest pair until the final triple approach. There does not appear to be any fundamental reason for this curious avoidance, since other similar initial conditions produce mutual encounters between all the bodies, albeit with some tendency for the heaviest mass to control the motion.

As we have seen from many examples, numerical integrations of three mass-points readily produce chaotic orbits. However, stable few-body systems have received remarkably little attention in spite of some interesting theoretical developments. Given a hierarchical configuration, the question of stability poses a big challenge. This is particularly relevant for the treatment of weakly perturbed systems in cluster simulations, as discussed in previous chapters (cf. sections 9.5 and 15.7).

The observational search for stable hierarchies has yielded a few examples where the elements may change on relatively short time-scales. We highlight the triple system HD 109648 which reveals evidence for precession of the nodes and apsidal advance, as well as eccentricity variation over an 8-yr base line [Jha *et al.*, 2000]. Given a period ratio of 22:1 and outer period 120 d, this triple should be stable, with a modulation time-scale $T_{\rm Kozai} \simeq 15\,{\rm yr}$ for small outer eccentricity (cf. (15.17)). Since the inner period is only 5 d, the orbit is expected to have been circularized, yet $e_{\rm in} \simeq 0.011$. This small but non-zero value appears to change in a cyclical way consistent with either a small inclination and/or strong tidal force. Direct integrations using the *TRIPLE* code have also been made, based on the derived elements. In particular, a small relative inclination of $\psi \geq 5°$ obtained from geometrical and observational constraints was used. The results of the inner eccentricity modulations over a decade were found to be consistent with the observed variation.

In general, binaries with periods below some cutoff value that depends on stellar type are expected to be circularized [Mermilliod & Mayor, 1992; Verbunt & Phinney, 1995]. It is therefore interesting to note some evidence of significant deviation from circular motion that supports the data presented above. Thus two cluster binaries with periods of 2.4 and 4.4 d in the Hyades and M67, respectively, have eccentricities 0.057 and 0.027 [Mazeh, 1990]; moreover, a 4.3 d halo binary has $e = 0.031$. Although the circularization time is much less than the age in each case, there is additional evidence for the presence of a third companion. Once again,

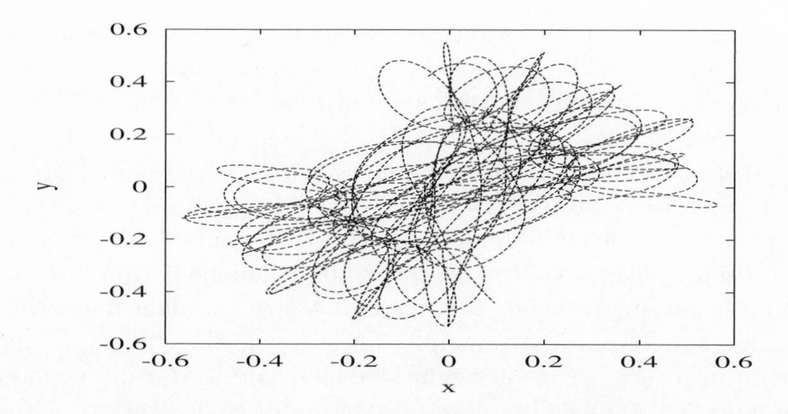

Fig. 18.2. Quasi-periodic three-body system

the relatively small eccentricities may provide a clue to the strength of the tidal dissipation which tends to have a stabilizing effect. The question of long-term stability of hierarchical stellar systems may therefore also depend on astrophysical processes such as mass transfer and tidal friction [Eggleton, Kiseleva & Hut, 1998].

On rare occasions even general three-body systems may exhibit regular motions. Remarkably, a sample of 5000 experiments in 2D yielded two kinds of regular orbits which were numerically stable over long times [Anosova *et al.*, 1984]. The first was of a chain type where a central body approached the other two successively on opposite sides without the latter getting close, while in the second case all bodies made close pair-wise encounters. Figure 18.2 illustrates the orbits in the former example until the rotating pattern has precessed by about 45°. In the absence of close triple encounters, the solution is quite accurate and displays the same pattern for much longer times.

The serendipitous discovery above shows that phase space contains hidden secrets. Another example was recently revealed by the existence proof for a periodic 2D orbit where three equal masses describe a figure eight solution [Chenciner & Montgomery, 2000]. Accurate numerical methods for finding such orbits based on minimization formulations have also been presented [Simó, 2001]. A variety of so-called 'choreographic solutions' have in fact been obtained for a range of particle numbers, including some cases without any symmetry.[††]

Finally, in this connection, we mention a practical algorithm for recognizing figure eight orbits with zero angular momentum [Heggie, 2000]. This scheme is based on identifying possible collinear configurations occurring in the sequence 123123123..., or its reverse, where each digit

[††] For entertaining animations see `http://www.maia.ub.es/dsg`.

denotes the middle body. Although the figure eight orbit appears special, it can be reproduced by scattering experiments in which a fourth incoming body produces two binaries, whereupon time reversal yields the desired result. Nevertheless, this new outcome of binary–binary scattering only has a low probability of occurring and hence the corresponding stability test in chain regularization is not needed.

The grand challenge of understanding the three-body problem has attracted much attention by celebrated mathematicians over the last few centuries, yet many mysteries remain. A new formalism has recently been developed from first principles for studying the energy and angular momentum exchange in hierarchical triples and scattering experiments [Mardling, 2001, 2003a,b], allowing a study of configurations with arbitrary initial conditions. The different scattering processes include fly-by, ionization, resonant capture and exchange. Moreover, stability boundaries for hierarchical systems can be determined for arbitrary masses and orbital elements and, in addition, the evolution of unstable triples may be obtained. A description of unstable behaviour is particularly useful for long-lived systems with large outer eccentricity that occur in star cluster simulations.

By analogy with the modes of oscillation of a *rotating* star that may be excited due to a close binary companion [Lai, 1997], the formalism introduces the concept of the *modes of oscillation of a binary.* The inner orbit is regarded as a pair of circular rings of material that spin in the same plane as the actual (inner) orbit. In a frame rotating with the spin frequency (i.e. the orbital frequency), the rings are stationary when the orbit is circular, but oscillate radially and azimuthally for an eccentric orbit. A related device was introduced by Gauss for the case of two planets orbiting the Sun. In this picture, the mass of the inner planet is smeared out along its orbit to form a single ring, such that the density is highest at apastron and lowest at periastron. The outer planet then responds to this *static* (non-oscillating) distribution of matter. This formulation predicts successfully secular evolution features such as the rate of apsidal motion of the outer orbit [Murray & Dermott, 1999].

The new formulation differs in two distinct ways from that of Gauss. First, a binary of arbitrary mass ratio is regarded as a *pair* of rings, and the third body responds to a *radially varying distribution of matter*, much like in the binary–tides problem where the second star responds to the moments of the mass distribution of the first. The other difference is that the rings interact *dynamically* with the outer orbit so that its *normal modes of oscillation*, given in terms of the mean anomaly, are excited by the outer body. Hence the rings oscillate as they exchange energy and angular momentum with the outer orbit, at all times remaining circular.

While Gauss's formulation involves averaging over the inner orbit to predict *secular* evolution, the new formalism involves averaging over the inner orbit to predict *dynamical* evolution. In effect, the averaging procedure isolates the dynamically important frequencies.

Dynamically active triple processes generally involve significant exchange of energy and angular momentum between the inner and outer orbits in contrast to, for example, stable (non-resonant) hierarchical triples in which the semi-major axis remains secularly constant, although the eccentricity may evolve secularly through angular momentum exchange. It is this aspect that the new formalism captures, an aspect particularly important for studying stability. Previous studies of stability have generally focused on the concept of *Hill stability*. A system is Hill stable if no close approaches of any pair of bodies is possible [Szebehely & Zare, 1977; Gladman, 1993]. In fact this is no guarantee of stability; it is possible for the outer eccentricity effectively to random-walk its way past unity without there ever being a close encounter. This introduces the more general concept of *Lagrange stability*: a system is Lagrange stable if in addition it remains bound for all time. This is very much associated with the question of whether or not significant energy can be transferred between the orbits, and this in turn is associated with the existence of internal resonances.

The interactions between non-linear oscillators and their chaotic behaviour have been studied extensively as mathematical dynamical systems [Chirikov, 1979 and references therein]. A particularly useful concept for predicting the onset of chaotic behavior is that of *resonance overlap* [Chirikov, 1979]. In the context of the three-body problem, this was first applied by Wisdom [1980] to the *restricted* problem, and it is this concept that is employed in the new formalism to study stability in the *general* three-body problem. Here, the dominant mode excited in the inner binary by the outer body is determined and it is then treated as an oscillator interacting with a Kepler orbit (the outer orbit). If the outer orbit is eccentric, the frequencies associated with it are the orbital frequency and all integer multiples of this. Hence there is the possibility of resonance between the inner orbit and any one of these frequencies, and if any of these resonances 'overlap', unstable behaviour ensues.

The energy pumped into or out of the 'oscillator' (the dominant mode) by the outer orbit represents the dynamical activity, both in unstable bound triples and unbound scattering events. Including this energy in the description allows one to study the general problem, and it is this aspect that is absent from other descriptions of stability and scattering.

Given such an analytical description of stability and scattering, it becomes possible to devise convenient fitting formulae for use in N-body simulations. These procedures are currently being formulated.

Questions concerning the stability of the Solar System have dominated the field of celestial mechanics since the days of Newton and Laplace. We consider briefly some developments involving numerical studies that are connected with chaotic motion. For a comprehensive review, see Lecar *et al.* [2001]. In the early days of direct N-body simulations, the asteroids presented a popular subject for applications [cf. Lecar & Franklin, 1973] and the results focused the attention on chaos. Although general planetary integrations are still unable to provide *observationally* correct solutions for time intervals of even 100 Myr because of the short Lyapunov times, much can be learnt about the long-term behaviour.

In view of the importance of the symplectic mapping method [Wisdom & Holman, 1991] for numerical studies, it is reassuring to note general agreement with conventional integrations for the outer planets over nearly 1 Gyr. The latter results were obtained using the special-purpose Digital Orrery computer [Sussman & Wisdom, 1988] which pointed the way to later hardware developments. In a new approach, 50 000 terms of the averaged secular equations were integrated by a 12th-order Adams method over 15 Gyr after comparison with direct solutions [Laskar, 1994]. The motion of the outer planets is always very regular, whereas Mercury is more susceptible to developing chaotic orbits. A number of successive solutions were adopted for the Earth, which maximized the eccentricity of Mercury and eventually large values were reached, indicating possible escape. Subsequently, it was found that the spacing of the inner planets show evidence of chaotic wanderings [Laskar, 1997]. However, these wanderings are limited by the so-called 'angular momentum deficit' (AMD) which contains the contributions from non-circular and non-planar motion. A simplified model of planetary accretion also made use of the AMD conservation to constrain the evolution following collisions [Laskar, 2000]. The final distribution of semi-major axes obeyed a square root relation reminiscent of the Titus–Bode rule which was also reproduced for the υ Andromedae system. Meanwhile, an application of the Wisdom–Holman method to asteroid orbits, with simultaneous solutions of the variational equations, yielded longer Lyapunov times in the main asteroid belt which may account for the apparent longevity [Mikkola & Innanen, 1995].

A new long-term study of planetary orbits [Ito & Tanikawa, 2002] can be seen as a culmination of the modern approach which seeks to answer the ultimate challenge in our field.[‡‡] Based on dedicated computations lasting several years, all the nine planets were integrated over time-spans of

[‡‡] Although direct integrations have so far only covered about 100 Myr, this objective might come closer to realization by more efficient methods. One promising alternative under consideration is the combination of Picard iteration with Chebyshev polynomial approximation, implemented on a vector computer, which may yield a large speed-up [Fukushima, 1997]. This method is well suited for dealing with regular motions, enabling fast convergence due to small perturbations, and further efficiency gain may also be possible by parallelization [Ito, personal communication, 2003].

$\pm 4\times 10^9$ yr using the Wisdom–Holman [1991] method [see also Mikkola & Saha, 2003]. The main outcome of this heroic effort is that the Solar System appears to be quite stable on this time-scale and no close encounters were recorded. However, the eccentricity of Mercury tends to show large amplitude variations which may indicate the onset of chaotic motion on a somewhat longer time-scale. Since the orbital motion of the outer planets is quite regular and the total AMD is conserved, the AMD of the inner four planets is expected to be approximately constant. The z-component of the normalized AMD is defined by

$$C_z = \sum_{j=1}^{4} \frac{m_j M_\odot}{m_j + M_\odot} [G(M_\odot + m_j)a_j]^{1/2} [1 - (1 - e_j^2)^{1/2} \cos \psi_j] \,. \quad (18.14)$$

The deviations from zero are relatively small, with Mercury again showing the largest effect. The authors concluded that the main reason for the long-term stability is that the terrestrial planets have larger separations measured in terms of their mutual Hill radii as compensation for the shorter orbital periods. Hence the stronger perturbations by the outer planets are less effective in inducing chaotic features in the inner Solar System. These results also indicate that much can be learnt by studying the behaviour of the Jovian planets in isolation.

So far, this book has been exclusively concerned with topics restricted to dynamics, with emphasis on the relevant integration methods. However, it is interesting to note that such methods have found applications in other fields. One problem studied is the semi-classical quantization of the quadratic Zeeman effect in the hydrogen atom [Krantzman, Milligan & Farrelly, 1992]. KS regularization was used effectively to avoid singularities associated with the classical separatrix. Moreover, this problem is an important example of a non-integrable Hamiltonian that is chaotic in the classical limit. Another investigation of classical electron motion of two-electron atoms [Richter, Tanner & Wintgen, 1992] employed the three-body regularization of section 5.2 to calculate stability properties of periodic orbits with two degrees of freedom. These systems display an amazingly rich structure and a variety of different types of motion. In particular, the behaviour for zero angular momentum where the orbits degenerate to collinear motion exhibit Poincaré maps analogous to the properties of the 1D systems discussed above [cf. Mikkola & Hietarinta, 1989]. Hence, in conclusion, the general theme of chaos and stability connects Newtonian dynamics with the atomic world.

Appendix A
Global regularization algorithms

A collection of the relevant formulae for Heggie's global regularization method is given, based on the reformulation of Mikkola [1985a]. In the second section, we summarize procedures for including external perturbations for the case $N = 3$.

A.1 Transformations and equations of motion

We begin with the algorithm for constructing the initial values, then derive the equations of motion and finally write expressions for obtaining the physical variables. In the following we use the Lagrangian time transformation, with $g = 1/(T + U)$ and assume the centre-of-mass frame.

Let the coordinates and velocities be given by $\mathbf{r}_i, \dot{\mathbf{r}}_i$. Define the index $k = (i-1)N - i(i+1)/2 + j$ for $1 \leq i < j \leq N$ by adding one to k when incrementing i and j. At the same time form the quantities

$$\begin{aligned}
\mathbf{q}_k &= \mathbf{r}_i - \mathbf{r}_j \,, \\
\mathbf{p}_k &= (m_i\dot{\mathbf{r}}_i - m_j\dot{\mathbf{r}}_j)/N \,, \\
M_k &= m_i m_j \,, \\
a_{ik} &= 1 \,, \quad a_{jk} = -1 \,,
\end{aligned} \tag{A.1}$$

and take the other components of a_{ij} as zero. Hence there are a total of $N(N-1)/2$ terms $\mathbf{q}_k$, $\mathbf{p}_k$. We define $\mathbf{q}_k = (X, Y, Z)^{\mathrm{T}}$ and write the standard KS transformation for $X \geq 0$ as

$$\begin{aligned}
\mathbf{Q}_1 &= [\tfrac{1}{2}(q_k + X)]^{1/2} \,, \\
\mathbf{Q}_2 &= \tfrac{1}{2}Y/Q_1 \,, \\
\mathbf{Q}_3 &= \tfrac{1}{2}Z/Q_1 \,, \\
\mathbf{Q}_4 &= 0 \,,
\end{aligned} \tag{A.2}$$

otherwise we choose

$$\begin{aligned}
\mathbf{Q}_2 &= [\tfrac{1}{2}(q_k - X)]^{1/2}\,, \\
\mathbf{Q}_1 &= \tfrac{1}{2}Y/Q_2\,, \\
\mathbf{Q}_3 &= 0\,, \\
\mathbf{Q}_4 &= \tfrac{1}{2}Z/Q_2\,.
\end{aligned} \tag{A.3}$$

The corresponding regularized momenta are then related to the relative physical momenta, $\mathbf{p}_k$, by

$$\mathbf{P}_k = 2\mathcal{L}_k^{\mathrm{T}}\mathbf{p}_k\,, \tag{A.4}$$

where the superscript T denotes the transpose. At this stage we also calculate $K = N(N-1)/2$ elements of the matrix T_{uv} from

$$T_{uv} = \tfrac{1}{2}\sum_{e=1}^{N} a_{eu}a_{ev}/m_e\,. \qquad (u = 1,\ldots,K) \qquad (v = 1,\ldots,K) \tag{A.5}$$

Given the KS variables $\mathbf{Q}_k$, $\mathbf{P}_k$, the respective derivatives are evaluated by the following sequential steps

$$\begin{aligned}
\mathbf{p}_k &= \tfrac{1}{2}\mathcal{L}_k\mathbf{P}_k/Q_k^2\,, \\
\mathbf{A}_k &= \sum_{v=1}^{K} T_{kv}\mathbf{p}_v\,, \\
D_k &= \mathbf{A}_k\cdot\mathbf{p}_k\,, \\
T &= \sum_{k=1}^{K} D_k\,, \\
U &= \sum_{k=1}^{K} M_k/Q_k^2\,, \\
T_{\mathbf{P}_k} &= \mathcal{L}_k^{\mathrm{T}}\mathbf{A}_k/Q_k^2\,, \\
\mathbf{A}_k^* &= (A_1, A_2, A_3, -A_4)_k^{\mathrm{T}}\,, \\
T_{\mathbf{Q}_k} &= [\mathcal{L}(\mathbf{P}_k)^{\mathrm{T}}\mathbf{A}_k^* - 4D_k\mathbf{Q}_k]/Q_k^2\,, \\
U_{\mathbf{Q}_k} &= -2M_k\mathbf{Q}_k/Q_k^4\,, \\
L &= T + U\,, \\
H &= T - U\,, \\
G &= (H - E_0)/L\,, \\
G_{\mathrm{T}} &= (1 - G)/L\,, \\
G_{\mathrm{U}} &= -(1 + G)/L\,.
\end{aligned} \tag{A.6}$$

Here the equation for $T_{\mathbf{Q}_k}$ contains the transpose Levi-Civita matrix with argument $\mathbf{P}_k$. Defining partial derivatives with subscripts, we write the equations of motion

$$\begin{aligned} \mathbf{Q}_k' &= G_{\mathrm{T}} T_{\mathbf{P}_k} , \\ \mathbf{P}_k' &= -G_{\mathrm{T}} T_{\mathbf{Q}_k} - G_{\mathrm{U}} U_{\mathbf{Q}_k} , \\ t' &= 1/L . \end{aligned} \tag{A.7}$$

With $k = k(i, j)$ as before, the transformations to physical variables take the form

$$\begin{aligned} \mathbf{q}_{ij} &= \mathbf{q}_k = \mathcal{L}_k \mathbf{Q}_k , \\ \mathbf{p}_{ij} &= \mathbf{p}_k = \tfrac{1}{2} \mathcal{L}_k \mathbf{P}_k / Q_k^2 , \\ \mathbf{r}_i &= \frac{1}{M} \sum_{j=i+1}^{N} m_j \mathbf{q}_{ij} - \frac{1}{M} \sum_{j=1}^{i-1} m_j \mathbf{q}_{ji} , \\ \dot{\mathbf{r}}_i &= \frac{1}{m_i} \sum_{j=i+1}^{N} \mathbf{p}_{ij} - \frac{1}{m_i} \sum_{j=1}^{i-1} \mathbf{p}_{ji} , \end{aligned} \tag{A.8}$$

with M the total mass. In order to obtain accurate solutions, a high-order integrator such as the Bulirsch–Stoer method [1966] is recommended. As is usual in canonical theory, the differential equations to be solved are of first order.

A.2 External perturbations

Some of the steps for including external effects in the three-body problem are outlined below. Analogous derivations for the general case $N > 3$ are given elsewhere [cf. Heggie, 1988]. Only the potential energy interaction term requires differentiation if we also introduce an equation for the energy itself. The former can be expressed in the form

$$\mathcal{R} = -\sum_{i=1}^{3} \sum_{j=4}^{N} \frac{m_i m_j}{|\mathbf{r}_i - \mathbf{r}_j|} . \tag{A.9}$$

By analogy with (5.30), we now obtain the desired perturbation term by partial differentiation

$$\frac{\partial \mathcal{R}}{\partial \mathbf{Q}_k} = \sum_{i=1}^{3} \frac{\partial \mathcal{R}}{\partial \mathbf{r}_i} \frac{\partial \mathbf{r}_i}{\partial \mathbf{q}_k} \frac{\partial \mathbf{q}_k}{\partial \mathbf{Q}_k} . \tag{A.10}$$

Explicit differentiation yields the perturbing force

$$\partial \mathcal{R} / \partial \mathbf{r}_i = -m_i \mathbf{F}_i , \tag{A.11}$$

where $\mathbf{F}_i$ has the usual meaning. Alternatively, we can obtain the perturbation directly in the form $\partial\mathcal{R}/\partial\mathbf{q}_k$. Its evaluation is facilitated by employing the cyclic notation defined in section 5.5,

$$\mathbf{r}_1 = (m_2\mathbf{q}_3 - m_3\mathbf{q}_2)/M + \mathbf{r}_0\,, \qquad (*) \tag{A.12}$$

where $\mathbf{r}_0$ denotes the local centre of mass and the asterisk indicates two more equations by cyclic interchange of indices. Consequently, the expressions for $\partial\mathbf{r}_i/\partial\mathbf{q}_k$ reduce to mass ratios. Lastly, we have

$$\partial\mathbf{q}_k/\partial\mathbf{Q}_k = 2\mathcal{L}_k^{\mathrm{T}}\,, \tag{A.13}$$

and by combining terms we recover the full derivative expression (A.10). By analogy with the perturbed equation of motion given by (5.29), this is substituted into the differentiated form of (5.56).

The last term of the Hamiltonian (5.56) represents the total energy which is no longer constant. Accordingly, we can obtain its changing value by direct integration rather than by explicit evaluation. Employing the known equations of motion for $\dot{\mathbf{q}}_i$ and $\dot{\mathbf{p}}_i$, we derive the simplified result [Heggie, 1974]

$$\dot{E} = -\sum_{i=1}^{3} \dot{\mathbf{q}}_i^{\mathrm{T}}\partial\mathcal{R}/\partial\mathbf{q}_i\,. \tag{A.14}$$

Finally, we convert to regularized derivatives by introducing primes and employ the standard KS relation $\mathbf{q}_k = \mathcal{L}_k\mathbf{Q}_k$ which, together with (4.21), simplifies to

$$E' = -2\sum_{i=1}^{3} \mathcal{L}_k\mathbf{Q}_k'\partial\mathcal{R}/\partial\mathbf{q}_i\,. \tag{A.15}$$

In conclusion, it may be remarked that the additional effort of using the global three-body method, compared with the Aarseth–Zare regularization, might be relatively modest when perturbations are included, since the similar terms $\partial\mathcal{R}/\partial\mathbf{q}_i$ could well dominate the cost. It appears that the perturbed case has not been attempted in any N-body simulation so far. Hence an implementation would provide a good programming exercise. However, an application to general N-body systems would still require considerable decision-making. Its introduction would therefore depend on whether several simultaneous chain regularizations are developed.

Appendix B
Chain algorithms

Below we provide a practical collection of the necessary expressions for chain regularization, based closely on the improved notation [Mikkola & Aarseth, 1993].

B.1 Transformations and switching

In the following we refer to the actual indices of the physical variables instead of those assigned while relabelling the particles. This notation facilitates the construction of a computer code. We first describe transformation to regularized variables and assume that the initial chain structure has been determined, as described in section 5.4.

Let the location of a particle with mass m_j be denoted by an index I_j, such that the actual index (or name) is also I_j. The c.m. and chain variables for N members are then defined by

$$
\begin{aligned}
\mathbf{r}_0 &= \sum_{j=1}^{N} m_j \mathbf{r}_j / M \,, \\
\dot{\mathbf{r}}_0 &= \sum_{j=1}^{N} m_j \dot{\mathbf{r}}_j / M \,, \\
\mathbf{p}_j &= m_{I_j} (\dot{\mathbf{r}}_{I_j} - \dot{\mathbf{r}}_0) \,, \qquad (j = 1, \ldots, N) \\
\mathbf{W}_1 &= -\mathbf{p}_1 \,, \\
\mathbf{W}_k &= \mathbf{W}_{k-1} - \mathbf{p}_k \,, \qquad (k = 2, \ldots, N-2) \\
\mathbf{W}_{N-1} &= -\mathbf{p}_N \,, \\
\mathbf{R}_k &= \mathbf{r}_{I_{k+1}} - \mathbf{r}_{I_k} \,. \qquad (k = 1, \ldots, N-1)
\end{aligned} \tag{B.1}
$$

Note the explicit absence of the term $\mathbf{p}_{N-1}$ from the chain momenta. This is connected with the c.m. condition since $\mathbf{W}_{N-1}$ consists of the negative sum of the first $N - 1$ physical momenta. Hence for $N = 3$ we have $\mathbf{W}_2 = -(\mathbf{p}_1 + \mathbf{p}_2)$, whereas the two momenta are treated symmetrically in the equivalent formulation of section 5.2.

We define $\mathbf{R}_k = (X, Y, Z)^{\mathrm{T}}$, whereupon the standard KS transformation for $X \geq 0$ can be written as

$$\begin{aligned} \mathbf{Q}_1 &= [\tfrac{1}{2}(R + X)]^{1/2}\,, \\ \mathbf{Q}_2 &= \tfrac{1}{2}Y/Q_1\,, \\ \mathbf{Q}_3 &= \tfrac{1}{2}Z/Q_1\,, \\ \mathbf{Q}_4 &= 0\,, \end{aligned} \tag{B.2}$$

otherwise we choose well-determined values by

$$\begin{aligned} \mathbf{Q}_2 &= [\tfrac{1}{2}(R - X)]^{1/2}\,, \\ \mathbf{Q}_1 &= \tfrac{1}{2}Y/Q_2\,, \\ \mathbf{Q}_3 &= 0\,, \\ \mathbf{Q}_4 &= \tfrac{1}{2}Z/Q_2\,. \end{aligned} \tag{B.3}$$

The corresponding momenta are obtained by

$$\mathbf{P}_k = 2\mathcal{L}_k^{\mathrm{T}}\mathbf{W}_k\,. \qquad (k = 1, \ldots, N-1) \tag{B.4}$$

The transformation from $\mathbf{Q}, \mathbf{P}$ to $\mathbf{r}, \dot{\mathbf{r}}$ begins by the inverse KS mapping of chain vectors and associated momenta,

$$\begin{aligned} \mathbf{R}_k &= \mathcal{L}_k\mathbf{Q}_k\,, \\ \mathbf{W}_k &= \tfrac{1}{2}\mathcal{L}_k\mathbf{P}_k/Q_k^2\,, && (k = 1, \ldots, N-1) \\ \mathbf{p}_1 &= -\mathbf{W}_1\,, \\ \mathbf{p}_k &= \mathbf{W}_{k-1} - \mathbf{W}_k\,, && (k = 2, \ldots, N-1) \\ \mathbf{p}_N &= \mathbf{W}_{N-1}\,, \\ \dot{\mathbf{r}}_{I_j} &= \mathbf{p}_j/m_j + \dot{\mathbf{r}}_0\,. && (j = 1, \ldots, N) \end{aligned} \tag{B.5}$$

To obtain the coordinates $\mathbf{r}_j$ we first form coordinates $\mathbf{q}_j$ in a system where $\mathbf{q}_1 = 0$, which are then converted to the physical coordinates. This gives rise to the expressions

$$\begin{aligned} \mathbf{q}_1 &= 0\,, \\ \mathbf{q}_{j+1} &= \mathbf{q}_j + \mathbf{R}_j\,, && (j = 1, \ldots, N-1) \\ \mathbf{q}_0 &= \sum_{j=1}^{N} m_j\mathbf{q}_j/M\,, \\ \mathbf{r}_{I_j} &= \mathbf{q}_j - \mathbf{q}_0 + \mathbf{r}_0\,. && (j = 1, \ldots, N) \end{aligned} \tag{B.6}$$

Switching becomes desirable when one of the non-chained distances would connect a dominant particle interaction. For each value of the index $\mu = 1, \ldots, N-1$, we find the indices k_0 and k_1 such that $I^{\rm old}_{k_0} = I^{\rm new}_{\mu}$ and $I^{\rm old}_{k_1} = I^{\rm new}_{\mu+1}$. This enables us to write

$$\mathbf{R}^{\rm new}_{\mu} = \sum_{\nu=1}^{N-1} B_{\mu\nu} \mathbf{R}^{\rm old}_{\nu}\,, \tag{B.7}$$

where $B_{\mu\nu} = 1$ if($k_1 > \nu$ and $k_0 \le \nu$) and $B_{\mu\nu} = -1$ if($k_1 \le \nu$ and $k_0 > \nu$), otherwise $B_{\mu\nu} = 0$.

To transform the chain momenta, we first obtain the physical momenta from the old chain by

$$\begin{aligned}
\mathbf{p}_{I^{\rm old}_1} &= -\mathbf{W}_1\,, \\
\mathbf{p}_{I^{\rm old}_k} &= \mathbf{W}_{k-1} - \mathbf{W}_k\,, \qquad (k = 2, \ldots, N-1) \\
\mathbf{p}_{I^{\rm old}_N} &= \mathbf{W}_{N-1}\,,
\end{aligned} \tag{B.8}$$

and evaluate the new chain momenta with the same notation by

$$\begin{aligned}
\mathbf{W}_1 &= -\mathbf{p}_{I^{\rm new}_1}\,, \\
\mathbf{W}_k &= \mathbf{W}_{k-1} - \mathbf{p}_{I^{\rm new}_k}\,, \qquad (k = 2, \ldots, N-2) \\
\mathbf{W}_{N-1} &= \mathbf{p}_{I^{\rm new}_N}\,.
\end{aligned} \tag{B.9}$$

Now the procedure follows the established path of performing the KS transformations given above, whereupon the integration is continued.

B.2 Evaluation of derivatives

Let us define the matrix T_{ij} and the mass products M_k, M_{ij} by

$$\begin{aligned}
T_{kk} &= \tfrac{1}{2}(1/m_k + 1/m_{k+1})\,, \\
T_{k\,k+1} &= -1/m_k\,, \\
M_k &= m_k m_{k+1}\,, \qquad (k = 1, \ldots, N-1) \\
M_{ij} &= m_i m_j\,. \qquad (i < j)
\end{aligned} \tag{B.10}$$

Further auxiliary variables are introduced by

$$\begin{aligned}
\mathbf{W}_k &= \tfrac{1}{2}\mathcal{L}_k \mathbf{P}_k / Q_k^2\,, \\
\mathbf{A}_k &= \sum \tfrac{1}{2}(T_{ki} + T_{ik})\mathbf{W}_i\,, \qquad (|i-k| \le 1) \quad (1 \le i \le N-1) \\
D_k &= \mathbf{A}_k \cdot \mathbf{W}_k\,, \\
T &= \sum D_k\,, \qquad (1 \le k \le N-1)
\end{aligned}$$

$$
\begin{aligned}
\mathbf{R}_{ij} &= \mathbf{q}_j - \mathbf{q}_i = \sum \mathbf{R}_{k'} \,, \qquad & (i \le k' \le j-1) \\
U_{nc} &= \sum M_{ij}/R_{ij} \,, \qquad & (1 \le i \le j-2) \\
U &= \sum_{k=1}^{N-1} M_k/R_k + U_{nc} \,, \\
g &= 1/(T+U) \,.
\end{aligned} \tag{B.11}
$$

When calculating the vectors $\mathbf{R}_{ij}$ above, the difference form is generally more efficient, but the summation form should be used if there are only two terms in the sum; this avoids some round-off problems. We express partial derivatives with subscripts and obtain

$$
\begin{aligned}
T_{\mathbf{P}_k} &= \mathcal{L}_k^{\mathrm{T}} \mathbf{A}_k / Q_k^2 \,, \\
\mathbf{A}_k^* &= (A_1, A_2, A_3, -A_4)_k^{\mathrm{T}} \,, \\
T_{\mathbf{Q}_k} &= [\mathcal{L}(\mathbf{P}_k)^{\mathrm{T}} \mathbf{A}_k^* - 4D_k \mathbf{Q}_k]/Q_k^2 \,, \\
U_{nc\mathbf{R}_k} &= -\sum M_{ij} \mathbf{R}_{ij}/R_{ij}^3 \,, \qquad & (i \le k \le j-1) \\
U_{nc\mathbf{Q}_k} &= 2\mathcal{L}_k^{\mathrm{T}} U_{nc\mathbf{R}_k} \,, \\
U_{\mathbf{Q}_k} &= -2M_k \mathbf{Q}_k/Q_k^4 + U_{nc\mathbf{Q}_k} \,.
\end{aligned} \tag{B.12}
$$

Here the third equation (B.12) contains the transpose Levi-Civita matrix with argument $\mathbf{P}_k$.

If perturbations are present, the physical coordinates (and velocities if needed) are evaluated according to the transformation formulae given above, whereupon the external acceleration, $\mathbf{F}_j$, acting on each particle can be calculated. Since these quantities appear as tidal terms, we first define the c.m. acceleration by

$$
\mathbf{F}_0 = \sum_{j=1}^{N} m_j \mathbf{F}_j / M \,. \tag{B.13}
$$

Given the perturbative contributions

$$
\begin{aligned}
\delta \dot{\mathbf{p}}_j &= m_j(\mathbf{F}_j - \mathbf{F}_0) \,, \qquad & (j = 1, \ldots, N) \\
\delta \dot{\mathbf{W}}_1 &= -\delta \dot{\mathbf{p}}_1 \,, \\
\delta \dot{\mathbf{W}}_k &= \delta \dot{\mathbf{W}}_{k-1} - \delta \dot{\mathbf{p}}_k \,, \qquad & (k = 2, \ldots, N-2) \\
\delta \dot{\mathbf{W}}_{N-1} &= \delta \mathbf{p}_N \,, \\
\delta \mathbf{P}_k' &= 2g \mathcal{L}_k^{\mathrm{T}} \delta \dot{\mathbf{W}}_k \,,
\end{aligned} \tag{B.14}
$$

the equations of motion for the KS variables are obtained by

$$
H = T - U \,,
$$

$$
\begin{aligned}
\Gamma &= g(H - E_0)\,, \\
\Gamma_{\mathrm{T}} &= (1 - \Gamma) g\,, \\
\Gamma_{\mathrm{U}} &= -(1 + \Gamma) g\,, \\
\Gamma_{\mathbf{P}_k} &= \Gamma_{\mathrm{T}} T_{\mathbf{P}_k}\,, \\
\Gamma_{\mathbf{Q}_k} &= \Gamma_{\mathrm{T}} T_{\mathbf{Q}_k} + \Gamma_{\mathrm{U}} U_{\mathbf{Q}_k}\,, \\
\mathbf{P}_k' &= -\Gamma_{\mathbf{Q}_k} + \delta \mathbf{P}_k'\,, \\
\mathbf{Q}_k' &= \Gamma_{\mathbf{P}_k}\,.
\end{aligned} \tag{B.15}
$$

The c.m. coordinates and velocities may be advanced by standard integration according to

$$
\begin{aligned}
d\mathbf{v}_0/d\tau &= g\mathbf{F}_0\,, \\
d\mathbf{r}_0/d\tau &= g\mathbf{v}_0\,,
\end{aligned} \tag{B.16}
$$

or in the more usual way taking derivatives with respect to the physical time, t, by omitting g.

Finally, we need the rates of change for the internal energy and time. The former may be derived in analogy with the two-body energy equation (4.38) and also with the expression (5.37) of the Aarseth–Zare [1974] three-body regularization. Hence with the last equation (B.14), the desired equations take the form

$$
\begin{aligned}
E' &= 2 \sum_{k=1}^{N-1} \mathbf{Q}_k'^{\mathrm{T}} \mathcal{L}_k^{\mathrm{T}} \delta \dot{\mathbf{W}}_k\,, \\
t' &= g\,.
\end{aligned} \tag{B.17}
$$

We remark that because tidal effects enter the energy equation, relatively few perturbers may need to be considered. Still, the perturbed treatment can be quite time-consuming when employing a high-order integrator and only compact subsystems are therefore suitable for special study since external effects are included out to a distance $\simeq \lambda R_{\mathrm{grav}}$.

B.3 Errata

It has been pointed out [Orlov, private communication, 1998] that the original chain paper [Mikkola & Aarseth, 1990] contains some errors in the explicit expressions for the regularized derivatives, whereas the Hamiltonian itself is correct. The corrections are as follows: (i) the terms $\mathbf{B}_1 \cdot \mathbf{B}_2$ in (45) and $\mathbf{B}_3 \cdot \mathbf{B}_2$ in (47) should be interchanged; (ii) a factor of 2 is missing from the term $R_1 R_2 R_3$ in (45), (46) and (47). This formulation has mostly been superseded by the 1993 version but the corresponding code is free from these errors.

Appendix C
Higher-order systems

C.1 Introduction

We describe some relevant algorithms for initializing and terminating higher-order systems which are selected for the merger treatment discussed in chapter 11. Such configurations may consist of a single particle or binary in bound orbit around an existing triple or quadruple, or be composed of two stable triples. There are some significant differences from the standard case, the main one being that the KS solution of the binary is not terminated at the initialization. In other words, the complexity of the structure is increased, and the whole process is reminiscent of molecular chemistry. However, once formed as a KS solution, the new hierarchy needs to be restored to its original constituents at the termination which is usually triggered by large perturbations or mass loss. Special procedures are also required for removing all the relevant components of escaping hierarchies, and here we include merged triples and quadruples since the treatment is similar to that for higher-order systems.

C.2 Initialization

Consider an existing hierarchy of arbitrary multiplicity and mass, m_i, which is to be merged with the mass-point m_j, representing any object in the form of a single particle or even another hierarchy. Again we adopt the convention of denoting the component masses of a KS pair by m_k and m_l, respectively. Some of the essential steps are listed in Algorithm C.1.

Nearly all of these steps also appear in the standard case and therefore do not require comment. We note one important difference here, in that there is no termination of the KS solution. Hence the new c.m. particle occupies the same position in the general tables as the original primary, assuming there has been no switching. Moreover, the merger table contains

Algorithm C.1. *Initialization of higher-order system.*

1	Ensure that the primary binary is a hierarchy
2	Save the component masses and neighbour or perturber list
3	Copy the basic KS elements of the primary binary
4	Obtain potential energy of primary components and neighbours
5	Retain current values $\mathbf{R}, \dot{\mathbf{R}}$ of the primary components
6	Evaluate tidal correction for primary mass m_i and neighbours
7	Set initial conditions for the c.m. body by $m_{\rm cm} = m_i + m_j$
8	Form basic KS quantities for the new outer binary m_i and m_j
9	Define initial conditions for ghost KS if required [case $j > N$]
10	Create ghost particle $m_l = 0$ with large values of $\mathbf{r}_0, \mathbf{r}$ and t_0
11	Remove the ghost from all neighbour or perturber lists
12	Initialize the c.m. and KS polynomials for the binary
13	Record the ghost name for identification at termination
14	Specify negative c.m. name for the decision-making
15	Update $E_{\rm merge}$ by the old binding energy and tidal correction

four locations for saving component masses, in case the outer member also represents a KS solution which may be standard or hierarchical. On the other hand, there is only one array entry containing KS quantities for each merger. Hence any secondary KS solution needs to be defined as a ghost pair, with zero component masses and perturber membership, as well as a large value of t_0. In order to distinguish between different levels of a hierarchy, we use the convention that the c.m. particle is assigned a name $\mathcal{N}_{\rm cm}$ which is effectively $-2N_0$ smaller than the previous one. Thus the initial particle number, N_0, is used as a fixed reference number, such that the corresponding names of single particles can at most reach this value. This simple device allows an arbitrary number of hierarchical levels referring to the same system as it increases in complexity.

C.3 Termination

At some stage, the higher-order hierarchy discussed above may cease to satisfy one of the stability conditions, whether it be increased outer eccentricity or large perturbation. Algorithm C.2 lists the steps involved in restoring the present configuration to its original state.

Again most of the listed steps are similar to the standard case. Note that the existing c.m. location (but *not* its value) is also used for the new inner binary, which is initialized in the standard way. The corresponding KS solution is also initialized after forming the appropriate perturber list. One new feature is the re-evaluation of the stability parameter (9.14), together

Algorithm C.2. *Termination of higher-order system.*

1	Locate the current index in the merger table
2	Save the neighbour or perturber list for corrections
3	Predict the c.m. motion to highest order
4	Define mass, coordinates and velocity for the new c.m.
5	Add the old outer component to the perturber list
6	Obtain the potential energy of inner binary and neighbours
7	Identify the location, j, of the corresponding ghost particle
8	Set initial conditions for ghost and new inner c.m.
9	Restore the basic quantities of the primary components
10	Copy the saved KS variables from the merger table
11	Form new force polynomials for the old outer component
12	Initialize the c.m. polynomials and corresponding KS
13	Find the merger index for the next hierarchical level
14	Re-evaluate the stability parameter, $R_{\rm p}^{\rm out}$, including inclination
15	Update the original name of the hierarchy by adding $2N_0$
16	Perform perturbed initialization of any outer binary [case $j > N$]
17	Subtract new binding energy and add tidal correction to $E_{\rm merge}$
18	Compress the merger table, including any escaped hierarchies

with the basic inclination modification. This is achieved by obtaining the inner semi-major axis with the specified merger index and the outer two-body elements in the standard way, whereas the inclination is determined from the saved relative coordinates and velocities together with those of the c.m. and outer component. Finally, regarding step 18, care is needed to retain any inner members of hierarchies that would otherwise appear to be associated with escapers; i.e. those for which the saved c.m. name equals $2nN_0 + \mathcal{N}_{\rm cm}$, where $n = 0, 1, 2, \ldots$ denotes the possible levels and $\mathcal{N}_{\rm cm} < 0$ represents the actual c.m. name.

C.4 Escape of hierarchies

It is not uncommon for hierarchical systems to acquire escape velocity and be ejected from the cluster. This requires special treatment of the data structure since removal of only the active binary itself would leave single ghost particles and even massless binaries as cluster members. Hierarchies of any order are identified by a negative c.m. name, and it is therefore natural to discuss such systems together.

Let us first consider the case of a stable triple that satisfies the standard escape criterion for the associated c.m. particle, i. After the active KS binary elements and c.m. variables have been updated in the usual way, we determine the merger index using the c.m. name. The total energy, E, and merger energy, $E_{\rm merge}$, are updated by subtracting the inner binding energy, $\mu_{\rm kl}h$, where the latter quantities are obtained from the merger table. This enables the global index of the corresponding ghost particle to be identified by comparing it with the saved ghost name, whereupon the large value of the x-coordinate is reduced to facilitate escape removal. Accordingly, this should be a single particle with zero mass which can be removed in the usual way together with reducing N and $N_{\rm tot}$. Again all the relevant lists must be updated, but only to the extent of reducing any larger indices since the ghost particle itself should not be a member.

An escaping quadruple is distinguished by the second or outer component being a c.m. particle; hence we have $\mathcal{N}_{2p} > N_0$, where the pair index is given by $p = i - N$. First the KS index of the ghost binary is determined by comparing the relevant names with $\mathcal{N}_{2p}$, since the second component is in fact another binary c.m. The ghost pair index, q (say), then specifies the corresponding ghost c.m. particle as $i_q = N + q$, which must also be removed. The index of the latter is obtained by comparing the relevant names with the corresponding one for i_q. Again the updating procedure above is carried out, whereby the present KS and corresponding c.m. variables are updated. At this stage we correct the energies E and $E_{\rm merge}$ by subtracting the original ghost contribution since the masses and specific binding energy are known if due allowance is made for table updating (i.e. $q > p$). At the next sequential search, an activated value of i_q is used to identify the necessary ghost binary by employing the specified name. Removal of the corresponding components can now be performed in the usual way, provided the large x-coordinate of the ghost c.m. has been re-assigned an intermediate value which ensures escape.

It is also possible to envisage higher-order systems escaping. Although this is extremely rare, examples of ejected quintuplets have been noted for realistic populations. The above escape algorithm has been generalized to include one more hierarchical level where the outermost component may be a single particle or another binary. Note that the energy correction procedure now needs to account for the interaction between the two inner levels, as well as that of the outer level. Consequently, a successful update should reduce $E_{\rm merge}$ to the value given by any remaining hierarchies (apart from small tidal corrections) and remove all the relevant ghost members. For this purpose we introduce a notation where the innermost binary is said to be 'hierarchical' and any other binary (excluding the current KS pair) is denoted as a 'quadruple', although such a usage is only intended to be descriptive. Accordingly, an escaping quintuplet of the type $[[B, S], B]$ will be dissected into four different configurations, where the outermost ghost binary is called a quadruple.

Appendix D
Practical algorithms

In the following we give some useful algorithms to assist in the construction of codes or explain more fully the present implementations.

D.1 Maxwellian distribution

A Maxwellian velocity distribution is often used for generating initial conditions. This is specifically needed for the kick velocities, as described in section 15.5. We adopt an elegant method due to Hénon [Heggie, private communication, 1997]. It is based on a pair-wise procedure as follows. First select two random numbers, X_1, X_2, and form the quantities

$$\begin{aligned} S &= \sigma[-2\ln(1 - X_1)]\,, \\ \theta &= 2\pi X_2\,, \end{aligned} \tag{D.1}$$

where σ is the velocity dispersion in 1D. The two corresponding velocity components are then given by $v_x = S\cos\theta$ and $v_y = S\sin\theta$. For a 3D distribution we perform the same procedure a second time and choose $v_z = S\cos\theta$.

D.2 Ghost particles

The general problem of recovering the actual mass, coordinates and velocity of a ghost particle is fairly simple for a triple system. However, quadruples also occur and the procedure is now more involved. Consequently, the general case is presented here, where only the first part is relevant for triples. Thus a quadruple system is defined by $\mathcal{N}_j > N_0$ for ghost particle j. Algorithm D.1 is needed if we are considering a c.m. particle with index $i > N$, mass $m_i > 0$, name $\mathcal{N}_i < 0$ and wish to obtain global quantities for any ghost components. The two mass components of the active KS solution are denoted by m_k, m_l as usual.

Algorithm D.1. *Identification of ghost particles.*

1	Locate the index p in the merger table using $\mathcal{N}_{\rm merge} = \mathcal{N}_i$
2	Determine the global index j of the ghost by $\mathcal{N}_j = \mathcal{N}_{\rm ghost}$
3	Copy component masses from m_1, m_2 of the merger table
4	Form $\mathbf{r}, \dot{\mathbf{r}}$ for binary components from m_k and old $\mathbf{R}_1, \dot{\mathbf{R}}_1$
5	Obtain the c.m. index, q, from $\mathcal{N}_q = \mathcal{N}_j$ [case $\mathcal{N}_j > N_0$]
6	Get the component masses from m_3, m_4 of merger index p
7	Transform $\mathbf{U}, \mathbf{U}'$ of any ghost binary pair q–N to $\mathbf{R}_2, \dot{\mathbf{R}}_2$
8	Derive global $\mathbf{r}, \dot{\mathbf{r}}$ for components using m_l and $\mathbf{R}_2, \dot{\mathbf{R}}_2$

The merger table holds the array $\{m_j\}$ for saving the masses of the first and any second binary components. However, only the relative quantities $\mathbf{R}_1, \dot{\mathbf{R}}_1$ and corresponding KS elements, including h, of the inner binary are recorded in the table since any second binary is preserved as a ghost. In addition, the table contains the names of the original c.m. as well as the ghost. Parts of this algorithm are required for at least six different purposes in the codes *NBODY*4 and *NBODY*6.

D.3 KS procedures for averaging

A variety of tasks need to be carried out during the averaging of hierarchical systems. Let us concentrate on those aspects that are connected with KS regularization. Given the appropriate merger index, the current values of the relative coordinates and velocity, $\mathbf{R}, \mathbf{V}$, are obtained by KS transformations, whereupon the Runge–Lenz and angular momentum vectors are constructed.

Following evaluation of $e_{\rm max}$ and other quantities, together with time-step selection, the next integration interval is carried out and new values of $\mathbf{R}, \mathbf{V}$ are determined from the orbital elements. Transformation to the corresponding KS variables now takes place (cf. (4.32) and (4.36)) and the physical values are recorded. The binding energy, h, is also an important quantity and is evaluated via the semi-major axis in terms of the specific angular momentum $\mathbf{j}$, binary mass $m_{\rm b}$ and eccentricity,

$$a = |\mathbf{j}|^2/m_{\rm b}(1 - |\mathbf{e}|^2)\,. \tag{D.2}$$

Finally, any change in energy given by $\mu(h - h_0)$ is added to the merger energy, $E_{\rm merge}$, and *subtracted* from $E_{\rm coll}$. Although there is no net change in the total energy budget, the corrected quantities will now be consistent at termination.

D.4 Determination of pericentre or apocentre

We now describe procedures for specifying KS variables at the pericentre or apocentre position. This is needed for several purposes, such as during tidal interactions or collisions. Thus unperturbed binaries are defined at the first point past apocentre, whereas the energy change in tidal interactions are implemented at pericentre. To facilitate decision-making, the two-body separation is compared with the semi-major axis and the sign of the radial velocity, $t'' = R'$, defines approach or recession.

By combining two algorithms, we can obtain any desired transformation. Thus the case of increasing the orbital phase by an angle θ yields the new values [Stiefel & Scheifele, 1971]

$$\begin{aligned} \mathbf{U} &= \mathbf{U}_0 \cos\theta + \mathbf{U}'_0 \sin\theta/\nu\,, \\ \mathbf{U}' &= \mathbf{U}'_0 \cos\theta - \mathbf{U}_0 \sin\theta\,\nu\,, \end{aligned} \tag{D.3}$$

where $\nu = (\frac{1}{2}|h|)^{1/2}$ is the regularized orbital frequency. This procedure assumes unperturbed motion and is used frequently for reflection by an angle $\theta = \pi/2$ which corresponds to *half* a physical period.

In order to consider an arbitrary orbital phase, a second algorithm is needed which should include integration for perturbed motion. We distinguish between near-collision, elliptic or hyperbolic cases and evaluate the pericentre time from an expansion or the two forms of Kepler's equation, respectively [Mikkola, private communication, 1991]. If the orbit is perturbed and $R > a$, we first perform a reflection by $\pi/2$, followed by integration back to the pericentre. Treating this as a provisional solution, improved values are determined by the inverted relations

$$\begin{aligned} \mathbf{U} &= \tilde{\mathbf{U}} x_{\rm c} - \tilde{\mathbf{U}}' y_{\rm s}\,, \\ \mathbf{U}' &= \tilde{\mathbf{U}}' x_{\rm c} + \tfrac{1}{4}\tilde{\mathbf{U}} y_{\rm s} m_{\rm b}/a\,, \end{aligned} \tag{D.4}$$

since $\mathbf{U}'' = -m_{\rm b}\mathbf{U}/4a$. If $R < a$, we adopt $x_{\rm c} = [\frac{1}{2} + \frac{1}{2}(1 - R/a)/e]^{1/2}$ and $y_{\rm s} = t''/em_{\rm b}x_{\rm c}$. The latter expressions are derived from $\cos(\theta/2)$ and $\sin(\theta/2)$ by inverting (D.3), with $R = a(1 - e\cos\theta)$, Kepler's equation and θ half the eccentric anomaly. However, these coefficients are not well behaved near the apocentre since $x_{\rm c} \to 0$ and alternative expressions should then be used [Mikkola, private communication, 1997]. From the relation $\xi^2 + \psi^2/a = e^2$, with $\xi = 1 - R/a$ and $\psi = t''/m_{\rm b}^{1/2}$, we have $\xi/e = -(1 - \psi^2/ae^2)^{1/2}$ if $\xi < 0$. After some manipulation we obtain

$$\begin{aligned} x_{\rm c} &= \frac{\psi}{ea^{1/2}(2 - 2\xi/e)^{1/2}}\,, \\ y_{\rm s} &= [\frac{a}{m_{\rm b}}(2 - 2\xi/e)]^{1/2}\,, \end{aligned} \tag{D.5}$$

where $-y_{\rm s}$ should be chosen if $\psi < 0$.

D.5 Partial unperturbed reflection

Highly eccentric orbits treated by KS regularization are characterized by small perturbations in the inner part. Hence it might be of interest to assume unperturbed motion and perform a partial reflection, provided the relative perturbation is sufficiently small. Accordingly, we give an algorithm for achieving this task.*

Algorithm D.2. *Partial two-body reflection.*

1	Determine osculating elements a and e
2	Specify the total reflection time, $t_{\rm ref}$, by Kepler's equation
3	Convert to regularized time units, $\delta\tau$
4	Skip reflection near pericentre if $\delta\tau < 4\,\Delta\tau$
5	Predict c.m. and obtain coordinates of the components
6	Generate reflected solutions of $\mathbf{U}$ and $\mathbf{U}'$ by (D.4)
7	Set next look-up time with $\Delta t = t_{\rm ref}$
8	Reverse sign of t'' and define unperturbed interval Δt
9	Obtain potential energy of perturbers and components
10	Transform to reflected coordinates and repeat step 9
11	Modify c.m. velocity by differential potential energy $\Delta\Phi$

Using ξ and ψ from the previous section, the total reflection time is obtained from Kepler's equation, which gives

$$t_{\rm ref} = 2\,(a^3/m_{\rm b})^{1/2}\,(\theta - |\psi|/m_{\rm b}^{1/2})\,, \tag{D.6}$$

where θ is the eccentric anomaly in $[0, \pi]$. The corresponding regularized time interval is obtained from the differential expression [Baumgarte & Stiefel, 1974]

$$\delta\tau = -(2\,ht_{\rm ref} - \Delta t'')/m_{\rm b}\,, \tag{D.7}$$

with $\Delta t'' = t''_1 - t''_0 = -2\,t''_0$ by symmetry (cf. (9.27)). Here we adopt simplified coefficients for the transformations (D.4), with $x_{\rm c} = \xi/e$ and $y_{\rm s} = 2\psi/em_{\rm b}^{1/2}$. Thus the full eccentric anomaly is used (instead of half) because of the time derivative $t' = \mathbf{U}\cdot\mathbf{U}$, and the squaring implies doubling of angles. Since unperturbed motion is defined, the new KS polynomials will be initialized at the next updating time, $t + t_{\rm ref}$. Finally, the c.m. velocity is modified to compensate for the small tidal energy change, with the correction factor

$$C = (1 + 2\Delta\Phi/m_{\rm b}v_{\rm cm}^2)^{1/2}\,, \tag{D.8}$$

which facilitates formal energy conservation (cf. Algorithm 12.1).

* This optional procedure is no longer used by the NBODYn codes.

Appendix E
KS procedures with GRAPE

When using HARP or GRAPE, force calculations involving a perturbed c.m. body require modification because the summation is performed over all interactions in the point-mass approximation. Thus in order to obtain the correct c.m. force given by (8.57), we need to perform certain corrections, and the same principle applies to a single particle in the neighbourhood of a c.m. body or composite particle. A description of the relevant procedures is contained in the sections below.

HARP holds the quantities $\mathbf{r}_0, \dot{\mathbf{r}}_0, \mathbf{F}_0, \dot{\mathbf{F}}_0, t_0$ (GRAPE has also $\ddot{\mathbf{F}}$) for all single particles and c.m. bodies, and all current values of $\mathbf{r}, \dot{\mathbf{r}}$ are predicted by the hardware at each block-step before the force is evaluated. To facilitate decision-making, a list of perturbed KS solutions is maintained and updated at the end of a KS cycle if the status has changed. Coordinates and velocities of all the KS components are predicted on the host during a block-step but some are done at no cost while HARP is busy.

E.1 Single particles

The force on a single particle, i, exerted by a nearby c.m. body of mass $m_{\rm b} = m_k + m_l$ is obtained by vector summation over both components. Thus if the c.m. approximation is *not* satisfied, the force is modified by

$$\tilde{\mathbf{F}}_i = \mathbf{F}_i + \frac{m_{\rm b}(\mathbf{r}_i - \mathbf{r}_{\rm cm})}{|\mathbf{r}_i - \mathbf{r}_{\rm cm}|^3} - \frac{m_k(\mathbf{r}_i - \mathbf{r}_k)}{|\mathbf{r}_i - \mathbf{r}_k|^3} - \frac{m_l(\mathbf{r}_i - \mathbf{r}_l)}{|\mathbf{r}_i - \mathbf{r}_l|^3}\,, \qquad \text{(E.1)}$$

otherwise no correction is made. A similar procedure is carried out for $\dot{\mathbf{F}}_i$ and any other perturbed c.m. bodies are treated in an analogous way.

Likewise, a differential force correction is made on the rare occasions when chain regularization is used. The same principle as used above applies with respect to the chain perturber list or the chain c.m. itself, which is distinguished by zero name.

E.2 Regularized KS pairs

The force on a c.m. particle is given by (8.57), except that the c.m. approximation is used when appropriate. Thus in the case of unperturbed two-body motion, the correction procedure is the same as for a single particle. Since $i > N$ here, the sequential perturber lists of active KS solutions are searched backwards.

The case of perturbed two-body motion is more complicated but the principle is the same. Each member, j, of the perturber list is considered in turn. If $j \leq N$ and $r_{\rm ij} < \lambda R$, the HARP contribution to the force and its derivative is first subtracted. This is followed by prediction of coordinates and velocity* before the individual interaction terms are evaluated.

In the case $j > N$, we need to check *two* c.m. approximations and obtain the appropriate force terms if required. For example, a binary may have small size such that the c.m. approximation with respect to another binary is satisfied but the reverse may not be true. When all the perturbers have been considered, the new component forces and first derivatives are combined vectorially according to (8.57) and its derivative.

The above procedure is less cumbersome than an earlier formulation based on creating a mask and sending zero masses of relevant perturbers and active KS solutions to HARP before the force evaluation. This was followed by addition of the remaining contributions on the host. Finally, the proper perturber masses were restored on HARP. With some effort, an arbitrary number of active KS solutions were treated together at the same time. This scheme was used successfully by *NBODY4* for some years until the present procedure was implemented (end of 1998).

Unfortunately, the gain in efficiency of the new scheme is made at the expense of accuracy. Thus the numerically inelegant practice of subtracting previously added terms also suffers from the fact that the precision on HARP is different from the host [Makino *et al.*, 1997]. This is a design feature to optimize the performance and especially the first derivative is less accurate. However, the force errors are still sufficiently small to be of no practical significance in most cases. A test calculation with $N = 5000$ showed typical relative force errors of 1×10^{-7} when compared with evaluation on the host, whereas the corresponding errors in the force derivatives were about 4×10^{-6}. Similar errors are also present when using GRAPE-6 for a *static* configuration. Note, however, that predictions to order $\ddot{\mathbf{F}}$ are only performed on the hardware in the latter case. Small additional errors are therefore introduced when making corrections during actual calculations. To compensate, explicit force summation is performed on the host if the first derivative is very large during polynomial initialization.

---

* All coordinates and velocities are predicted on HARP and GRAPE but only the current block members and active KS solutions are updated on the host.

Appendix F
Alternative simulation method

Realistic N-body simulations of star cluster evolution require a substantial programming effort. Since it takes time to develop suitable software, published descriptions tend to lag behind or be non-existent. However, one large team effort has reached a degree of development that merits detailed comments, especially since many results have been described in this book. In the following we highlight some aspects relating to the integration method as well as the treatment of stellar evolution, based on one available source of information [Portegies Zwart *et al.*, 2001].

F.1 N-body treatment

The `kira` integrator* advances the particle motions according to the standard Hermite method [Makino, 1991a] using hierarchical (or quantized) time-steps [McMillan, 1986]. An efficient scheme was realized with the construction of the high-precision GRAPE computers which calculate the force and force derivative and also include predictions on the hardware.

One special feature here is the use of hierarchical Jacobi coordinates which is reminiscent of an earlier binary tree formulation [Jernigan & Porter, 1989]. This representation is equivalent to the data structure used in KS and chain regularization. Increased numerical accuracy is achieved by evaluating nearby contributions on the host, whereas the traditional brute force way involves the subtraction of two larger distances. On the other hand, binaries and multiple close encounters are not studied by regularization methods which also employ local coordinates. Hence direct integration of binaries requires significantly shorter time-steps in order to maintain adequate accuracy even for circular orbits. In compensation, direct integration offers advantages of simplicity.

* Available within the STARLAB software package on `http://www.manybody.org`.

As in the KS formulation, unperturbed binaries are treated in the c.m. approximation, whereas weak perturbations are included by a slow-down procedure (cf. section 5.8). Large numerical terms are avoided at small pericentre distances by the adoption of partial unperturbed motion (cf. (D.5)). On the other hand, long-lived stable hierarchies are only defined by means of the perturbations. Although the `kira` scheme exploits the GRAPE facility to full advantage, the lack of regular solutions for studying strong interactions may be a cause for concern. This situation is exemplified by the so-called 'terrible triples' which must be integrated over many inner orbits with small time-steps when the stability condition is not satisfied.† So far simulations of clusters containing primordial binaries have been discussed for $N_s = 1\,\mathrm{K}$ single stars and 50% binaries.

F.2 Stellar evolution

Single star and binary evolution are handled by the `SeBa` module, which is also part of *STARLAB*. The treatment of single stars is currently based on standard fitting functions for solar metallicities [cf. Eggleton *et al.*, 1989; Tout *et al.*, 1997]. Further prescriptions for the core mass and mass loss by stellar winds have been added [Portegies Zwart & Verbunt, 1996]. A randomized velocity kick [Hartman, 1997] is assigned when stars in the initial range 8–25 $M_\odot$ become neutron stars. An upper limit of 100 $M_\odot$ is assumed for the IMF, with the heaviest stars losing mass rapidly before becoming Wolf–Rayet stars, which eventually form black holes.

An extensive network of processes must be considered when binaries are included [Portegies Zwart, 1996]. The look-up interval is 1% of the typical evolution time for each stage in the HR diagram, with further reductions during Roche-lobe mass transfer. Sequential circularization without synchronizing spin effects is implemented if $5r_1^* > a(1-e)$, with r_1^* the largest stellar radius (cf. (11.34)).‡ The angular momentum loss associated with stellar winds is also modelled, as is gravitational radiation for compact stars in short-period orbits. Unstable mass transfer is implemented in the standard way, including prescriptions for the spiral-in process when the envelope is ejected [Webbink, 1984]. Again, stable mass transfer is treated on different time-scales, where the details depend on the period and evolutionary state of the components.

Collisions are implemented when the stellar radii overlap, with coalescence due to mass transfer or common-envelope phases, followed by system mass loss. Finally, the scheme for assigning collision products includes a full range of astrophysically interesting objects.

† In one large simulation, five out of 50 stable hierarchies had $R_\mathrm{p}/R_\mathrm{p}^\mathrm{crit} < 2$ (cf. (9.14)).

‡ Note the use of 1/5 and missing square root of equation (A2) in the original paper.

Appendix G
Table of symbols

G.1 Introduction

In this table we present a list of the most commonly used symbols. Depending on the context, some of these quantities may be in physical or N-body units but this should not cause confusion since relevant scaling factors are provided. Inevitably some symbols are not unique, but adopting the traditional notation for clarity has been given precedence. In any case, such multiplicity does not usually occur within the same section. A few duplicated definitions are listed in footnotes to the table.

Table G.1. Frequently used symbols.

Symbol	Definition
a	Semi-major axis
$a_{\rm hard}$	Hard binary semi-major axis
$\mathbf{D}^k$	Divided difference
e	Binary eccentricity
E	Energy
$E_{\rm b}$	Binary energy
$E_{\rm ch}$	Energy of chain members
$E_{\rm sub}$	Subsystem energy
$f_{\rm b}$	Binary fraction
$\mathbf{F}$	Force per unit mass
$\mathbf{F}^{(k)}$	Force derivative (also $\dot{\mathbf{F}}$)
$\mathbf{F}_{\rm I}$	Irregular force
$\mathbf{F}_{\rm R}$	Regular force
G	Gravitational constant
h	Two-body energy per unit mass

Table G.1. (*cont.*) Frequently used symbols.

H	Hamiltonian function
$\mathbf{J}$	Angular momentum
K	Kilo byte unit, 1024
L	Neighbour list*
$\mathcal{L}$	Levi-Civita matrix
$m_{\rm b}$	Binary mass
m_i	Particle mass
m_k	Mass of first binary component
m_l	Mass of second binary component
M	Total cluster mass
$M_{\rm S}$	Mean stellar mass ($M_\odot$)
$n_{\rm max}$	Maximum neighbour number
N	Particle number
$N_{\rm b}$	Number of binaries
$N_{\rm c}$	Particle number in the core
$N_{\rm p}$	Number of regularized pairs
$N_{\rm s}$	Number of single stars
$\mathcal{N}_i$	Particle name
$\mathbf{P}$	Physical perturbation
$\mathbf{P}_k$	Regularized momenta
$\mathbf{Q}_k$	Regularized coordinates
$Q_{\rm vir}$	Virial theorem ratio
$r_{\rm c}$	Core radius (cluster or stellar)
$\mathbf{r}_{\rm d}$	Density centre
$r_{\rm h}$	Half-mass radius
$\mathbf{r}_i$	Global coordinates
r_k^*	Stellar radius
$r_{\rm t}$	Tidal radius
R	Two-body separation
R_0	Initial two-body separation
$R_{\rm cl}$	Close encounter distance
$R_{\rm grav}$	Gravitational radius
$R_{\rm p}$	Pericentre distance
$R_{\rm s}$	Neighbour sphere radius
$R_{\rm V}$	Virial cluster radius (pc)
t	Time in N-body units
$t_{\rm block}$	Block boundary time
$t_{\rm cr}$	Crossing time

* Also Lagrangian energy

Table G.1. (*cont.*) Frequently used symbols.

$t_{\rm ev}$	Stellar evolution time
$t_{\rm rh}$	Half-mass relaxation time
T	Total kinetic energy
T^*	Time scaling factor (Myr)
U	Potential energy
$\mathbf{U}$	Regularized coordinates (also $\mathbf{u}$)
$\mathbf{U}'$	Regularized velocities (also $\mathbf{u}'$)
$\mathbf{v}_i$	Global velocities
V	Virial energy
V^*	Velocity scaling factor ($\rm km\,s^{-1}$)
W	Tidal energy
α	IMF power-law exponent[†]
γ	Relative two-body perturbation[‡]
$\gamma_{\rm min}$	Limit for unperturbed motion
Γ	Regularized Hamiltonian
$\Delta t_{\rm cl}$	Close encounter time-step
Δt_i	Individual time-step
$\Delta\tau$	Regularized time-step
ϵ	Softening parameter
$\epsilon_{\rm hard}$	Hard binary energy
η	Standard time-step parameter
η_U	KS time-step parameter
θ	Eccentric anomaly[§]
κ	Slow-down factor
λ	C.m. approximation factor
μ	Reduced mass of binary
ρ	Spatial density
σ	Velocity dispersion
τ	Regularized time
Φ	Potential or potential energy
ψ	Inclination angle
ω	General purpose angle
Ω	Angular velocity[¶]

[†] Also relative energy error
[‡] Also in Coloumb logarithm
[§] Also opening angle
[¶] Also energy in Stumpff method

Appendix H
Hermite integration method

Here we give the FORTRAN listing of standard Hermite integration for a test particle. The accuracy is controlled by the tolerance denoted ETA. Results are given at times scaled by the initial period.

```
*             H E R M I T
*             ***********
*
*         Standard Hermite integration.
*         -----------------------------
*
      IMPLICIT REAL*8  (A-H,O-Z)
      REAL*8  X(3),XDOT(3),F(3),FDOT(3),FI(3),FD(3),D2(3)
      DATA TIME,NSTEPS /0.0,0/
*
      READ (5,*)  ETA, DELTAT, TCRIT
      READ (5,*) (X(K),K=1,3), (XDOT(K),K=1,3)
      TNEXT = DELTAT
      RI2 = X(1)**2 + X(2)**2 + X(3)**2
      RI = SQRT(RI2)
      VI2 = XDOT(1)**2 + XDOT(2)**2 + XDOT(3)**2
      SEMI0 = 2.0/RI - VI2
      SEMI0 = 1.0/SEMI0
      RDOT = X(1)*XDOT(1) + X(2)*XDOT(2) + X(3)*XDOT(3)
      ECC0 = SQRT((1.0 - RI/SEMI0)**2 + RDOT**2/SEMI0)
      TWOPI = 8.0*ATAN(1.0D0)
      ONE3 = 1.0/3.0D0
      ONE12 = 1.0/12.0D0
      P = TWOPI*SEMI0*SQRT(SEMI0)
      DELTAT = DELTAT*P
```

```
      TNEXT = DELTAT
      TCRIT = TCRIT*P
*
*       Evaluate the initial force and first derivative.
      RIN3 = 1.0/(RI2*SQRT(RI2))
      RDOT = 3.0*(X(1)*XDOT(1) + X(2)*XDOT(2) +
     &                           X(3)*XDOT(3))/RI2
      DO 5 K = 1,3
          F(K) = -X(K)*RIN3
          FDOT(K) = -(XDOT(K) - RDOT*X(K))*RIN3
    5 CONTINUE
*
*       Choose initial time-step by force criterion.
      FF = SQRT(F(1)**2 + F(2)**2 + F(3)**2)
      STEP = SQRT(ETA*RI/FF)
*
*       Predict coordinates and velocity to order FDOT.
   10 DT = STEP
      DT2 = 0.5*DT
      DT3 = ONE3*DT
      DO 20 K = 1,3
          X(K) = ((FDOT(K)*DT3 + F(K))*DT2 + XDOT(K))*DT +
     &                                        X(K)
          XDOT(K) = (FDOT(K)*DT2 + F(K))*DT + XDOT(K)
   20 CONTINUE
*
*       Obtain force and first derivative at end of step.
      RI2 = X(1)**2 + X(2)**2 + X(3)**2
      RIN3 = 1.0/(RI2*SQRT(RI2))
      RDOT = 3.0*(X(1)*XDOT(1) + X(2)*XDOT(2) +
     &                           X(3)*XDOT(3))/RI2
      DO 30 K = 1,3
          FI(K) = -X(K)*RIN3
          FD(K) = -(XDOT(K) - RDOT*X(K))*RIN3
   30 CONTINUE
*
*       Set corrector time factors and advance the time.
      DTSQ = DT**2
      DT2 = 2.0/DTSQ
      DT6 = 6.0/(DT*DTSQ)
      DT13 = ONE3*DT
      DTSQ12 = ONE12*DTSQ
      TIME = TIME + DT
```

```
*
*       Include the Hermite corrector and update F & FDOT.
      DO 40 K = 1,3
          DF = F(K) - FI(K)
          SUM = FDOT(K) + FD(K)
          AT3 = 2.0*DF + SUM*DT
          BT2 = -3.0*DF - (SUM + FDOT(K))*DT
          X(K) = X(K) + (0.6*AT3 + BT2)*DTSQ12
          XDOT(K) = XDOT(K) + (0.75*AT3 + BT2)*DT13
          F(K) = FI(K)
          FDOT(K) = FD(K)
          D2(K) = BT2*DT2 + DT*AT3*DT6
   40 CONTINUE
*
*       Determine next step from F & F2DOT in Taylor series.
      FF = SQRT(F(1)**2 + F(2)**2 + F(3)**2)
      FF2 = SQRT(D2(1)**2 + D2(2)**2 + D2(3)**2)
      STEP = SQRT(ETA*FF/FF2)
*
*       Increase step counter and check output time.
      NSTEPS = NSTEPS + 1
      IF (TIME.LT.TNEXT) GO TO 10
*
*       Print errors in semi-major axis & eccentricity.
      RI2 = X(1)**2 + X(2)**2 + X(3)**2
      RI = SQRT(RI2)
      VI2 = XDOT(1)**2 + XDOT(2)**2 + XDOT(3)**2
      SEMI = 2.0/RI - VI2
      SEMI = 1.0/SEMI
      RDOT = X(1)*XDOT(1) + X(2)*XDOT(2) + X(3)*XDOT(3)
      ECC = SQRT((1.0 - RI/SEMI)**2 + RDOT**2/SEMI)
      DA = (SEMI - SEMI0)/SEMI0
      DE = ECC0 - ECC
      WRITE (6,50)  TIME/P, NSTEPS, ECC, DA, DE
   50 FORMAT ('  T/P =',F8.1,'  # =',I7,'  E =',F8.4,
     &                 '  DA/A =',1P,E9.1,'  DE =',E9.1)
      TNEXT = TNEXT + DELTAT
      IF (TIME.LT.TCRIT) GO TO 10
*
      END
```

References

Aarseth, S. J. [1963a], *Dynamics of Galaxies*, Ph.D. Thesis (University of Cambridge).

[1963b], 'Dynamical evolution of clusters of galaxies, I', *Mon. Not. R. Astron. Soc.* **126**, 223–55.

[1966a], 'Dynamical evolution of clusters of galaxies, II', *Mon. Not. R. Astron. Soc.* **132**, 35–65.

[1966b], 'Third integral of motion for high-velocity stars', *Nature*, **212**, 57–8.

[1967], 'On a collisionless method in stellar dynamics, I', *Bull. Astron.* **2**, 47–57.

[1968], 'Dynamical evolution of simulated N-body systems', *Bull. Astron.* **3**, 105–25.

[1969], 'Dynamical evolution of clusters of galaxies, III', *Mon. Not. R. Astron. Soc.* **144**, 537–48.

[1970], 'Perturbation treatment of close binaries in the N-body problem', *Astron. Astrophys.* **9**, 64–9.

[1972a], 'Binary evolution in stellar systems', in *Gravitational N-Body Problem*, ed. M. Lecar (Reidel, Dordrecht), 88–98.

[1972b], 'Direct integration methods of the N-body problem', in *Gravitational N-Body Problem*, ed. M. Lecar (Reidel, Dordrecht), 373–87.

[1973], 'Computer simulations of star cluster dynamics', *Vistas Astron.* **15**, 13–37.

[1974], 'Dynamical evolution of simulated star clusters', *Astron. Astrophys.* **35**, 237–50.

[1975], 'N-body simulations', in *Dynamics of Stellar Systems*, ed. A. Hayli (Reidel, Dordrecht), 57–9.

[1976], 'A note on stabilization in three-body regularization', in *Long-Time Predictions in Dynamics*, ed. V. Szebehely & B. D. Tapley (Reidel, Dordrecht), 173–7.

[1980], 'Dynamics of initial binaries in open clusters', in *Star Clusters*, ed. J. E. Hesser (Reidel, Dordrecht), 325–6.

[1985a], 'Direct methods for N-body simulations', in *Multiple Time Scales*, ed. J. U. Brackbill & B. I. Cohen (Academic Press, Orlando), 377–418.

[1985b], 'Direct N-body calculations', in *Dynamics of Star Clusters*, ed. J. Goodman & P. Hut (Reidel, Dordrecht), 251–8.

[1988a], 'Integration methods for small N-body systems', in *The Few Body Problem*, ed. M. J. Valtonen (Reidel, Dordrecht), 287–307.

[1988b], 'Dynamics of molecular clouds', *Bol. Acad. Nac. Cienc. Cordoba* **58**, 201–8.

[1994], 'Direct methods for N-body simulations', in *Lecture Notes in Physics*, ed. G. Contopoulos, N. K. Spyrou, & L. Vlahos (Springer-Verlag, New York), **433**, 277–312.

[1996a], 'Star cluster simulations on HARP', in *Dynamical Evolution of Star Clusters*, ed. P. Hut & J. Makino (Kluwer, Dordrecht), 161–70.

[1996b], 'N-body simulations of open clusters with binary evolution', in *The Origins, Evolution, and Destinies of Binary Stars in Clusters*, ed. E. F. Milone & J.-C. Mermilliod (ASP, San Francisco), **90**, 423–30.

[1999a], 'From NBODY1 to NBODY6: the growth of an industry', *Publ. Astron. Soc. Pacific* **111**, 1333–46.

[1999b], 'Star cluster simulations: the state of the art', in *Impact of Modern Dynamics in Astronomy*, ed. J. Henrard & S. Ferraz-Mello (Kluwer, Dordrecht), 127–37.

[2001a], 'The formation of hierarchical systems', in *Dynamics of Star Clusters and the Milky Way*, ed. S. Deiters, B. Fuchs, A. Just, R. Spurzem & R. Wielen (ASP, San Francisco), **228**, 111–16.

[2001b], 'NBODY2: a direct N-body integration code', *New Astron.* **6**, 277–91.

[2001c], 'Regularization methods for the N-body problem', in *The Restless Universe*, ed. B. A. Steves & A. J. Maciejewski (Institute of Physics, Bristol), 93–108.

[2003a], 'Regularization tools for binary interactions', in *Astrophysical Supercomputing using Particle Simulations*, ed. J. Makino & P. Hut (ASP, San Francisco), 295–304, `astro-ph/0110148`.

[2003b], 'Black hole binary dynamics', in *Fred Hoyle's Universe*, ed. G. Burbidge, J. V. Narlikar & N. C. Wickramasinghe, `astro-ph/0210116`, *Astrophys. Sp. Sci.* **285**.

Aarseth, S. J. & Fall, S. M. [1980], 'Cosmological N-body simulations of galaxy merging', *Astrophys. J.* **236**, 43–57.

Aarseth, S. J. & Heggie, D.C. [1976], 'The probability of binary formation by three-body encounters', *Astron. Astrophys.* **53**, 259–65.

[1993], 'Core collapse for $N = 6000$', in *The Globular Cluster–Galaxy Connection*, ed. G. H. Smith & J. P. Brodie (ASP, San Francisco), **48**, 701–4.

[1998], 'Basic N-body modelling of the evolution of globular clusters, I. Time scaling', *Mon. Not. R. Astron. Soc.* **297**, 794–806.

Aarseth, S. J. & Hills, J. G. [1972], 'The dynamical evolution of a stellar cluster with initial subclustering', *Astron. Astrophys.* **21**, 255–63.

Aarseth, S. J. & Hoyle, F. [1964], 'An assessment of the present state of the gravitational N-body problem', *Astrophysica Norvegica* **9**, 313–21.

Aarseth, S. J. & Inagaki, S. [1986], 'Vectorization of N-body codes', in *The Use of Supercomputers in Stellar Dynamics*, ed. P. Hut & S. McMillan (Springer-Verlag, New York), 203–5.

Aarseth, S. J. & Lecar, M. [1975], 'Computer simulations of stellar systems', *Ann. Rev. Astron. Astrophys.* **13**, 1–21.

[1984], 'The formation of the terrestrial planets from lunar sized planetesimals', in *Planetary Rings*, ed. A. Brahic (Cepadues, Toulouse), 661–74.

Aarseth, S. & Mardling, R. [1997], 'Neutron star binaries in open clusters', in *Pulsars: Problems and Progress*, ed. S. Johnston, M.A. Walker & M. Bailes (ASP, San Francisco), **105**, 541–2.

[2001], 'The formation and evolution of multiple star systems', in *Evolution of Binary and Multiple Star Systems; A Meeting in Celebration of Peter Eggleton's 60th Birthday*, ed. P. Podsiadlowski, S. Rappaport, A. R. King, F. D'Antona & L. Burderi (ASP, San Francisco), **229**, 77–88.

Aarseth, S. J. & Saslaw, W. C. [1972], 'Virial mass determinations of bound and unstable groups of galaxies', *Astrophys. J.* **172**, 17–35.

[1982], 'Formation of voids in the galaxy distribution', *Astrophys. J.* **258**, L7–L10.

Aarseth, S. J. & Wolf, N. J. [1972], 'Depletion of low-mass stars in clusters', *Astrophys. Lett.* **12**, 159–64.

Aarseth, S. J. & Zare, K. [1974], 'A regularization of the three-body problem', *Celes. Mech.* **10**, 185–205.

Aarseth, S. J., Gott, J. R. & Turner, E. L. [1979], 'N-body simulations of galaxy clustering, I. Initial conditions and galaxy collapse times', *Astrophys. J.* **228**, 664–83.

Aarseth, S. J., Hénon, M. & Wielen, R. [1974], 'A comparison of numerical methods for the study of star cluster dynamics', *Astron. Astrophys.* **37**, 183–7.

Aarseth, S. J., Lin, D. N. C. & Palmer, P. L. [1993], 'Evolution of planetesimals, II. Numerical simulations', *Astrophys. J.* **403**, 351–76.

Aarseth, S. J., Lin, D. N. C. & Papaloizou, J. C. B. [1988], 'On the collapse and violent relaxation of protoglobular clusters', *Astrophys. J.* **324**, 288–310.

Aarseth, S. J., Anosova, J. P., Orlov, V. V. & Szebehely, V. G. [1994a], 'Global chaoticity in the Pythagorean three-body problem', *Celes. Mech. Dyn. Ast.* **58**, 1–16.

[1994b], 'Close triple approaches and escape in the three-body problem', *Celes. Mech. Dyn. Ast.* **60**, 131–7.

Adams, F. C. & Laughlin, G. [2001], 'Constraints on the birth aggregate of the Solar System', *Icarus* **150**, 151–62.

Agekian, T. A. & Anosova, J. P. [1968a], 'Investigation of the dynamics of triple systems by the method of statistical tests, II', *Astrophys.* **4**, 11–16.

[1968b], 'Evolution of binary systems in the galactic field', *Astrophys.* **4**, 196–8.

[1968c], 'A study of the dynamics of triple systems by means of statistical sampling', *Soviet Astron.* **11**, 1006–14.

[1971], 'Probability of binary-system formation through triple encounters', *Soviet Astron.* **15**, 411–14.

Aguilar, L. A. & White, S. D. M. [1985], 'Tidal interactions between spherical galaxies', *Astrophys. J.* **295**, 374–87.

Ahmad, A. & Cohen, L. [1973], 'A numerical integration scheme for the N-body gravitational problem', *J. Comput. Phys.* **12**, 389–402.

[1974], 'Integration of the N-body gravitational problem by separation of the force into a near and a far component', in *Lecture Notes in Mathematics*, ed. D. G. Bettis (Springer-Verlag, New York), **362**, 313–6.

Aitken, R. G. [1914], 'Measures of double stars', *Publ. Lick Obs.* **12**, 35–6.

Aksnes, K. & Standish, E. M. [1969], 'A numerical test of the relaxation time', *Astrophys. J.* **158**, 519–27.

Alexander, M. E. [1986], 'Simulation of binary–single star and binary–binary scattering', *J. Comput. Phys.* **64**, 195–219.

Allen, C. & Poveda, A. [1972], 'On the reproducibility of run-away stars formed in collapsing clusters', in *Gravitational N-Body Problem*, ed. M. Lecar (Reidel, Dordrecht), 114–23.

Ambartsumian, V. A. [1938], *Ann. Leningrad State Univ.*, No. 22, 19.

[1985], 'On the dynamics of open clusters', in *Dynamics of Star Clusters*, ed. J. Goodman & P. Hut (Reidel, Dordrecht), 521–4.

Anosova, J. P. [1986], 'Dynamical evolution of triple systems', *Astrophys. Sp. Sci.* **124**, 217–41.

[1991], 'Strong triple interactions in the general three-body problem', *Celes. Mech. Dyn. Ast.* **51**, 1–15.

Anosova, J. P., Bertov, D. I. & Orlov, V. V. [1984], 'Influence of rotation on the evolution of triple systems', *Astrophys.* **20**, 177–84.

Anosova, J. P., Orlov, V. V. & Aarseth, S. J. [1994], 'Initial conditions and dynamics of triple systems', *Celes. Mech. Dyn. Ast.* **60**, 365–72.

Arabadjis, J. S. & Richstone, D. O. [1996], 'The evolution of dense rotating N-body systems', *Bull. Am. Astron. Soc.* **28**, 1310.

Bacon, D., Sigurdsson, S. & Davies, M. B. [1996], 'Close approach during hard binary–binary scattering', *Mon. Not. R. Astron. Soc.* **281**, 830–46.

Bailey, V. C. & Davies, M. B. [1999], 'Red giant collisions in the Galactic Centre', *Mon. Not. R. Astron. Soc.* **308**, 257–70.

Bailyn, C. D. [1987], Ph.D. Thesis (Yale University).

[1995], 'Blue stragglers and other stellar anomalies: implications for the dynamics of globular clusters', *Ann. Rev. Astron. Astrophys.* **33**, 133–62.

Barnes, J. [1984], 'N-body studies of compact groups of galaxies dominated by dark matter', *Mon. Not. R. Astron. Soc.* **208**, 873–85.

Barnes, J. & Hut, P. [1986], 'A hierarchical $O(N \log N)$ force-calculation algorithm', *Nature* **324**, 446–9.

Baumgardt, H. [2001], 'Scaling of N-body calculations', *Mon. Not. R. Astron. Soc.* **325**, 1323–31.

[2002], 'Lifetimes of star clusters and dynamical relaxation', in *Dynamics of Star Clusters and the Milky Way*, ed. S. Deiters, B. Fuchs, A. Just, R. Spurzem & R. Wielen (ASP, San Francisco), **228**, 125–30.

Baumgardt, H. & Makino, J. [2003] 'Dynamical evolution of star clusters in tidal fields', *Mon. Not. R. Astron. Soc.* **340**, 227–46.

Baumgardt, H., Hut, P. & Heggie, D.C. [2002], 'Long-term evolution of isolated N-body systems', *Mon. Not. R. Astron. Soc.* **336**, 1069–81.

Baumgardt, H., Hut, P., Makino, J., McMillan, S. & Portegies Zwart, S. [2003], 'On the central structure of M15', *Astrophys. J.* **582**, L21–L24.

Baumgarte, J. [1973], 'Numerical stabilization of all laws of conservation in the many body problem', *Celes. Mech.* **8**, 223–8.

Baumgarte, J. & Stiefel, E. L. [1974], 'Examples of transformations improving the numerical accuracy of the integration of differential equations', in *Lecture Notes in Mathematics*, ed. D. G. Bettis (Springer-Verlag, New York), **362**, 207–36.

Beaugé, C. & Aarseth, S. J. [1990], 'N-body simulations of planetary formation', *Mon. Not. R. Astron. Soc.* **245**, 30–9.

Benz, W. & Hills, J. G. [1992], 'Three-dimensional hydrodynamical simulations of colliding stars, III. Collisions and tidal captures of unequal-mass main-sequence stars', *Astrophys. J.* **389**, 546–57.

Bettis, D. G. & Szebehely, V. [1972], 'Treatment of close approaches in the numerical integration of the gravitational problem of N bodies', in *Gravitational N-Body Problem*, ed. M. Lecar (Reidel, Dordrecht), 388–405.

Bettwieser, E. & Sugimoto, D. [1984], 'Post-collapse evolution and gravothermal oscillations of globular clusters', *Mon. Not. R. Astron. Soc.* **208**, 493–509.

Binney, J. & Tremaine, S. [1987], *Galactic Dynamics* (Princeton University Press).

Blaes, O., Lee, M. H. & Socrates, A. [2002], 'The Kozai mechanism and the evolution of binary supermassive black holes', *Astrophys. J.* **578**, 775–86.

Boily, C. M. [2000], 'The impact of rotation on cluster dynamics', in *Massive Stellar Clusters*, ed. A. Lançon & C. M. Boily (ASP, San Francisco), **211**, 190–6.

Boily, C. M. & Spurzem, R. [2000], 'N-body modelling of rotating star clusters', in *The Galactic Halo*, ed. A. Noels *et al.* (Liege, Belgium), 607–12.

Bolte, M. [1992], 'CCD photometry in the globular cluster NGC 288, I. Blue stragglers and main-sequence binary stars', *Astrophys. J. Suppl.* **82**, 145–65.

Bolte, M., Hesser, J. E. & Stetson, P. B. [1993], 'Canada–France–Hawaii telescope observations of globular cluster cores: blue straggler stars in M3', *Astrophys. J.* **408**, L89–L92.

Bonnell, I. A. & Davies, M. B. [1998], 'Mass segregation in young stellar clusters', *Mon. Not. R. Astron. Soc.* **295**, 691–8.

Bonnell, I. A., Smith, K. W., Davies, M. B. & Horne, K. [2001], 'Planetary dynamics in stellar clusters', *Mon. Not. R. Astron. Soc.* **322**, 859–65.

Bouvier, P. & Janin, G. [1970], 'Disruption of star clusters through passing interstellar clouds investigated by numerical experiments', *Astron. Astrophys.* **9**, 461–5.

Boyd, P. T. & McMillan, S. L. W. [1993], 'Chaotic scattering in the gravitational three-body problem', *Chaos* **3**, 507–23.

Brehm, M., Bader, R., Heller, H. & Ebner, R. [2000], 'Pseudovectorization, SMP and message passing on the Hitachi SR8000-F1', in *Lecture Notes in Computer Science*, ed. A. Bode *et al.* (Springer-Verlag, New York), **1900**, 1351–62.

Brouwer, D. & Clemence, G. M. [1961], *Methods of Celestial Mechanics*, (Academic Press, New York).

Bulirsch, R. & Stoer, J. [1966], 'Numerical treatment of ordinary differential equations by extrapolation methods', *Num. Math.* **8**, 1–13.

Burdet, C. A. [1967], 'Regularization of the two-body problem', *Z. Angew. Math. Phys.* **18**, 434–8.

[1968], 'Theory of Kepler motion: the general perturbed two body problem', *Z. Angew. Math. Phys.* **19**, 345–68.

Burrau, C. [1913], 'Numerische Berechnung eines Spezialfalles des Dreikörper-Problems', *Astron. Nach.* **195**, 113–18.

Casertano, S. & Hut, P. [1985], 'Core radius and density measurements in N-body experiments: connections with theoretical and observational definitions', *Astrophys. J.* **298**, 80–94.

Cazenave, A., Lago, B. & Dominh, K. [1982], 'Numerical experiment applicable to the latest stage of planet growth', *Icarus* **51**, 133–48.

Chambers, J. E. [1999], 'A hybrid symplectic integrator that permits close encounters between massive bodies', *Mon. Not. R. Astron. Soc.* **304**, 793–9.

Chambers, J. E. & Wetherill, G. W. [1998], 'Making the terrestrial planets: N-body integrations of planetary embryos in three dimensions', *Icarus* **136**, 304–27.

Chandrasekhar, S. [1942], *Principles of Stellar Dynamics* (Dover, New York).

Chenciner, A. & Montgomery, R. [2000], 'A remarkable periodic solution of the three-body problem in the case of equal masses', *Ann. Math.* **152**, 881–901.

Chernoff, D. & Weinberg, M. [1990], 'Evolution of globular clusters in the Galaxy', *Astrophys. J.* **351**, 121–56.

Chirikov, B. V. [1979], 'A universal instability of many-dimensional oscillator systems', *Phys. Rep.* **52**, 263–379.

Cohn, H. [1979], 'Numerical integration of the Fokker–Planck equation and the evolution of star clusters', *Astrophys. J.* **234**, 1036–53.

[1980], 'Late core collapse in star clusters and the gravothermal instability', *Astrophys. J.* **242**, 765–71.

Cohn, H., Hut, P. & Wise, M. [1989], 'Gravothermal oscillations after core collapse in globular cluster evolution', *Astrophys. J.* **342**, 814–22.

Cool, A., Piotto, G. & King, I. [1996], 'The main sequence and a white dwarf sequence in the globular cluster NGC 6397', *Astrophys. J.* **468**, 655–62.

Couchman, H. M. P. [1991], 'Mesh-refined P^3M: a fast adaptive N-body algorithm', *Astrophys. J.* **368**, L23–L26.

Cox, L. P. & Lewis, J. S. [1980], 'Numerical simulation of the final stages of terrestrial planet formation', *Icarus* **44**, 706–21.

Daisaka, H. & Makino, J. [2003], 'The formation process and dynamics of the Uranian elliptical rings', *Astrophys. J.* , submitted.

Daisaka, H., Tanaka, H. & Ida, S. [2001], 'Viscosity in a dense planetary ring with self-gravitating particles', *Icarus* **154**, 296–312.

Danby, J. M. A. [1992], *Fundamentals of Celestial Mechanics* (Willmann–Bell, Richmond).

Davies, M. B. [2001], 'Stellar collisions in the Galactic centre', in *Dynamics of Star Clusters and the Milky Way*, ed. S. Deiters, B. Fuchs, A. Just, R. Spurzem & R. Wielen (ASP, San Francisco), **228**, 147–52.

Davies, M. B., Benz, W. & Hills, J. G. [1991], 'Stellar encounters involving red giants in globular cluster cores', *Astrophys. J.* **381**, 449–61.

Davies, M. B., Blackwell, R., Bailey, V. C. & Sigurdsson, S. [1998], 'The destructive effects of binary encounters on red giants in the Galactic centre', *Mon. Not. R. Astron. Soc.* **301**, 745–53.

de la Fuente Marcos, C. & de la Fuente Marcos, R. [1997], 'Eccentric giant planets in open star clusters', *Astron. Astrophys.* **326**, L21–L24.

[1999], 'Runaway planets', *New Astron.* **4**, 21–32.

[2001], 'Reshaping the outskirts of planetary systems', *Astron. Astrophys.* **371**, 1097–206.

de la Fuente Marcos, R. [1995], 'The initial mass function and the dynamical evolution of open clusters, I. Conservative systems', *Astron. Astrophys.* **301**, 407–18.

[1996a], 'The initial mass function and the dynamical evolution of open clusters, II. With mass loss', *Astron. Astrophys.* **308**, 141–50.

[1996b], 'The initial mass function and the dynamical evolution of open clusters, III. With primordial binaries', *Astron. Astrophys.* **314**, 453–64.

[1997], 'The initial mass function and the dynamical evolution of open clusters, IV. Realistic systems', *Astron. Astrophys.* **322**, 764–77.

[1998], 'Searching for open cluster remnants', *Astron. Astrophys.* **333**, L27–L30.

de la Fuente Marcos, R. & de la Fuente Marcos, C. [2000], 'On the dynamical evolution of the brown dwarf population in open clusters', *Astrophys. Sp. Sci.* **271**, 127–44.

[2002], 'Dynamics of very rich open clusters', *Astrophys. Sp. Sci.* **280**, 381–404.

de la Fuente Marcos, R., Aarseth, S. J., Kiseleva, L. G. & Eggleton, P. P. [1997], 'Hierarchical systems in open clusters', in *Visual Double Stars: Formation, Dynamics and Evolutionary Tracks*, ed. J. A. Docobo, A. Elipe & H. McAlister (Kluwer, Dordrecht), 165–78.

Dehnen, W. [2001], 'Towards optimal softening in three-dimensional N-body codes, I. Minimizing the force error', *Mon. Not. R. Astron. Soc.* **324**, 273–91.

Dejonghe, H. & Hut, P. [1986], 'Round-off sensitivity in the N-body problem', in *The Use of Supercomputers in Stellar Dynamics*, ed. P. Hut & S. McMillan (Springer-Verlag, New York), 212–18.

Dekel, A. & Aarseth, S. J. [1984], 'The spatial correlation function of galaxies confronted with theoretical scenarios', *Astrophys. J.* **283**, 1–23.

Dorband, E. N., Hemsendorf, M. & Merritt, D. [2003], 'Systolic and hyper-systolic algorithms for the gravitational N-body problem with an application to Brownian motion', astro-ph/0112092, *J. Comput. Phys.* **185**, 484–511.

Dubinski, J. [1996], 'A parallel tree code', *New Astron.* **1**, 133–47.

Duchêne, G. [1999], 'Binary fraction in low-mass star forming regions: a reexamination of the possible excesses and implications', *Astron. Astrophys.* **341**, 547-52.

Duquennoy, A. & Mayor, M. [1991], 'Multiplicity among solar-type stars in the solar neighbourhood, II. Distribution of the orbital elements in an unbiased sample', *Astron. Astrophys.* **248**, 485–524.

Dyer, C. C. & Ip, P. S. S. [1993], 'Softening in N-body simulations of collisionless systems', *Astrophys. J.* **409**, 60–7.

Ebisuzaki, T., Makino, J., Tsuru, T. G., Funato, Y., Portegies Zwart, S., Hut, P., McMillan, S., Matsushita, S., Matsumoto, H. & Kawabe, R. [2001], 'Missing link found? The "runaway" path to supermassive black holes', *Astrophys. J.* **562**, L19–L22.

Efstathiou, G. P. & Eastwood, J. [1981], 'On the clustering of particles in an expanding universe', *Mon. Not. R. Astron. Soc.* **194**, 503–25.

Eggleton, P. P. [1983], 'Approximations to the radii of Roche lobes', *Astrophys. J.* **268**, 368–9.

Eggleton, P. & Kiseleva, L. [1995], 'An empirical condition for stability of hierarchical triple systems', *Astrophys. J.* **455**, 640–5.

Eggleton, P. P., Fitchett, M. J. & Tout, C. A. [1989], 'The distribution of visual binaries with two bright components', *Astrophys. J.* **347**, 998–1011.

Eggleton, P. P., Kiseleva, L. G. & Hut, P. [1998], 'The equilibrium tide model for tidal friction', *Astrophys. J.* **499**, 853–70.

Einsel, C. & Spurzem, R. [1999], 'Dynamical evolution of rotating stellar systems, I. Pre-collapse, equal-mass system', *Mon. Not. R. Astron. Soc.* **302**, 81–95.

Elson, R. A. W., Sigurdsson, S., Davies, M., Hurley, J. & Gilmore, G. [1998], 'The binary star population of the young cluster NGC 1818 in the Large Magellanic Cloud', *Mon. Not. R. Astron. Soc.* **300**, 857–62.

Farouki, R. T. & Salpeter, E. E. [1982], 'Mass segregation, relaxation and the Coulomb logarithm in N-body systems', *Astrophys. J.* **253**, 512–19.

Fekel, F. C., Scarfe, C. D., Barlow, D. J., Hartkopf, W. I., Mason, B. & McAlister, H.A. [2002], 'The quadruple system μ Orionis: three-dimensional orbit and physical parameters', *Astron. J.* **123**, 1723–40.

Fellhauer, M., Kroupa, P., Baumgardt, H., Bien, R., Boily, C. M., Spurzem, R. & Wassmer, N. [2000], 'SUPERBOX – An efficient code for collisionless galactic dynamics', *New Astron.* **5**, 305–26.

Fellhauer, M., Lin, D. N. C, Bolte, M., Aarseth, S. J. & Williams, K. A. [2003], 'The white dwarf deficit in open clusters: dynamical processes', *Astrophys. J. Letters*, in press.

Ferraro, F. R., Sills, A., Rood, R. T., Paltrinieri, B. & Buonanno, R. [2003], 'Blue straggler stars: a direct comparison of star counts and population ratios in six Galactic globular clusters', *Astrophys. J.* **558**, 464–77.

Findlay, D. A. [1983], *Cosmological N-Body Simulations of Galaxy Clustering: Tidal Interactions and Merging*, Ph.D. Thesis (University of Cambridge).

Ford, E. B., Havlickova, M. & Rasio, F. [2001], 'Dynamical instabilities in extrasolar planetary systems containing two giant planets', *Icarus* **150**, 303–13.

Ford, E. B., Kozinsky, B. & Rasio, F. A. [2000], 'Secular evolution of hierarchical triple star systems', *Astrophys. J.* **535**, 385–401.

Freitag, M. & Benz, W. [2001], 'A new Monte Carlo code for star cluster simulations, I. Relaxation', *Astron. Astrophys.* **375**, 711–38.

[2002], 'A new Monte Carlo code for star cluster simulations, II. Central black hole and stellar collisions', *Astron. Astrophys.* **394**, 345–74.

Fukushige, T. & Heggie, D.C. [1995], 'Pre-collapse evolution of galactic globular clusters', *Mon. Not. R. Astron. Soc.* **276**, 206–18.

[2000], 'The time-scale of escape from star clusters', *Mon. Not. R. Astron. Soc.* **318**, 753–61.

Fukushima, T. [1997], 'Vector integration of dynamical motions by the Picard–Chebyshev method', *Astron. J.* **113**, 2325–8.

Fullerton, L. W. & Hills, J. G. [1982], 'Computer simulations of close encounters between binary and single stars: the effect of the impact velocity and the stellar masses', *Astron. J.* **87**, 175–83.

Funato, Y., Hut, P., McMillan, S. & Makino, J. [1996], 'Time-symmetrized Kustaanheimo–Stiefel regularization', *Astron. J.* **112**, 1697–708.

Gerhard, O. E. [1981], 'N-body simulations of disc-halo galaxies: isolated systems, tidal interactions and merging', *Mon. Not. R. Astron. Soc.* **197**, 179–208.

[1982], *Dynamical Effects of Galaxy Encounters*, Ph.D. Thesis (University of Cambridge).

[2000], 'Dynamical masses, time-scales and evolution of star clusters', in *Massive Stellar Clusters*, ed. A. Lançon & C. M. Boily (ASP, San Francisco), **211**, 12–24.

Giannone, G. & Molteni, D. [1985], 'The role of initial binaries in cluster evolution: direct N-body calculations', *Astron. Astrophys.* **143**, 321–6.

Giersz, M. [1996], 'Monte-Carlo simulations', in *Dynamical Evolution of Star Clusters*, ed. P. Hut & J. Makino (Kluwer, Dordrecht), 101–10.

[2001a], 'Monte Carlo simulations of star clusters, II. Tidally limited, multi-mass systems with stellar evolution', *Mon. Not. R. Astron. Soc.* **324**, 218–30.

[2001b], 'Monte Carlo simulations of star clusters', in *Dynamics of Star Clusters and the Milky Way*, ed. S. Deiters, B. Fuchs, A. Just, R. Spurzem & R. Wielen (ASP, San Francisco), **228**, 61–6.

Giersz, M. & Heggie, D.C. [1994a], 'Statistics of N-body simulations, I. Equal masses before core collapse', *Mon. Not. R. Astron. Soc.* **268**, 257–75.

[1994b], 'Statistics of N-body simulations, II. Equal masses after core collapse', *Mon. Not. R. Astron. Soc.* **270**, 298–324.

[1996], 'Statistics of N-body simulations, III. Unequal masses', *Mon. Not. R. Astron. Soc.* **279**, 1037–56.

[1997], 'Statistics of N-body simulations, IV. Unequal masses with a tidal field', *Mon. Not. R. Astron. Soc.* **286**, 709–31.

Giersz, M. & Spurzem, R. [1994], 'Comparing direct N-body integration with anisotropic gaseous models of star clusters', *Mon. Not. R. Astron. Soc.* **269**, 241–56.

[2000], 'A stochastic Monte Carlo approach to model real star cluster evolution, II. Self-consistent models and primordial binaries', *Mon. Not. R. Astron. Soc.* **317**, 581–606.

[2003], 'A stochastic Monte Carlo approach to model real star cluster evolution, III. Direct integrations of three- and four-body interactions', *Mon. Not. R. Astron. Soc.* in press.

Gladman, B. [1993], 'Dynamics of systems of two close planets', *Icarus* **106**, 247–63.

Gnedin, O. Y. & Ostriker, J. P. [1997], 'Destruction of the galactic globular cluster system', *Astrophys. J.* **474**, 223–55.

Goldreich, P. & Tremaine, S. D. [1978], 'The velocity dispersion in Saturn's rings', *Icarus* **34**, 227–39.

Gonzalez, C. C. & Lecar, M. [1968], 'Encounters and escapes', *Bull. Astron.* **3**, 209–11.

Goodman, J. [1987], 'On gravothermal oscillations', *Astrophys. J.* **313**, 576–95.

Goodman, J., Heggie, D.C. & Hut, P. [1993], 'On the exponential instability of N-body systems', *Astrophys. J.* **415**, 715–33.

Goodwin, S. P. [1997], 'Residual gas expulsion from young globular clusters', *Mon. Not. R. Astron. Soc.* **284**, 785–802.

Gott, J. R., Turner, E. L. & Aarseth, S. J. [1979], 'N-body simulations of galaxy clustering, III. The covariance function', *Astrophys. J.* **234**, 13–26.

Griffith, J. S. & North, R. D. [1974], 'Escape or retention in the three-body problem', *Celes. Mech.* **8**, 473–9.

Grillmair, C. J. [1998], 'Probing the Galactic Halo with globular cluster tidal tails', in *Galactic Halos*, ed. D. Zaritsky (ASP, San Francisco), **136**, 45–52.

Grindlay, J. E. [1996], 'High resolution studies of compact binaries in globular clusters with HST and ROSAT', in *Dynamical Evolution of Star Clusters*, ed. P. Hut & J. Makino (Kluwer, Dordrecht), 171–80.

Guhathakurta, P., Webster, Z. T., Yanny, B., Schneider, D. P. & Bahcall, J. N. [1998], 'Globular cluster photometry with the Hubble Space Telescope, VII. Color gradients and blue stragglers in the central region of M30', *Astron. J.* **116**, 1757–74.

Gunn, J. E. & Griffin, R. F. [1979], 'Dynamical studies of globular clusters based on photometric radial velocities of individual stars, I. M3', *Astron. J.* **84**, 752–73.

Hansen, B. M. S. & Phinney, E. S. [1997], 'The pulsar kick velocity distribution', *Mon. Not. R. Astron. Soc.* **291**, 569–77.

Harrington, R. S. [1972], 'Stability criteria for triple stars', *Celes. Mech.* **6**, 322–7.

[1974], 'The dynamical decay of unstable 4-body systems', *Celes. Mech.* **9**, 465–70.

[1975], 'Production of triple stars by the dynamical decay of small stellar systems', *Astron. J.* **80**, 1081–86.

Lecar, M. & Cruz-González, C. [1972], 'A numerical experiment on relaxation times in stellar dynamics', in *Gravitational N-Body Problem*, ed. M. Lecar (Reidel, Dordrecht), 131–5.

Lecar, M. & Franklin, F. A. [1973], 'On the original distribution of the asteroids, I', *Icarus* **20**, 422–36.

Lecar, M., Loeser, R., & Cherniack, J. R. [1974], 'Numerical integration of gravitational N-body systems with the use of explicit Taylor series', in *Lecture Notes in Mathematics*, ed. D. G. Bettis (Springer-Verlag, New York), **362**, 451–70.

Lecar, M., Franklin, F. A., Holman, M. J. & Murray, N. W. [2001], 'Chaos in the Solar System', *Ann. Rev. Astron. Astrophys.* **39**, 581–631.

Lee, H. M., Fahlman, G. B. & Richer, H. B. [1991], 'Multicomponent models for the dynamic evolution of globular clusters', *Astrophys. J.* **366**, 455–63.

Lee, M. H. [1993], 'N-body evolution of dense clusters of compact stars', *Astrophys. J.* **418**, 147–62.

Leinhardt, Z. M. & Richardson, D. C. [2002], 'N-body simulations of planetesimal evolution: effect of varying impactor mass ratio', *Icarus* **159**, 306–13.

Leinhardt, Z. M., Richardson, D. C. & Quinn, T. [2000], 'Direct N-body simulations of rubble pile collisions', *Icarus* **146**, 133–51.

Lemaître, G. [1955], 'Regularization of the three-body problem', *Vistas Astron.* **1**, 207–15.

Leonard, P. J. T. [1988], *The Dynamics of Open Star Clusters*, Ph.D. Thesis (University of Toronto).

[1989], 'Stellar collisions in globular clusters and the blue straggler problem', *Astron. J.* **98**, 217–26.

[1991], 'The maximum possible velocity of dynamically ejected runaway stars', *Astron. J.* **101**, 562–71.

[1995], 'Merged dynamically ejected OB runaway stars', *Mon. Not. R. Astron. Soc.* **277**, 1080–6.

[1996], 'The implications of the binary properties of the M67 blue stragglers', *Astrophys. J.* **470**, 521–7.

Leonard, P. J. T. & Duncan, M. J. [1988], 'Runaway stars from young star clusters containing initial binaries, I. Equal-mass, equal-energy binaries', *Astron. J.* **96**, 222–32.

[1990], 'Runaway stars from young star clusters containing initial binaries, II. A mass spectrum and a binary energy spectrum', *Astron. J.* **99**, 608–16.

Leonard, P. J. T. & Linnell, A. P. [1992], 'On the possibility of a collisional origin for the blue stragglers and contact binaries in the old open clusters M67 and NGC 188', *Astron. J.* **103**, 1928–44.

Levi-Civita, T. [1920], 'Sur la régularisation du problème des trois corps', *Acta Math.* **42**, 99–44.

Levison, H. F. & Duncan, M. J. [1994], 'The long-term dynamical behaviour of short-period comets', *Icarus* **108**, 18–36.

Lissauer, J. J & Rivera, E. J. [2001], 'Stability analysis of the planetary system orbiting υ Andromedae, II. Simulations using new Lick Observatory fits', *Astrophys. J.* **554**, 1141–50.

Lombardi, J. C., Rasio, F. A. & Shapiro, S. L. [1996], 'Collisions of main-sequence stars and the formation of blue stragglers in globular clusters', *Astrophys. J.* **468**, 797–818.

Louis, P. D. & Spurzem, R. [1991], 'Anisotropic gaseous models for the evolution of star clusters', *Mon. Not. R. Astron. Soc.* **251**, 408–26.

Lupton, R. H. & Gunn, J. E. [1987], 'Three-integral models of globular clusters', *Astron. J.* **93**, 1106–13.

Lynden-Bell, D. [1967], 'Statistical mechanics of violent relaxation in stellar systems', *Mon. Not. R. Astron. Soc.* **136**, 101–21.

Lynden-Bell, D. & Eggleton, P. P. [1980], 'On the consequences of the gravothermal catastrophe', *Mon. Not. R. Astron. Soc.* **191**, 483–98.

Magorrian, J. & Tremaine, S. [1999], 'Rates of tidal disruption of stars by massive central black holes', *Mon. Not. R. Astron. Soc.* **309**, 447–60.

Makino, J. [1990], 'Comparison of two different tree algorithms', *J. Comput. Phys.* **88**, 393–408.

[1991a], 'Optimal order and time-step criterion for Aarseth-type N-body integrators', *Astrophys. J.* **369**, 200–12.

[1991b], 'A modified Aarseth code for GRAPE and vector processors', *Publ. Astron. Soc. Japan* **43**, 859–76.

[1996a], 'Gravothermal oscillations', in *Dynamical Evolution of Star Clusters*, ed. P. Hut & J. Makino (Kluwer, Dordrecht), 151–60.

[1996b], 'Postcollapse evolution of globular clusters', *Astrophys. J.* **471**, 796–803.

[1997], 'Merging of galaxies with central black holes, II. Evolution of the black hole binary and the structure of the core', *Astrophys. J.* **478**, 58–65.

[2002], 'An efficient parallel algorithm for $O(N^2)$ direct summation method and its variations on distributed-memory parallel machines', *New Astron.* **7**, 373–84.

[2003], 'The GRAPE project: current status and future outlook', in *Astrophysical Supercomputing Using Particle Simulations*, ed. P. Hut & J. Makino (ASP, San Francisco), 13–24.

Makino, J. & Aarseth, S. J. [1992], 'On a Hermite integrator with Ahmad–Cohen scheme for gravitational many-body problems', *Publ. Astron. Soc. Japan* **44**, 141–51.

Makino, J. & Ebisuzaki, T. [1996], 'Merging of galaxies with central black holes, I. Hierarchical mergings of equal-mass galaxies', *Astrophys. J.* **465**, 527–33.

Makino, J. & Fukushige, T. [2002], 'A 11.55 Tflops simulation of black holes in a galactic center on GRAPE-6', Gordon Bell Prize, in *Proceedings IEE/ACM SC2002 Conference*, (CD-ROM).

Makino, J. & Hut, P. [1988], 'Performance analysis of direct N-body calculations', *Astrophys. J. Suppl.* **68**, 833–56.

[1990], 'Bottlenecks in simulations of dense stellar systems', *Astrophys. J.* **365**, 208–18.

Makino, J. & Sugimoto, D. [1987], 'Effect of suprathermal particles on gravothermal oscillation', *Publ. Astron. Soc. Japan* **39**, 589–603.

Makino, J. & Taiji, M. [1998], *Scientific Simulations with Special-Purpose Computers: The GRAPE systems* (John Wiley & Sons, Chichester).

Makino, J., Fukushige, T. & Namura, K. [2003], 'GRAPE-6: the massively-parallel special-purpose computer for astrophysical particle simulations', Publ. Astron. Soc. Japan, submitted.

Makino, J., Kokubo, E. & Taiji, M. [1993], 'HARP: a special-purpose computer for N-body problem', *Publ. Astron. Soc. Japan* **45**, 349–60.

Makino, J., Tanekusa, J. & Sugimoto, D. [1986], 'Gravothermal oscillation in gravitational many-body systems', *Publ. Astron. Soc. Japan* **38**, 865–77.

Makino, J., Taiji, M., Ebisuzaki, T. & Sugimoto, D. [1997], 'GRAPE-4: a massively parallel special-purpose computer for collisional N-body simulations', *Astrophys. J.* **480**, 432–46.

Mann, P. J. [1987], 'Finite difference methods for the classical particle–particle gravitational N-body problem', *Comp. Phys. Comm.* **47**, 213–28.

Marchal, C., Yoshida, J. & Sun, Y.-S. [1984], 'A test of escape valid even for very small mutual distances, I. The acceleration and the escape velocities of the third body', *Celes. Mech.* **33**, 193–207.

Mardling, R. A. [1991], *Chaos in Binary Star Systems*, Ph.D. Thesis (Monash University).

[1995a], 'The role of chaos in the circularization of tidal capture binaries, I. The chaos boundary', *Astrophys. J.* **450**, 722–31.

[1995b], 'The role of chaos in the circularization of tidal capture binaries, II. Long-time evolution', *Astrophys. J.* **450**, 732–47.

[1996a], 'Tidal capture binaries in clusters: the chaos model for tidal evolution', in *The Origins, Evolution, and Destinies of Binary Stars in Clusters*, ed. E. F. Milone & J.-C. Mermilliod (ASP, San Francisco), 399–408.

[1996b], 'Chaos and tidal capture', in *Evolutionary Processes in Binary Stars*, ed. R. A. M. J. Wijers, M. B. Davies & C. A. Tout (Kluwer, Dordrecht), 81–99.

[1996c], 'Tidal capture in star clusters', in *Dynamical Evolution of Star Clusters*, ed. P. Hut & J. Makino (Kluwer, Dordrecht), 273–82.

[2001], 'Stability in the general three-body problem', in *Evolution of Binary and Multiple Star Systems; A Meeting in Celebration of Peter Eggleton's 60th Birthday*, ed. P. Podsiadlowski, S. Rappaport, A. R. King, F. D'Antona & L. Burderi (ASP, San Francisco), **229**, 101–16.

[2003a], 'A new formalism for studying three-body interactions', in *Astrophysical Supercomputing using Particle Simulations*, ed. J. Makino & P. Hut (ASP, San Francisco), 123–30.

[2003b], 'A new three-body formalism', in preparation.

Mardling, R. & Aarseth, S. [1999], 'Dynamics and stability of three-body systems', in *The Dynamics of Small Bodies in the Solar System*, ed. B. A. Steves & A. E. Roy (Kluwer, Dordrecht), 385–92.

[2001], 'Tidal interactions in star cluster simulations', *Mon. Not. R. Astron. Soc.* **321**, 398–420.

Mardling, R. A. & Lin, D. N. C. [2002], 'Calculating the tidal, spin, and dynamical evolution of extrasolar planetary systems', *Astrophys. J.* **573**, 829–44.

[2003], 'On the interaction between three planets and their host star Upsilon Andromedae', *Astrophys. J.* , submitted.

Mathieu, R. D. [1983], *The Structure, Internal Kinematics and Dynamics of Open Star Clusters*, Ph.D. Thesis (University of California, Berkeley).

Mathieu, R. D., Latham, D. W. & Griffin, R. F. [1990], 'Orbits of 22 spectroscopic binaries in the open cluster M67', *Astron. J.* **100**, 1859–81.

Mazeh, T. [1990], 'Eccentric orbits in samples of circularized binary systems: the fingerprint of a third star', *Astron. J.* **99**, 675–7.

McCrea, W. H. [1964], 'Extended main-sequence of some clusters', *Mon. Not. R. Astron. Soc.* **128**, 147–55.

McMillan, S. L. W. [1986], 'The vectorization of small-N integrators', in *The Use of Supercomputers in Stellar Dynamics*, ed. P. Hut & S. McMillan (Springer-Verlag, New York), 156–61.

McMillan, S. L. W. & Aarseth, S. J. [1993], 'An $O(N \log N)$ integration scheme for collisional stellar systems', *Astrophys. J.* **414**, 200–12.

McMillan, S. L. W. & Hut. P. [1994], 'Star cluster evolution with primordial binaries, III. Effect of the galactic tidal field', *Astrophys. J.* **427**, 793–807.

[1996], 'Binary–single-star scattering, VI. Automatic determination of interaction cross sections', *Astrophys. J.* **467**, 348–58.

McMillan, S. L. W. & Lightman, A. P. [1984a], 'A unified N-body and statistical treatment of stellar dynamics, I. The hybrid code', *Astrophys. J.* **283**, 801–12.

[1984b], 'A unified N-body and statistical treatment of stellar dynamics, II. Applications to globular cluster cores', *Astrophys. J.* **283**, 813–24.

McMillan, S., Hut. P. & Makino, J. [1990], 'Star cluster evolution with primordial binaries, I. A comparative study', *Astrophys. J.* **362**, 522–37.

[1991], 'Star cluster evolution with primordial binaries, II. Detailed analysis', *Astrophys. J.* **372**, 111–24.

Mermilliod, J.-C. & Mayor, M. [1992], 'Distribution of orbital elements for red giant spectroscopic binaries in open clusters', in *Binaries as Tracers of Stellar Formation*, ed. A. Duquennoy & M. Mayor (Cambridge University Press), 183–201.

Merritt, D. [1996], 'Optimal smoothing for N-body codes', *Astron. J.* **111**, 2462–4.

Merritt, D. & Quinlan, G. D. [1998], 'Dynamical evolution of elliptical galaxies with central singularities', *Astrophys. J.* **498**, 625–39.

Meusinger, H., Scholz, R. D. & Irwin, M. J. [2001], 'A proper motion search for stars escaping from a globular cluster with high velocity', in *Dynamics of Star Clusters and the Milky Way*, ed. S. Deiters, B. Fuchs, A. Just, R. Spurzem & R. Wielen (ASP, San Francisco), **228**, 520–2.

Meylan, G. & Heggie, D. C. [1997], 'Internal dynamics of globular clusters', *Astron. Astrophys. Rev.* **8**, 1–143.

Michie, R. W. [1963], 'On the distribution of high energy stars in spherical stellar systems', *Mon. Not. R. Astron. Soc.* **125**, 127–39.

Mikkola, S. [1983], 'Encounters of binaries, I. Equal energies', *Mon. Not. R. Astron. Soc.* **203**, 1107–21.

[1984a], *Dynamics of Interacting Binaries*, Ph.D. Thesis (University of Turku).

[1984b], 'Encounters of binaries, II. Unequal energies', *Mon. Not. R. Astron. Soc.* **207**, 115–26.

[1984c], 'Encounters of binaries, III. Fly-bys', *Mon. Not. R. Astron. Soc.* **208**, 75–82.

[1985a], 'A practical and regular formulation of the N-body equations', *Mon. Not. R. Astron. Soc.* **215**, 171–7.

[1985b], 'Numerical simulations of encounters of hard binaries', in *Dynamics of Star Clusters*, ed. J. Goodman & P. Hut (Reidel, Dordrecht), 335–8.

[1988], 'On the effects of unequal masses in the statistics of three- and four-body interactions', in *The Few Body Problem*, ed. M. J. Valtonen (Reidel, Dordrecht), 261–4.

[1994], 'A numerical exploration of the phase-space structure of chaotic three-body scattering', *Mon. Not. R. Astron. Soc.* **269**, 127–36.

[1997a], 'Practical symplectic methods with time transformation for the few-body problem', *Celes. Mech. Dyn. Ast.* **67**, 145–65.

[1997b], 'Numerical treatment of small stellar systems with binaries', *Celes. Mech. Dyn. Ast.* **68**, 87–104.

Mikkola, S. and Aarseth, S. J. [1990], 'A chain regularization method for the few-body problem', *Celes. Mech. Dyn. Ast.* **47**, 375–90.

[1993], 'An implementation of N-body chain regularization', *Celes. Mech. Dyn. Ast.* **57**, 439–59.

[1996], 'A slow-down treatment for close binaries', *Celes. Mech. Dyn. Ast.* **64**, 197–208.

[1998], 'An efficient integration method for binaries in N-body simulations', *New Astron.* **3**, 309–20.

[2002], 'A time-transformed leapfrog scheme', *Celes. Mech. Dyn. Ast.* **84**, 343–54.

Mikkola, S. & Hietarinta, J. [1989], 'A numerical investigation of the one-dimensional Newtonian three-body problem', *Celes. Mech. Dyn. Ast.* **46**, 1–18.

[1990], 'A numerical investigation of the one-dimensional Newtonian three-body problem, II. Positive energies', *Celes. Mech. Dyn. Ast.* **47**, 321–31.

[1991], 'A numerical investigation of the one-dimensional Newtonian three-body problem, III. Mass dependence in the stability of motion', *Celes. Mech. Dyn. Ast.* **51**, 379–94.

Mikkola, S. & Innanen, K. [1995], 'Solar system chaos and the distribution of asteroid orbits', *Mon. Not. R. Astron. Soc.* **277**, 497–501.

Mikkola, S. & Saha, P. [2003], *Numerical Celestial Mechanics* (Kluwer, Dordrecht), in preparation.

Mikkola, S. & Tanikawa, K. [1999a], 'Explicit symplectic algorithms for time-transformed Hamiltonians', *Celes. Mech. Dyn. Ast.* **74**, 287–95.

[1999b], 'Algorithmic regularization of the few-body problem', *Mon. Not. R. Astron. Soc.* **310**, 745–9.

Miller, M. C. & Hamilton, D. C. [2002], 'Four-body effects in globular cluster black hole coalescence', *Astrophys. J.* **576**, 894–8.

Miller, R. H. [1964], 'Irreversibility in small stellar dynamical systems', *Astrophys. J.* **140**, 250–6.
[1974], 'Numerical difficulties with the gravitational N-body problem', in *Lecture Notes in Mathematics*, ed. D. G. Bettis (Springer-Verlag, New York), **362**, 260–75.
Milosavljević, M. & Merritt, D. [2001], 'Formation of galactic nuclei', *Astrophys. J.* **563**, 34–62.
Miyamoto, M. & Nagai, R. [1975], 'Three-dimensional models for the distribution of mass in galaxies', *Publ. Astron. Soc. Japan* **27**, 533–43.
Monaghan, J. J. [1976], 'A statistical theory of the disruption of three-body systems, II. High angular momentum', *Mon. Not. R. Astron. Soc.* **177**, 583–94.
Montgomery, K. A., Marschall, L. A. & Janes, K. A. [1993], 'CCD photometry of the old open cluster M67', *Astron. J.* **106**, 181–219.
Murray, C. D. & Dermott, S. F. [1999], *Solar System Dynamics* (Cambridge University Press).
Nacozy, P. E. [1972], 'The use of integrals in numerical integrations of the N-body problem', in *Gravitational N-Body Problem*, ed. M. Lecar (Reidel, Dordrecht), 153–64.
Nakano, T. & Makino, J. [1999], 'On the origin of density cusps in elliptical galaxies', *Astrophys. J.* **510**, 155–66.
Nash, P. E. & Monaghan, J. J. [1978], 'A statistical theory of the disruption of three-body systems, III. Three-dimensional motion', *Mon. Not. R. Astron. Soc.* **184**, 119–25.
Oh, K. S., Lin, D. N. C. & Aarseth, S. J. [1992], 'Tidal evolution of globular clusters, I. Method', *Astrophys. J.* **386**, 506–18.
[1995], 'On the tidal disruption of dwarf spheroidal galaxies around the Galaxy', *Astrophys. J.* **442**, 142–58.
Oort, J. H. [1965], 'Stellar dynamics', in *Galactic Structure*, ed. A. Blaauw & M. Schmidt (University of Chicago Press), 455–511.
Ostriker, J. P., Spitzer, L. & Chevalier, R. A. [1972], 'On the evolution of globular clusters', *Astrophys. J.* **176**, L51–L56.
Paczynski, B. [1976], 'Common envelope binaries', in *Structure and Evolution of Close Binary Systems*, ed. P. Eggleton, S. Mitton & J. Whelan, (Reidel, Dordrecht), 75–9.
Palmer, P. L., Lin, D. N. C. & Aarseth, S. J. [1992], 'Evolution of planetesimals, I. Dynamics: relaxation in a thin disk', *Astrophys. J.* **403**, 336–50.
Palmer, P. L., Aarseth, S. J., Mikkola, S. & Hashida, Y. [1998], 'High precision integration methods for orbit propagation', *J. Astronaut. Sci.* **46**, 329–42.
Peters, C. F. [1968a], *Treatment of Close Approaches in Stellar Dynamics*, Ph.D. Thesis (Yale University).
[1968b], 'Numerical Regularization', *Bull. Astron.* **3**, 167–75.
Peters, P. C. [1964], 'Gravitational radiation and the motion of two point masses', *Phys. Rev.* **136**, B1224–32.
Pfahl, E., Rappaport, S. & Podsiadlowski, P. [2002], 'A comprehensive study of neutron star retention in globular clusters', *Astrophys. J.* **573**, 283–305.

Pfahl, E., Rappaport, S., Podsiadlowski, P. & Spruit, H. [2002], 'A new class of high-mass X-ray binaries: implications for core collapse and neutron star recoil', *Astrophys. J.* **574**, 364–76.

Plummer, H. C. [1911], 'On the problem of distribution in globular star clusters', *Mon. Not. R. Astron. Soc.* **71**, 460–70.

Podsiadlowski, P. [1996], 'The response of tidally heated stars', *Mon. Not. R. Astron. Soc.* **279**, 1104–10.

Portegies Zwart, S. F. [1996], *Interacting Stars*, Ph.D. Thesis (University of Utrecht).

Portegies Zwart, S. F. & McMillan, S. L. W. [2002], 'The runaway growth of intermediate-mass black holes in dense star clusters', *Astrophys. J.* **576**, 899–907.

Portegies Zwart, S. F. & Meinen, A. T. [1993], 'Quick method for calculating energy dissipation in tidal interaction', *Astron. Astrophys.* **280**, 174–6.

Portegies Zwart, S. F. & Verbunt, F. [1996], 'Population synthesis of high-mass binaries', *Astron. Astrophys.* **309**, 179–96.

Portegies Zwart, S. F., Hut, P., McMillan, S. L. W. & Verbunt, F. [1997], 'Star cluster ecology, II. Binary evolution with single-star encounters', *Astron. Astrophys.* **328**, 143–57.

Portegies Zwart, S. F., Hut, P., Makino, J. & McMillan, S. L. W. [1998], 'On the dissolution of evolving star clusters', *Astron. Astrophys.* **337**, 363–71.

Portegies Zwart, S. F., Makino, J., McMillan, S. L. W. & Hut, P. [1999], 'Star cluster ecology, III. Runaway collisions in young compact star clusters', *Astron. Astrophys.* **348**, 117–26.

Portegies Zwart, S. F., McMillan, S. L. W., Hut, P. & Makino, J. [2001], 'Star cluster ecology, IV. Dissection of an open star cluster: photometry', *Mon. Not. R. Astron. Soc.* **321**, 199–226.

Portegies Zwart, S. F., Makino, J., McMillan, S. L. W. & Hut, P. [2002], 'The lives and deaths of star clusters near the Galactic center', *Astrophys. J.* **565**, 265–79.

Poveda, A., Ruiz, J. & Allen, C. [1967], 'Run-away stars as the result of the gravitational collapse of proto-stellar clusters', *Bol. Obs. Tonantzintla y Tacubaya* **28**, 86–90.

Press, W. H. [1986], 'Techniques and tricks for N-body computation', in *The Use of Supercomputers in Stellar Dynamics*, ed. P. Hut & S. McMillan (Springer-Verlag, New York), 184–92.

Press, W. H. & Spergel, D. N. [1988], 'Choice of order and extrapolation method in Aarseth-type N-body algorithms', *Astrophys. J.* **325**, 715–21.

Press, W. H. & Teukolsky, S. [1977], 'On the formation of close binaries by two-body tidal capture', *Astrophys. J.* **213**, 183–92.

Preto, M. & Tremaine, S. [1999], 'A class of symplectic integrators with adaptive time step for separable Hamiltonian systems', *Astron. J.* **118**, 2532–41.

Pryor, C., McClure, R. D., Hesser, J. E. & Fletcher, J. M. [1989], 'The frequency of primordial binary stars', in *Dynamics of Dense Stellar Systems*, ed. D. Merritt (Cambridge University Press), 175–81.

Quinlan, G. D. & Hernquist, L. [1997], 'The dynamical evolution of massive black hole binaries, II. Self-consistent N-body integrations', *New Astron.* **2**, 533–54.

Quinlan, G. D. & Tremaine, S. [1990], 'Symmetric multistep methods for the numerical integration of planetary orbits', *Astron. J.* **100**, 1694–700.

[1992], 'On the reliability of gravitational N-body integrations', *Mon. Not. R. Astron. Soc.* **259**, 505–18.

Raboud, R. & Mermilliod, J.-C. [1998a], 'Investigation of the Pleiades cluster, IV. The radial structure', *Astron. Astrophys.* **329**, 101–14.

[1998b], 'Evolution of mass segregation in open clusters: some observational evidences', *Astron. Astrophys.* **333**, 897–909.

Rasio, F. A., McMillan, S. & Hut, P. [1995], 'Binary–binary interactions and the formation of the PSR B1620–26 triple system in M4', *Astrophys. J.* **438**, L33–L36.

Richardson, D. C. [1993a], *Planetesimal Dynamics*, Ph.D. Thesis (University of Cambridge).

[1993b], 'A new tree code method for simulation of planetesimal dynamics', *Mon. Not. R. Astron. Soc.* **261**, 396–414.

[1994], 'Tree code simulations of planetary rings', *Mon. Not. R. Astron. Soc.* **269**, 493–511.

[1995], 'A self-consistent numerical treatment of fractal aggregate dynamics', *Icarus* **115**, 320–35.

Richardson, D. C., Asphaug. E. & Benner, L. [1995], 'Comet Shoemaker–Levy 9: a 'rubble pile' model with dissipative collisions and gravitational perturbations', *Bull. Am. Astron. Soc.* **27**, 1114.

Richardson, D. C., Bottke, W. F. & Love, S. G. [1998], 'Tidal distortion and disruption of Earth-crossing asteroids', *Icarus* **134**, 47–76.

Richardson, D. C., Quinn, T., Stadel, J. & Lake, G. [2000], 'Direct large-scale N-body simulations of planetesimal dynamics', *Icarus* **143**, 45–59.

Richardson, D. C., Leinhardt, Z. M., Melosh, H. J., Bottke, W. F. & Asphaug, E. [2003], 'Gravitational aggregates: evidence and evolution', in *Asteroids III*, ed. W. F. Bottke *et al.* (University of Arizona Press), 501–15.

Richter, K., Tanner, G. & Wintgen, D. [1993], 'Classical mechanics of two-electron atoms', *Phys. Rev.* **A48**, 4182–96.

Risken, H. [1984], *The Fokker–Planck Equation* (Springer-Verlag, Berlin).

Rivera, E. J. & Lissauer, J. J. [2000], 'Stability analysis of the planetary system orbiting υ Andromedae', *Astrophys. J.* **530**, 454–63.

Roos, N. & Aarseth, S. J. [1982], 'Evolution of rich clusters of galaxies', *Astron. Astrophys.* **114**, 41–52.

Roos, N. & Norman, C. A. [1979], 'Galaxy collisions and their influence on the dynamics and evolution of groups and clusters of galaxies', *Astron. Astrophys.* **76**, 75–85.

Ross, D. J., Mennim, A. & Heggie, D. C. [1997], 'Escape from a tidally limited star cluster', *Mon. Not. R. Astron. Soc.* **284**, 811–14.

Rosseland, S. [1928], 'On the time of relaxation of closed stellar systems', *Mon. Not. R. Astron. Soc.* **88**, 208–12.

Roy, A. E. [1988], *Orbital Motion* (Adam Hilger, Bristol).

Saar, S. H., Nordström, B. & Andersen, J. [1990], 'Physical parameters for three chromospherically active binaries', *Astron. Astrophys.* **235**, 291–304.

Safronov, V. S. [1969], *Evolution of the Protoplanetary Cloud and Formation of the Earth and the Planets* (Nauka, Moscow; NASA TT–F–667).

Salo, H. [1992], 'Gravitational wakes in Saturn's rings', *Nature* **359**, 619–21.

Salpeter, E. E. [1955], 'The luminosity function and stellar evolution', *Astrophys. J.* **121**, 161–7.

Sandage, A. R. [1953], 'The color–magnitude diagram for the globular cluster M3', *Astron. J.* **58**, 61–75.

Saslaw, W. C., Valtonen, M. & Aarseth, S. J. [1974], 'The gravitational slingshot and the structure of extragalactic radio sources', *Astrophys. J.* **190**, 253–70.

Scally, A. & Clarke, C. [2001], 'Destruction of protoplanetary discs in the Orion Nebula Cluster', *Mon. Not. R. Astron. Soc.* **325**, 449–56.

Schubart, J. [1956], 'Numerische Aufsuchung periodischer Lösungen im Dreikörperproblem', *Astron. Nach.* **283**, 17–22.

Schwarzschild, M. [1958], *Structure and Evolution of the Stars*, (Princeton University Press).

Shara, M. M. & Hurley, J. R. [2002], 'Star clusters as type Ia supernova factories', *Astrophys. J.* **571**, 830–42.

Shara, M. M., Hurley, J. R. & Mardling, R. A. [2003], 'Planet migration and solar system survival in open cluster', in preparation.

Shara, M. M., Saffer, R. A. & Livio, M. [1997], 'The first direct measurement of the mass of a blue straggler in the core of a globular cluster: BSS 19 in 47 Tucanae', *Astrophys. J.* **489**, L59–L62.

Shara, M. M., Fall, S. M., Rich, R. M. & Zurek, D. [1998], 'Hubble Space Telescope observations of NGC 121: first detection of blue stragglers in an extragalactic globular cluster', *Astrophys. J.* **508**, 570–5.

Sigurdsson, S. & Phinney, E. S. [1993], 'Binary–single star interactions in globular clusters', *Astrophys. J.* **415**, 631–51.

Sills, A., Lombardi, J. C., Bailyn, C. D., Demarque, P., Rasio, F. A. & Shapiro, S. L. [1997], 'Evolution of stellar collision products in globular clusters, I. Head-on collisions', *Astrophys. J.* **487**, 290–303.

Simó, C. [2001], 'Periodic orbits of the planar N-body problem with equal masses and all bodies on the same path', in *The Restless Universe*, ed. B. A. Steves & A. J. Maciejewski (Institute of Physics, Bristol), 265–84.

Smith, H. [1974], 'Integration errors and their effects on macroscopic properties of calculated N-body systems', in *Lecture Notes in Mathematics*, ed. D. G. Bettis (Springer-Verlag, New York), **362**, 360–73.

[1977], 'The validity of statistical results from N-body calculations', *Astron. Astrophys.* **61**, 305–12.

Smith, K. W. & Bonnell, I. A. [2001], 'Free-floating planets in stellar clusters?', *Mon. Not. R. Astron. Soc.* **322**, L1–L4.

Soffel, M. H. [1989], *Relativity in Astrometry, Celestial Mechanics and Geodesy* (Springer-Verlag, Berlin).

Spitzer, L. [1940], 'The stability of isolated clusters', *Mon. Not. R. Astron. Soc.* **100**, 396–413.

[1958], 'Disruption of galactic clusters', *Astrophys. J.* **127**, 17–27.

[1969], 'Equipartition and the formation of compact nuclei in spherical stellar systems', *Astrophys. J.* **158**, L139–L43.

[1987], *Dynamical Evolution of Globular Clusters* (Princeton University Press).

Spitzer, L. & Hart, M. H. [1971a], 'Random gravitational encounters and the evolution of spherical systems, I. Method', *Astrophys. J.* **164**, 399–409.

[1971b], 'Random gravitational encounters and the evolution of spherical systems, II. Models', *Astrophys. J.* **166**, 483–511.

Spitzer, L. & Mathieu, R. D. [1980], 'Random gravitational encounters and the evolution of spherical systems, VIII. Clusters with an initial distribution of binaries', *Astrophys. J.* **241**, 618–36.

Spitzer, L. & Shull, J. M. [1975a], 'Random gravitational encounters and the evolution of spherical systems, VI. Plummer's model', *Astrophys. J.* **200**, 339–42.

[1975b], 'Random gravitational encounters and the evolution of spherical systems, VII. Systems with several mass groups', *Astrophys. J.* **201**, 773–82.

Spitzer, L. & Thuan, T. X. [1972], 'Random gravitational encounters and the evolution of spherical systems, IV. Isolated systems of identical stars', *Astrophys. J.* **175**, 31–61.

Springel, V., Yoshida, N. & White, S. D. M. [2001], 'GADGET: a code for collisionless and gasdynamical cosmological simulations', *New Astron.* **6**, 79–117.

Spurzem, R. [1999], 'Direct N-body simulations', *J. Comp. Applied Maths.* **109**, 407–32.

Spurzem, R. & Aarseth, S. J. [1996], 'Direct collisional simulation of 10 000 particles past core collapse', *Mon. Not. R. Astron. Soc.* **282**, 19–39.

Spurzem, R. & Giersz, M. [1996], 'A stochastic Monte Carlo approach to modelling of real star cluster evolution, I. The model', *Mon. Not. R. Astron. Soc.* **283**, 805–10.

Spurzem, R. & Takahashi, K. [1995], 'Comparison between Fokker–Planck and gaseous models of star clusters in the multi-mass case revisited', *Mon. Not. R. Astron. Soc.* **272**, 772–84.

Spurzem, R., Baumgardt, H. & Ibold, N. [2003], 'A parallel implementation of an N-body integrator on general and special purpose computers', *Mon. Not. R. Astron. Soc.*, submitted, `astro-ph/0103410`.

Spurzem, R., Makino, J., Fukushige, T., Lienhart, G., Kugel, A., Männer, R., Wetzstein, M., Burkert, A. & Naab, T. [2002], 'Collisional stellar dynamics, gas dynamics and special purpose computing', Fifth JSPS/CSE Symposium, Tokyo, in press, `astro-ph/0204326`.

Stadel, J., Wadsley, J. & Richardson, D. C. [2002], 'High performance computational astrophysics with pkdgrav', in *High Performance Computing Systems and Applications*, ed. N.J. Dimopoulos & K. F. Lie (Kluwer, Boston), 501–23.

Standish, E. M. [1968a], 'The numerical integration of collapsing clusters', *Bull. Astron.* **3**, 135–45.

[1968b], *Numerical Studies of the Gravitational Problem of N Bodies*, Ph.D. Thesis (Yale University).

[1971], 'Sufficient conditions for escape in the three-body problem', *Celes. Mech.* **4**, 44–8.

[1972], 'The dynamical evolution of triple star systems: a numerical study', *Astron. Astrophys.* **21**, 185–91.

Sterzik, M. F. & Durisen, R. H. [1995], 'Escape of T Tauri stars from young stellar systems', *Astron. Astrophys.* **304**, L9–L12.

[1998], 'The dynamic decay of young few-body stellar systems, I. The effect of a mass spectrum for $N = 3$, 4, and 5', *Astron. Astrophys.* **339**, 95–112.

Sterzik, M. F. & Tokovinin, A. A. [2002], 'Relative orientation of orbits in triple stars', *Astron. Astrophys.* **384**, 1030–7.

Stiefel, E. L. & Scheifele, G. [1971], *Linear and Regular Celestial Mechanics* (Springer-Verlag, Berlin).

Stodółkiewicz, J. S. [1982], 'Dynamical evolution of globular clusters, I', *Acta Astronomica* **32**, 63–91.

[1985], 'Monte-Carlo simulations', in *Dynamics of Star Clusters*, ed. J. Goodman & P. Hut (Reidel, Dordrecht), 361–72.

Stoer, J. & Bulirsch, R. [1980], *Introduction to Numerical Analysis* (Springer-Verlag, New York).

Strömgren, E. [1900], 'Über mekanische Integration und deren Verwendung für numerische Rechnungen auf dem Gebiete des Dreikörper-Problems', *Meddelanden från Lunds Astron. Obs.* No. 13, 1–9.

[1909], 'Ein numerisch gerechneter Spezialfall des Dreikörper-Problems mit Massen und Distanzen von derselben Grössenordnung', *Astron. Nach.* **182**, 181–92.

Stumpff, K. [1962], *Himmelsmechanik*, Vol. I (VEB, Berlin).

Sugimoto, D., Chikada, Y., Makino, J., Ito, T., Ebisuzaki, T. & Umemura, M. [1990], 'A special-purpose computer for gravitational many-body problems', *Nature* **345**, 33–35.

Sundman, K. F. [1912], 'Mémoire sur le probléme des trois corps', *Acta Math.* **36**, 105–79.

Sussman, G. J. & Wisdom, J. [1988], 'Numerical evidence that the motion of Pluto is chaotic', *Science* **241**, 433–7.

Sweatman, W. L. [1994], 'The development of a parallel N-body code for the Edinburgh concurrent supercomputer', *J. Comput. Phys.* **111**, 110–19.

[2002a], 'The symmetrical one-dimensional Newtonian four-body problem: a numerical investigation', *Celes. Mech. Dyn. Ast.* **82**, 179–201.

[2002b], 'Vector rational extrapolation methods in Aarseth-type N-body algorithms', unpublished.

Szebehely, V. [1967], *Theory of Orbits* (Academic Press, New York).

[1971], 'Classification of the motions of three bodies in a plane', *Celes. Mech.* **4**, 116–18.

[1972], 'Mass effects in the problem of three bodies', *Celes. Mech.* **6**, 84–107.

Szebehely V. & Peters, C. F. [1967], 'Complete solution of a general problem of three bodies', *Astron. J.* **72**, 876–83.

Szebehely, V. & Zare, K. [1977], 'Stability of classical triplets and of their hierarchy', *Astron. Astrophys.* **58**, 145–52.

Takahashi, K. [1995], 'Fokker–Planck models of star clusters with anisotropic velocity distributions, I. Pre-collapse evolution', *Publ. Astron. Soc. Japan* **47**, 561–73.

[1997], 'Fokker–Planck models of star clusters with anisotropic velocity distributions, III. Multi-mass clusters', *Publ. Astron. Soc. Japan* **49**, 547–60.

Takahashi, K. & Portegies Zwart, S. F. [1998], 'The disruption of globular star clusters in the Galaxy: a comparative analysis between Fokker–Planck and N-body models', *Astrophys. J.* **503**, L49–L52.

Tamaki, S. *et al.*, [1999], 'Node architecture and performance evaluation of the Hitachi super technical server SR8000', in *Proceedings ISCA 12th International Conference on Parallel and Distributed Computing Systems*, 487–93.

Terlevich, E. [1983], *Evolution of Open Clusters*, Ph.D. Thesis (University of Cambridge)

[1987], 'Evolution of N-body open clusters', *Mon. Not. R. Astron. Soc.* **224**, 193–225.

Teuben, P. [1995], 'The stellar dynamics toolbox NEMO', in *Astronomical Data Analysis Software and Systems IV*, ed. R. Shaw, H. E. Payne & J. J. E. Heyes (ASP, San Francisco), **77**, 398–401.

Theis, C. [1998], 'Two-body relaxation in softened potentials', *Astron. Astrophys.* **330**, 1180–9.

Theuns, T. [1992a], 'Hydrodynamics of encounters between star clusters and molecular clouds, I. Code validation and preliminary results', *Astron. Astrophys.* **259**, 493–502.

[1992b], 'Hydrodynamics of encounters between star clusters and molecular clouds, II. Limits on cluster lifetimes', *Astron. Astrophys.* **259**, 503–9.

Thorne, K. S. & Żytkow, A. N. [1977], 'Stars with degenerate neutron cores, I. Structure of equilibrium models', *Astrophys. J.* **212**, 832–58.

Thüring, B. [1958], 'Der Gravitations-Stoss', *Astron. Nach.* **284**, 263–8.

[1968], 'New numerical results on gravitational impact', *Bull. Astron.* **3**, 177–88.

Tomisaka, K. [1986], 'Formation of giant molecular clouds by coagulation of small clouds and spiral structure', *Publ. Astron. Soc. Japan* **38**, 95–9.

Toomre, A. [1977], 'Mergers and some consequences', in *The Evolution of Galaxies and Stellar Populations*, ed. B. M. Tinsley & R. B. Larson (Yale University Observatory, New Haven), 401–26.

Tout, C. A., Aarseth, S. J., Pols, O. R. & Eggleton, P. P. [1997], 'Rapid binary star evolution for N-body simulations and population synthesis', *Mon. Not. R. Astron. Soc.* **291**, 732–48.

Trimble, V. [1980], 'Binary stars in globular and open clusters', in *Star Clusters*, ed. J. E. Hesser (Reidel, Dordrecht), 259–79.

Turner, E. L., Aarseth, S. J., Gott, J. R., Blanchard, N. T. & Mathieu, R. D. [1979], 'N-body simulations of galaxy clustering, II. Groups of galaxies', *Astrophys. J.* **228**, 684–95.

Umehara, H. & Tanikawa, K. [2000], 'Slingshot-escape condition in the planar three-body problem', in *The Chaotic Universe*, ed. V. G. Gurzadyan & R. Ruffini (World Scientific, Singapore), 568–78.

Valtonen, M. J. [1974], 'Statistics of three-body experiments', in *The Stability of the Solar System and of Small Stellar Systems*, ed. Y. Kozai (Reidel, Dordrecht), 211–23.

[1975], 'Statistics of three-body experiments: probability of escape and capture', *Mem. R. Astron. Soc.* **80**, 77–91.

[1976], 'Statistics of three-body experiments: escape energy, velocity and final binary eccentricity', *Astrophys. Sp. Sci.* **42**, 331–47.

[1988], 'The general three-body problem in astrophysics', *Vistas Astron.* **32**, 23–48.

Valtonen, M. J., Mikkola, S. & Pietilä, H. [1995], 'Burrau's three-body problem in the post-Newtonian approximation', *Mon. Not. R. Astron. Soc.* **273**, 751–4.

van Albada, T. S. [1967], 'The evolution of small stellar systems and the formation of double stars', *Bull. Astron.* **2**, 59–65.

[1968], 'Numerical integrations of the N-body problem', *Bull. Astron. Inst. Neth.* **19**, 479–99.

van Albada, T. S. & van Gorkom, J. H. [1977], 'Experimental stellar dynamics for systems with axial symmetry', *Astron. Astrophys.* **54**, 121–6.

van Leeuwen, F. [1983], *The Pleiades: An Astrometric and Photometric Study of an Open Cluster*, Ph.D. Thesis (Leiden University).

van Leeuwen, F. & van Gendern, A. M. [1997], 'The discovery of a new massive O-type close binary: τ CMa (HD 57061), based on Hipparcos and Walraven photometry', *Astron. Astrophys.* **327**, 1070–6.

Verbunt, F. & Phinney, E. S. [1995], 'Tidal circularization and the eccentricity of binaries containing giant stars', *Astron. Astrophys.* **296**, 709–21.

Vesperini, E. & Chernoff, D. F. [1996], 'Truncation of the binary distribution function in globular cluster formation', *Astrophys. J.* **458**, 178–93.

Vesperini, E. & Heggie, D. C. [1997], 'On the effects of dynamical evolution on the initial mass function of globular clusters', *Mon. Not. R. Astron. Soc.* **289**, 898–920.

Villumsen, J. V. [1982], 'Simulation of galaxy mergers', *Mon. Not. R. Astron. Soc.* **199**, 493–516.

von Hippel, T. [1998], 'Contribution of white dwarfs to cluster masses', *Astron. J.* **115**, 1536–42.

von Hoerner, S. [1957], 'The internal structure of globular clusters', *Astrophys. J.* **125**, 451–69.

[1960], 'Die numerische Integration des n-Körper-Problems für Sternhaufen, I', *Z. Astrophys.* **50**, 184–214.

[1963], 'Die numerische Integration des n-Körper-Problems für Sternhaufen, II', *Z. Astrophys.* **57**, 47–82.

[2001], 'How it all started', in *Dynamics of Star Clusters and the Milky Way*, ed. S. Deiters, B. Fuchs, A. Just, R. Spurzem & R. Wielen (ASP, San Francisco), **228**, 11–15.

Waldvogel, J. [1972], 'A new regularization of the planar problem of three bodies', *Celes. Mech.* **6**, 221–31.

Webbink, R. F. [1984], 'Double white dwarfs as progenitors of R Coronae Borealis stars and Type I supernovae', *Astrophys. J.* **277**, 355–60.

Wetherill, G. W. & Stewart, G. R. [1989], 'Accumulation of a swarm of small planetesimals', *Icarus* **77**, 330–57.

White, S. D. M. [1976], 'The dynamics of rich clusters of galaxies', *Mon. Not. R. Astron. Soc.* **177**, 717–33.

[1978], 'Simulations of merging galaxies', *Mon. Not. R. Astron. Soc.* **184**, 185–203.

[1979], 'Further simulations of merging galaxies', *Mon. Not. R. Astron. Soc.* **189**, 831–52.

[1983], 'Simulations of sinking satellites', *Astrophys. J.* **274**, 53–61.

Whittaker, E. T. [1904], *A Treatise on Analytical Dynamics of Particles and Rigid Bodies; with an Introduction to the Problem of Three Bodies* (Cambridge University Press).

Wickes, W. C. [1975], 'Orbits and masses of Hyades visual binaries', *Astron. J.* **80**, 1059–64.

Wielen, R. [1967], 'Dynamical evolution of star cluster models, I', *Veröffent. Astron. Rechen-Inst. Heidelberg* No. 19, 1–43.

[1968], 'On the escape rate of stars from clusters', *Bull. Astron.* **3**, 127–33.

[1972], 'On the lifetimes of galactic clusters', in *Gravitational N-Body Problem*, ed. M. Lecar (Reidel, Dordrecht), 62–70.

[1974], 'On the numerical integration of the N-body problem for star clusters', in *Lecture Notes in Mathematics*, ed. D. G. Bettis (Springer-Verlag, New York), **362**, 276–90.

[1975], 'Dynamics of star clusters: comparison of theory with observations and simulations', in *Dynamics of Stellar Systems*, ed. A. Hayli (Reidel, Dordrecht), 119–31.

Wilkinson. M. I., Hurley, J. R., Mackey, A. D., Gilmore, G. & Tout, C. A. [2003], 'Core radius evolution of star clusters', *Mon. Not. R. Astron. Soc.* in press.

Williamson, R. E. & Chandrasekhar, S. [1941], 'The time of relaxation of stellar systems, II', *Astrophys. J.* **93**, 305–22.

Wisdom, J. [1980], 'The resonance overlap criterion and the onset of stochastic behavior in the restricted three-body problem', *Astron. J.* **85**, 1122–33.

Wisdom, J. & Holman, M. [1991], 'Symplectic maps for the N-body problem', *Astron. J.* **102**, 1528–38.

Wisdom, J. & Tremaine, S. [1988], 'Local simulations of planetary rings', *Astron. J.* **95**, 925–40.

Worral, G. [1967], 'The few-body problem, close binary stars and stellar associations', *Mon. Not. R. Astron. Soc.* **135**, 83–98.

Yabushita, S. [1966], 'Lifetime of binary stars', *Mon. Not. R. Astron. Soc.* **133**, 133–43.

Yoshida, H. [1982], 'A new derivation of the Kustaanheimo–Stiefel variables', *Celes. Mech.* **28**, 239–42.

Zare, K. [1974], 'A regularization of multiple encounters in gravitational N-body problems', *Celes. Mech.* **10**, 207–15.

[1976], *The Effects of Integrals on the Totality of Solutions of Dynamical Systems*, Ph.D. Thesis (University of Texas, Austin).

[1977], 'Bifurcation points in the planar problem of three bodies', *Celes. Mech.* **16**, 35–8.

Zare, K. & Szebehely, V. [1975], 'Time transformations in the extended phase-space', *Celes. Mech.* **11**, 469–82.

Zhao, H. [1996], 'Analytical models for galactic nuclei', *Mon. Not. R. Astron. Soc.* **278**, 488–96.

Index

Aarseth–Zare method, 67
AC neighbour scheme, 164
accretion time-scale, 314
accuracy dependence, 325
adiabatic invariance, 192
Ahmad–Cohen method, 33
alternative data structure, 127
angular momentum, 7
angular momentum deficit, 348
anisotropic formulation, 273
anisotropy radius, 123
apsidal motion, 291
asteroid belt, 348
automated scattering package, 334
average neighbour number, 251
averaged solutions, 268
averaging method, 290

Barnes–Hut scheme, 96
Beowulf PC clusters, 249
BH binary, 232
bilinear relation, 54
binary formation, 332
binary fraction, 125, 294
binary–binary collisions, 260, 269
binary–binary scattering, 283, 335
binary–single star scattering, 334
binary–tides problem, 150
bisecting principle, 263
black hole binaries, 90, 179, 302
black hole formation, 280
block time-step algorithm, 100
block-step scheme, 171
blue stragglers, 257, 284
Bode's Law, 307
boot-strapping, 20, 39
braking, 288
Brownian motions, 177
Bulirsch–Stoer method, 91, 230
Burdet–Heggie scheme, 61

canonical variables, 68
capture theory, 283
cascade, 331
category scheme, 25
cell time-step, 100
central singularity, 268
chain regularization, 86, 341
chain vectors, 82
chaos boundary, 203, 225
chaotic behaviour, 343
chaotic evolution, 200
chaotic exchange of energy, 203
chaotic wanderings, 348
choreographic solutions, 345
circularization, 289
circularization time, 202
close encounter distance, 144
close triple approaches, 327
cluster simulations, 264
clusters of galaxies, 303
coalescence, 205
coefficient of restitution, 48
coefficients of restitution, 317

collision criterion, 157, 308
collision procedure, 157
collision remnants, 284
collision time, 315
collisional tree code, 102
collisionless calculations, 273
collisionless systems, 106
commensurability condition, 309
common-envelope, 156, 206, 228
COMOVE, 106, 306
comoving formulation, 38
compact subsystem, 147, 207
comparison of regularization, 336
computer movies, 258
contact binaries, 257
contact binary, 206
coordinate transformation, 54
core collapse, 268, 273
core oscillations, 269
core radius, 265, 288
correlation function, 305
cosmological simulation, 305
Coulomb factor, 245
Coulomb logarithm, 9
CPU times, 250
CRAY T3E, 173
cross sections, 331
crossing time, 8, 260, 342
cumulative cross section, 339

data bank, 254, 259
data structure, 117
deflection angle, 144
degenerate objects, 293
degenerate triple, 222
density centre, 265
depletion rate, 272
differential corrections, 162, 198
differential cross sections, 334
Digital Orrery computer, 348
disc potential, 135
disk shocking, 272
distribution function, 123
divided difference, 19, 61, 131
domain decomposition, 175
dominant two-body motion, 214
doubly degenerate binary, 205
drag force, 312
dwarf spheroidal galaxies, 300
dynamical friction, 10, 243, 314

eccentricity, 58, 201, 289
eccentricity distributions, 326
eccentricity evolution, 321
eccentricity growth, 308
effective particle number, 272
eigenevolution, 277
ejected quintuplets, 362
Encke's method, 65
energy budget, 161, 260
energy conservation, 257
energy equipartition, 313
energy errors, 238, 262, 332
energy stabilization, 45, 237
epoch of initialization, 224
equilibrium eccentricity, 326
equilibrium spin, 289
equilibrium tides, 288
equipartition time, 11
error growth, 2, 234
error tolerance, 163
escape criterion, 153
escape probability, 272
escape rate, 129, 271
escaping quadruple, 362
events, 255
exact binary motion, 92
exchange, 331
exchange criterion, 197
exchange cross section, 332
explicit differentiation, 39
exponential growth, 239
extended impulse approximation, 130
extended phase space, 69
external tidal field, 153, 170, 184
extra-solar planets, 319

few-body systems, 328
fictitious particle, 58
figure eight solution, 345
fitting functions, 203, 279
flattened superclusters, 306
Fokker–Planck method, 268

Fokker–Planck models, 244
force polynomial, 20, 138
FORTRAN listing, 374
fossil fuel, 163, 260
four-body scattering, 334
fractal structure, 343
fragmentation, 46, 309
free-floating planets, 322
fuzzy boundary, 196

GADGET, 176
galaxy cluster model, 303
galaxy collisions, 301
gaseous model, 244, 268
generating function, 68
ghost particle, 47, 199, 362
global regularization method, 336
globular cluster evolution, 273
globular clusters, 124, 292
GR coalescence, 233
GRAPE-4, 246, 270, 314
GRAPE-6, 107, 247, 319, 368
graphics, 258
gravitational clustering picture, 306
gravitational radiation, 156, 232
gravitational radius, 147
gravitational slingshot, 326
gravothermal oscillations, 270
grid perturbation method, 46, 308

half-mass relaxation time, 8, 271
hard and soft configurations, 335
hard binary, 111, 176, 256, 331
hardening rate, 176
harmonic oscillator, 53
HARP-1, 246, 313
HARP-2, 106, 246, 313
heap-sort algorithm, 142
Heggie's global formulation, 78
Hermite integration, 29, 374
Hermite rule, 187
Hermite scheme, 27, 146
hierarchical formation, 260, 329
hierarchical levels, 28, 360
hierarchical space, 197, 286
hierarchical stability, 221, 325
hierarchical systems, 150
hierarchical time-steps, 28
hierarchical triple formation, 335
hierarchical triples, 260, 275, 337
high-order prediction, 237
high-velocity escapers, 238, 275
high-velocity particles, 168
higher-order system, 222, 286
Hill radii, 311, 349
homology model, 268
hyper-elastic encounter, 335
hyperbolic capture, 257

IAU Comparison Problem, 242
idealized models, 267
impact parameter, 333
implicit midpoint method, 92
improved Euler method, 14
impulse approximation, 130
inclined systems, 151, 196
individual time-step scheme, 22
inelastic coalescence, 303
inelastic collisions, 314
inertial reference frame, 131
initial segregation, 245
initialization procedure, 20
input parameters, 112
integration cycle, 165
integration errors, 238
integration variables, 114
interacting galaxies, 301
interaction energy, 163, 335
intermediate-mass black holes, 295
interstellar clouds, 129, 272
ionization, 334
irregular time-step, 169, 251

Jacobi integral, 134
Jupiter-like planets, 320

Kepler's equation, 159
kick velocity, 280
kinematical cooling, 277
King model, 123, 244
King–Michie models, 123
kira, 369
Kozai cycles, 238, 286
KS initialization, 127, 149

KS termination, 188
Kustaanheimo–Stiefel method, 56

Lagrange points, 128
Lagrange stability, 347
large-N simulations, 295
latent energy, 260
Legendre coefficients, 97
Legendre polynomials, 40
Levi-Civita matrix, 54
linearized equations of motion, 47
linearized tidal force, 136
load balance, 256
logarithmic Hamiltonian, 93
logarithmic potential, 133, 300
long-lived hierarchy, 262
long-lived unstable triples, 330
long-term stability, 345
loss-cone depletion, 302
loss-cone replenishment, 177
low-order prediction, 22, 237
lunar-size planetesimals, 308
Lyapunov times, 348

M67, 284
magnetic braking, 156, 288
mass generating function, 121
mass loss, 154, 228, 279
mass segregation, 245, 313
mass spectrum, 120, 245, 272
mass-loss correction, 162, 281
mass-transfer hypothesis, 283
maximum binary fractions, 277
maximum neighbour number, 113, 263
mean escape time, 325
merger, 151
merger procedure, 198
molecular clouds, 46, 298
Monte Carlo method, 243, 340
Monte Carlo models, 269
Monte Carlo sampling, 332
multiple regularization, 66, 106, 147
multipole expansion, 40

narrow ring of particles, 313
NBODY1, 106, 236
NBODY2, 106
NBODY3, 106, 192
NBODY4, 106, 247, 270
NBODY5, 106, 192, 266
NBODY6, 107, 274
NBODY7, 107, 230
neighbour list, 35, 115, 248
neighbour sphere radius, 34
NEMO, 257
neutron star formation, 293
neutron star retention, 293
new binary formation, 259
Newton's Law, 6
Newton–Raphson iteration, 188
Newtonian cosmology, 37
non-solar metallicity, 279
normal tidal motion, 225
numerical experiments, 323
numerical singularities, 268

oblateness effect, 290
oligarchic growth, 315
onset of chaos, 225
onset of chaotic motion, 204
Oort's constants, 128
open clusters, 129, 153, 271
opening angle, 95, 104
orbital shrinkage, 228, 291

pancake simulations, 306
parallelization, 173
partial reflection, 366
partial unperturbed motion, 191
particle in box method, 312
particle in box scheme, 103
pericentre parameters, 227
pericentre passage time, 193
pericentre time, 159
periodic boundary, 103
perturbed subsystem, 74
perturbed three-body problem, 80
perturber selection, 185
perturbing force, 55
perturbing function, 74
perturbing potential, 85
PGPLOT, 258
phase-space maps, 343
physical collisions, 283, 293

pipeline efficiency, 247
PKDGRAV, 315
planetary embryos, 310
planetary formation, 46, 307
planetary integrations, 348
planetary orbits, 321, 348
planetary rings, 103, 317
planetesimal simulations, 315
planetesimal tree code, 104
Plummer model, 121
Plummer sphere, 21
point-mass approximation, 167
point-mass potential, 300
post-collapse phase, 197, 286
post-collapse solutions, 273
post-Newtonian acceleration, 232
post-Newtonian terms, 327
practical regularization scheme, 92
practical stability, 152
pre-main-sequence evolution, 125
predictor–corrector methods, 30
preferential escape, 272, 274
primordial binaries, 124, 269, 276
probability of binary formation, 332
Pythagorean Problem, 238, 324

quadruple system, 260, 275
quadrupole moments, 97
quiescent phase, 206

rapid binary evolution, 279
rational extrapolation, 26
realistic simulations, 248, 271
recoil velocity, 331
rectification, 185, 218
recursive procedure, 95
recursive relations, 85
regular orbits, 345
regular time-step, 137, 169
regularized Hamiltonian, 69, 239
regularized polynomials, 138
regularized time-step, 140
regularized two-body Hamiltonian, 70
relative energy error, 163, 240, 325
relative force criterion, 241
relative perturbation, 217
relativistic effects, 91
relativistic precession, 232, 290, 321
relativistic simulations, 328
relaxation time, 8, 268, 322
reproducibility, 238, 341
resonance, 311, 331, 334
resonance overlap, 347
resonance scattering, 343
retrograde motion, 196, 285
ring formulation, 49
Roche lobe, 155, 204
Rosetta Stone, 66
rotating coordinates, 132
round-off errors, 93
round-off sensitivity, 325
rubble piles, 316
runaway growth, 294, 313
runaway stars, 276, 338
Runge–Kutta method, 289
Runge–Lenz vector, 290
Runge–Lenz–Laplace vector, 61

Saturn's B-ring, 319
scaling problem, 270
scattering experiments, 321, 331
scheduling, 142, 185, 209
Schwarzschild radii, 233
SeBa, 370
second-order Bulirsch–Stoer, 93
secular oscillations, 311
secular perturbations, 288
secular resonances, 311
self-force, 42
self-similar approximation, 312
semi-iteration, 17, 20
semi-major axis, 58
sensitivity of solutions, 239
sequential circularization, 200, 225
sequential ordering, 119, 142
serendipitous discovery, 345
shadow orbits, 235
shell method, 178
similarity solutions, 268
simulation of $M67$, 293
single star–binary encounters, 330
singularity, 69
sliding box configuration, 47

slingshot, 177, 231, 332
slow-down factor, 87, 192
softened potential, 21
solar system models, 322
sorting algorithm, 143
special-purpose computers, 246
SPH, 249
spin synchronization, 289
spin–orbit coupling, 290, 321
stability boundary, 197, 320, 330
stability criterion, 150, 285
stable hierarchies, 194, 285, 344
standard N-body units, 110
standard KS formulation, 59
standard leapfrog equations, 90
star cluster model, 181, 266
STARLAB, 257
stellar encounters, 40, 321
stellar evolution, 279
stellar oscillations, 288
stellar rotation, 287
strong interactions, 207, 335
Stumpff coefficients, 140
Stumpff functions, 63
Stumpff KS method, 65, 237
subsystem energy, 74
Sundman's time smoothing, 326
Superfast particles, 261
superhard binary, 206
supermassive objects, 294
supernova event, 156
supernova mass loss, 281
survey of N-body models, 274
symplectic integration, 93
symplectic integrator, 311
symplectic mapping method, 348
synchronization time-scale, 289
synthetic HR diagram, 258, 281
synthetic stellar evolution, 154, 279
systematic errors, 145, 180, 236

terrestrial planets, 311, 349
terrestrial region, 311
test problem, 243
theories of escape, 271
thermal distribution, 125
thermodynamic theory, 306
three-body problem, 67, 324
three-body scattering, 330
three-body studies, 326
tidal capture, 282
tidal circularization, 155, 289
tidal evolution, 202
tidal interaction, 321
tidal radius, 129, 300
tidal shocks, 292
tidal torque, 127, 306
time quantization, 25, 211
time reversal, 238, 325
time smoothing, 37, 51, 331
time transformation, 52, 80, 337
time-step, 23, 185, 236
time-symmetric formulation, 31
time-symmetric method, 188
time-transformed leapfrog, 90, 229
total energy, 7, 161
tree code, 315
tree construction, 98
tree repair, 104
tree structure, 95
TRIPLE, 327, 341, 343
triple collision, 73, 80
two-body encounters, 144, 189
two-body energy, 55, 159
two-body perturbation, 45
two-body regularization, 54, 143

UNIX, 258
unperturbed binary, 152, 190
unperturbed orbits, 191
Uranian rings, 319

VAX 11-780, 306
velocity escape criterion, 210
velocity kick, 162, 168, 281
virial radius, 8, 110
virial ratio, 11, 111
voids, 306

water-shed effect, 330
wheel-spoke regularization, 78, 106, 177, 337
Wisdom–Holman method, 348